Ceramic Processing

Ceramic Processing

Mohamed N. Rahaman
University of Missouri-Rolla, USA

Taylor & Francis
Taylor & Francis Group
Boca Raton   London   New York

CRC is an imprint of the Taylor & Francis Group,
an informa business

CRC Press
Taylor & Francis Group
6000 Broken Sound Parkway NW, Suite 300
Boca Raton, FL 33487-2742

© 2007 by Taylor & Francis Group, LLC
CRC Press is an imprint of Taylor & Francis Group, an Informa business

Library of Congress Cataloging-in-Publication Data

Rahaman, M. N. 1950-
  Ceramic processing / Mohamed Rahaman.
     p. cm.
  Includes bibliographical references and index.
  ISBN 0-8493-7285-2 (acid-free paper)
  1. Ceramics. 2. Ceramic engineering. 3. Polycrystals. 4. Sintering. I. Title.

TP807.R278 2006
666--dc22   2006043889

Visit the Taylor & Francis Web site at
http://www.taylorandfrancis.com

and the CRC Press Web site at
http://www.crcpress.com

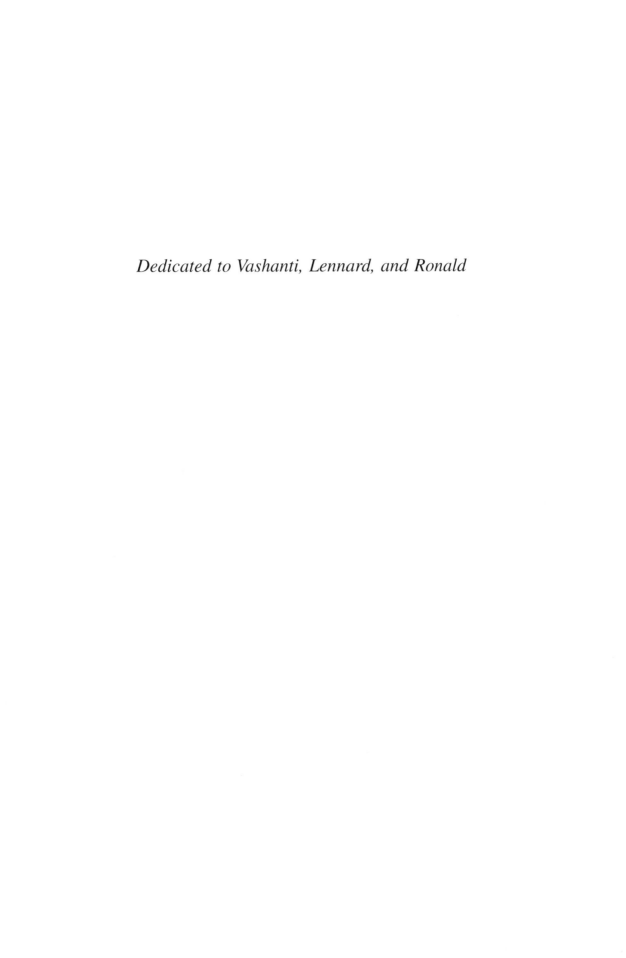

Dedicated to Vashanti, Lennard, and Ronald

Preface

The production and application of ceramics are among the oldest technological skills. The field of ceramic materials has its roots in the more traditional aspects of the subject such as clay-based ceramics and glasses. During the past few decades, new developments in the use of ceramics in more advanced technological applications have attracted considerable attention. The discovery of ceramic superconductors, as well as the use of ceramics for heat-resistant tiles in the space shuttle, optical fibers, components in high temperature engines, and the electrolyte in solid oxide fuel cells, has generated considerable interest in the field.

The increasing use of ceramics in more advanced technological applications has resulted in a heightened demand for improvements in properties and reliability. In recent years, there has been the realization that such improvements can be achieved only through careful attention to the fabrication process. The engineering properties of polycrystalline ceramics are controlled by the microstructure, which, in turn, depends on the processing method used to fabricate the body. Therefore, the fabrication process governs the production of microstructures with the desired properties. It is often stated that materials science is a field at the interface between the physical sciences (physics, chemistry, and mathematics) and engineering (such as electrical, mechanical, and civil). In this view, the approach to ceramic processing is concerned with the understanding of fundamentals and the application of the knowledge to the production of microstructures with useful properties.

This book is concerned primarily with the processing of polycrystalline ceramics. Because of its importance and widespread use, the fabrication of ceramics by the consolidation and sintering of powders forms the focus of this book. A brief treatment of the production of ceramics (and glasses) by the less conventional sol–gel route is also included. The approach is to outline the fundamental issues of each process and to show how they are applied to the practical fabrication of ceramics. Each fabrication route involves a number of processing steps, and each step has the potential for producing microstructural flaws that degrade the properties of the fabricated material. An important feature of this treatment is the attempt to show the importance of each step as well as the interconnection between the various steps in the overall fabrication route. Chapter 1 provides an introductory overview of the various methods that can be used for the production of ceramic materials. The overview also forms a basis for the more detailed treatment of powder processing and sol–gel processing, which is discussed later in the book. Chapter 2 to Chapter 9 form a logical development from the start of the fabrication process to the final fabricated microstructure.

My intention has been to prepare a textbook that is suitable for a one-semester (or two-quarter) course in ceramic processing at the upper undergraduate level or the introductory graduate level. A background in the concepts and processing of traditional ceramics, typically obtained in lower-level undergraduate courses, is assumed. It is hoped that the book will also be useful to researchers in industry who are involved in the production of ceramics or who would wish to develop a background in the processing of ceramics.

I wish to thank the many authors and publishers who have allowed me permission to reproduce their figures in this book. Last but not least, I wish to thank my wife, Vashanti, for her encouragement and support when I was preoccupied with the completion of this book.

The Author

Mohamed N. Rahaman is professor of ceramics in the department of materials science and engineering, University of Missouri-Rolla. He earned a B.A. (Hons.) and an M.A. from the University of Cambridge, England, and a Ph.D. from the University of Sheffield, England. Prior to joining the University of Missouri in 1986, Dr. Rahaman held positions at the University of Leeds, England; the University of the West Indies, Trinidad; and the Lawrence Berkeley National Laboratory, Berkeley, California. Dr. Rahaman is the author of two books, and the author or coauthor of over 125 publications, most of them in the area of processing and sintering of ceramics.

Contents

1 Ceramic Fabrication Processes: An Introductory Overview

1.1 INTRODUCTION

The subject of ceramics covers a wide range of materials. Recent attempts have been made to divide it into two parts: traditional ceramics and advanced ceramics. The usage of the term *advanced* has, however, not received general acceptance and other forms, including technical, special, fine and engineering will also be encountered. Traditional ceramics bear a close relationship to those materials that have been developed since the time of the earliest civilizations. They are pottery, structural clay products, and clay-based refractories, with which we may also group cements, concretes, and glasses. Whereas traditional ceramics still represent a major part of the ceramics industry, the interest in recent years has focused on advanced ceramics, ceramics that, with minor exceptions, have been developed within the last 50 years or so. Advanced ceramics include ceramics for electrical, magnetic, electronic, and optical applications (sometimes referred to as *functional ceramics*) and ceramics for structural applications at ambient as well as at elevated temperatures (*structural ceramics*). Although the distinction between traditional and advanced ceramics may be referred to in this book occasionally for convenience, we do not wish to overemphasize it. There is much to be gained through continued interaction between the traditional and the advanced sectors.

Chemically, with the exception of carbon, ceramics are nonmetallic, inorganic compounds. Examples are the silicates such as kaolinite ($Al_2Si_2O_5(OH)_4$) and mullite ($Al_6Si_2O_{13}$); simple oxides such as alumina (Al_2O_3) and zirconia (ZrO_2); complex oxides other than the silicates such as barium titanate ($BaTiO_3$) and the superconducting material $YBa_2Cu_3O_{6+\delta}$ ($0 \leq \delta \leq 1$). In addition, there are nonoxides, including carbides, such as silicon carbide (SiC) and boron carbide (B_4C); nitrides such as silicon nitride (Si_3N_4), and boron nitride (BN); borides such titanium diboride (TiB_2); silicides such as molybdenum disilicide ($MoSi_2$); and halides such as lithium fluoride (LiF). There are also compounds based on nitride-oxide or oxynitride systems (e.g., β'-sialons with the general formula $Si_{6-z}Al_zN_{8-z}O_z$ where $0 < z < \sim4$).

Structurally, all materials are either crystalline or amorphous (also referred to as *glassy*). The difficulty and expense of growing single crystals means that, normally, crystalline ceramics (and metals) are actually polycrystalline; they are made up of a large number of small crystals, or grains, separated from one another by grain boundaries. In ceramics as well as in metals, we are concerned with two types of structure, both of which have a profound effect on properties. The first type of structure is at the atomic scale: the type of bonding and the crystal structure (for a crystalline ceramic) or the amorphous structure (if it is glassy). The second type of structure is at a larger scale: the microstructure, which refers to the nature, quantity, and distribution of the structural elements or phases in the ceramic (e.g., crystals, glass, and porosity).

It is sometimes useful to distinguish between the intrinsic properties of a material and the properties that depend on the microstructure. The intrinsic properties are determined by the structure at the atomic scale and are properties that are not susceptible to significant change by modification of the microstructure. These properties include the melting point, elastic modulus, and coefficient of thermal expansion, and factors such as whether or not the material is brittle, magnetic, ferroelectric, or semiconducting. In contrast, many of the properties critical to the engineering applications of materials are strongly dependent on the microstructure (e.g., mechanical strength, dielectric constant, and electrical conductivity).

Intrinsically, ceramics usually have high melting points and are therefore generally described as refractory. They are also usually hard, brittle, and chemically inert. This chemical inertness is usually taken for granted, for example, in ceramic and glass tableware and in the bricks, mortar, and glass of our houses. However, when used at high temperatures, as in the chemical and metallurgical industries, this chemical inertness is severely tried. The electrical, magnetic, and dielectric behaviors cover a wide range, for example, in the case of electrical behavior, from insulators to conductors. The applications of ceramics are many. Usually, for a given application one property may be of particular importance but, in fact, all relevant properties need to be considered. We are, therefore, usually interested in combinations of properties. For traditional ceramics and glasses, familiar applications include structural building materials (e.g., bricks and roofing tile); refractories for furnace linings, tableware, and sanitaryware; electrical insulation (e.g., electrical porcelain and steatite); glass containers; and glasses for building and transportation vehicles. The applications for which advanced ceramics have been developed or proposed are already very diverse, and this area is expected to continue to grow at a reasonable rate. Table 1.1 illustrates some of the applications for advanced ceramics [1].

The important relationships between chemical composition, atomic structure, fabrication, microstructure, and properties of polycrystalline ceramics are illustrated in Figure 1.1. The intrinsic properties must be considered at the time of materials selection. For example, the phenomenon of ferroelectricity originates in the perovskite crystal structure of which $BaTiO_3$ is a good example. For the production of a ferroelectric material, we may therefore wish to select $BaTiO_3$. The role of the fabrication process, then, is to produce microstructures with the desired engineering properties. For example, the measured dielectric constant of the fabricated $BaTiO_3$ will depend significantly on the microstructure (grain size, porosity, and the presence of any secondary phases). Normally, the overall fabrication method can be divided into a few or several discrete steps, depending on the complexity of the method. Although there is no generally accepted terminology, we will refer to these discrete steps as *processing steps*. The fabrication of a ceramic body, therefore, involves a number of processing steps. In the next section, we examine, in general terms, some of the commonly used methods for the fabrication of ceramics.

1.2 CERAMIC FABRICATION PROCESSES

Ceramics can be fabricated by a variety of methods, some of which have their origins in early civilization. Our normal objective is the production, from suitable starting materials, of a solid product with the desired shape such as a film, fiber, or monolith, and with the desired microstructure. As a first attempt, we divide the main fabrication methods into three groups, depending on whether the starting materials involve a gaseous phase, a liquid phase, or a solid phase (Table 1.2). In the following sections, we examine briefly the main features of the processing steps involved in the ceramic fabrication methods and their main advantages and disadvantages from the point of view of ease of processing.

1.2.1 GAS-PHASE REACTIONS

By far the most important reactions are obtained by vapor deposition methods, in which the desired material is formed by chemical reaction between gaseous species. Creating a reaction between a liquid and a gas is generally impractical but has been developed recently into an elegant technique, referred to as *directed metal oxidation*. Reaction between a gas and a solid, commonly referred to

TABLE 1.1
Application of Advanced Ceramics Classified by Function

Function	Ceramic	Application
Electric	Insulation materials (Al_2O_3, BeO, MgO)	Integrated circuit substrate, package, wiring substrate, resistor substrate, electronics interconnection substrate
	Ferroelectric materials ($BaTiO_3$, $SrTiO_3$)	Ceramic capacitor
	Piezoelectric materials (PZT)	Vibrator, oscillator, filter, etc.
		Transducer, ultrasonic humidifier, piezoelectric spark generator, etc.
	Semiconductor materials ($BaTiO_3$, SiC, $ZnO–Bi_2O_3$, V_2O_5 and other transition metal oxides)	NTC thermistor: temperature sensor, temperature compensation, etc.
		PTC thermistor: heater element, switch, temperature compensation, etc.
		CTR thermistor: heat sensor element
		Thick film sensor: infrared sensor
		Varistor: noise elimination, surge current absorber, lightning arrestor, etc.
		Sintered CdS material: solar cell
		SiC heater: electric furnace heater, miniature heater, etc.
	Ion-conducting materials (β-Al_2O_3, ZrO_2)	Solid electrolyte for sodium battery
		ZrO_2 ceramics: oxygen sensor, pH meter, fuel cells
Magnetic	Soft ferrite	Magnetic recording head, temperature sensor, etc.
	Hard ferrite	Ferrite magnet, fractional horse power motors, etc.
Optical	Translucent alumina	High-pressure sodium vapor lamp
	Translucent Mg–Al spinel, mullite, etc.	Lighting tube, special-purpose lamp, infrared transmission window materials
	Translucent Y_2O_3–ThO_2 ceramics	Laser materials
	PLZT ceramics	Light memory element, video display and storage system, light modulation element, light shutter, light valve
Chemical	Gas sensor (ZnO, Fe_2O_3, SnO_2)	Gas leakage alarm, automatic ventilation alarm, hydrocarbon, fluorocarbon detectors, etc.
	Humidity sensor ($MgCr_2O_4$–TiO_2)	Cooking control element in microwave oven, etc.
	Catalyst carrier (cordierite)	Catalyst carrier for emission control
	Organic catalysts	Enzyme carrier, zeolites
	Electrodes (titanates, sulfides, borides)	Electrowinning aluminum, photochemical processes, chlorine production
Thermal	ZrO_2, TiO_2	Infrared radiator
Mechanical	Cutting tools (Al_2O_3, TiC, TiN, others)	Ceramic tool, sintered CBN; cermet tool, artificial diamond; nitride tool
	Wear-resistant materials (Al_2O_3, ZrO_2)	Mechanical seal, ceramic liner, bearings, thread guide, pressure sensors
	Heat-resistant materials (SiC, Al_2O_3, Si_3N_4, others)	Ceramic engine, turbine blade, heat exchangers, welding burner nozzle, high-frequency combustion crucibles)
Biological	Al_2O_3, MgO-stabilized ZrO_2, hydroxyapatite, bioactive glass	Artificial tooth root, bone, and joint implantation.
Nuclear	UO_2, UO_2–PuO_2	Nuclear fuels
	C, SiC, B_4C	Cladding materials
	SiC, Al_2O_3, C, B_4C	Shielding materials

Source: Kenney, G.B. and Bowen, H.K., High tech ceramics in Japan: current and future markets, *Am. Ceram. Soc. Bull.*, 62, 590, 1983. With permission.

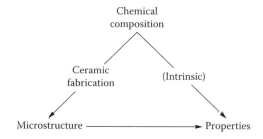

FIGURE 1.1 The important relationships in ceramic fabrication.

TABLE 1.2
Common Ceramic Fabrication Methods

Starting Materials	Method	Product
Gases	Chemical vapor deposition	Films, monoliths
Gas–liquid	Directed metal oxidation	Monoliths
Gas–solid	Reaction bonding	Monoliths
Liquid–solid	Reaction bonding	Monoliths
Liquids	Sol–gel process	Films, fibers
	Polymer pyrolysis	Fibers, films
Solids (powders)	Melt casting	Monoliths
	Sintering of powders	Monoliths, films

as *reaction bonding* (*or reaction forming*) has been used mainly for the production of Si_3N_4 but is now also being applied to the production of oxide ceramics. Reaction bonding (by a solid–liquid reaction) is also an important fabrication route for SiC.

1.2.1.1 Chemical Vapor Deposition

Chemical vapor deposition (CVD) is a process by which reactive molecules in the gas phase are transported to a surface at which they chemically react and form a solid film [2–9]. It is a well-established technique that can be used to deposit all classes of materials, including metals, ceramics, and semiconductors, for a variety of applications. Large areas can be coated, and the process is amenable to mass production. Thick films or even monolithic bodies can also be produced by basically prolonging the deposition process so that the desired thickness is achieved. Table 1.3 shows some of the important reactions used for the fabrication of ceramics, together with the temperature range of the reactions and the applications of the fabricated articles.

The apparatus used for CVD depends on the reaction being used, the reaction temperature, and the configuration of the substrate. Figure 1.2 shows examples of reactors for the deposition of films on substrates such as Si wafers. The general objective for any design is to provide uniform exposure of the substrate to the reactant gases. CVD has a number of process variables that must be manipulated to produce a deposit with the desired properties. These variables include the flow rate of the reactant gases, the nature and flow rate of any carrier gases, the pressure in the reaction vessel, and the temperature of the substrate. Substrate heating is required in CVD reactors because the films are produced preferably by endothermic reactions. The temperature of the substrate influences the deposition rate and is the main factor controlling the structure of the deposit. In general, high temperatures will yield crystalline deposits whereas low temperatures result in amorphous materials. Between these two extremes a polycrystalline deposit will be formed.

TABLE 1.3
Some Important CVD Reactions for the Fabrication of Ceramics

Reaction	Temperature (°C)	Application
$2C_xH_y \rightarrow 2xC + yH2$	900–2400	Pyrolytic carbon and graphite
$CH_3Cl_3Si \rightarrow SiC + 3HCl$	1000–1300	Composites
$W(CO)_6 \rightarrow WC + CO_2 + 4CO$	400–800	Coatings
$TiCl_4 + O_2 \rightarrow TiO_2 + 2Cl_2$	900–1200	Films for electronic devices
$SiCl_4 + 2CO_2 + 2H_2 \rightarrow SiO_2 + 4HCl + 2CO$	800–1000	Films for electronic devices, optical fibers
$SiCl_4 + 2H_2O \rightarrow SiO_2 + 4HCl$	500–1000	Films for electronic devices, optical fibers
$SiCl_4 + 2H_2 \rightarrow Si + 4HCl$	500–800	Films for electronic devices
$TiCl_4 + 2BH_3 \rightarrow TiB_2 + 4HCl + H_2$	1000–1300	Monoliths, composites
$SiH_4 + CH_4 \rightarrow SiC + 4H_2$	1000–1400	Coatings
$3SiH_4 + 4NH_3 \rightarrow Si_3N_4 + 12H_2$	800–1500	Films for semiconductor devices
$3HSiCl_3 + 4NH_3 \rightarrow Si_3N_4 + 9HCl + 3H_2$	800–1100	Composites
$BCl_3 + NH_3 \rightarrow BN + 3HCl$	700–1000	Monoliths

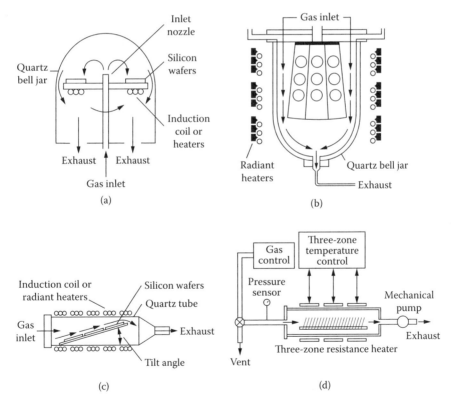

FIGURE 1.2 Typical reactors used in chemical vapor deposition: (a) pancake reactor; (b) barrel reactor; (c) horizontal reactor; (d) low-pressure (LPCVD) reactor. (From Jensen, K.F., Modeling of chemical vapor deposition reactors for the fabrication of microelectronic devices, *Am. Chem. Soc. Symp. Series,* 237, 197, 1984.)

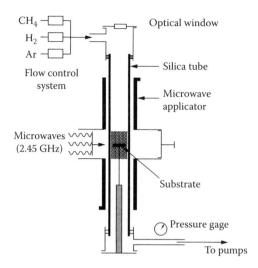

FIGURE 1.3 Schematic diagram of microwave-plasma-assisted chemical vapor deposition (MPACVD) diamond growth system. (From Spear, K.E., Diamond — ceramic coating of the future, *J. Am. Ceram. Soc.*, 72, 171, 1989. With permission.)

The pressure in the reaction vessel influences the concentration of the reactant gases, the diffusion of reactants toward the substrate, and the diffusion of the products away from the surface. The higher diffusivity at lower pressure leads to the formation of films with better uniformity, so most CVD reactors are operated in the pressure range of 1 to 15 kPa. The reactant gases, also referred to as *precursor molecules*, are chosen to react and produce a specific film. Properties necessary for a good precursor include thermal stability at its vaporization temperature and sufficient vapor pressure (at least ~125 Pa) at a reasonable temperature (~300°C) for effective gas-phase delivery to the growth surface. In addition, the molecules must be obtainable at high purity and must not undergo parasitic or side reactions which would lead to contamination or degradation of the film [10]. Examples of the classes of precursor molecules (e.g., hydrides, halides, carbonyls, hydrocarbons, and organometallics) and the types of chemical reaction (pyrolysis, oxidation/hydrolysis, reduction, carbidization/nitridation, and disproportionation) are summarized in Table 1.3.

CVD technology has been attracting much interest recently for the production of diamond films or coatings [11]. The diamond has several attractive properties but, in the past, high pressures and high temperatures were required to produce synthetic diamond. In contrast, a plasma-assisted CVD process allows the production of diamond films at relatively low temperatures and low pressures (Figure 1.3). The deposition process is complex and is not understood clearly at present. The basic reaction involves the pyrolysis of a carbon-containing precursor such as methane:

$$CH_4(g) \rightarrow C(diamond) + 2H_2(g) \qquad (1.1)$$

The typical process consists of the reactant gas at less than atmospheric pressure and containing > 95% H_2. The gas is activated by passing it through a plasma or past a heated filament (at ~2000°C) before deposition on a substrate at 800 to 1000°C.

CVD technology has also been attracting significant interest as a fabrication route for ceramic composites [12]. For fiber-reinforced ceramics, one approach that has shown considerable promise is chemical vapor infiltration (CVI). The fibers, preformed into the shape and dimensions of the finished body, are placed into the reactant gases and held at the desired temperature so that the deposited material is formed in the interstices between the fibers. Significant effort has been devoted to SiC matrix composites reinforced with SiC or C fibers. The SiC matrix is typically deposited

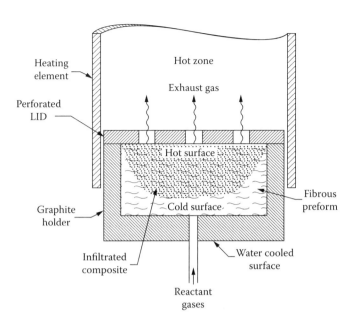

FIGURE 1.4 Schematic diagram of chemical vapor infiltration process exploiting forced flow of the infiltrating gas. (From Stinton, D.P., Besmann, T.M., and Lowder, R.A., Advanced ceramics by chemical vapor deposition techniques, *Am. Ceram. Soc. Bull.*, 67, 350, 1988. With permission.)

from methyltrichlorosilane, CH_3Cl_3Si, at temperatures of ~1200°C and pressures of ~3 kPa. The process is slow, and a serious problem is the tendency for most of the reaction to occur near the surface of the fiber preform, leading to density gradients and the sealing off of the interior. A promising route involves the exploitation of forced flow of the reacting gas into the preform using pressure and temperature gradients (Figure 1.4). Matrices with reasonably high density (typically ~10% porosity) have been produced. The CVI route has an inherent advantage over conventional ceramic powder processing routes that commonly require higher temperatures and high pressures for fabrication: mechanical and chemical degradation of the composite during fabrication is not severe. Composites containing as high as 45 volume percent (vol%) of fibers have been fabricated with an open porosity of ~10%. The measured fracture toughness remained unchanged at ~30 $MPa.m^{1/2}$ up to 1400°C which is considerably better than unreinforced SiC with a fracture toughness of ~3 $MPa.m^{1/2}$ [13].

Table 1.3 indicates that the reaction temperatures for the CVD fabrication of most of the highly refractory ceramics listed are rather low. CVD methods therefore provide a distinct advantage of fairly low fabrication temperatures for ceramics and composites with high melting points that are difficult to fabricate by other methods or which require very high fabrication temperatures. The low reaction temperatures also increase the range of materials that can be coated by CVD, especially for the highly refractory coatings. However, a major disadvantage is that the material deposition rate by CVD is very slow, typically in the range of 1 to 100 μm/h. The production of monolithic bodies can therefore be very time consuming and expensive. Another problem that is normally encountered in the fabrication of monolithic bodies by CVD is the development of a microstructure consisting of fairly large columnar grains, which leads to fairly low intergranular strength. These difficulties limit CVD methods primarily to the formation of thin films and coatings.

1.2.1.2 Directed Metal Oxidation

Fabrication routes involving reactions between a gas and a liquid are generally impractical for the production of ceramic bodies because the reaction product usually forms a solid protective coating,

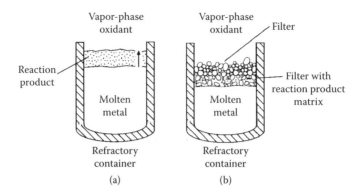

FIGURE 1.5 Schematic diagram of (a) the formation of a matrix of oxide and unreacted metal by directed oxidation of molten metal and (b) oxidation in the presence of a filler (From Newkirk, M.S. et al., Preparation of Lanxide™ ceramic matrix composites: matrix formation by directed metal oxidation, *Ceram. Eng. Sci., Proc.*, 8, 879, 1987. With permission.)

thereby separating the reactants and effectively stopping the reaction. However, a novel method employing directed oxidation of a molten metal by a gas has been developed by the Lanxide Corporation for the production of porous and dense materials, as well as composites [14–19]. Figure 1.5 shows a schematic of the reaction process. In Figure 1.5a, a molten metal (e.g., an Al alloy) is being oxidized by a gas (e.g., air). If the temperature is in the range of 900 to 1350°C, and the Al alloy contains a few percent of Mg and a group IVA element (e.g., Si, Ge, Sn, or Pb), the oxide coating is no longer protective. Instead, it contains small pores through which molten metal is drawn up to the top surface of the film, thereby continuing the oxidation process. As long as the molten metal and the oxidizing gas are available to sustain the process, and the temperature is maintained, the reaction product continues to grow at a rate of a few centimeters per day until the desired thickness is obtained. The material produced in this way consists of two phases: the oxidation product (e.g., Al_2O_3) that is continuous and interconnected, and unreacted metal (Al alloy). The amount of unreacted metal (typically 5 to 30 vol%) depends on the starting materials and processing parameters (e.g., the temperature).

For the production of composites, a filler material (e.g., particles, platelets, or fibers) is shaped into a preform of the size and shape desired of the product. The filler and the metal alloy are then heated to the growth temperature in which the oxidation process occurs outward from the metal surface and into the preform (Figure 1.5b), so that the oxidation product becomes the matrix of the composite. A micrograph of a SiC fiber preform that has been filled with an Al_2O_3/Al matrix by directed oxidation of molten aluminum is shown in Figure 1.6.

The term *directed metal oxidation* is taken to include all reactions in which the metal gives up or shares its electrons. The method has been used to produce composites with not only matrices of oxides but also nitrides, borides, carbides, and titanates. Composite systems produced by the method include matrices of Al_2O_3/Al, AlN/Al, ZrN/Zr, TiN/Ti, and ZrC/Zr, and fillers of Al_2O_3, SiC, $BaTiO_3$, AlN, B_4C, TiB_2, ZrN, ZrB_2, and TiN. A distinct advantage of the method is that growth of the matrix into the preform involves little or no change in dimensions. The problems associated with shrinkage during densification in other fabrication routes (e.g., powder processing) are therefore avoided. Furthermore, large components can be produced readily with good control of the component dimensions.

1.2.1.3 Reaction Bonding

The term *reaction bonding* (or *reaction forming*) is commonly used to describe fabrication routes in which a porous solid preform reacts with a gas (or a liquid) to produce the desired chemical

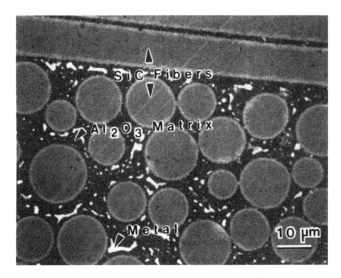

FIGURE 1.6 Optical micrograph of an Al_2O_3/Al matrix reinforced with SiC fibers produced by directed metal oxidation (From Newkirk, M.S. et al., Preparation of Lanxide™ ceramic matrix composites: matrix formation by directed metal oxidation, *Ceram. Eng. Sci., Proc.*, 8, 879, 1987. With permission.)

compound and bonding between the grains. Commonly, the process is accompanied by little or no shrinkage of the perform, thereby providing the benefit of near-net-shape fabrication. Reaction bonding is used on a large scale as one of the fabrication routes for Si_3N_4 and SiC [20,21].

1.2.1.4 Silicon Nitride

Si_3N_4 is the most widely known reaction-bonded system involving the reaction between a solid and a gas phase. In the formation of reaction-bonded Si_3N_4 (RBSN), Si powder is consolidated by one of the common ceramic forming methods (e.g., die pressing, isostatic pressing, slip casting, or injection molding) to form a billet or a shaped article. This is then preheated in argon at ~1200°C to develop some strength, after which it can be machined to the required component shape and dimensions. Finally, the component is heated, usually in N_2 gas at atmospheric pressure and at temperatures in the region of 1250 to 1400°C, when reaction bonding occurs to produce RBSN. Although the mechanism is fairly complex [22], the overall reaction can be written:

$$3Si(s) + 2N_2(g) \rightarrow Si_3N_4(s) \qquad (1.2)$$

The densities of Si and Si_3N_4 are 2.33 g/cm³ and 3.18 g/cm³, respectively, so the reaction of a silicon particle to form Si_3N_4 involves a volume expansion of 22%. However, very little change in the external dimensions of the component occurs during the nitridation. This means that the nitridation occurs by a mechanism in which new mass that has been added to the body expands into the surrounding pore space. As the pore sizes and the number of pores decrease, the pore channels close off and the reaction effectively stops. In preforms with relatively high density, pore channel closure commonly occurs prior to complete reaction. The RBSN has a porosity of 15 to 20% and some residual, unreacted Si. The Si_3N_4 consists of 60 to 90 wt% of the α-phase, the remainder being β-Si_3N_4. Several factors influence the reaction kinetics and the resulting microstructure, including the Si particle size, the composition and pressure of the nitriding gas, the reaction temperature, and impurities in the Si starting powder [22]. Because of the high porosity, the strength of RBSN is inferior to that of dense Si_3N_4 produced by other methods (e.g., hot pressing). On the other hand, RBSN bodies with a high degree of dimensional accuracy and with complex shapes can be prepared fairly readily without the need for expensive machining after firing.

1.2.1.5 Oxides

A reaction bonding route involving both gas–solid and gas–liquid reactions is the reaction-bonded aluminum oxide (RBAO) process developed by Claussen and coworkers [23–27]. The RBAO process utilizes the oxidation of powder mixtures containing a substantial amount of Al (30 to 65 vol%). A mixture of Al (particle size ~20 μm), α-Al_2O_3 (0.5 to 1.0 μm), and ZrO_2 (~0.5 μm) is milled vigorously in an attrition mill, dried and compacted to produce a green article (porosity $\approx$ 30 to 40%). During heat treatment in an oxidizing atmosphere (commonly air), the Al oxidizes to nanometer-sized γ-Al_2O_3 crystals below ~900°C that undergo a phase transformation to α-Al_2O_3 at temperatures up to ~11~00°C. The volume expansion (28 vol%) associated with the oxidation of Al to α-Al_2O_3 is used to partially compensate for the shrinkage due to sintering, so that dense RBAO ceramics can be achieved with lower shrinkage than conventionally sintered Al_2O_3 ceramics. Figure 1.7 shows a micrograph of a reaction-bonded Al_2O_3/ZrO_2 ceramic produced from a starting mixture of 45 vol% Al, 35 vol% Al_2O_3, and 20 vol% ZrO_2. Successful application of the RBAO process depends on several variables, such as the characteristics of the starting powders (e.g., particle size and volume fraction of the Al), the milling parameters, the green density of the compacted mixture, and the heating (oxidation) schedule. ZrO_2 is known to aid the microstructure development during sintering, but its role is not clear. In addition to Al_2O_3, the RBAO process has been applied to the fabrication of mullite ceramics [28] and composites [29].

A gas–solid reaction involving the oxidation of a combination of an alkaline earth metal and another metal has been used recently to produce ceramics containing an alkaline earth element [30–35]. An unusual feature of most alkaline earth metals is the reduction in solid volume accompanying oxidation. For example, the molar volume of MgO is 19% smaller than that of Mg. In contrast, most other metals tend to expand during oxidation. The reduction in volume due to the oxidation of an alkaline earth metal can be used to offset the volume expansion of accompanying the oxidation of another metal. In this way, dense ceramics containing an alkaline earth element can be produced with little change in external dimensions from dense preforms of metal-bearing precursors.

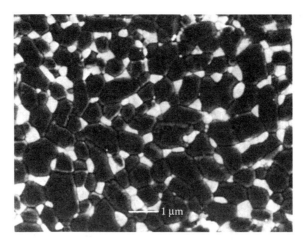

FIGURE 1.7 Scanning electron micrograph showing the microstructure of a reaction-bonded aluminum oxide (RBAO) sample. The white phase is ZrO_2 (~20 vol%) and the dark phase is Al_2O_3. (Courtesy of M.P. Harmer.)

1.2.1.6 Silicon Carbide

Reaction-bonded silicon carbide (RBSC) represents the most important example of the fabrication route based on the reaction between a solid and a liquid [36,37]. Commonly, a mixture of SiC particles (5 to 10 μm), carbon, and a polymeric binder is formed into a green article by pressing, extrusion, or injection molding. In some variations, silicon carbide particles and a carbon-forming resin are used as the starting mixture. The binder or resin is burned off or converted to microporous carbon by pyrolysis, after which the porous preform is infiltrated with liquid Si at temperatures somewhat above the melting point of Si (1410°C). Reaction between the carbon and Si occurs according to:

$$Si(l) + C(s) \rightarrow SiC(s) \qquad (1.3)$$

The reaction product crystallizes on the original SiC grains and bonds them together.

The infiltration and reaction processes occur simultaneously. Capillary pressure provides the driving force for infiltration, and good wetting of the surfaces by liquid Si is a key requirement. The kinetics of the infiltration are complex. However, if the pore structure of the preform is simplified as a set of parallel cylinders, the kinetics can be determined from Poiseuille's equation for viscous flow through a tube:

$$\frac{dV}{dt} = \frac{\pi r^4}{8\eta}\left(\frac{\Delta p}{l}\right) \qquad (1.4)$$

where dV/dt is the rate of liquid flow, r is the radius of the tube with a length l, Δp is the pressure difference across the length of the tube, and η is the viscosity of the liquid. The height infiltrated in time t is found to be:

$$h^2 = \left(\frac{r\gamma_{LV}\cos\theta}{2\eta}\right)t \qquad (1.5)$$

where γ_{LV} is the surface tension of the liquid and θ is the contact angle for wetting. Equation 1.5 shows that the rate of infiltration is proportional to the pore size. As the pore size scales as the particle size, preforms made with larger particles are more readily infiltrated. On the other hand, the strength of the material decreases with larger grain sizes. The reaction given by Equation 1.3 is exothermic and is accompanied by a large heat of reaction. The development of stresses due to thermal gradients can lead to cracking if the reaction kinetics are too fast.

Theoretically, it is possible to calculate the green density of the preform required to produce a fully dense RBSC but in practice it is necessary to leave additional porosity to allow the infiltrating Si to move freely. Consequently, the fabricated material contains an interconnected network of unreacted Si. Commonly, there is no unreacted carbon or porosity, and the principal microstructural features are the original SiC grains coated with the SiC formed in the reaction and the residual Si [38,39]. Figure 1.8 shows the microstructure of an RBSC. The optimum RBSC composition is ~90 vol% SiC and 10 vol% Si but special products with 5 to 25 vol% Si are sometimes made. The presence of Si leads to a deterioration of the mechanical strength at temperatures above ~1200°C [40]. The use of Si alloys (e.g., Si + 2% Mo), instead of pure Si, has been investigated to eliminate the residual Si in the fabricated body [41]. The alloying elements form chemically stable, refractory silicides (e.g., $MoSi_2$) with the excess Si. The reaction bonding process has also been applied to

(a)

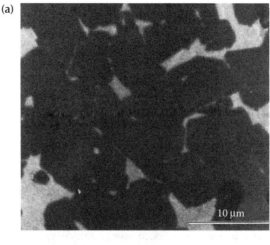

(b)

FIGURE 1.8 The same area of reaction-bonded SiC observed with (a) reflected light showing the light Si phase due to its high reflectivity and (b) secondary electrons in which the Si now appears black and the outer impure regions of SiC grains are lighter than the dark gray cores. (From Sawyer, G.R. and Page, T.F., Microstructural characterization of "REFEL" (reaction bonded) silicon carbides, *J. Mater. Sci.*, 13, 885, 1978. With permission.)

the fabrication of composites in which a reinforcing phase is incorporated into the preform prior to infiltration and reaction [42].

1.2.2 LIQUID PRECURSOR METHODS

Fabrication routes in which a solution of metal compounds is converted into a solid body are sometimes referred to as liquid precursor methods. The sol–gel process has attracted considerable interest since the mid-1970s and forms the most important liquid precursor route for the production of simple or complex oxides. The pyrolysis of suitable polymers to produce ceramics (mainly nonoxides such as SiC and Si_3N_4) has attracted a fair degree of interest in the past 20 years. It is an important route for the production of SiC fibers.

1.2.2.1 Sol–Gel Processing

In the sol–gel process [43–47], a solution of metal compounds or a suspension of very fine particles in a liquid (referred to as a *sol*) is converted into a highly viscous mass (the *gel*). Two different

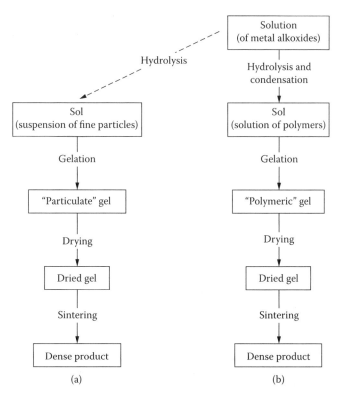

FIGURE 1.9 Basic flowcharts for sol–gel processing using (a) a suspension of fine particles and (b) a solution.

sol–gel processes can be distinguished, depending on whether a sol (Figure 1.9a) or a solution (Figure 1.9b) is used. Starting with a sol, the gelled material consists of identifiable colloidal particles that have been joined together by surface forces to form a network (Figure 1.10a). When a solution is used, typically a solution of metal–organic compounds (such as metal alkoxides), the gelled material in many cases may consist of a network of polymer chains formed by hydrolysis and condensation reactions (Figure 1.10b). Whereas this "solution sol–gel process" is receiving considerable research interest, the sol–gel process based on the gelling of suspensions sees more widespread industrial application.

We shall consider the sol–gel process in more detail later (Chapter 5). Here, we outline the main sequence of steps in the solution sol–gel route as a basis for comparison with the processing steps in other fabrication routes. The starting material normally consists of a solution of metal alkoxides in an appropriate alcohol. Metal alkoxides have the general formula, $M(OR)_x$ and can be considered as either a derivative of an alcohol, ROH, where R is an alkyl group, in which the

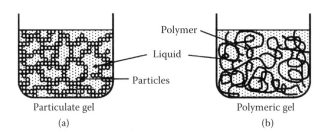

FIGURE 1.10 Schematic diagram of the structure of (a) a particulate gel formed from a suspension of fine particles and (b) a polymeric gel from a solution.

hydroxyl proton is replaced by a metal, M, or a derivative of a metal hydroxide, $M(OH)_x$. To this solution water is added, either in the pure state or diluted with more alcohol. Under constant stirring at temperatures slightly above room temperature (normally ~50 to 90°C) and with suitable concentration of reactants and pH of the solution, hydrolysis and condensation reactions may occur, leading to the formation of polymer chains. Taking the example of a tetravalent metal (e.g., M = Si), the reactions may be expressed as:

Hydrolysis

$$M(OR)_4 + H_2O \rightarrow M(OR)_3OH + ROH \tag{1.6}$$

Condensation

$$M(OR)_3OH + M(OR)_4 \rightarrow (RO)_3 - M - O - M - (OR)_3 + ROH \tag{1.7}$$

Polymerization of the species formed by the hydrolysis and condensation reactions together with interlinking and cross-linking of the polymer chains eventually leads to a marked increase in the viscosity of the reaction mixture and the production of a gel. Normally, excess water and alcohol are used in the reactions so that the amount of solid matter in the gel (i.e., the solids content of the gel) can be quite low, being < 5 to 10 vol% in many cases. The remainder of the volume consists of liquid that must be removed prior to sintering.

Drying of the gels can be the most time consuming and difficult step in the overall fabrication route, especially when a monolithic material is required directly from the gel. Normally, the liquid is present in fine channels, typically ~2 to 50 nm in diameter. Removal of the liquid by evaporation generates large capillary stresses which can cause cracking, warping, and considerable shrinkage of the gel [48,49]. If the liquid-filled pore channels in the gel are simplified as a set of parallel cylinders of radius a, then the maximum capillary stress exerted on the solid network of the gel is:

$$p = \frac{2\gamma_{LV}\cos\theta}{a} \tag{1.8}$$

where γ_{LV} is the specific surface energy of the liquid–vapor interface and θ is the contact angle. For an alkoxide-derived gel with $\gamma_{LV}\cos\theta \approx 0.02$ to 0.07 J/m^2 and $a \approx 1$ to 10 nm, the maximum capillary stress is ≈ 4 to 150 MPa, indicating that the gel can be subjected to quite large stresses. If the evaporation is carried out slowly to control the vapor pressure of the liquid, drying can take weeks for a gel with a thickness of a few centimeters. Aging before drying helps to strengthen the gel network and thereby reduce the risk of fracture [50]. The addition of certain chemical agents, referred to as *drying control chemical additives* (DCCAs), to the solution prior to gelation has been reported to speed up the drying process considerably [51]. However, this may cause serious problems during sintering because some DCCAs are difficult to burn off. Another approach is the use of supercritical drying, in which the problems associated with the capillary stresses are avoided by removing the liquid above the critical temperature and critical pressure [52,53].

Despite drying, the gel contains a small amount of adsorbed water and organic groups such as residual alkyl groups chemically attached to the polymer chains. These are removed below ~500°C prior to densification of the gels at a higher temperature. In general, this densification takes place at a much lower temperature than would be required to make an equivalent material by a more conventional fabrication route such as sintering of powders. This lower temperature densification results primarily from the amorphous nature of the gel and the very fine porosity. Crystallization of the gel prior to significant densification can severely reduce the ease of sintering.

As a fabrication route for ceramics, the sol–gel process has a number of advantages, such as the production of materials with high purity and good chemical homogeneity, and lower temperature sintering. However, the disadvantages are also real. The starting materials (e.g., the metal alkoxides) can be fairly expensive, and conventional drying often presents severe difficulties. Mainly because of the problems in drying, the sol–gel route has seen little use for the fabrication of monolithic ceramics. Instead, it is more suitable to the fabrication of small or thin articles such as films, fibers, and powders.

1.2.2.2 Polymer Pyrolysis

Polymer pyrolysis refers to the pyrolytic decomposition of metal–organic polymeric compounds to produce ceramics. The polymers used in this way are sometimes referred to as *preceramic polymers* in that they form the precursors to ceramics. Unlike conventional organic polymers (e.g., polyethylene) that contain a chain of carbon atoms, the chain backbone in preceramic polymers contains elements other than carbon (e.g., Si, B, and N) or in addition to it. The pyrolysis of the polymer produces a ceramic containing some of the elements present in the chain. Polymer pyrolysis is an extension of the well-known route for the production of carbon materials (e.g., fibers from pitch or polyacrylonitrile) by the pyrolysis of carbon-based polymers [54].

Whereas the possibility of preparing ceramics from metal–organic polymers was recognized for many years [55], heightened interest in the process was generated in the mid-1970s when the formation of fibers with high SiC content was reported by Yajima et al. [56]. Although potentially a large field, the polymer pyrolysis route has been applied most effectively to the production of fibers of nonoxide ceramics, particularly SiC and Si_3N_4 [57–63]. We shall, therefore, focus most of our attention on these two nonoxide ceramics.

The characteristics of the ceramic product formed by polymer pyrolysis depend on the structure and composition of the polymer, and on the pyrolysis conditions. The composition and structure of the polymer are dictated by the chemical synthesis of the monomers and their polymerization reactions. Furthermore, the usefulness of the polymer is determined by its processing characteristics, the ceramic yield on pyrolysis (percentage by weight of ceramic product formed from a given mass of polymer), the purity and microstructure of the ceramic product, and the manufacturing cost. Key requirements for the polymer include: (a) the ability to be synthesized from low-cost starting materials and polymerization reactions, (b) fusibility at moderate temperatures or solubility in solvents for formation into the required shape (e.g., fibers), (c) high ceramic yield on pyrolysis (greater than ~75 wt%) to minimize volume changes, porosity, and shrinkage stresses, and (d) on pyrolysis gives the desired chemical composition and crystalline microstructure of the ceramic product for property optimization (e.g., tensile strength and Young's modulus). Based on these requirements, only a limited number of synthetic routes have been developed. Examples of the types of polymers and the ceramics produced from them are shown in Table 1.4.

1.2.2.3 Silicon Carbide

Polymer precursor routes based on the synthesis of polycarbosilanes have been studied most extensively. The chain backbone of these polymers contains the Si–C bond. The polycarbosilane route, used by Yajima et al. [56] to produce fibers with high SiC content, formed an important contribution to the polymer pyrolysis route for Si-based ceramics. It is used here as an example to illustrate basic steps in the process. The initial step (Equation 1.9) is the condensation reaction between Na and dimethyldichlorosilane, $(CH_3)_2SiCl_2$, in xylene to produce an insoluble poly(dimethylsilane), $[(CH_3)_2Si]_n$, where n ≈ 30.

Poly(dimethylsilane) is not a useful preceramic polymer in that it gives a low ceramic yield on pyrolysis. When treated in an autoclave at 450 to 470°C, complex reactions take place, and a new polymer is produced that no longer has a backbone consisting of just Si atoms. This polymer has Si–C–Si bonds in the chain and is referred to by the general term: *polycarbosilane.*

TABLE 1.4

Some Precursor Polymers and the Ceramics Produced by Their Pyrolysis

Polymer Precursor	Ceramic Yield (Wt%/Atmosphere)	Ceramic Product	Reference
Polycarbosilanes	55–60 (N_2/vacuum)	SiC + amorphous SiC_xO_y	63
Polymethysilanes	~80 (Ar)	SiC	67,68
Polysilazanes	80–85 (N_2)	Amorphous SiC_xN_y	69
	60–65 (N_2)	Amorphous Si_3N_4 + carbon	70
	~75 (N_2)	Amorphous SiC_xN_y	71
Polysiladiazanes	70–80 (Ar)	Amorphous Si_3N_4 + SiC_xN_y	72
Polyborasilazanes	~90 (Ar)	Amorphous $BSi_xC_yN_z$	74
	~75 (NH_3)	Amorphous BSi_xN_y	74

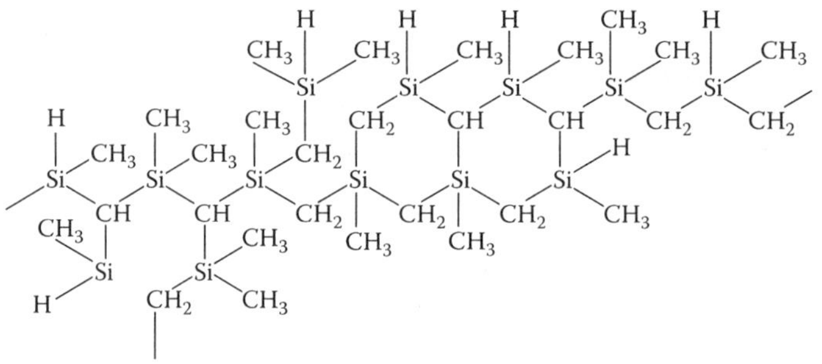

FIGURE 1.11 Constitutional chemical formula of polycarbosilane.

Alternatively, a catalyst, 3 to 5 wt% poly(borodiphenyl siloxane) allows the reaction to take place at lower temperatures (~350°C) and at atmospheric pressure. The purified polycarbosilanes have a complex structure with components consisting of the $CH_3(H)SiCH_2$ unit and the $(CH_3)_2SiCH_2$ unit. The polycarbosilane structure suggested by Yajima [64] is shown in Figure 1.11. The polycarbosilanes are glassy solids with a number average molecular weight, M_n, of 1000 to 2000 and, on pyrolysis in an inert atmosphere, give a ceramic yield of 55 to 60 wt%. They are soluble in common organic solvents and the melt can be spun into fibers with small diameters (10 to 20 μm).

In the formation of fibers, after melt spinning at 250 to 350°C, the fibers undergo a rigidization step, referred to as *curing*, prior to conversion to the ceramic. Curing is accomplished by heating in air at 100 to 200°C to cross-link the polymer chains and serves to maintain geometrical integrity of the fibers. During pyrolysis in an inert atmosphere (e.g., N_2), an amorphous product is produced between 800 and 1100°C which crystallizes above ~1200°C. The ceramic yield increases to ~80 wt% and this increase in the yield (when compared to the uncured material) correlates with cross-linking of the polymer chains by oxygen atoms during the curing step. The ceramic is not stoichiometric SiC but contains excess carbon and some oxygen (presumably from the curing step). In a variation of the process, Yajima et al. [65] heated the polycarbosilane with titanium tetrabutoxide, $Ti(OC_4H_9)_4$, to give a more highly cross-linked polymer, referred to as polytitano-carbosilane, containing Si, Ti, C, O, and H. Pyrolysis in an inert atmosphere produces a ceramic consisting mainly of SiC, a small amount of TiC, excess carbon and some oxygen. The ceramic yield is ~75 wt%.

$$(1.9)$$

The procedures developed by Yajima et al. [56,65] form the basis for the industrial production of fibers with high SiC content. Fibers derived from polycarbosilane are produced by Nippon Carbon Company (Japan) under the trade name of Nicalon whereas Ube Industries (Japan) produces fibers having the trade name Tyranno, which are derived from polytitanocarbosilane. Although nominally described as silicon carbide fibers, Nicalon fibers contain excess carbon and oxygen with a nominal composition of $SiC_{1.25}O_{0.33}$. The microstructure consists of microcrystalline β-SiC grains in an amorphous silicon oxycarbide matrix as well as significant microporosity. As the only high performance fibers available for several years, Nicalon fibers have been studied extensively and have been incorporated into more ceramic composites than any other fiber. The ceramic grade (CG) Nicalon fibers commonly used in ceramic composites have a tensile strength of ~3 GPa and a Young's modulus of ~200 GPa [66].

The nonstoichiometric composition coupled with the microporosity and amorphous phase in the microstructure reduces the thermal stability of the fibers. Depending on the magnitude and duration of the applied stress, the use of the fibers may be limited to temperatures lower than 1000°C. To overcome this deficiency, recent efforts have been devoted to producing dense, polycrystalline, stoichiometric SiC through modification of the synthesis and processing conditions or through the synthesis of more useful preceramic polymers [63,67,68].

1.2.2.4 Silicon Nitride

When compared to SiC, less work has been reported on the production of Si_3N_4 by the polymer pyrolysis route. Most efforts have focused on polymer precursors based on polysilazanes, a class of polymers having Si–N bonds in the main chain [58–61]. The reactions to produce the Si–N bond in the chain backbone are based on the ammonolysis of methylchlorosilanes. As indicated in Equation 1.10, a preceramic polymer can be prepared by the ammonolyis of methyldichlorosilane, followed by the polymerization of the silazane product catalyzed by potassium hydride [69]. Silazanes with an average molecular weight of ~300 are produced in the first step. Pyrolysis of this material gives a ceramic yield of ~20 wt%. The second step results in the production of polysilazanes with molecular weight ranging from 600 to 1800. The suggested structure is shown in Figure 1.12. Pyrolysis of these polysilazanes in N_2 at ~1000°C gives a ceramic yield of 80 to 85 wt%. The constitution of the product is not clear. Based on elemental chemical analysis, the constitution of the product has been suggested as an amorphous mixture of Si_3N_4 (~66 wt%), SiC (~27 wt%), C (~5 wt%) and, possibly, SiO_2 (~2 wt%). However, x-ray diffraction and nuclear

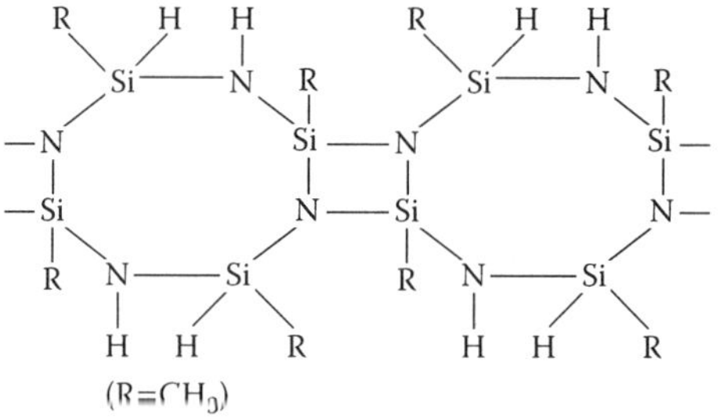

FIGURE 1.12 Example of the linkage of eight-membered rings via Si_2N_2 bridges in a polysilazane.

magnetic resonance indicate that it may be an amorphous mixture silicon–carbon–nitride and free carbon. Polysilazanes with a repeating unit of $[H_2SiNCH_3]$ that is isostructural and chemically identical with the $[CH_3HSiNH]$ repeating unit in Equation 1.10 have also been studied as a precursor to silicon nitride [70]. In this case, pyrolysis in N_2 at 1273 K gives a ceramic yield of 63 wt% with the product consisting of predominantly of amorphous Si_3N_4 (77 wt%) and free carbon (18 wt%). These results indicate that the structure of the repeating unit in the preceramic polymer may have a significant influence on the composition and microstructure of the ceramic product.

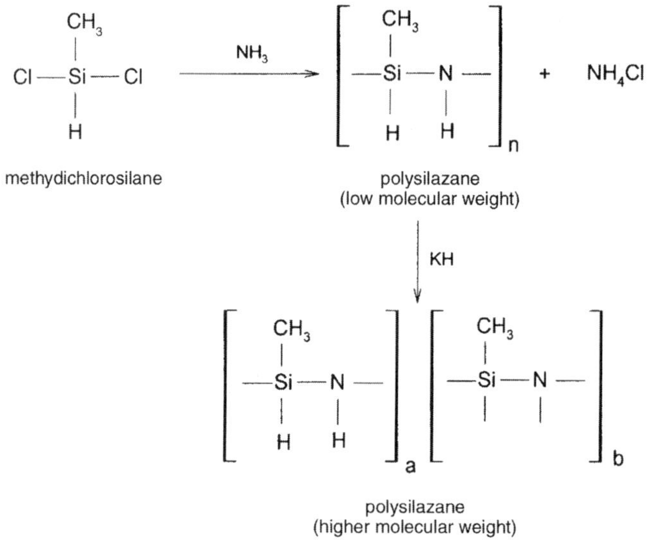

(1.10)

Attempts have been made to improve the Si_3N_4 content and reduce the amount of carbon in the ceramic product through modification of the precursor chemistry. They include the synthesis of precursors based on hydridopolysilazane [71] and poly(methysiladiazane) [72].

1.2.2.5 Boron Nitride and Boron Carbide

Several routes exist for the preparation of polymer precursors to BN and B_4C [58–61,73]. The reaction between $H_3B \cdot S(CH_3)_2$ and the silazane product, $(CH_3HSiNH)_n$, formed in the first step of Equation 1.10, produces borasilazane polymers (molecular weight ~800) that are useful precursors to boron-containing ceramics [74]. The polymers are soluble in many common organic solvents and can be processed into fibers. On pyrolysis in argon at 1000°C, polymers (formed with a Si/B reactant ratio of ~2) give a high yield (~90 wt%) of an amorphous boron–silicon–carbon–nitride ceramic with a composition of $B_{1.0}Si_{1.9}C_{1.7}N_{2.5}$. Pyrolysis in NH_3 at 1000°C gives an amorphous boron–silicon–nitride ceramic with the composition BSi_3N_5 with a yield of ~75 wt%. For B_4C/BN ceramics, high yield synthesis has been reported from polymer precursors synthesized by the reaction of decaborane ($B_{10}H_{14}$) with a diamine (e.g., $H_2NCH_2CH_2NH_2$) in an organic solvent [75].

Whereas the polymer pyrolysis route has key advantages, such as ease of processing into the desired shape and relatively low conversion temperature to nonoxide ceramics, the pyrolysis step is accompanied by a large volume change that makes the fabrication of monolithic ceramics difficult. Preceramic polymers have been investigated for use as binders in the processing of nonoxide ceramic powders [76–78] but being very expensive, they increase the fabrication costs significantly. The polymer pyrolysis route finds its most important use in the production of thin objects, in particular fibers and, to a more limited extent, coatings [79].

1.2.3 Fabrication from Powders

This route involves the production of the desired body from an assemblage of finely divided solids (powders) by the action of heat. It gives rise to the two most widely used methods for the fabrication of ceramics: melting followed by casting (or forming) into shape, referred to simply as *melt casting,* and sintering of compacted powders.

1.2.3.1 Melt Casting

In its simplest form, this method involves melting a mixture of raw materials (commonly in the form of powders), followed by forming into shape by one of several methods, including casting, rolling, pressing, blowing, and spinning. For ceramics that crystallize relatively easily, solidification of the melt is accompanied by rapid nucleation and growth of crystals (i.e., grains). Uncontrolled grain growth is generally a severe problem that leads to the production of ceramics with undesirable properties (e.g., low strength). Also, obtaining a melt is rather difficult because many ceramics have high melting points (e.g., ZrO_2 with a melting point of ~2600°C) or decompose prior to melting (e.g., Si_3N_4). The melt casting method is therefore limited to the fabrication of glasses [80–83].

An important variation of glass processing is the glass–ceramic route [84]. Here, the raw materials are first melted and formed into shape in the glassy state by the conventional glass fabrication methods outlined above. The glass is then crystallized using a two-step heat treatment consisting of a lower temperature hold to induce nucleation of crystals, followed by one or more higher-temperature holds to promote growth of the crystals throughout the glass. Glass ceramics are by definition ≥ 50% crystalline by volume and most are > 90% crystalline. The glass–ceramic route has the advantage of economical fabrication by the conventional glass manufacturing methods. Furthermore, the composite structure, consisting of small crystals (~0.1 to 10 μm) held together in a glassy matrix, provides glass–ceramics with improved properties when compared to the original glass. Generally, glass–ceramics have higher strength, chemical durability, and electrical resistance, and can be made with very low thermal expansion coefficients, giving excellent thermal shock resistance [85]. The highest volume applications are cookware and tableware, architectural cladding, stove tops, and stove windows.

The potential range of glass ceramic compositions is fairly broad because the method depends only on the ability to form a glass and to control its crystallization. However, almost all glass ceramics developed so far are based on silicate glass compositions, such as the $Li_2O–Al_2O_3–SiO_2$ (LAS) and $MgO–Al_2O_3–SiO_2$ (MAS) systems. A typical commercial cordierite glass ceramic (Corning Code 9606) has the composition (in wt%): SiO_2 (56.0), Al_2O_3 (19.7), MgO (14.7), TiO_2 (9.0), CaO (0.11) and As_2O_3 (0.5). Compared to the stoichiometric cordierite composition ($2MgO.2Al_2O_3.5SiO_2$), this commercial composition has an excess of SiO_2 and utilizes 9 wt% TiO_2 for the nucleation of crystals. The phases produced by heat treatment above ~1200°C are cordierite, magnesium aluminum titanate, and cristobalite. The microstructure of a cordierite glass ceramic (Corning Code 9606) is shown in Figure 1.13. The predominant phase with somewhat rounded grains is cordierite, and the angular, needle-shaped crystals are magnesium aluminum titanate.

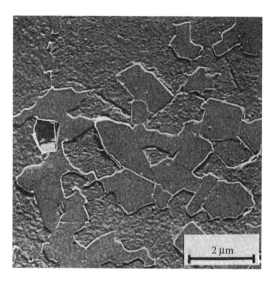

FIGURE 1.13 The microstructure of a commercial cordierite glass ceramic (Corning Code 9606). The predominant phase with somewhat rounded grains is cordierite and the angular, needle-shaped crystals are magnesium aluminum titanate. (From Grossman, D.G., Glass ceramics, in *Concise Encyclopedia of Advanced Ceramics*, Brook, R.J., Ed., The MIT Press, Cambridge, MA, 1991, p. 173. With permission.)

1.2.3.2 Sintering of Compacted Powders

Whereas in principle this route can be used for the production of both glasses and polycrystalline ceramics, in practice it is hardly ever used for glasses because of the availability of more economical fabrication methods (e.g., melt casting). It is, however, by far the most widely used method for the production of polycrystalline ceramics. The various processing steps are shown in Figure 1.14. In its simplest form, this method involves the consolidation of a mass of fine particles (i.e., a powder) to form a porous, shaped article (also referred to as a *green body* or *powder compact*), which is then sintered (i.e., heated) to produce a dense product. Because of its importance and widespread use, the fabrication of polycrystalline ceramics from powders will form the main focus of this book.

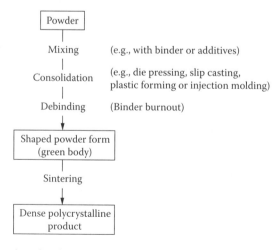

FIGURE 1.14 Basic flowchart for the production of polycrystalline ceramics by sintering of consolidated powders.

In the next section, we provide an overview of the fabrication of polycrystalline ceramics from powders, which will form the basis for the more detailed considerations in subsequent chapters.

1.3 PRODUCTION OF POLYCRYSTALLINE CERAMICS FROM POWDERS: AN OVERVIEW

In the flowchart shown in Figure 1.14, we can divide the processing steps into two parts: processes prior to the sintering of the green body and those that occur during sintering. Until recently, most emphasis was placed on the processes that occur during sintering, and the understanding gained from these studies is considerable. However, the increased attention given recently to powder synthesis and forming methods has yielded clear benefits. In general, each processing step in the fabrication route has the potential for producing undesirable microstructural flaws in the body which can limit properties and reliability. Therefore, close attention should be paid to each step if the very specific properties required of many ceramics are to be achieved. The important issues in each step and how these influence or are influenced by the other steps are outlined in the following sections.

1.3.1 POWDER SYNTHESIS AND POWDER CHARACTERIZATION

In most cases, the fabrication process starts from a mass of powder obtained from commercial sources. Nevertheless, knowledge of powder synthesis methods is very important. Equally important are methods that can be used to determine the physical, chemical, and surface characteristics of the powder. The characteristics of the powder depend strongly on the method used to synthesize it, and these, in turn, influence the subsequent processing of the ceramic. The powder characteristics of greatest interest are the size, size distribution, shape, degree of agglomeration, chemical composition, and purity. Many methods are available for the synthesis of ceramic powders. These range from mechanical methods involving predominantly grinding or milling for the reduction in size of a coarse, granular material (referred to as *comminution*) to the chemical methods involving chemical reactions under carefully controlled conditions. Some of the chemical methods, although often more expensive than the mechanical methods, offer unprecedented control of the powder characteristics. Figure 1.15 shows an example of a chemically synthesized powder consisting of spherical particles of approximately the same size [86].

One of the more troublesome issues in the production of advanced ceramics is the effect of minor variations in the chemical composition and purity of the powder on processing and properties.

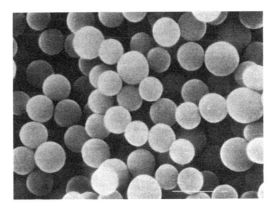

FIGURE 1.15 An example of a submicrometer TiO_2 powder prepared by controlled chemical precipitation from a solution. (From Matijevic, E., Monodispersed colloidal metal oxides, sulfides, and phosphates, in *Ultrastructure Processing of Ceramics, Glasses, and Composites*, Hench, L.L. and Ulrich, D.R., Eds., John Wiley & Sons, New York, 1984, p. 334. With permission.)

These variations, produced by insufficient control of the synthesis procedure or introduced during subsequent handling, often go unrecorded or undetected but their effects on the microstructure and properties of the fabricated material can often be quite profound. For ceramics that must satisfy very demanding property requirements, one of the major advances made in the last 20 years has been the attention paid to powder quality. This has resulted in greater use of chemical methods for powder synthesis coupled with careful handling during subsequent processing.

A continuing trend is toward the preparation of fine powders. In particular, very fine powders with a particle size smaller than 50 to 100 nm, often referred to as *nanoscale* or *nanosize* powders, have been the subject of considerable recent research (see Chapter 2). Although the enhanced activity of fine powders is beneficial for achieving high density at lower sintering temperatures [87], the benefits of fine powders are normally realized only when extreme care is taken in their handling and subsequent consolidation. Generally, as the size decreases below ~1 µm, the particles exhibit a greater tendency to interact, leading to the formation of agglomerates. One consequence of the presence of agglomerates is that the packing of the consolidated powder can be quite nonuniform. The overall effect is that during the sintering stage little benefit is achieved over a coarse powder with a particle size corresponding to the agglomerate size of the fine powder. The use of fine powders, therefore, requires proper control of the handling and consolidation procedures in order to minimize the deleterious effects due to the presence of agglomerates. Such procedures may be quite demanding and expensive.

1.3.2 POWDER CONSOLIDATION

The consolidation of ceramic powders to produce a shaped article is commonly referred to as *forming*. The main forming methods include: (1) dry or semidry pressing of the powder (e.g., in a die), (2) mixing of the powder with water or organic polymers to produce a plastic mass that is shaped by pressing or deformation (referred to as *plastic forming*), and (3) casting from a concentrated suspension or slurry (e.g., slip casting and tape casting). These methods have been in use for a long time and most have originated in the traditional ceramics industry for the manufacture of clay-based materials. More recently, considerable interest has been devoted to a group of forming methods referred to as *solid freeform fabrication* or *rapid prototyping* which employ computer-assisted techniques and do not require the use of conventional tools such as dies and molds (see Chapter 7).

Perhaps the greatest advance made in the last 20 years or so has been the realization that the green body microstructure has a profound influence on the subsequent sintering stage. If severe variations in packing density occur in the green body, then under conventional sintering conditions, the fabricated body will usually have a heterogeneous microstructure that limits the engineering properties and reliability. This realization has led to the increasing use of colloidal techniques for the consolidation of powders from a suspension [88]. Figure 1.16 shows an example of the uniform arrangement formed from almost spherical, nearly monosized particles [89]. Although their benefits have been demonstrated, colloidal methods have not made inroads into many industrial applications in which mass production is desired and fabrication cost is a serious consideration.

1.3.3 THE SINTERING PROCESS

In this stage, the green body is heated to produce the desired microstructure. The changes occurring during this stage may be fairly complex, depending on the complexity of the starting materials. Two terms will be encountered for the heating stage: firing and sintering. Generally, the term *firing* is often used when the processes occurring during the heating stage are fairly complex, as in many traditional ceramics produced from clay-based materials. In less complex cases, the term *sintering* is often used. We will not overemphasize the distinction in this book.

The less complex nature of sintering allows it to be analyzed theoretically in terms of idealized models. The theoretical analyses combined with experimental investigations during the last 50 years

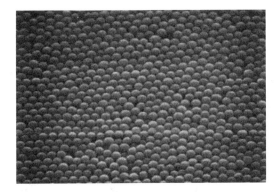

FIGURE 1.16 Uniformly packed submicrometer TiO_2 powder produced by consolidation from a stable suspension. (From Barringer, E.A. and Bowen, H.K., Formation, packing and sintering of monodispersed TiO_2 powders, *J. Am. Ceram. Soc.*, 65, C-199, 1982. With permission.)

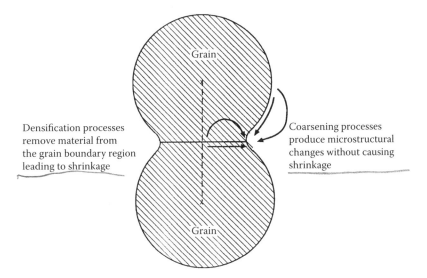

FIGURE 1.17 Schematic indication of the distinction between densifying and nondensifying microstructural changes resulting from atom transport during the sintering of ceramic powders.

or so have provided a considerable understanding of sintering. The simplest case is that for a pure, single-phase material (e.g., Al_2O_3). The system is heated to a temperature that is in the range of 0.5 to 0.75 of the melting temperature. (For Al_2O_3 with a melting temperature of 2073°C, the sintering temperature is commonly 1400 to 1650°C.) The powder does not melt; instead, the joining together of the particles and the reduction in the porosity (i.e., densification) of the body, as required in the fabrication process, occurs by atomic diffusion in the solid state. This type of sintering is usually referred to as *solid-state sintering*. Whereas solid-state sintering is the simplest case of sintering, the processes occurring and their interaction can be fairly complex.

The driving force for sintering is the reduction in surface free energy of the consolidated mass of particles. This reduction in energy can be accomplished by atom diffusion processes that lead to either densification of the body (by transport matter from inside the grains into the pores) or coarsening of the microstructure (by rearrangement of matter between different parts of the pore surfaces without actually leading to a decrease in the pore volume). The diffusion paths for densification and coarsening are shown in Figure 1.17 for an idealized situation of two spherical particles in contact [90]. From the point of view of achieving high density during sintering, a major

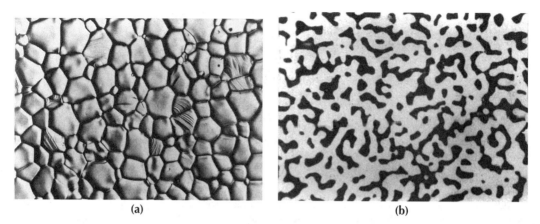

(a)   (b)

FIGURE 1.18 (a) The surface of an alumina ceramic from which all porosity has been removed during the firing of the powder; the microstructure consists of the cystalline grains and the boundaries (interfaces) between them. (b) The sintering of silicon results in the formation of a continuous network of solid material (white) and porosity (black); this microstructural change in not accompanied by any shrinkage. (From Brook, R.J., Advanced ceramic materials: an overview, in *Concise Encyclopedia of Advanced Ceramic Materials*, Brook, R.J., Ed., The MIT Press, Cambridge, MA, 1991, p. 3. With permission.)

problem is that the coarsening process reduces the driving force for densification. This interaction is sometimes expressed by the statement that sintering involves a competition between densification and coarsening. The domination of densifying diffusion processes will favor the production of a dense body (Figure 1.18a). When coarsening processes dominate, the production of a highly porous body will be favored (Figure 1.18b).

The effects of key material and processing parameters such as temperature, particle (or grain) size, applied pressure, and gaseous atmosphere on the densification and coarsening processes are well understood. The rates of these processes are enhanced by higher sintering temperature and by fine particle size. Densification is further enhanced by the application of an external pressure. A key issue that has received increasing attention in recent years is the effect of microstructural inhomogeneities present in the green body (e.g., density, grain size and compositional variations). It is now well recognized that inhomogeneities can seriously hinder the ability to achieve high density and to adequately control the microstructure. As outlined earlier, a consequence of this realization has been the increased attention paid to powder quality and to powder consolidation by colloidal methods.

A common difficulty in solid-state sintering is that coarsening may dominate the densification process, with the result that high density is difficult to achieve. This difficulty is especially common in highly covalent ceramics (e.g., Si_3N_4 and SiC). One solution is the use of an additive that forms a small amount of liquid phase between the grains at the sintering temperature. This method is referred to as *liquid-phase sintering*. The liquid phase provides a high diffusivity path for transport of matter into the pores to produce densification but is insufficient, by itself, to fill up the porosity. A classic example of liquid-phase sintering in ceramics is the addition of 5 to 10 wt% of MgO to Si_3N_4 (Figure 1.19). Whereas the presence of the liquid phase adds a further complexity to the sintering process, the benefits can be significant, as demonstrated by the widesread use of liquid-phase sintering in industry. Another solution to the difficulty of inadequate densification is the application of an external pressure to the body during heating in either case of solid-state or liquid-phase sintering. This method is referred to as *pressure sintering* of which hot pressing and hot isostatic pressing are well-known examples. The applied pressure has the effect of increasing the driving force for densification without significantly affecting the rate of coarsening. A common drawback of pressure sintering, however, is the increase in fabrication costs.

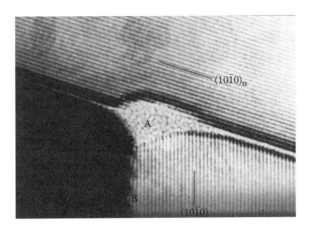

FIGURE 1.19 Microstructure produced by liquid-phase sintering, under an applied pressure, of Si_3N_4 with MgO additive. A continuous glassy phase, ~0.8-nm thick, separates the crystalline grains. (Courtesy of D.R. Clarke.)

1.3.4 Ceramic Microstructures

The microstructure of the fabricated article is significantly dependent on the processing methods, so therefore an examination of the microstructure may serve as a test of successful processing. Equally important, the microstructural observations may lead to inferences about the way in which the processing methods must be modified or changed in order to obtain the desired characteristics. As we have seen from the previous section, ceramic microstructures cover a wide range. For solid-state sintering in which all the porosity is successfully removed, a microstructure consisting of crystalline grains separated from one another by grain boundaries is obtained (Figure 1.18a). However, most ceramics produced by solid-state sintering contain some residual porosity (Figure 1.20). The use of liquid-phase sintering leads to the formation of an additional phase at the grain boundaries. Depending on the nature and amount of liquid, the grain boundary phase may be continuous, thereby separating each grain from the neighboring grains (Figure 1.19), or may be discontinuous, e.g., at the corners of the grains.

Generally, the advanced ceramics that must meet exacting property requirements tend to have relatively simple microstructures. A good reason for this is that the microstructure is more amenable to control when the system is less complex. Even so, the attainment of these relatively simple

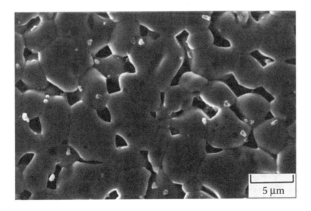

FIGURE 1.20 Incomplete removal of the porosity during solid-state sintering of CeO_2 results in a microstructure consisting of grains, grain boundaries and pores.

FIGURE 1.21 A commercial sanitaryware body produced by firing a mixture of clay, feldspar and quartz, showing a fairly complex microstructure containing some residual feldspar, F, porosity, P, and quartz, Q. (From Chu, G.P.K., Microstructure of complex ceramics, in *Ceramic Microstructures*, Fulrath, R.M. and Pask, J.A., Eds., John Wiley & Sons, New York, 1968, p. 828, chap. 38. With permission.)

microstructures in advanced ceramics can be a difficult task. For the traditional clay-based ceramics for which the properties achieved are often less critical than the cost or shape of the fabricated article, the microstructures can be fairly complex, as shown in Figure 1.21 for a ceramic used for sanitaryware [16].

Ceramic composites are a class of materials that have undergone rapid development in recent years because of their promising properties for structural applications at high temperatures (e.g., heat engines). A reinforcing phase (e.g., SiC fibers) is deliberately added to the ceramic (e.g., Al_2O_3) to make it mechanically tougher (i.e., reduce its brittleness). For these materials, the complexity of the microstructure is controlled by having an ordered distribution of the reinforcing phase in the ceramic matrix (see Figure 1.6).

We discussed earlier the strong consequences of insufficient attention to the quality of the powder and to the consolidation of the powder in the production of the green body. Microstructural flaws that limit the properties of the fabricated body normally originate in these stages of processing. Large voids and foreign objects such as dust or milling debris are fairly common. The control of the powder quality, including the particle size distribution, the shape, and the composition, along with minor constituents, has a major influence on the development of the microstructure. Impurities can lead to the presence of a small amount of liquid phase at the sintering temperature, which causes selected growth of large individual grains (Figure 1.22). In such a case, the achievement of

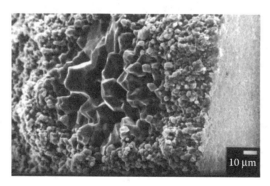

FIGURE 1.22 Large-grained region of microstructural heterogeneity resulting from an impurity in hot-pressed Al_2O_3. (Courtesy of B.J. Dalgleish.)

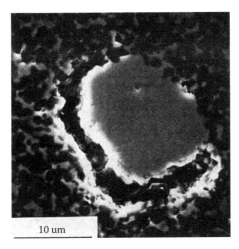

FIGURE 1.23 Crack-like void produced by a ZrO_2 agglomerate shrinking away from the surrounding Al_2O_3/ZrO_2 matrix during sintering. (Courtesy of F.F. Lange.)

a fine uniform grain size would be impossible. The homogeneity of particle packing in the consolidated material is also very important. In general, any significant microstructural heterogeneity in the green body is exaggerated in the sintering process, leading to the development of crack-like voids or large pores between relatively dense regions (Figure 1.23).

Assuming proper precautions were taken in the processing steps prior to firing, further microstructural manipulation must be performed during sintering. Unless high porosity is a deliberate requirement, we normally wish to achieve as high a density, as small a grain size, and as uniform a microstructure as possible. Insufficient control of the sintering conditions (e.g., sintering temperature, time at sintering temperature, rate of heating, and sintering atmosphere) can lead to flaws and the augmentation of coarsening, which make attainment of the desired microstructure impossible. The sintering of many materials (e.g., Si_3N_4, lead-based ferroelectric ceramics and β-alumina) also requires control of the atmosphere in order to prevent decomposition or volatilization.

Most ceramics are not single-phase solids. For a material consisting of two solid phases, our general requirement is for a homogeneous distribution of one phase in a homogeneously packed matrix of the other phase. Close attention must be paid to the processing steps prior to sintering (e.g., during the mixing and consolidation stages) if this requirement is to be achieved in the fabricated body. The consequences of inhomogeneous mixing for the microstructure control of an Al_2O_3 powder containing fine ZrO_2 particles are illustrated in Figure 1.24. As in the case for single-phase ceramics, we must also pay close attention to the firing stage. An additional effect now is the possibility of chemical reactions between the phases. As we have outlined earlier, Al_2O_3 can be made mechanically tougher (i.e., less brittle) by the incorporation of SiC fibers. However, oxidation of the SiC fibers leads to the formation of an SiO_2 layer on the fiber surfaces. The SiO_2 can then react with the Al_2O_3 to form aluminosilicates with a consequent deterioration of the fibers. Prolonged exposure of the system to oxygen will eventually lead to the deterioration and practical elimination of the fibers from the system. A primary aim therefore would be the control of the atmosphere during densification in order to prevent oxidation.

To summarize at this stage, although sufficient attention must be paid to the sintering stage, defects produced in the processing steps prior to sintering cannot normally be reduced or eliminated. In most cases, these defects are enhanced during the sintering stage. In general properties are controlled by the microstructure (e.g., density and grain size), but the flaws in the fabricated body have a profound effect on those properties that depend on failure of the material (e.g., mechanical strength, dielectric strength, and thermal shock resistance). Failure events are almost always initiated

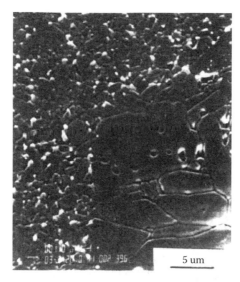

FIGURE 1.24 The nonuniform distribution of fine ZrO_2 particles (light phase) in Al_2O_3 (dark phase) is seen to result in a region of uncontrolled grain growth during sintering. (Courtesy of F.F. Lange.)

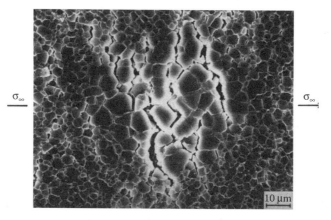

FIGURE 1.25 Crack nucleation at a large-grained heterogeneity during deformation of a hot-pressed Al_2O_3 body. (Courtesy of B.J. Dalgleish.)

at regions of physical or chemical heterogeneity. An example is shown in Figure 1.25 in which mechanical failure originates at a large-grained heterogeneity [92].

1.4 CASE STUDY IN PROCESSING: FABRICATION OF TiO₂ FROM POWDERS

Earlier, we outlined the potential benefits that may be obtained from careful control of the powder quality and the powder consolidation method. To further illustrate these benefits, we consider the case of TiO_2. One fabrication route for the production of TiO_2 bodies, referred to as the *conventional route* [93], is summarized in the flow diagram of Figure 1.26a. Typically, TiO_2 powder available from commercial sources is mixed with small amounts of additives (e.g., Nb_2O_5 and $BaCO_3$), which aid the sintering process, and calcined for 10 to 20 h at ~900°C in order to incorporate the additives into solid solution with the TiO_2. The calcined material is mixed with a small amount of binder (to aid the powder compaction process) and milled in a ball mill to break down agglomerates present

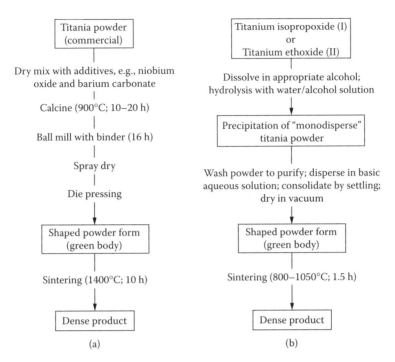

FIGURE 1.26 Flowchart for the production of TiO_2 by (a) a conventional powder route and (b) the powder route used by Barringer and Bowen. (From Barringer, E.A. and Bowen, H.K., Formation, packing and sintering of monodispersed TiO_2 powders, *J. Am. Ceram. Soc.*, 65, C-199, 1982. With permission.)

in the material. The milled powder in the form of a slurry is sprayed into a drying apparatus (available commercially and referred to as a *spray dryer*). The spray-drying process serves to produce a dried powder in the form of spherically-shaped agglomerates. After compaction, the powder is sintered for 10 h at ~1400°C to produce a body with a density of ~93% of the theoretical density of TiO_2.

In the second fabrication route used by Barringer and Bowen [89], powders with controlled characteristics are prepared and consolidated, by colloidal methods, into uniformly packed green bodies (e.g., similar to that shown in Figure 1.16). An abbreviated flow diagram of the fabrication route is shown in Figure 1.26b. Powders are prepared by controlled hydrolysis of titanium tetra-isopropoxide (denoted powder I) or titanium tetraethoxide (powder II). The powder particles are nearly spherical and almost monosized. After washing, each powder is dispersed in a basic aqueous solution to produce a stable suspension. Consolidation of the suspension is accomplished by gravitational or centrifugal settling. The consolidated body is dried in a vacuum and sintered for ~90 min at 800°C (for powder I) or 1050°C (for powder II) to produce bodies with a density greater than 99% of the theoretical.

Table 1.5 gives a summary of the main results of the preceding two fabrication methods. It is easily recognized that the benefits achieved by the fabrication route of Barringer and Bowen are very extraordinary. These include a substantial reduction in sintering temperature (by 350 to 600°C) and sintering time, a higher final density and a much smaller grain size. Also, because no milling, binder or sintering aids are required, the purity of the fabricated bodies can be expected to be higher than for the conventional route.

The experiments of Barringer and Bowen clearly demonstrate that substantial benefits can be achieved in sintering and microstructure control when the powder quality is carefully controlled and the powder packing is homogeneous, i.e., when careful attention is paid to the processing steps that precede sintering. However, this fabrication route is not currently used in many industrial applications in which mass production and low cost are important considerations.

TABLE 1.5
Processing Data for the Production of Titanium Oxide by a Conventional Route [93] and by the Method of Barringer and Bowen [89]

Process Parameter	Conventional	Barringer and Bowen
Powder particle size (μm)	1	0.1 (I)
		0.4 (II)
Sintering temperature (°C)	1400	800 (I)
		1050 (II)
Sintering time (h)	10	1.5
Final relative density (%)	93	>99
Final grain size (μm)	50–100	0.15 (I)
		1.2 (II)

Source: From Yan, M.F. and Rhodes, W.W., Efficient process for volume production of TiO_2 varistor powder, *Am. Ceram. Soc. Bull.*, 63, 1484, 1984; Barringer, E.A. and Bowen, H.K., Formation, packing and sintering of monodispersed TiO_2 powders, *J. Am. Ceram. Soc.*, 65, C-199, 1982. With permission.

1.5 CONCLUDING REMARKS

In this chapter we have examined, in general terms, the common methods for the fabrication of ceramics. By far the most widely used method for the production of polycrystalline ceramics is the sintering of consolidated powders. Solution-based methods (e.g., sol–gel processing) are attracting significant research interest and their use is expected to grow substantially in the future. Although the powder processing route has its origins in early civilization, its use in the production of advanced ceramics with the desired microstructure can still involve severe difficulties. Each processing step has the potential for producing microstructural flaws in the fabricated body that cause a deterioration in properties. Careful attention must therefore be paid to each processing step in order to minimize the microstructural flaws. Finally, we considered the fabrication of titania and showed the enormous benefits that can be achieved by careful attention to the processing.

PROBLEMS

1.1

Silicon powder is compacted to form a green body with a density equal to 65% of the theoretical value and heated in N_2 to form reaction-bonded silicon nitride (RBSN). If the reaction goes to completion and there is no change in the external dimensions of the compact, determine the porosity of the RBSN.

1.2

Consider the formation of Al_2O_3/ZrO_2 composites by the reaction-bonded aluminum oxide (RBAO) process. A mixture of 45 vol% Al, 35 vol% Al_2O_3, and 20 vol% ZrO_2 (monoclinic phase) is attrition-milled, compacted to form a green body with 35% porosity, and heated slowly to oxidize the Al prior to sintering. If the volume fraction of Al oxidized during the milling step is zero, determine the porosity of the compact after the oxidation step, assuming that the Al is completely oxidized.

If the compact reaches full density during the sintering step, determine the linear shrinkage of the body. State any assumptions that you make.

Repeat the calculations for the case when 10% of the Al is oxidized during the milling step, assuming that the porosity of the green body is the same.

1.3

One approach to the formation of mullite by the reaction-bonded aluminum oxide (RBAO) process uses a starting mixture of Al and SiC powders. After attrition milling, the mixture is compacted to form a green body with a relative density of ρ_o. Oxidation of the green body leads to the formation of a mixture of Al_2O_3 and SiO_2, which is converted to mullite at higher temperatures. If the relative density of the final mullite body is ρ, derive a condition for zero shrinkage of the compact in terms of ρ_o, ρ, and the volume fractions of Al and SiC in the starting powder mixture (V_{Al} and V_{SiC}, respectively). Assume that the volume fraction of the starting powders oxidized during the attrition milling step is zero.

Show the condition for zero shrinkage on a graph, using axes corresponding to the final density and the green density.

1.4

Repeat the exercise in Problem 1.3 for the case when the starting mixture consists of Si and Al_2O_3 powders.

1.5

Compare the production of dense $BaTiO_3$ tapes by the following two methods:

    a. The oxidation of dense sheets of a solid metallic precursor consisting of Ba and Ti (see Schmutzler, H.J. et al., *J. Am. Ceram. Soc.*, 77, 721, 1994)
    b. Tape casting of $BaTiO_3$ powder slurries (see Chapter 7)

1.6

Derive Equation 1.5 for the height of a cylindrical pore infiltrated by a liquid.

1.7

Determine the linear and volume shrinkages that occur during the drying and sintering of a gel containing 5 vol% solids if the dried gel and the sintered gel have a solids content (relative density) of 50% and 100%, respectively.

1.8

Assuming that a polycarbosilane consists of equal numbers of the two structural components given in Equation 1.9 that on pyrolysis give SiC, estimate the ceramic yield (in wt%) on pyrolysis.

1.9

Taking 1 cm³ of a polycarbosilane with a density of ~1.2 g/cm³, estimate the volume of SiC formed on pyrolysis if the ceramic yield is 60 wt%.

1.10

Discuss the key factors involved with using preceramic polymers for the following applications:

a. Production of high-strength ceramic fibers
b. A binder in ceramic green bodies
c. The formation of a fine reinforcing phase in ceramic composites
d. The production of ceramic thin films

REFERENCES

1. Kenney, G.B. and Bowen, H.K., High tech ceramics in Japan: current and future markets, *Am. Ceram. Soc. Bull.*, 62, 590, 1983.
2. Pierson, H.O., *Handbook of Chemical Vapor Deposition (CVD): Principles, Technology, and Applications*, Noyes, Park Ridge, NJ, 1992.
3. Hitchman, M.L. and Jensen, K.F., Eds., *Chemical Vapor Deposition: Principles and Applications*, Academic Press, London, 1993.
4. Ohring, M., *The Materials Science of Thin Films*, Academic Press, San Diego, CA, 1992.
5. Schuegraf, K.L., *Handbook of Thin Film Deposition Processes and Techniques: Principles, Methods, Equipment, and Techniques*, Noyes, Park Ridge, NJ, 1998.
6. Sherman, A., *Chemical Vapor Deposition for Microelectronics*, Noyes, Park Ridge, NJ, 1987.
7. Hess, D.W. and Jensen, K.F., Eds., *Microelectronics Processing: Chemical Engineering Aspects*, American Chemical Society, Washington, D.C., 1989.
8. Galasso, F.S., *Chemical Vapor Deposited Materials*, CRC Press, Boca Raton, FL, 1991.
9. Hess, D.W., Jensen, K.F., and Anderson, T.J., Chemical vapor deposition: a chemical engineering perspective, *Rev. Chem. Eng.*, 3, 130, 1985.
10. Aardahl, C.L., and Rogers, J.W. Jr., Chemical vapor deposition, in *Inorganic Reactions and Methods*, Vol. 18, Atwood, J.D., Ed., Wiley-VCH, New York, 1999, p. 83.
11. Spear, K.E., Diamond — ceramic coating of the future, *J. Am. Ceram. Soc.*, 72, 171, 1989.
12. Stinton, D.P., Besmann, T.M., and Lowder, R.A., Advanced ceramics by chemical vapor deposition techniques, *Am. Ceram. Soc. Bull.*, 67, 350, 1988.
13. Lamicq, P.J., Bernhart, G.A., Dauchier, M.M., and Mace, J.G., SiC/SiC composite ceramics, *Am. Ceram. Soc. Bull.*, 65, 336, 1986.
14. Newkirk, M.S., Urquhart, A.W., Zwicker, H.R., and Breval, E., Formation of Lanxide™ ceramic composite materials, *J. Mater. Res.*, 1, 81, 1986.
15. Newkirk, M.S., Lesher, H.D., White, D.R., Kennedy, C.R., Urquhart, A.W., and Claar, T.D., Preparation of Lanxide™ ceramic matrix composites: matrix formation by directed metal oxidation, *Ceram. Eng. Sci., Proc.*, 8, 879, 1987.
16. Claussen, N. and Urquhart, A.W., Directed oxidation of molten metals, in *Encyclopedia of Materials Science and Engineering*, Supplementary Vol. 2, Cahn, R.W., Ed., Pergamon Press, Oxford, 1990, p. 1111.
17. Urquhart, A.W., Directed metal oxidation, in *Engineered Materials Handbook*, Vol. 4: Ceramics and Glasses, ASM International, Materials Park, OH, 1991, p. 232.
18. Antolin, S., Nagelberg, A.S., and Creber, D.K., Formation of Al_2O_3/metal composites by the directed metal oxidation of molten aluminum-magnesium-silicon alloys: Part I, microstructural development, *J. Am. Ceram. Soc.*, 75, 447, 1992.
19. Nagelberg, A.S., Antolin, A., and Urquhart, A.W., Formation of Al_2O_3/metal composites by the directed metal oxidation of molten aluminum-magnesium-silicon alloys: Part II, growth kinetics, *J. Am. Ceram. Soc.*, 75, 455, 1992.
20. Washburn, M.E. and Coblenz, W.S., Reaction-formed ceramics, *Am. Ceram. Soc. Bull.*, 67, 356, 1988.
21. Haggerty, J.S. and Chiang, Y.-M., Reaction-based processing methods for ceramics and composites, *Ceram. Eng. Sci. Proc.*, 11, 757, 1990.
22. Moulson, A.J., Reaction-bonded silicon nitride: its formation and properties, *J. Mater. Sci.*, 14, 1017, 1979.
23. Claussen, N., Le, T., and Wu, S., Low-shrinkage reaction-bonded alumina, *J. Eur. Ceram. Soc.*, 5, 29, 1989.
24. Claussen, N., Wu, S., and Holz, D., Reaction bonding of aluminum oxide (RBAO) composites: processing, reaction mechanisms, and properties, *J. Eur. Ceram. Soc.*, 14, 97, 1994.

25. Wu, S., Holz, D., and Claussen, N., Mechanisms and kinetics of reaction-bonded aluminum oxide ceramics, *J. Am. Ceram. Soc.*, 76, 970, 1993.
26. Holz, D. et al., Effect of processing parameters on phase and microstructure evolution in RBAO ceramics, *J. Am. Ceram. Soc.*, 77, 2509, 1994.
27. Claussen, N., Janssen, R., and Holz, D., Reaction bonding of aluminum oxide — science and technology, *J. Ceram. Soc. Japan*, 103, 749, 1995.
28. Wu, S. and Claussen, N., Fabrication and properties of low-shrinkage reaction-bonded mullite, *J. Am. Ceram. Soc.*, 74, 2460, 1991.
29. Wu, S. and Claussen, N., Reaction bonding and mechanical properties of mullite/silicon carbide composites, *J. Am. Ceram. Soc.*, 77, 2898, 1994.
30. (a) Sandhage, K.H., Electroceramics and process for making the same, U.S. Patent No. 5,318,725, June 7, 1994.
    (b) Sandhage, K.H., Processes for fabricating structural ceramic bodies and structural ceramic-bearing composite bodies, U.S. Patent No. 5,447,291, September 5, 1995.
31. Schmutzler, H.J. and Sandhage, K.H., Transformation of Ba-Al-Si precursor to celsian by high-temperature oxidation and annealing, *Metall. Trans. B*, 26B, 135, 1995.
32. Schmutzler, H.J., Sandhage, K.H., Nava, J.C., Fabrication of dense, shaped barium cerate by the oxidation of solid metal-bearing precursors, *J. Am. Ceram. Soc.*, 79, 1575, 1996.
33. Ward, G.A. and Sandhage, K.H., Synthesis of barium hexaferrite by the oxidation of a metallic barium-iron precursor, *J. Am. Ceram. Soc.*, 80, 1508, 1997.
34. Allameh, S.M. and Sandhage, K.H., The oxidative transformation of solid, barium-metal-bearing precursors into monolithic celsian with a retention of shape, dimensions and relative density, *J. Mater. Res.*, 13, 1271, 1998.
35. Kumar, P. and Sandhage, K.H., The fabrication of near net-shaped spinel bodies by the oxidative transformation of Mg/Al_2O_3 precursors, *J. Mater. Res.*, 13, 3423, 1998.
36. Forrest, C.W. Kennedy, P., and Shennan, J.V., The fabrication and properties of self-bonded silicon carbide bodies, *Special Ceramics*, 5, 99, 1972.
37. Kennedy, P., Effect of microstructural features on the mechanical properties of REFEL self-bonded silicon carbide, in *Non-Oxide Technical and Engineering Ceramics*, Hampshire, S., Ed., Elsevier, New York, 1986, 301.
38. Sawyer, G.R. and Page, T.F., Microstructural characterization of "REFEL" (reaction bonded) silicon carbides, *J. Mater. Sci.*, 13, 885, 1978.
39. Ness, J.N. and Page, T.F., Microstructural evolution in reaction bonded silicon carbide, *J. Mater. Sci.*, 21, 1377, 1986.
40. Larsen, D.C., *Ceramic Materials for Heat Engines: Technical and Economic Evaluation*, Noyes Publications, Park Ridge, NJ, 1985.
41. Messner, R.P. and Chiang, Y.-M., Liquid-phase reaction-bonding of silicon carbide using alloyed silicon-molybdenum melts, *J. Am. Ceram. Soc.*, 73, 1193, 1990.
42. Singh, M. and Levine, S.R., Low Cost Fabrication of Silicon Carbide Based Ceramics and Fiber Reinforced Composites, NASA-TM-107001, 1995.
43. Brinker, C.J. and Scherer, G.W., *Sol-Gel Science*, Academic Press, New York, 1990.
44. Klein, L.C., Ed., *Sol-Gel Technology for Thin Films, Fibers, Preforms, Electronics and Specialty Shapes*, Noyes Publications, Park Ridge, NJ, 1988.
45. Zelinski, B.J.J. and Uhlmann, D.R., Gel technology in ceramics, *J. Phys. Chem. Solids*, 45, 1069, 1984.
46. *Materials Research Society Symposium Proceedings: Better Ceramics through Chemistry*, Materials Research Society, Pittsburg, PA.
47. Conference Series, *Ceramic Transactions*: Vol. 55, 1995; Vol. 81, 1998, The American Ceramic Society, Westerville, OH.
48. Scherer, G.W., Drying gels: VIII, revision and review, *J. Non-Cryst. Solids*, 109, 171, 1989.
49. Scherer, G.W., Theory of drying, *J. Am. Ceram. Soc.*, 73, 3, 1990.
50. Zarzycki, J., Prassas, M., and Phalippou, J., Synthesis of glasses from gels: the problem of monolithic gels, *J. Mater. Sci.*, 17, 3371, 1982.
51. Hench, L.L., Use of drying control chemical additives (DCCAs) in controlling sol-gel processing, in *Science of Ceramic Chemical Processing*, Hench, L.L. and Ulrich, D.R., Eds., John Wiley & Sons, New York, 1986, p. 52.

52. Kistler, S.S., Coherent expanded aerogels, *J. Phys. Chem.*, 36, 52, 1932.

53. Tewari, P.H., Hunt, A.J., and Lofftus, K.D., Ambient-temperature supercritical drying of transparent silica aerogels, *Mater. Lett.*, 3, 363, 1985.

54. Jenkins, G.M. and Kawamura, K., *Polymer Carbons — Carbon Fiber, Glass and Char*, Cambridge University Press, London, 1976.

55. Chantrell, P.G. and Popper, P., Inorganic polymers and ceramics, in *Special Ceramics* 1964, Popper, P., Ed., Academic Press, New York, 1965, p. 87.

56. Yajima, S., Hayashi, J., Omori, M., and Okamura, K., Development of a silicon carbide fiber with high tensile strength, *Nature*, 261, 683, 1976.

57. Rice, R.W., Ceramics from polymer pyrolysis, opportunities and needs — a materials perspective, *Am. Ceram. Soc. Bull.*, 62, 889, 1983.

58. Wynne, K.J. and Rice, R.W., Ceramics via polymer pyrolysis, *Annu. Rev. Mater. Sci.*, 14, 297, 1984.

59. Baney, R. and Chandra, G., Preceramic polymers, *Encycl. Polym. Sci. Eng.*, 13, 312, 1985.

60. Pouskouleli, G. Metallorganic compounds as preceramic materials I. Non-oxide ceramics, *Ceram. Int.*, 15, 213, 1989.

61. Toreki, W., Polymeric precursors to ceramics — a review, *Polym. News*, 16, 6, 1991.

62. Zeldin, M., Wynne, K.J., and Allcock, H.R., Eds., *Inorganic and Organometallic Polymers*, ACS Symposium Series, Vol. 360, American Chemical Society, Washington, D.C., 1988.

63. Laine, R.M. and Babonneau, F., Preceramic polymer routes to silicon carbide, *Chem. Mater.*, 5, 260, 1993.

64. Yajima, S., Special heat-resisting materials from organometallic polymers, *Am. Ceram. Soc. Bull.*, 62, 893, 1983.

65. Yajima, S., Iwai, T., Yamamura, T., Okamura, K., Hasegawa, Y., Synthesis of a polytitanocarbosilane and its conversion into inorganic compounds, *J. Mater. Sci.*, 16, 1349, 1981.

66. Mai, Y.-W., Guest editorial, *Composites Sci. Technol.*, 51, 124, 1994.

67. Zhang, Z-F., Babonneau, F., Laine, R.M., Mu, Y., Harrod, J.F., and Rahn, J.A., Poly(methylsilane) — a high ceramic yield precursor to silicon carbide, *J. Am. Ceram. Soc.*, 74, 670, 1991.

68. Zhang, Z.-F., Scotto, C.S., and Laine, R.M., Processing stoichiometric silicon carbide fibers from polymethylsilane. Part I, Precursor fiber processing, *J. Mater. Chem.*, 8, 2715, 1998.

69. Seyferth, D. and Wiseman, G.H., High-yield synthesis of Si_3N_4/SiC ceramic materials by pyrolysis of a novel polyorganosilane, *J. Am. Ceram. Soc.*, 67, C-132, 1984.

70. Laine, R.M., Babonneau, F., Blowhowiak, K.Y., Kennish, R.A., Rahn, J.A., Exarhos, G.J., and Waldner, K., The evolutionary process during pyrolytic transformation of poly(N-methylsilazane) from a preceramic polymer into an amorphous silicon nitride/carbon composite, *J. Am. Ceram. Soc.*, 78, 137, 1995.

71. Legrow, G.E., Lim, T.F., Lipowitz, J., and Reaoch, R.S., Ceramics from hydridopolysilazane, *Am. Ceram. Soc. Bull.*, 66, 363, 1987.

72. He, J., Scarlete, M., and Harrod, J.E., Silicon nitride and silicon carbonitride by the pyrolysis of poly(methylsiladiazane), *J. Am. Ceram. Soc.*, 78, 3009, 1995.

73. Paine, R.T. and Narula, C.K., Synthetic routes to boron nitride, *Chem. Rev.*, 90, 73, 1990.

74. Seyferth, D. and Plenio, H., Borasilazane polymeric precursors for borosilicon nitride, *J. Am. Ceram. Soc.*, 73, 2131, 1990.

75. Rees, W.R. Jr. and Seyferth, D., High yield synthesis of B_4C/BN ceramic materials by pyrolysis of Lewis base adducts of decaborane, *J. Am. Ceram. Soc.*, 71, C196, 1988.

76. Schwartz, K.B., Rowcliffe, D.J., and Blum, Y.D., Microstructural development in Si_3N_4/polysilazane bodies during heating, *Adv. Ceram. Mater.*, 3, 320, 1988.

77. Schwab, S.T. and Blanchard-Ardid, C.R., The use of organometallic precursors to silicon nitride as binders, *Mater. Res. Soc. Symp. Proc.*, 121, 345, 1989.

78. Mohr, D.L., Desai, P., and Starr, T.L., Production of silicon nitride/silicon carbide fibrous composite using polysilazanes as preceramic binders, *Ceram. Eng. Sci. Proc.*, 11, 920, 1990.

79. Mucalo, M.R., Milestone, N.B., Vickridge, I.C., and Swain, M.V., Preparation of ceramic coatings from preceramic precursors. Part 1, SiC and Si_3N_4/Si_2N_2O coatings on alumina substrates, *J. Mater. Sci.*, 29, 4487, 1994.

80. Tooley, F.V., Ed., *The Handbook of Glass Manufacture*, 3rd ed., Vol. 1 and Vol. 2. Ashlee Publishing Co., New York, 1984.

81. Boyd, D.C. and McDowell, J.F., Eds., *Commercial Glasses*. Advances in Ceramics, Vol. 18, The American Ceramic Society, Westerville, OH, 1986.

82. Varshneya, A.K., *Fundamentals of Inorganic Glasses*, Academic Press, San Diego, CA, 1994, chap. 20.

83. (a) Uhlmann, D.R. and Kreidl, N.J., Eds., *Glass: Science and Technology*, Vol. 1: Glass Forming Systems, Academic Press, San Diego, CA, 1983.
    (b) Uhlmann, D.R. and Kreidl, N.J., Eds., *Glass: Science and Technology*, Vol. 2, Pt. 1: Processing, Academic Press, San Diego, CA, 1983.

84. McMillan, P.W., *Glass Ceramics*, 2nd ed., Academic Press, New York, 1979.

85. Grossman, D.G., Glass ceramics, in *Concise Encyclopedia of Advanced Ceramics*, Brook, R.J., Ed., The MIT Press, Cambridge, MA, 1991, p. 170.

86. Matijevic, E., Monodispersed colloidal metal oxides, sulfides, and phosphates, in *Ultrastructure Processing of Ceramics, Glasses, and Composites*, Hench, L.L. and Ulrich, D.R., Eds., John Wiley & Sons, New York, 1984, p. 334.

87. Zhou, Y.C. and Rahaman, M.N., Hydrothermal synthesis and sintering of ultrafine CeO_2 powders, *J. Mater. Res.*, 8, 1680, 1993.

88. Lange, F.F., Powder processing science and technology for increased reliability, *J. Am. Ceram. Soc.*, 72, 3, 1989.

89. Barringer, E.A. and Bowen, H.K., Formation, packing and sintering of monodispersed TiO_2 powders, *J. Am. Ceram. Soc.*, 65, C-199, 1982.

90. Brook, R.J., Advanced ceramic materials: an overview, in *Concise Encyclopedia of Advanced Ceramic Materials*, Brook, R.J., Ed., The MIT Press, Cambridge, MA, 1991, p. 1.

91. Chu, G.P.K., Microstructure of complex ceramics, in *Ceramic Microstructures*, Fulrath, R.M. and Pask, J.A., Eds., John Wiley & Sons, New York, 1968, chap. 38.

92. Dalgleish, B.J. and Evans, A.G., Influence of shear bands on creep rupture in ceramics, *J. Am. Ceram. Soc.*, 68, 44, 1985.

93. Yan, M.F. and Rhodes, W.W., Efficient process for volume production of TiO_2 varistor powder, *Am. Ceram. Soc. Bull.*, 63, 1484, 1984.

2 Synthesis of Powders

2.1 INTRODUCTION

The powder characteristics, as outlined in the previous chapter, have a pronounced effect on subsequent processing. As a result, powder synthesis forms the first important step in the overall fabrication of ceramics. In this chapter, we shall first define, in general terms, the desirable characteristics that a powder should possess for the production of successful ceramics and then consider some of the main synthesis methods. In practice, the choice of a powder preparation method will depend on the production cost and the capability of the method for providing a certain set of desired characteristics. For convenience, we shall divide the powder synthesis methods into two categories: mechanical methods and chemical methods. Powder synthesis by chemical methods is an area of ceramic processing which has received a high degree of interest and has seen considerable developments since the early 1980s. Further new developments are expected in the future, particularly in the area of very fine powders with particle sizes smaller than 50 to 100 nm, referred to as *nanoscale* or *nanosize* powders.

2.2 POWDER CHARACTERISTICS

Traditional ceramics generally must meet less specific property requirements than advanced ceramics. They can be chemically inhomogeneous and can have complex microstructures. In these, unlike in the case of advanced ceramics, chemical reaction during firing is often a requirement. The starting materials for traditional ceramics, therefore, consist of mixtures of powders with a chosen reactivity. For example, the starting powders for an insulating porcelain can, typically, be a mixture of clay (~50 wt%), feldspar (~25 wt%), and silica (~25 wt%). Fine particle size is desirable for good chemical reactivity. The powders must also be chosen to give a reasonably high packing density, which serves to limit the shrinkage and distortion of the body during firing. Clays form the major constituent and therefore provide the fine particle size constituent in the starting mixture for most traditional ceramics. Generally, low-cost powder preparation methods are used for traditional ceramics.

Advanced ceramics must meet very specific property requirements and therefore their chemical composition and microstructure must be well controlled. Careful attention must therefore be paid to the quality of the starting powders. The important powder characteristics are the size, size distribution, shape, state of agglomeration, chemical composition, and phase composition. The structure and chemistry of the particle surfaces are also important.

The size, size distribution, shape, and state of agglomeration have an important influence on both the powder consolidation step and the microstructure of the sintered body. A particle size greater than 5 to 10 μm generally precludes the use of colloidal consolidation methods because the settling time of the particles is fairly short. The most profound effect of the particle size, however, is on the sintering. As described in Chapter 9, the rate at which a body densifies increases strongly with a decrease in particle size. Normally, if other factors do not cause severe difficulties during sintering, a particle size of less than ~1 μm allows the achievement of high density within a reasonable time (e.g., a few hours). A powder with a wide distribution of particle sizes (sometimes referred to as a *polydisperse* powder) generally leads to higher packing density in the green body,

TABLE 2.1
Desirable Powder Characteristics for Advanced Ceramics

Powder Characteristic	Desired Property
Particle size	Fine (< ~1 μm)
Particle size distribution	Narrow or monodisperse
Particle shape	Spherical or equiaxial
State of agglomeration	No agglomeration or soft agglomerates
Chemical composition	High purity
Phase composition	Single phase

but this benefit is often vastly outweighed by difficulties in microstructural control during sintering. A common problem is that if the particles are not packed homogeneously to give uniformly distributed fine pores, densification in the later stages can be slow. The large grains coarsen rapidly at the expense of the smaller grains, making the attainment of high density with controlled grain size impossible. Homogeneous packing of a narrow size distribution powder (i.e., a nearly *monodisperse* powder) generally allows better control of the microstructure. A spherical or equiaxial particle shape is beneficial for controlling the homogeneity of the packing.

Agglomerates lead to heterogeneous packing in the green body which, in turn, leads to differential sintering, whereby different regions of the body shrink at different rates. This can lead to serious problems such as the development of large pores and crack-like voids in the sintered body (see Figure 1.23). Furthermore, the rate at which the body densifies is roughly that for a coarse-grained body with a particle size equal to the agglomerate size. An agglomerated powder therefore has serious limitations for the fabrication of ceramics when high density coupled with a fine-grained microstructure is required. Agglomerates are classified into two types: *soft* agglomerates, in which the particles are held together by weak van der Waals forces, and *hard* agglomerates, in which the particles are chemically bonded together by strong bridges. The ideal situation is the avoidance of agglomeration in the powder. However, in most cases this is not possible. We would, then, prefer to have soft agglomerates rather than hard agglomerates, as soft agglomerates can be broken down relatively easily by mechanical methods (e.g., pressing or milling) or by dispersion in a liquid. Hard agglomerates cannot be easily broken down and therefore should be avoided or removed from the powder.

Whereas surface impurities may have a significant influence on the dispersion of the powder in a liquid, the most serious effects of such variations in chemical composition are encountered in sintering. Impurities may lead to the formation of a small amount of liquid phase at the sintering temperature, which causes selected growth of large individual grains (Figure 1.22). In such a case, the achievement of a fine uniform grain size would be impossible. Chemical reactions between incompletely reacted phases can also be a source of problems. We would, therefore, like to have no chemical change in the powder during sintering. For some materials, polymorphic transformation between different crystalline structures can also be a source of severe difficulties for microstructure control. Unless it is stabilized, ZrO_2 undergoes a martensitic tetragonal to monoclinic phase transformation on cooling from the sintering temperature, which leads to severe cracking. In γ-Al_2O_3, the transformation to the α-phase results in rapid grain growth and a severe retardation in the densification rate. The desirable powder characteristics for the fabrication of advanced ceramics are summarized in Table 2.1.

2.3 POWDER SYNTHESIS METHODS

A variety of methods exist for the synthesis of ceramic powders. We shall divide them into two categories: *mechanical methods* and *chemical methods*. Powders of traditional ceramics are

TABLE 2.2
Common Powder Preparation Methods for Ceramics

Powder Preparation Method	Advantages	Disadvantages
Mechanical:		
Comminution	Inexpensive, wide applicability	Limited purity, limited homogeneity, large particle size
Mechanochemical synthesis	Fine particle size, good for nonoxides, low temperature route	Limited purity, limited homogeneity
Chemical:		
Solid-state reaction	Simple apparatus, inexpensive	Agglomerated powder, limited
Decomposition; reaction between solids		homogeneity for multicomponent powders
Liquid Solutions:		
Precipitation or coprecipitation; solvent vaporization (spray drying, spray pyrolysis, freeze drying); gel routes (sol–gel, Pechini, citrate gel, glycine nitrate)	High purity, small particle size, composition control, chemical homogeneity	Expensive, poor for nonoxides, powder agglomeration commonly a problem
Nonaqueous liquid reaction	High purity, small particle size	Limited to nonoxides
Vapor-Phase Reaction:		
Gas–solid reaction	Commonly inexpensive for large particle size	Commonly low purity; expensive for fine powders
Gas–liquid reaction	High purity, small particle size	Expensive, limited applicability
Reaction between gases	High purity, small particle size, inexpensive for oxides	Expensive for nonoxides, agglomeration commonly a problem

generally prepared by mechanical methods from naturally occurring raw materials. Powder preparation by mechanical methods is a fairly mature area of ceramic processing, in which the scope for new developments is rather small. However, in recent years, the preparation of fine powders of some advanced ceramics by mechanical methods involving milling at high speeds has received some interest.

Powders of advanced ceramics are generally synthesized by chemical methods either from synthetic materials or from naturally occurring raw materials that have undergone a considerable degree of chemical refinement. Some of the chemical methods include a mechanical milling step, which is often required for breaking down agglomerates and for producing the desired particle size characteristics. The introduction of impurities during this milling step should be avoided. Table 2.2 gives a summary of the common powder preparation methods for ceramics.

2.4 POWDER PREPARATION BY MECHANICAL METHODS

2.4.1 COMMINUTION

The process in which small particles are produced by reducing the size of larger ones by mechanical forces is usually referred to as *comminution*. It involves operations such as crushing, grinding, and milling. For traditional clay-based ceramics, machines such as jaw, gyratory, and cone crushers are used for coarse size reduction of the mined raw material, to produce particles in the size range of 0.1 to 1 mm or so. The equipment and the processes involved in the production of these coarse particles are well described elsewhere [1–3]. Here we will assume that a stock of coarse particles (with sizes < 1 mm or so) is available and consider the processes applicable to the subsequent size

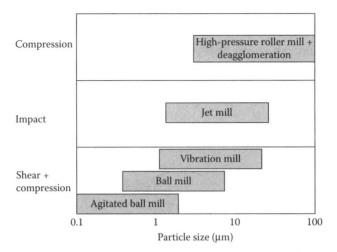

FIGURE 2.1 Range of particle sizes reached with different types of mills. (From Polke, R. and Stadler, R., Grinding, in *Concise Encyclopedia of Advanced Ceramic Materials*, Brook, R.J., Ed., MIT Press, Cambridge, MA, 1991, p. 189. With permission.)

reduction to produce a fine powder. The most common way to achieve this size reduction is by milling. One or more of a variety of mills may be used, including high compression roller mills, jet mills (also referred to as fluid energy mills), and ball mills [3,4]. Ball mills are categorized into various types, depending on the method used to impart motion to the balls (e.g., tumbling, vibration, and agitation).

We define the *energy utilization* of the comminution method as the ratio of the new surface area created to the total mechanical energy supplied. The *rate of grinding* is defined as the amount of new surface area created per unit mass of particles per unit time. Obviously, there is a connection between the two terms. A comminution method that has a high energy utilization will also have a high rate of grinding, so the achievement of a given particle size will take a shorter time. For a given method, we will also want to understand how the rate of grinding depends on the various experimental factors.

In milling, the particles experience mechanical stresses at their contact points due to compression, impact, or shear with the mill medium or with other particles. The mechanical stresses lead to elastic and inelastic deformation and, if the stress exceeds the ultimate strength of the particle, to fracture of the particles. The mechanical energy supplied to the particle is used not only to create new surfaces but also to produce other physical changes in the particle (e.g., inelastic deformation, increase in temperature, and lattice rearrangements within the particle). Changes in the chemical properties (especially the surface properties) can also occur, especially after prolonged milling or under very vigorous milling conditions. Consequently, the energy utilization of the process can be fairly low, ranging from < 20% for milling produced by compression forces, to < 5% for milling by impact. Figure 2.1 summarizes the stress mechanisms and the range of particle sizes achieved with different types of mills.

2.4.1.1 High Compression Roller Mills

In the high compression roller mill, the material is stressed between two rollers. In principle, the process is similar to a conventional roller mill but the contact pressure is considerably higher (in the range of 100 to 300 MPa). The stock of coarse particles is comminuted and compacted. This process must, therefore, be used in conjunction with another milling process (e.g., ball milling) to produce a powder. Whereas the process is unsuitable for the production of particle sizes below ~10 μm, the energy utilization is fairly good because the mechanical energy supplied to the rollers goes directly into comminuting the particles. For the production of the same size of particles from a stock of coarse particles, the use of a high energy roller mill in conjunction with a ball mill is more efficient than the use of a ball mill only. As only a small amount of material makes contact with the rolls, the wear can be fairly low (e.g., much lower than in ball milling).

2.4.1.2 Jet Mills

Jet mills are manufactured in a variety of designs. Generally, the operation consists of the interaction of one or more streams of high-speed gas bearing the stock of coarse particles with another high-speed stream. Comminution occurs by particle–particle collisions. In some designs, comminution is achieved by collisions between the particles in the high-speed stream and a wall (fixed or movable) within the mill. The milled particles leave the mill in the emergent fluid stream and are usually collected in a cyclone chamber outside the mill. The gas for the high-speed stream is usually compressed air, but inert gases such as nitrogen or argon may be used to reduce oxidation of certain nonoxide materials. The average particle size and the particle size distribution of the milled powder depend on several factors, including the size, size distribution, hardness and elasticity of the feed particles, and the pressure at which the gas is injected. The dimensions of the milling chamber and the utilization of particle classification in conjunction with the milling also influence the particle size and distribution.

Multiple gas inlet nozzles are incorporated into some jet mill designs in order to provide multiple collisions between the particles, enhancing the comminution process. In some cases the flow of the particles in the high-speed gas stream can be utilized for their classification in the milling chamber. The feed particles remain in the grinding zone until they are reduced to a sufficiently fine size and then removed from the milling chamber. An advantage of jet mills is that when combined with a particle classification device, they provide a rapid method for the production of a powder with a narrow size distribution for particle sizes down to ~1 μm. A further advantage is that for some designs, the particles do not come into contact with the surfaces of the milling chamber, so contamination is not a problem.

2.4.1.3 Ball Mills

The high compression roller mills and jet mills just described achieve comminution without the use of grinding media. For mills that incorporate grinding media (balls or rods), comminution occurs by compression, impact, and shear (friction) between the moving grinding media and the particles. Rod mills are not suitable for the production of fine powders, but ball milling can be used to produce particle sizes from ~10 μm to as low as a fraction of a micrometer. Ball milling is suitable for wet or dry milling.

The rate of grinding depends on several factors, such as the mill parameters, the properties of the grinding medium, and the properties of the particles to be ground [3]. Generally, ball mills that run at low speeds contain large balls because most of the mechanical energy supplied to the particle is in the form of potential energy. Those mills that run at high speeds contain small balls because, in this case, most of the energy supplied to the particle is in the form of kinetic energy. For a given size of grinding medium, as the mass is proportional to the density, the grinding medium should consist of materials with as high a density as possible. In practice, cost often determines the choice of the grinding medium.

The size of the grinding medium is an important consideration. Small grinding media are generally better than large ones. For a given volume, the number of balls increases inversely as the cube of the radius. Assuming that the rate of grinding depends on the number of contact points between the balls and the powder, and that the number of contact points, in turn, depends on the surface area of the balls, then the rate of grinding will increase inversely as the radius of the balls. However, the balls cannot be too small because they must impart sufficient mechanical energy to the particles to cause fracture.

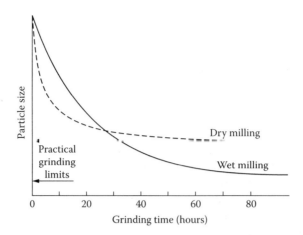

FIGURE 2.2 Particle size vs. grinding time for ball milling. (Courtesy of L.C. De Jonghe.)

The rate of grinding also depends on the particle size. The rate decreases with decreasing particle size and, as the particles become fairly fine (e.g., ~1 µm to a few micrometers), it becomes more and more difficult to achieve further reduction in size. A practical grinding limit is approached (Figure 2.2). This limit depends on several factors. First, the increased tendency for the particles to agglomerate with decreasing particle size means that a physical equilibrium is set up between the agglomeration and comminution processes. Second, the probability for the occurrence of a comminution event decreases with decreasing particle size. Third, the probability of a flaw with a given size existing in the particle decreases with decreasing particle size, i.e., the particle becomes stronger. The limiting particle size may be made smaller by wet milling as opposed to dry milling (Figure 2.2), the use of a dispersing agent during wet milling (Figure 2.3), and by performing the milling in stages [3]. For staged milling, as the particles gets finer, they are transferred to another compartment of the mill or to another mill operating with smaller balls.

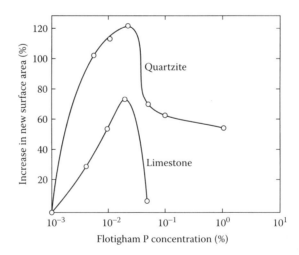

FIGURE 2.3 Effect of Flotigam P, an organic dispersing agent, on grinding of quartzite and limestone in a rod mill. (From Somasundaran, P., Theories of grinding, in *Ceramic Processing Before Firing*, Onoda, G.Y. and Hench, L.L., Eds., John Wiley & Sons, New York, 1978, p. 105, chap. 11. With permission.)

TABLE 2.3
Commercially Available Grinding
Media for Ball Milling

Grinding Media	Density (g/cm3)
Porcelain	2.3
Silicon nitride	3.1
Silicon carbide	3.1
Alumina	
Lower than 95% purity	3.4–3.6
Greater than 99% purity	3.9
Zirconia	
MgO stabilized	5.5
High-purity Y_2O_3-stabilized	6.0
Steel	7.7
Tungsten carbide	14.5

A disadvantage of ball milling is that wear of the grinding medium can be fairly high. For advanced ceramics, the presence of impurities in the powder is a serious problem. The best solution is to use balls with the same composition as the powder itself. However, this is only possible in very few cases and, even for these, at fairly great expense. Another solution is to use a grinding medium that is chemically inert at the sintering temperature of the body (e.g., ZrO_2 balls) or can be removed from the powder by washing (e.g., steel balls). A common problem is the use of porcelain balls or low purity Al_2O_3 balls that wear easily and introduce a fair amount of SiO_2 into the powder. Silicate liquids will form at the sintering temperature and make microstructural control more difficult. A list of grinding balls available commercially and the approximate density of each is given in Table 2.3.

Tumbling ball mills, usually referred to simply as *ball mills*, consist of a slowly rotating horizontal cylinder that is partly filled with grinding balls and the particles to be ground. In addition to the factors discussed earlier, the rotation speed of the mill is an important variable as it influences the trajectory of the balls and the mechanical energy supplied to the powder. The *critical speed of rotation*, defined as the speed (in revolutions per unit time) required to just take the balls to the apex of revolution (i.e., to the top of the mill where the centrifugal force just balances the force of gravity), is equal to $(g/R_m)^{1/2}/2\pi$, where R_m is the radius of the mill and g is the acceleration due to gravity. In practice, ball mills are operated at ~75% of the critical speed, so the balls do not reach the top of the mill (Figure 2.4).

Ball milling is a complex process that does not lend itself easily to rigorous theoretical analysis. We therefore have to be satisfied with empirical relationships. One such empirical relationship is:

$$\text{Rate of milling} \approx AR_m^{1/2}\rho x / a \tag{2.1}$$

where A is numerical constant that is specific to the mill being used and the powder being milled, R_m is the radius of the mill, ρ is the density of the balls, x is the particle size of the powder, and a is the radius of the balls. According to Equation 2.1, the rate decreases with decreasing particle size, but this holds up to a certain point because a practical grinding limit is reached after a certain time. The variation of the rate of grinding with the radius of the balls must also be taken with caution: the balls will not possess sufficient energy to cause particle fracture if they are too small.

In ball milling, the objective is to have the balls fall onto the particles at the bottom of the mill rather than onto the mill liner itself. For a mill operating at ~75% of its critical speed, this occurs,

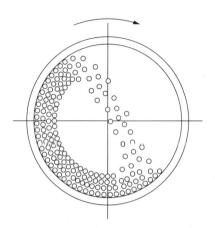

FIGURE 2.4 Schematic of a ball mill in cataracting motion. (From Polke, R. and Stadler, R., Grinding, in *Concise Encyclopedia of Advanced Ceramic Materials*, Brook, R.J., Ed., MIT Press, Cambridge, MA,1991, p. 191. With permission.)

for dry milling, for a quantity of balls filling ~50% of the mill volume and for a charge of particles filling ~25% of the mill volume. For wet milling, a useful guide is for the balls occupying ~50% of the mill volume and the slurry ~40% of the mill volume, with the solids content of the slurry equal to ~25 to 40%.

Wet ball milling has a somewhat higher energy utilization (by ~10 to 20%) than dry milling and the ability to produce a higher fraction of finer particles. On the other hand, wet milling suffers from increased wear of the grinding media, the need for drying the powder after milling, and possible contamination of the powder by adsorbed molecules from the liquid medium.

Vibratory ball mills or *vibro-mills* consist of a drum, almost filled with a well-packed arrangement of grinding media and the charge of particles, that is vibrated fairly rapidly (10 to 20 Hz) in three dimensions. The grinding medium, usually cylindrical in shape, occupies more than 90% of the mill volume, and the amplitude of the vibrations is controlled so as not to disrupt the well-packed arrangement of the grinding media. The three-dimensional motion helps in the distribution of the stock of particles and, in the case of wet milling, to minimize segregation of the particles in the slurry. Because of the fairly rapid vibratory motion, the impact energy is much greater than the energy supplied to the particles in a tumbling ball mill. Vibratory ball mills therefore provide a more rapid comminution process than tumbling ball mills. They are also more energy efficient than tumbling ball mills.

Attrition mills, also referred to as *agitated ball mills* or *stirred media mills*, differ from tumbling ball mills in that the milling chamber does not rotate. Instead, the stock of particles and the grinding medium are stirred rather vigorously with a stirrer rotating continuously at frequencies of 1 to 10 Hz. The grinding chamber is aligned either vertically or horizontally (Figure 2.5), with the stirrer located at the center of the chamber. The grinding media consist of small spheres (~0.2 to 10 mm) that make up ~60 to 90% of the available volume of the mill. Although it can be used for dry milling, most attrition milling is carried out with slurries. Most attrition ball milling is also carried out continuously, with the slurry of particles to be milled fed in at one end and the milled product removed at the other end. For milling where the agitation is very intense, considerable heat is produced, so the milling chamber must be cooled.

The energy utilization in attrition mills is significantly higher than that in tumbling ball mills or in vibratory ball mills, and the fine grinding media gives an improved rate of milling. Attrition mills also have the ability to handle a higher solids content in the slurry to be milled. The high efficiency of the process coupled with the short duration required for milling means that contamination of the milled powder is less serious than in the case for tumbling ball mills or vibratory

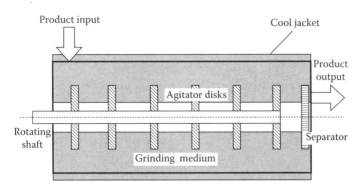

FIGURE 2.5 Schematic of an agitated ball mill. (From Polke, R. and Stadler, R., Grinding, in *Concise Encyclopedia of Advanced Ceramic Materials*, Brook, R.J., Ed., MIT Press, Cambridge, MA, 1991, p. 192. With permission.)

ball mills. Contamination in agitated ball milling can be further reduced by lining the mill chamber with a ceramic or plastic material and using ceramic stirrers and grinding media.

2.4.2 MECHANOCHEMICAL SYNTHESIS

In comminution, our interest lies mainly in achieving certain physical characteristics, such as particle size and particle size distribution. In recent years, chemical changes during milling have also been exploited for the preparation of some powders. Grinding enhances the chemical reactivity of powders. Rupture of the bonds during particle fracture results in surfaces with unsatisfied valences. This, combined with a high surface area favors reaction between mixed particles or between the particles and their surroundings.

Powder preparation by high energy ball milling of elemental mixtures is referred to as *mechanochemical synthesis*, but other terms such as *mechanosynthesis*, *mechanical driven synthesis*, *mechanical alloying*, and *high energy milling* will also be encountered. The method has attracted a fair amount of interest since the early 1980s for the production of metal and alloy powders [6–8]. Whereas less attention has been paid to ceramic systems, the method has been investigated for the preparation of several oxide, carbide, nitride, boride, and silicide powders [9–12].

Mechanochemical synthesis can be carried out in small mills, such as the Spex mill, for synthesizing a few grams of powder or in attrition mills for larger quantities. In the Spex mill, a cylindrical vial containing the milling balls and the charge of particles undergoes large amplitude vibrations in three dimensions at a frequency of ~20 Hz. The charge occupies ~20% of the volume of the vial, and the amount of milling media (in the form of balls 5 to 10 mm in diameter) makes up 2 to 10 times the mass of the charge. The milling is normally carried out for a few tens of hours for the set of conditions indicated here. The method therefore involves high-intensity vibratory milling for extended periods.

An advantage of mechanochemical synthesis is the ease of preparation of some powders that can otherwise be difficult to produce, such as those of the silicides and carbides. For example, most metal carbides are formed by the reaction between metals or metal hydrides and carbon at high temperatures (in some cases as high as 2000°C). Some carbides and silicides have a narrow compositional range that is difficult to produce by other methods. A disadvantage is the incorporation of impurities from the mill and milling medium into the powder.

The mechanism of mechanochemical synthesis is not clear. The occurrence of the reaction by a solid-state diffusion mechanism appears to be unlikely because the temperature in the mill is significantly lower than that required for a true solid-state mechanism. Evidence for compound formation by local melting during mechanochemical synthesis is unclear. A strong possibility is the occurrence of the reaction by a form of self-propagating process at high temperature. In highly

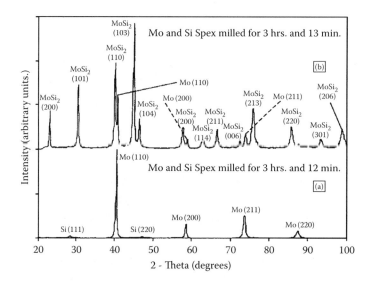

FIGURE 2.6 X-ray diffraction pattern of a stoichiometric mixture of Mo and Si powders after high-speed milling for (a) 3 h 12 min and (b) 3 h 13 min, showing a fairly abrupt formation of $MoSi_2$. (From Patankar, S.N. et al., The mechanism of mechanical alloying of $MoSi_2$, *J. Mater. Res.*, 8, 1311, 1993. With permission.)

exothermic reactions, such as the formation of molybdenum and titanium silicides from their elemental mixtures, the heat that is liberated is often sufficient to sustain the reaction [13,14]. However, for the reaction to first occur, a source of energy must be available to raise the adiabatic temperature of the system to that required for it to become self-sustaining. A possible source is the enormous surface energy, together with the stored strain energy of the very fine powders prior to any extensive reaction. For example, the average particle sizes of the Mo and Si powders prior to extensive formation of $MoSi_2$ have been reported as ~20 nm and ~10 nm, respectively [13]. The surface energy of the particles alone is estimated as 5 to 10% of the heat of formation of $MoSi_2$ (131.9 kJ/mol at 298K).

A critical step for the formation reaction appears to be the generation of a fine enough particle size so that the available surface and strain energy is sufficient to make the reaction self-sustaining. The reaction for $MoSi_2$ and other silicides shows features that are characteristic of a self-propagating process. After an induction period, the reaction occurs quite abruptly (Figure 2.6). It is likely that after a small portion of the elemental powders has reacted, the heat liberated by the reaction ignites the unreacted portion until the bulk of the elemental powders is converted to the product. Immediately following the reaction, the product is highly agglomerated. For $MoSi_2$, the agglomerate size was found to be $\approx$100 μm, made up of primary particles of ~0.3 μm in diameter (Figure 2.7).

2.5 POWDER SYNTHESIS BY CHEMICAL METHODS

A wide range of chemical methods exists for the synthesis of ceramic powders and several reviews of the subject are available in the literature [15–20]. We will consider these methods in three broad categories: (1) solid-state reactions, (2) synthesis from liquid solutions, and (3) vapor-phase reactions.

2.5.1 SOLID-STATE REACTIONS

Chemical decomposition reactions, in which a solid reactant is heated to produce a new solid plus a gas, are commonly used for the production of powders of simple oxides from carbonates,

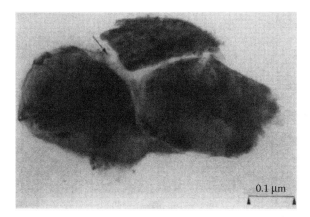

FIGURE 2.7 TEM image of $MoSi_2$ in the sample milled for 3 h 13 min showing three particles separated by Mo (arrow). (From Patankar, S.N. et al., The mechanism of mechanical alloying of $MoSi_2$, *J. Mater. Res.*, 8, 1311, 1993. With permission.)

hydroxides, nitrates, sulfates, acetates, oxalates, alkoxides and other metal salts. An example is the decomposition of calcium carbonate (calcite) to produce calcium oxide and carbon dioxide gas:

$$CaCO_3(s) \rightarrow CaO(s) + CO_2(g) \tag{2.2}$$

Chemical reactions between solid starting materials, usually in the form of mixed powders, are common for the production of powders of complex oxides such as titanates, ferrites, and silicates. The reactants normally consist of simple oxides, carbonates, nitrates, sulfates, oxalates, or acetates. An example is the reaction between zinc oxide and alumina to produce zinc aluminate:

$$ZnO(s) + Al_2O_3(s) \rightarrow ZnAl_2O_4(s) \tag{2.3}$$

These methods, involving decomposition of solids or a chemical reaction between solids, are referred to in the ceramic literature as *calcination*.

2.5.1.1 Decomposition

Because of the industrial and scientific interest, a large body of literature exists on the principles, kinetics, and chemistry of decomposition reactions [21–23]. The most widely studied systems are $CaCO_3$, $MgCO_3$ and $Mg(OH)_2$. We will focus on the basic thermodynamics, reaction kinetics and mechanisms, and process parameters pertinent to the production of powders.

Considering the thermodynamics, for the decomposition of $CaCO_3$ defined by Equation 2.2, the standard heat (enthalpy) of reaction at 298K, ΔH_R°, is 44.3 kcal/mol [24]. The reaction is strongly *endothermic* (i.e., ΔH_R° is positive), which is typical for most decomposition reactions. This means that heat must be supplied to the reactant to sustain the decomposition. The Gibbs free energy change associated with any reaction is given by:

$$\Delta G_R = \Delta G_R^o + RT \ln K \tag{2.4}$$

where ΔG_R° is the free energy change for the reaction when the reactants are in their standard state, R is the gas constant, T is the absolute temperature, and K is the equilibrium constant for the reaction. For the reaction defined by Equation 2.2,

$$K = \frac{a_{CaO} a_{CO_2}}{a_{CaCO_3}} = p_{CO_2} \qquad (2.5)$$

where a_{CaO} and a_{CaCO_3} are the activities of the pure solids CaO and $CaCO_3$, respectively, taken to be unity, and a_{CO_2} is the activity of CO_2, taken to be the partial pressure of the gas. At equilibrium, $\Delta G_R = 0$, and combining Equation 2.4 and Equation 2.5, we get:

$$\Delta G_R^o = -RT \ln p_{CO_2} \qquad (2.6)$$

The standard free energy for the decomposition of $CaCO_3$, $MgCO_3$, and $Mg(OH)_2$ is plotted in Figure 2.8, along with the equilibrium partial pressure of the gas for each of the reactions [25]. Assuming that the compounds become unstable when the partial pressure of the gaseous product above the solid equals the partial pressure of the gas in the surrounding atmosphere, we can use the values in Figure 2.8 to determine the temperatures at which the compounds become unstable when heated in air. For example, $CaCO_3$ becomes unstable above ~810K, $MgCO_3$ above 480K, and depending on the relative humidity, $Mg(OH)_2$ becomes unstable above 445 to 465K. Furthermore, acetates, sulfates, oxalates, and nitrates have essentially zero partial pressure of the product gas in the ambient atmosphere, so they are predicted to be unstable. The fact that these compounds are observed to be stable at much higher temperatures indicates that their decomposition is controlled by kinetic factors and not by thermodynamics.

Kinetic investigations of decomposition reactions can provide information about the reaction mechanisms and the influence of the process variables such as temperature, particle size, mass of reactant, and the ambient atmosphere. They are conducted isothermally or at a fixed heating rate. In isothermal studies, the maintenance of a constant temperature represents an ideal that cannot be achieved in practice, as a finite time is required to heat the sample to the required temperature. However, isothermal decomposition kinetics are easier to analyze. The progress of the reaction is commonly measured by the weight loss, and the data are plotted as the fraction of the reactant decomposed, α, vs. time, t, with α defined as:

$$\alpha = \Delta W / \Delta W_{max} \qquad (2.7)$$

where ΔW and ΔW_{max} are the weight loss at time t and the maximum weight loss according to the decomposition reaction, respectively.

There is no general theory of decomposition reactions. However, a generalized α vs. time curve similar to that shown in Figure 2.9 is often observed [22]. The stage A is an initial reaction, sometimes associated with the decomposition of impurities or unstable superficial material. B is an induction period that is usually regarded as terminated by the development of stable nuclei, whereas C is the acceleratory period of growth of such nuclei, perhaps accompanied by further nucleation, which extends to the maximum rate of reaction at D. Thereafter, the continued expansion of nuclei is no longer possible due to impingement and the consumption of the reactant, and this leads to a decay period, E, that continues until the completion of the reaction, F. In practice, one or more of these features (except D) may be absent or negligible.

The molar volume of the solid product is commonly smaller than that for the reactant, so that very often the product forms a porous layer around the nonporous core of reactant (Figure 2.10). Like most solid-state reactions, the reaction is *heterogeneous*, occurring at a sharply defined interface. The kinetics may be controlled by any one of three processes: (1) the reaction at the interface between the reactant and the solid product, (2) heat transfer to the reaction surface, or (3) gas diffusion or permeation from the reaction surface through the porous product layer. As seen

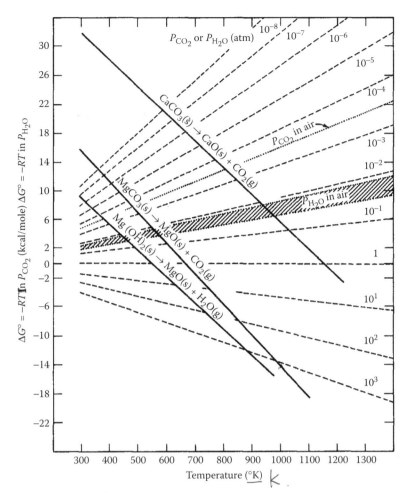

FIGURE 2.8 Standard free energy of reaction as a function of temperature. The dashed lines are the equilibrium gas pressure above the oxide/carbonate and oxide/hydroxide. (From Kingery, W.D., Bowen, H.K., and Uhlmann, D.R., *Introduction to Ceramics*, 2nd ed., John Wiley & Sons, New York, 1976, p. 415. With permission.)

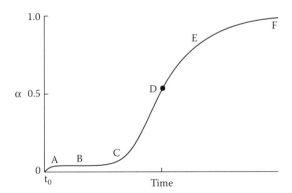

FIGURE 2.9 Generalized α vs. time plot summarizing characteristic kinetic behavior observed for isothermal decomposition of solids. α represents the weight loss divided by the maximum weight loss.

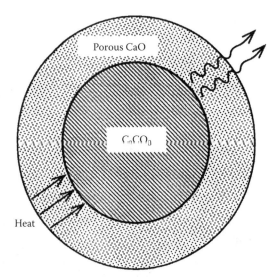

FIGURE 2.10 Schematic of the decomposition of calcium carbonate.

from Table 2.4, several expressions have been developed to analyze the reaction kinetics. It is generally assumed that the interface moves inward at a constant rate, so for a spherical reactant of initial radius a_o, the radius of the unreacted core at time t is given by:

$$a = a_o - Kt \tag{2.8}$$

where K is a constant. When the reaction at the interface is rate controlling, the different expressions reflect the different assumptions on the nucleation and growth of stable particles of the product from the reactant (*nucleation* equations in Table 2.4). If the nucleation step is fast, the equations depend only on the geometry of the model (*geometrical models*). For thin $CaCO_3$ (calcite) crystals, Figure 2.11 shows that the decomposition kinetics, measured under conditions such that the reaction interface advances in one dimension, follow a linear equation in accordance with Equation 7 in Table 2.4.

When large samples are used, the increasing thickness of the porous product layer may provide a barrier to the escape of the product gas. Kinetic equations for the decomposition of $CaCO_3$ controlled by the rate of removal of the product gas (CO_2), or the rate of heat transfer to the reaction interface, have been developed by Hills [26]. Table 2.4 includes equations for the reaction rate controlled by diffusion of the chemical components of the reactant.

The kinetics of chemical reactions are frequently classified with respect to reaction order. Taking the simple case where a reactant is decomposed:

$$\mathfrak{R} \rightarrow \text{Product} \tag{2.9}$$

The rate of the reaction can be written:

$$-\frac{dc}{dt} = Kc^{\beta} \tag{2.10}$$

TABLE 2.4
Rate Equations for the Analysis of Kinetic Data in Decomposition Reactions*

	Equation	Number
Nucleation		
Power law	$\alpha^{1/n} = Kt$	(1)
Exponential law	$\ln \alpha = Kt$	(2)
Avrami–Erofe'ev	$[-\ln(1-\alpha)]^{1/2} = Kt$	(3)
	$[-\ln(1-\alpha)]^{1/3} = Kt$	(4)
	$[-\ln(1-\alpha)]^{1/4} = Kt$	(5)
Prout–Tompkins	$\ln[\alpha / (1-\alpha)] = Kt$	(6)
Geometrical Models		
Contracting thickness	$\alpha = Kt$	(7)
Contracting area	$1 - (1-\alpha)^{1/2} = Kt$	(8)
Contracting volume	$1 - (1-\alpha)^{1/3} = Kt$	(9)
Diffusion		
One dimensional	$\alpha^2 = Kt$	(10)
Two dimensional	$(1-\alpha)\ln(1-\alpha) + \alpha = Kt$	(11)
Three dimensional	$[1 - (1-\alpha)^{1/3}]^2 = Kt$	(12)
Ginstling–Brounshtein	$[1 - (2\alpha / 3)] - (1-\alpha)^{2/3} = Kt$	(13)
Reaction Order		
First order	$-\ln(1-\alpha) = Kt$	(14)
Second order	$(1-\alpha)^{-1} = Kt$	(15)
Third order	$(1-\alpha)^{-2} = Kt$	(16)

* The reaction rate constants, K, are different in each expression and the times, t, are assumed to have been corrected for any induction period, t_0.

Source: From Bamford, C.H. and Tipper, C.F.H., Eds., *Comprehensive Chemical Kinetics, Vol. 22: Reactions in the Solid State*, Elsevier, Oxford, 1980. With permission.

---

where c is the concentration of the reactant $\Re$ at time t, K is a reaction rate constant, and β is an exponent that defines the order of the reaction. The reaction is first order if $\beta = 1$, second order if $\beta = 2$, and so on. The kinetic equations for first-, second-, and third-order reactions are included in Table 2.4.

When comparing experimental decomposition data with the theoretical equations, the best fit to a particular equation is consistent with, but not proof of, the mechanism on which the equation is based. It is often found that the data can be fitted with equal accuracy by two or more rate equations based on different mechanisms. In seeking mechanistic information, the kinetic analysis should commonly be combined with other techniques, such as structural observations by electron microscopy.

The observed rate of decomposition and the characteristics of the powder produced by the decomposition reaction depend on a number of material and processing factors, including the chemical nature of the reactants, the initial size and size distribution of the reactant particles, the atmospheric conditions, the temperature, and the time. Isothermal rate data measured at several different temperatures show that the rate of decomposition obeys the Arrhenius relation. The rate constant, K, in the kinetic equations is given by:

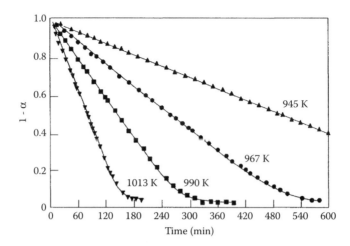

FIGURE 2.11 Isothermal decomposition kinetics of calcite (CaCO$_3$) single crystal. (From Beruto, D. and Searcy, A.W., Use of the Langmuir method for kinetic studies of decomposition reactions: calcite (CaCO$_3$), *J. Chem. Soc., Faraday Trans. I*, 70, 2145, 1974. With permission.)

$$K = A \exp\left(\frac{-Q}{RT}\right) \tag{2.11}$$

where *A* is a constant referred to as the *pre-exponential factor* or the *frequency factor*, *Q* is the activation energy, *R* is the gas constant, and *T* is the absolute temperature. Most reported activation energy values for CaCO$_3$ decompositions are close to the enthalpy of reaction [27]. Equation 2.11 has been established for gas-phase processes by the collision theory of reaction rates. The reason for its validity for decomposition reactions where a reactant is immobilized in the lattice of a solid phase has been the subject of some discussion [22].

Using the reaction described by Equation 2.2 as an example, decomposition reactions are commonly carried out under conditions where the equilibrium is driven far to the right-hand side. However, it is recognized that the decomposition kinetics of CaCO$_3$ will depend on the partial pressure of the CO$_2$ gas in the ambient atmosphere. High ambient CO$_2$ pressure drives the equilibrium to the left-hand side, so the decomposition rate decreases as the partial pressure of CO$_2$ increases [28].

In addition to the kinetics, the microstructure of the solid product particles is also dependent on the decomposition conditions. A feature of decomposition reactions is the ability to produce very fine particle size from a normally coarse reactant when the reaction is carried out under controlled conditions. In vacuum, the decomposition reaction is often *pseudomorphic* (i.e., the product particle often maintains the same size and shape as the reactant particle). As its molar volume is lower than that of the reactant, the product particle contains internal pores. Often the product particle consists of an aggregate of fine particles and fine internal pores. In the decomposition of 1 to 10 μm CaCO$_3$ particles performed at ~923K [29], the specific surface area of the CaO product formed in vacuum is as high as ~100 m^2/g with particle and pore sizes smaller than 10 nm (Figure 2.12a). If the reaction is carried out in an ambient atmosphere rather than in a vacuum, then high surface area powders are not obtained. In 1 atmosphere N$_2$, the surface area of the CaO particles is only 3 to 5 m^2/g (Figure 2.12b). Sintering of fine CaO particles during decomposition is also catalyzed by increasing partial pressure of CO$_2$ in the ambient atmosphere [30]. The atmospheric gas catalyzes the sintering of the fine particles, leading to larger particles and a reduction of the surface area. The sintering of fine MgO particles produced by decomposition of MgCO$_3$ or Mg(OH)$_2$ is catalyzed by water vapor in the atmosphere [31]. High decomposition

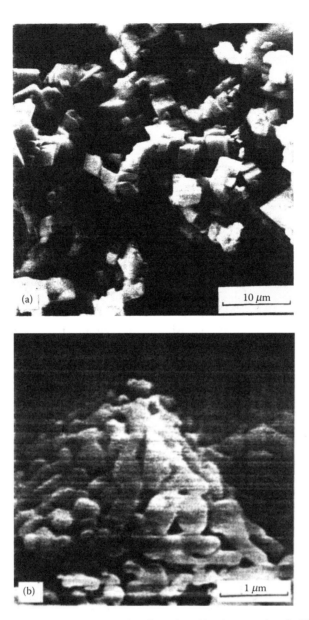

FIGURE 2.12 Scanning electron micrographs of CaO produced by decomposing $CaCO_3$ powder at 650°C in (a) vacuum (particles are same apparent size and shape as parent $CaCO_3$ particles) and (b) dry N_2 at atmospheric pressure. (From Ewing, J., Beruto, D., and Searcy, A.W., The nature of CaO produced by calcite powder decomposition in vacuum and in CO_2, *J. Am. Ceram. Soc.*, 62, 580, 1979. With permission.)

temperatures and long decomposition times promote sintering of the fine product particles, giving an agglomerated mass of low surface area powder. Although attempts are usually made to optimize the decomposition temperature and time schedule, agglomerates are invariably present, so a milling step is required to produce powders with controlled particle size characteristics.

2.5.1.2 Chemical Reaction between Solids

The simplest system involves the reaction between two solid phases, A and B, to produce a solid solution C (e.g., Equation 2.3). After the initiation of the reaction, A and B are separated by the

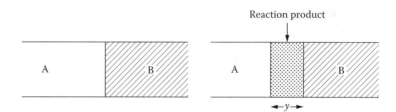

FIGURE 2.13 Schematic of solid-state reaction in single crystals.

reaction product C (Figure 2.13), and further reaction involves the transport of atoms, ions or molecules by several possible mechanisms through the phase boundaries and the reaction product. Whereas reactions between mixed powders are technologically important for powder synthesis, the study of reaction mechanisms is greatly facilitated by the use of single crystals because of the simplified geometry and boundary conditions.

The spinel formation reaction $AO + B_2O_3 = AB_2O_4$ is one of the most widely studied reactions [32]. Figure 2.14 shows a few of the possible reaction mechanisms. Included are (1) mechanisms in which O_2 molecules are transported through the gaseous phase, and electroneutrality is maintained by electron transport through the product layer (Figure 2.14a and Figure 2.14b), (2) mechanism involving counter diffusion of the cations with the oxygen ions remaining essentially stationary (Figure 2.14c), and (3) mechanisms in which O^{2-} ions diffuse through the product layer (Figure 2.14d and Figure 2.14e).

In practice, the diffusion coefficients of the ions differ widely. For example, in spinels, diffusion of the large O^{-2} ions is rather slow when compared to cationic diffusion, so the mechanisms in Figure 2.14d and Figure 2.14e can be eliminated. If ideal contact occurs at the phase boundaries, making the transport of O_2 molecules slow, then the mechanisms in Figure 2.14a and Figure 2.14b are unimportant. Under these conditions, the most likely mechanism is the counter diffusion of cations (Figure 2.14c), in which the cation flux is coupled to maintain electroneutrality. When the rate of product formation is controlled by diffusion through the product layer, the product thickness is observed to follow a parabolic growth law:

$$y^2 = Kt \hspace{4cm} (2.12)$$

where K is a rate constant that obeys the Arrhenius relation.

Several investigations have reported a parabolic growth rate for the reaction layer, which is usually taken to mean that the reaction is diffusion controlled [32]. The reaction between ZnO and Fe_2O_3 to from $ZnFe_2O_4$ is reported to occur by the counter-diffusion mechanism in which the cations migrate in opposite directions and the oxygen ions remain essentially stationary [33,34]. The reaction mechanism for the formation of $ZnAl_2O_4$ by Equation 2.3 is not as clear. The reaction rate is reported to be controlled by the diffusion of zinc ions through the product layer [35] but, as described below, for reactions between powders, the kinetics can also be described by a gas–solid reaction between ZnO vapor and Al_2O_3.

For powder reactions (Figure 2.15), a complete description of the reaction kinetics must take into account several parameters, thereby making the analysis very complicated. Simplified assumptions are commonly made in the derivation of kinetic equations. For isothermal reaction conditions, an equation derived by by Jander [36] is frequently used. In the derivation, it is assumed that equal-sized spheres of reactant A are embedded in a quasi-continuous medium of reactant B and that the reaction product forms coherently and uniformly on the A particles. The volume of unreacted material at time t is:

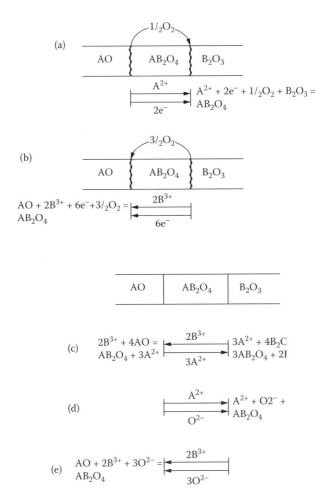

FIGURE 2.14 Reaction mechanisms and corresponding net phase boundary reactions for the spinel formation reaction $AO + B_2O_3 = AB_2O_4$. (From Schmalzried, H., *Solid State Reactions*, 2nd ed., Verlag Chemie, Weinheim, Germany, 1981. With permission.)

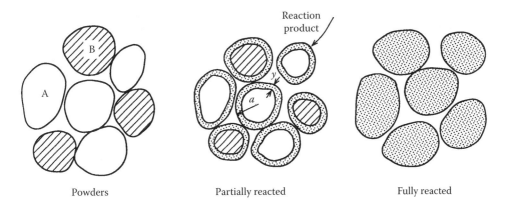

Powders   Partially reacted   Fully reacted

FIGURE 2.15 Schematic of solid-state reaction in mixed powders.

$$V = (4/3)\pi(a-y)^3 \tag{2.13}$$

where a is the initial radius of the spherical particles of reactant A and y is the thickness of the reaction layer. The volume of unreacted material is also given by:

$$V = (4/3)\pi a^3(1-\alpha) \tag{2.14}$$

where α is the fraction of the volume that has already reacted. Combining Equation 2.13 and Equation 2.14,

$$y = a[1-(1-\alpha)^{1/3}] \tag{2.15}$$

Assuming that y grows according to the parabolic relationship given by Equation 2.12, the reaction kinetics are given by:

$$[1-(1-\alpha)^{1/3}]^2 = Kt/a^2 \tag{2.16}$$

Equation 2.16, referred to as the *Jander equation*, suffers from two simplifications that limit its applicability and the range over which it adequately predicts reaction rates. First, the parabolic growth law assumed for the thickness of the reaction layer is valid for one-dimensional reaction across a planar boundary and not for a system with spherical geometry. At best, it is expected to be valid only for the initial stages of the powder reaction when $y \ll a$. Second, any change in molar volume between the reactant and the product is neglected. These two simplifications have been taken into account by Carter [37], who derived the following equation:

$$[1+(Z-1)\alpha]^{2/3} + (Z-1)(1-\alpha)^{2/3} = Z + (1-Z)\frac{Kt}{a^2} \tag{2.17}$$

where Z is the volume of the reaction product formed from unit volume of the reactant A. Equation 2.17, referred to as the *Carter equation*, is applicable to the formation of $ZnAl_2O_4$ by the reaction between ZnO and Al_2O_3 even up to 100% of reaction (Figure 2.16).

For a solid-state diffusion mechanism, the growth of the reaction product in powder systems occurs at the contact points and for nearly equal-sized spheres, the number of contact points is small. Nevertheless, for many systems, the Jander equation and the Carter equation give a good description of the reaction kinetics for at least the initial stages of the reaction. It appears that rapid surface diffusion provides a uniform supply of one of the reactants over the other. Alternatively, if the vapor pressure of one of the reactants is high enough (e.g., ZnO in Equation 2.3), condensation on the surface of the other reactant can also provide a uniform supply of the other reactant. In this case, the powder reaction can be better described as a gas–solid reaction rather than a solid-state reaction [32].

In practical systems, solid-state reaction in powder systems depends on several parameters, such as the chemical nature of the reactants and the product, the size, size distribution and shape of the particles, the relative sizes of the reactant particles in the mixture, the uniformity of the mixing, the reaction atmosphere, the temperature, and the time. The reaction rate will decrease with an increase in particle size of the reactants because, on average, the diffusion distances will increase. For coherent reaction layers and nearly spherical particles, the dependence of the reaction kinetics on particle size is given by Equation 2.16 or Equation 2.17. The reaction rate will increase

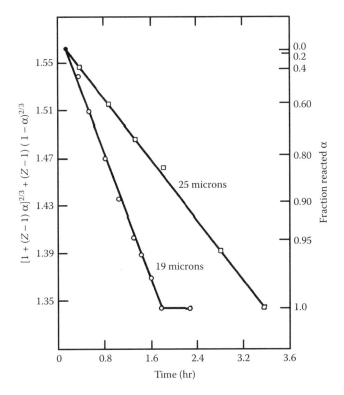

FIGURE 2.16 Kinetics of reaction between spherical particles of ZnO and Al_2O_3 to form $ZnAl_2O_4$ at 1400°C in air, showing the validity of the Carter equation. (From Schmalzried, H., *Solid State Reactions*, 2nd ed., Verlag Chemie, Weinheim, Germany, 1981. With permission.)

with temperature according to the Arrhenius relation. Commonly, the homogeneity of mixing is one of the most important parameters. It influences the diffusion distance between the reactants and the relative number of contacts between the reactant particles, and thus the ability to produce homogeneous, single-phase powders.

Powder preparation by solid-state reactions generally has an advantage in terms of production cost. However, the powder quality is also important for advanced ceramics. A grinding step is almost always required to break up the agglomerates present in the powder, often leading to contamination of the powder. The particle shape of ground powders is usually difficulty to control. Incomplete reactions, especially in poorly mixed powders, produce undesirable phases.

2.5.1.3 Reduction

The reduction of silica by carbon is used industrially to produce silicon carbide powders:

$$SiO_2 + 3C \rightarrow SiC + 2CO \tag{2.18}$$

This reaction should occur somewhat above 1500°C, but it is usually carried out at much higher temperatures, at which the SiO_2 is actually a liquid. The industrial process is referred to as the *Acheson process*. The mixture is self-conducting and is heated electrically to temperatures of ~2500°C. Side reactions occur, so the reaction is more complex that that indicated in Equation 2.18. The product obtained after several days of reaction consists of an aggregate of black or green crystals, which is crushed, washed, ground, and classified to produce the desired powder sizes. One disadvantage of the Acheson process is that the powder quality is often too poor for demanding applications such as high-temperature structural ceramics. Because the reactants exist as mixed particles, the extent of the reaction is limited by the contact area and inhomogeneous mixing between reactant particles, with the result that the SiC product contains some unreacted SiO_2 and C. These limitations have been surmounted recently by a process in which SiO_2 particles are coated with C prior to reduction [38,39], leading to the production of relatively pure SiC powders with fine particle sizes (<0.2 μm).

2.5.2 LIQUID SOLUTIONS

There are two general routes for the production of a powdered material from a solution: (1) evaporation of the liquid and (2) precipitation by adding a chemical reagent that reacts with the solution. The reader may be familiar with these two routes as they are commonly used in inorganic chemistry laboratories; e.g., the production of common salt crystals from a solution by evaporation of the liquid or of $Mg(OH)_2$ by the addition of NaOH solution to $MgCl_2$ solution. For the production of powders with controlled characteristics, an understanding of the principles of precipitation from solution is useful.

2.5.2.1 Precipitation from Solution

2.5.2.1.1 Principles

The kinetics and mechanism of precipitation are well covered in a textbook [40] and several review articles on the principles and procedures of precipitation from solution are available in the literature [41–46]. Precipitation from solution consists of two basic steps: (1) nucleation of fine particles and (2) their growth by addition of more material to the surfaces. In practice, control of the powder characteristics is achieved by controlling the reaction conditions for nucleation and growth and the extent to which these two processes are coupled.

2.5.2.1.1.1 Nucleation

The type of nucleation we consider here is referred to as *homogeneous nucleation*, taking place in a completely homogeneous phase with no foreign inclusions in the solution or on the walls of the reaction vessel. When these inclusions are present and act to assist the nucleation, the process is called *heterogeneous nucleation*. The occurrence of heterogeneous nucleation makes it difficult to obtain well-controlled particle sizes and must normally be avoided, but as we shall see later, it can be used to good advantage for the synthesis of coated particles.

Homogeneous nucleation of solid particles in solution is generally analyzed in terms of the classical theories developed for vapor→liquid and vapor→solid transformations [47]. We briefly outline the main features of the classical theories for vapor→liquid transformation and then examine how they are applied to nucleation of solid particles from solution. In a supersaturated vapor consisting of atoms (or molecules), random thermal fluctuations give rise to local fluctuations in density and free energy of the system. Density fluctuations produce clusters of atoms referred to as *embryos*, which can grow by addition of atoms from the vapor phase. A range of embryo sizes will be present in the vapor with vapor pressure p assumed to obey the Kelvin equation:

$$\ln \frac{p}{p_0} = \frac{2\gamma v_l}{kTa} \tag{2.19}$$

where p_0 is the saturated vapor pressure over a flat (reference) surface, γ is the specific surface energy of the cluster, v_l is the volume per molecule in a liquid drop formed by condensation of the vapor, k is the Boltzmann constant, T is the absolute temperature, and a is the radius of the embryo (assumed to be spherical). Because of their higher vapor pressure, small embryos evaporate back

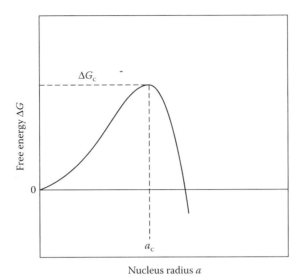

FIGURE 2.17 Schematic variation of the free energy vs. radius for a spherical droplet. Some critical size must be exceeded before a nucleus becomes stable.

to the vapor phase. Embryos with a radius, a, less than a critical radius, a_c, cannot grow whereas embryos with $a > a_c$ can. However, the formation of nuclei (i.e., embryonic droplets), requires an energy barrier to be surmounted. This may be illustrated by considering the free energy change in forming a spherical nucleus of radius a. The increase in the free energy can be written as:

$$\Delta G_n = 4\pi a^2 \gamma - (4/3)\pi a^3 \Delta G_v \qquad (2.20)$$

The first term on the right is the intrinsically positive contribution of the surface free energy. The second term represents the contribution by the bulk free energy change. Considering unit volume of liquid, the free energy decrease, ΔG_v, in going from vapor to liquid is given by [48]:

$$\Delta G_v = (kT / v_l) \ln(p / p_0) \qquad (2.21)$$

where v_l is the volume per molecule in the liquid. Substituting for ΔG_v in Equation 2.20 gives:

$$\Delta G_n = 4\pi a^2 \gamma - (4/3)\pi a^3 (kT / v_l) \ln(p / p_o) \qquad (2.22)$$

For the limiting case when the supersaturation ratio $S = p/p_0 = 1$, the bulk term vanishes and ΔG_n increases monotonically as a parabola. For $S < 1$, the ΔG_n curve rises more steeply because a fractional S makes the second term on the right go positive, reinforcing the effect due to the surface free energy barrier. For $S > 1$, the second term is negative and this assures the existence of a maximum in ΔG_n at some critical radius, a_c, as shown schematically in Figure 2.17. The critical radius, a_c, is obtained by putting $d(\Delta G_n)/da = 0$, giving:

$$a_c = \frac{2\gamma v_l}{kT \ln(p / p_0)} \qquad (2.23)$$

Substituting for a_c in Equation 2.22, the height of the free energy activation barrier is:

$$\Delta G_c = \frac{16\pi\gamma^3 v_l^2}{3[kT\ln(p/p_0)]^2} = (4/3)\pi a_c^2\gamma \tag{2.24}$$

To summarize at this stage, sufficient increase in the supersaturation ratio $S = p/p_0$ finally serves to increase the atomic (molecular) bombardment rate in the vapor and to reduce ΔG_v and a_c to such an extent that the probability of a subcritical embryo growing to supercritical size in a short time approaches unity. Homogeneous nucleation to form liquid droplets now becomes an effective process.

The nucleation rate, I, refers to the rate of formation of critical nuclei as only these can grow to produce liquid droplets. A pseudothermodynamic treatment of vapor→liquid transformation gives the result that I is proportional to exp $(-\Delta G_c/kT)$, where k is the Boltzmann constant and ΔG_c is given by Equation 2.24. The rate at which the nuclei grow will also depend on the frequency with which atoms join it, and this may be written as [ν exp$(-\Delta G_m/kT)$], where ν is the characteristic frequency, and ΔG_m is the activation energy for atom migration. Putting $\nu = kT/h$, where h is Planck's constant, an approximate expression for the nucleation rate is [47]:

$$I \approx \frac{NkT}{h}\exp\left(\frac{-\Delta G_m}{kT}\right)\exp\left\{\frac{-16\pi\gamma^3 v_l^2}{3kT[kT\ln(p/p_0)]^2}\right\} \tag{2.25}$$

where N is the number of atoms per unit volume in the phase undergoing transformation.

Homogeneous nucleation of particles from solution occurs in many techniques for the synthesis of ceramic powders. General aspects of nucleation from liquids and solutions are discussed by Walton [49]. In aqueous solution, metal ions are hydrated [50]. Embryos of hydrated metal ions are assumed to form by progressive addition of ions to one another by a polymerization process. These polynuclear ions are the precursors to nucleation. When the concentration of the polynuclear ions increases above some minimum supersaturation concentration, homogeneous nucleation to form solid nuclei becomes an effective process. The nucleation rate of particles from solution can be expressed as [49]:

$$I \approx \frac{2Nv_s(kT\gamma)^{1/2}}{h}\exp\left(\frac{-\Delta G_a}{kT}\right)\exp\left(\frac{-16\pi\gamma^3 v_s^2}{3k^3T^3[\ln(C_{ss}/C_s)]^2}\right) \tag{2.26}$$

where N is the number of ions per unit volume in the solution, v_s is the volume of a molecule in the solid phase, γ is the specific energy of the solid–liquid interface, ΔG_a is the activation energy for the transport of an ion to the solid surface, C_{ss} is the supersaturated concentration, and C_s the saturated concentration of the ions in the solution. The nucleation rate is strongly dependent on the supersaturation ratio C_{ss}/C_s.

2.5.2.1.1.2 Particle Growth by Solute Precipitation
Nuclei are normally very small but even during a short nucleation stage they may have grown to somewhat different sizes. Therefore, the starting system for growth is not monodisperse. Nuclei formed in a supersaturated solution can grow by transport of solute species (ions or molecules) through the solution to the particle surface, desolvation, and alignment on the particle surface. The rate-determining step in the growth of the particles can be:

1. Diffusion toward the particle
2. Addition of new material to the particle by a form of surface reaction.

The occurrence of the specific mechanisms and their interplay control the final particle size characteristics.

Diffusion-controlled growth: Assuming that the particles are far apart so that each can grow at its own rate, the diffusion of solute species toward the particle (assumed to be spherical of radius a) can be described by Fick's first law. The flux, J, through any spherical shell of radius x is given by:

$$J = 4\pi x^2 D (dC / dx) \tag{2.27}$$

where D is the diffusion coefficient for the solute through the solution, and C is its concentration. Assuming that the saturation concentration, C_s, is maintained at the particle surface and that the concentration of the solute far from the particle is C_∞, a concentration gradient is set up that approaches a stationary state in times of the order of a^2/D. In this stationary state, J does not depend on x, and integration of Equation 2.27 gives:

$$J = 4\pi a D (C_\infty - C_s) \tag{2.28}$$

The rate of growth of the particle radius is then:

$$\frac{da}{dt} = \frac{JV_s}{4\pi a^2} = \frac{DV_s(C_\infty - C_s)}{a} \tag{2.29}$$

where V_s is the molar volume of the solid precipitating on the particle. Equation 2.29 can also be written:

$$\frac{d(a^2)}{dt} = 2DV_s(C_\infty - C_s) \tag{2.30}$$

showing that irrespective of the original size of the particle, the square of the radius of all particles increases at the same constant rate. The treatment leading to Equation 2.30 is oversimplified. It has been shown more rigorously that for diffusion-controlled growth, $d(a^2)/dt$ is the same for particles of any size but not necessarily constant in time [51].

If the absolute width of the particle size distribution is Δa for a mean radius a and Δa_o for the mean radius a_o of the initial system, we can deduce from Equation 2.30 that:

$$\frac{\Delta a}{\Delta a_o} = \frac{a_o}{a} \ and \ \frac{\Delta a}{a} = \left(\frac{a_o}{a}\right)^2 \frac{\Delta a_0}{a_0} \tag{2.31}$$

Equation 2.31 shows that the absolute width of the size distribution becomes narrower in the ratio a_o/a and the relative width decreases even faster in the ratio $(a_o/a)^2$.

Surface reaction–controlled growth: Each new layer around the particle has to be nucleated first by a process that is different from the homogeneous nucleation discussed earlier. Two types of growth mechanisms can be distinguished, referred to as *mononuclear* growth and *polynuclear* growth (Figure 2.18). In the mononuclear growth mechanism, once a nucleation step is formed on

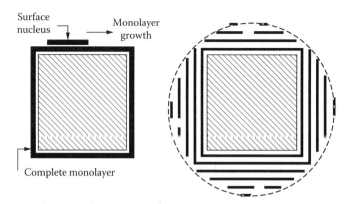

FIGURE 2.18 Nucleation of layers around a particle for (a) mononuclear growth and (b) polynuclear growth.

the particle surface, a layer has the time to achieve its completion before a new step appears. Growth proceeds, therefore, layer by layer and the particle surface may appear faceted on a macroscopic scale. Particle growth is described by [40]:

$$\frac{da}{dt} = K_1 a^2$$ (2.32)

where K_1 is a constant. The relative width of the size distribution is given by:

$$\frac{\Delta a}{a} = \left(\frac{a}{a_o} \right) \frac{\Delta a_o}{a_o}$$ (2.33)

and increases in the ratio a/a_o. In the *polynuclear* growth mechanism, formation of nucleation steps on the particle surface is fast enough to create a new layer before the previous one been completed. The growth rate is independent of the surface area of the existing particles and is given by [40]:

$$\frac{da}{dt} = K_2$$ (2.34)

where K_2 is a constant. In this case, the relative width of the distribution decreases according to:

$$\frac{\Delta a}{a} = \left(\frac{a_o}{a} \right) \frac{\Delta a_o}{a_o}$$ (2.35)

2.5.2.1.1.3   Controlled Particle Size Distribution

The basic principles for obtaining particles with a fairly uniform size by precipitation from solution were put forward nearly 50 years ago by LaMer and Dinegar [52]. The main features may be represented in terms of the diagram shown in Figure 2.19, generally referred to as a *LaMer diagram*. As the reaction proceeds, the concentration of the solute to be precipitated, C_x, increases to, or above, the saturation value, C_s. If the solution is free of foreign inclusions and the container walls are clean and smooth, then it is possible for C_x to exceed C_s by a large amount to give a supersaturated solution. Eventually a critical supersaturation concentration, C_{ss}, will be reached after some time

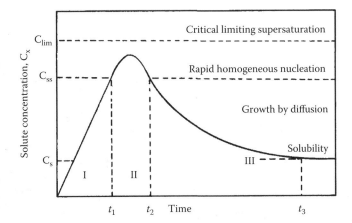

FIGURE 2.19 Schematic representation of the solute concentration vs. time in the nucleation and growth of particles from a solution (From LaMer, V.K. and Dinegar, R.H., Theory, production and mechanism of formation of monodispersed hydrosols, *J. Am. Chem. Soc.*, 72, 4847, 1950. With permission.)

t_1 and homogeneous nucleation and growth of solute particles occur, leading to a decrease in C_x to a value below C_{ss} after a time t_2. Further growth of the particles occurs by diffusion of solute through the liquid and precipitation on to the particle surfaces. Finally, particle growth stops after a time t_3 when $C_x = C_s$.

It is clear that if we wish to produce particles with a fairly uniform size, then one short burst of nucleation should occur in a short time interval, $t_2 - t_1$. One way of achieving this is through the use of a fairly low reactant concentration. Furthermore, uniform growth of the particles requires that the solute be released slowly to allow diffusion to the particles without build up of the solute concentration and further bursts of nucleation. This mechanism of nucleation followed by diffusion controlled growth does not apply to the formation of particles that are aggregates of finer primary particles. Instead, it may apply only to the primary particles.

2.5.2.1.1.4   Particle Growth by Aggregation

High-resolution electron micrographs of particles synthesized by several routes involving precipitation from solution show that the particles consist of aggregates of much finer primary particles. Titania particles prepared by the Stober process show primary particle features that are smaller than ~10 nm (Figure 2.20a). TEM of CeO_2 particles synthesized by hydrolysis of cerium nitrate salts in the presence of sulfate ions [53] show that the hexagonal particles consist of smaller primary particles with a spherical shape (Figure 2.20b). Bogush et al. [54,55] suggested that growth of SiO_2 particles in the Stober process occurs by aggregation of fine particles rather than by diffusion of solute to existing particles. Using the DLVO theory for colloid stability (see Chapter 4), they showed that under identical surface charge densities, the barrier to aggregation increases with size for two particles with equal size and that the rate of aggregation decreases exponentially. On the other hand, fine particles aggregate more quickly with large particles than they do with themselves. During a precipitation reaction, the first nuclei grow rapidly by aggregation to a colloidally stable size. These particles then sweep through the suspension, picking up freshly formed nuclei and smaller aggregates. The formation of particles with uniform size is thus achieved through size-dependent aggregation rates.

2.5.2.1.1.5   Particle Growth by Ostwald Ripening

Particles in a liquid can also grow by a process in which the smaller particles dissolve and the solute precipitates on the larger particles. This type of particle growth or coarsening is referred to as *Ostwald ripening* (Chapter 9). Matter transport from the smaller particles to the larger particles can be controlled by (1) diffusion through the liquid or (2) an interface reaction (dissolution of the

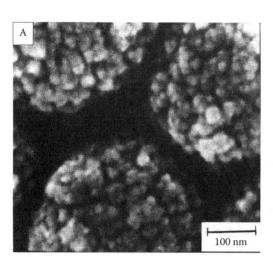
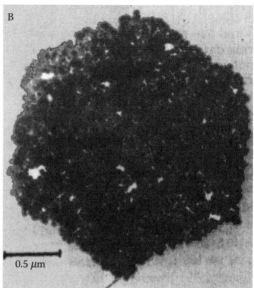

100 nm

0.5 μm

FIGURE 2.20 (a) Scanning electron micrograph of TiO$_2$ particles prepared by the Stober process showing that the particles consist of much finer primary particles (Courtesy of A.M. Glaeser.). (b) Transmission electron micrograph of hexagonal CeO$_2$ particles synthesized by hydrolysis of cerium nitrate salts in the presence of sulfate ions showing that the hexagonal particles consist of much smaller primary particles with a spherical shape. (From Hsu, W.P., Ronnquist, L., and Matijevic, E., Preparation and properties of monodispersed colloidal particles of lanthanide compounds. 2. Cerium (IV), *Langmuir*, 4, 31, 1988. With permission.)

solid or deposition of the solute onto the particle surfaces). The average radius, $\langle a \rangle$, of the particles (assumed to be spherical) is predicted to increase with time, t, according to:

$$\langle a \rangle^m = \langle a_o \rangle^m + Kt \tag{2.36}$$

where $\langle a_o \rangle$ is the initial average radius of the particles, K is a constant that obeys the Arrhenius relation, and m is an exponent the depends on the mechanism: $m = 2$ for interface reaction control and $m = 3$ for diffusion control. Regardless of the initial size distribution, the particle size distribution reaches a self-similar distribution as it depends only on $a / \langle a \rangle$ and is independent of time. The maximum radius of the distribution is $2 \langle a \rangle$ for the interface reaction mechanism and $3/2 \langle a \rangle$ for the diffusion mechanism. Ostwald ripening, by itself, cannot therefore lead to a monodisperse system of particles.

2.5.2.1.2 Procedures for Precipitation from Solution

The most straightforward use of precipitation is the preparation of simple oxides or hydrous oxides (hydroxides or hydrated oxides) by hydrolysis. Two main routes can be distinguished: (1) hydrolysis of metal-organic compounds (e.g., metal alkoxides) in alcoholic solution, generally referred to as the Stober process and (2) hydrolysis of aqueous solutions of metal salts, pioneered by Matijevic.

2.5.2.1.2.1 Hydrolysis of Solutions of Metal Alkoxides

Metal alkoxides have the general formula M(OR)$_z$, where z is an integer equal to the valence of the metal, M, and R is an alkyl chain (see Chapter 5). They can be considered as derivatives of either an alcohol, ROH, in which the hydrogen is replaced by the metal M, or of a metal hydroxide, M(OH)$_z$, in which the hydrogen is replaced by an alkyl group. The reactions involve hydrolysis:

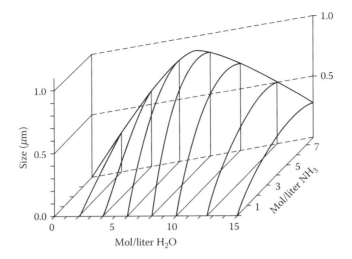

FIGURE 2.21 Correlation between particle size and the concentrations of water and ammonia in the hydrolysis of a solution of 0.28 mol/dm³ silicon tetraethoxide in ethanol. (From Stober, W., Fink, A., and Bohn, E., Controlled growth of monodisperse silica spheres in the micron size range, *J. Colloid Interface Sci.*, 26, 62, 1968. With permission.)

$$M(OR)_z + xH_2O \rightarrow M(OR)_{z-x}(OH)_x + xROH \tag{2.37}$$

followed by condensation and polymerization:

$$-M - OH + HO - M - \rightarrow -M - O - M - +H_2O \tag{2.38}$$

Stober et al. [56] carried out a systematic study of the factors that control the preparation of fine, uniform SiO_2 particles by the hydrolysis of silicon alkoxides in the presence of NH_3. The NH_3 served to produce pH values in the basic range. For the hydrolysis of silicon tetraethoxide, $Si(OC_2H_5)_4$, referred to commonly as TEOS, with ethanol as the solvent, the particle size of the powder was dependent on the ratio of the concentration of H_2O to TEOS and on the concentration of NH_3 but not on the TEOS concentration (in the range of 0.02 to 0.50 mol/dm³). For a TEOS concentration of 0.28 mol/dm³, Figure 2.21 shows the general correlation between particle size and the concentrations of H_2O and NH_3. The particle sizes varied between 0.05 to 0.90 μm and were very uniform, as shown in Figure 2.22. Different alcoholic solvents or silicon alkoxides were also found to have an effect. The reaction rates were fastest with methanol and slowest with n-butanol. Likewise, under comparable conditions, the particle sizes were smallest in methanol and largest in *n*-butanol.

The controlled hydrolysis of metal alkoxides has since been used to prepare fine powders of several simple oxides. We mentioned in Chapter 1 the work of Barringer and Bowen [57] for the preparation, packing, and sintering of monodisperse TiO_2 powders. In the formation of TiO_2 by hydrolysis of $Ti(OC_2H_5)_4$, the alkoxide reacts with water to produce a monomeric hydrolysis species [58]:

$$Ti(OC_2H_5)_4 + 3H_2O \leftrightarrow Ti(OC_2H_5)(OH)_3 + 2C_2H_5OH \tag{2.39}$$

However, the presence of dimers and trimers of the hydrolysis species cannot be excluded. Polymerization of the monomer to produce the hydrated oxide is represented by

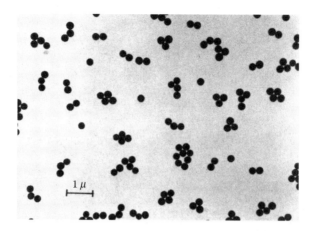

FIGURE 2.22 Silica spheres produced by the hydrolysis of a solution of silicon tetraethoxide in ethanol. (From Stober, W., Fink, A., and Bohn, E., Controlled growth of monodisperse silica spheres in the micron size range, *J. Colloid Interface Sci.*, 26, 62, 1968. With permission.)

$$Ti(OC_2H_5)(OH)_3 \leftrightarrow TiO_2 \cdot xH_2O + (1 - x)H_2O + C_2H_5OH \qquad (2.40)$$

The overall reaction can therefore be represented as

$$Ti(OC_2H_5)_4 + (2 + x)H_2O \leftrightarrow TiO_2 \cdot xH_2O + 4C_2H_5OH \qquad (2.41)$$

The value of x was found by thermogravimatric analysis to be between 0.5 and 1.

Because most metal alkoxides hydrolyze readily in the presence of water, stringent conditions must be maintained to achieve powders with controlled characteristics. The reactions are sensitive to the concentration of the reactants, the pH and the temperature. Oxide or hydrated oxide powders are produced. The precipitated particles are often amorphous and may consist of agglomerates of much finer particles (Figure 2.20a).

2.5.2.1.2.2 Hydrolysis of Solutions of Metal Salts

Procedures for the preparation of uniform particles by the hydrolysis of metal salt solutions have been developed and reviewed by Maitjevic [45,46,59–61]. Compared with the Stober process, the method has the ability for producing a wider range of chemical compositions, including oxides or hydrous oxides, sulfates, carbonates, phosphates and sulfides. On the other hand, the number of experimental parameters that must be controlled to produce uniform particles is generally higher. They include the concentration of the metal salts, the chemical composition of the salts used as starting materials, the temperature, the pH of the solution, and the presence of anions and cations that form intermediate complexes. A variety of particle sizes and shapes can be produced (Figure 2.23) but the morphology of the final particles can rarely be predicted. Whereas amorphous as well as crystalline particles can be produced, the factors that determine the crystalline vs. amorphous structure of the product are not clear [61].

Metal ions are normally hydrated in aqueous solution [50]. The conditions for homogeneous precipitation of uniform particles can be achieved by a *forced hydrolysis* technique based on promoting the *deprotonation* of hydrated cations by heating the solution at elevated temperatures (90 to 100 °C). For a metal, M, with a valence z, the reaction can be written as:

$$[M(OH_2)_n]^{z+} \leftrightarrow [M(OH)_y(OH_2)_{n-y}]^{(z-y)} + yH^+ \qquad (2.42)$$

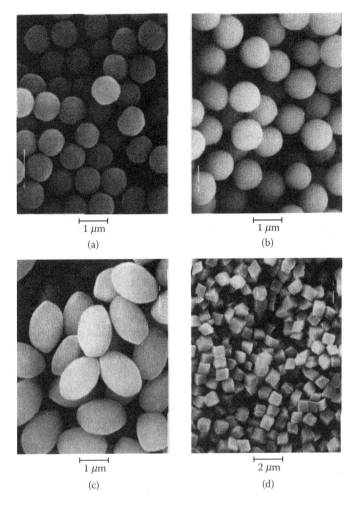

FIGURE 2.23 Examples of the sizes, shapes, and chemical compositions for powders prepared by precipitation form metal salt solutions, showing particles of (a) hematite (α-Fe$_2$O$_3$), (b) cadmium sulfide, (c) iron (III) oxide, and (d) calcium carbonate. (From Matijevic, E., Monodispersed colloidal metal oxides, sulfides, and phosphates, in *Ultrastructure Processing of Ceramics, Glasses, and Composites,* Hench, L.L. and Ulrich, D.R., Eds., Wiley, New York, 1984, chap. 27. With permission.)

The soluble hydroxylated complexes produced by the hydrolysis reaction form the precursors to the nucleation of particles. They can be generated at the proper rate to achieve nucleation and growth of uniform particles by adjustment of the temperature and pH. In principle, it is only necessary to age the solutions at elevated temperatures but, in practice, the process is very sensitive to minor changes in conditions. Anions other than hydroxide ions can also play a decisive role in the reaction. Some anions are strongly coordinated to the metal ions and thus end up in the precipitated solid of fixed stoichiometric composition. In other cases, the anions can be readily removed from the product by leaching or they can affect the particle morphology without being incorporated in the precipitated solid. The specific conditions for the precipitation of uniform particles must therefore be adjusted from case to case.

As an example of the sensitivity of the reactions to changes in conditions, we can consider the synthesis of spherical hydrated aluminum oxide particles with narrow size distribution [62]. Solutions of Al$_2$(SO$_4$)$_3$, KAl(SO$_4$)$_2$, a mixture of Al(NO$_3$)$_3$ and Al$_2$(SO$_4$)$_3$ or a mixture of Al$_2$(SO$_4$)$_3$ and Na$_2$SO$_4$ were aged in sealed Pyrex tubes at 98 $\pm$ 2°C for up to 84 h. The pH of the freshly

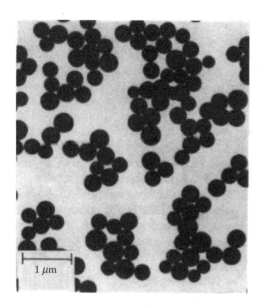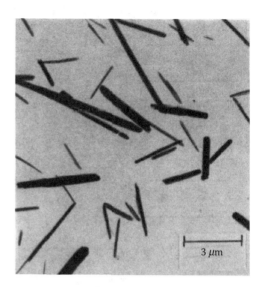

FIGURE 2.24 Particles obtained (a) by aging for 2.5 h at 90°C a solution of 1.5×10^{-2} mol/dm^3 YCl$_3$ and 0.5 mol/dm^3 urea and (b) by aging for 18 h at 115°C a solution of 3.0×10^{-2} mol/dm^3 YCl$_3$ and 3.3 mol/dm^3 urea. (From Aiken, B., Hsu, W.P., and Matijevic, E., Preparation and properties of monodispersed colloidal particles of lanthanide compounds, *J. Am. Ceram. Soc.*, 71, 845, 1988. With permission.)

prepared solutions was 4.1, and after aging and cooling to room temperature it was 3.1. Particles with uniform size were produced only when the Al concentration was between 2×10^{-4} and 5×10^{-3} mol/dm^3 provided that the [Al^{3+}] to [SO$_4^{2-}$] molar ratio was between 0.5 and 1. For a constant Al concentration, the particle size increased with increasing sulfate concentration. The temperature of aging was a critical parameter; no particles were produced below 90°C, whereas the best results were obtained at 98°C. The particles had reasonably constant chemical composition, which indicates that one or more well-defined aluminum basic sulfate complexes were the precursors to the nucleation of particles.

The conditions for nucleation and growth of uniform particles in solution can also be met by the slow release of anions from organic molecules such as urea or formamide. An example is the precipitation of yttrium basin carbonate particles from a solution of yttrium chloride, YCl$_3$, and urea, (NH$_2$)$_2$CO [63]. Particles of uniform size were produced by aging for 2.5 h at 90°C, a solution of 1.5×10^{-2} mol/dm^3 YCl$_3$ and 0.5 mol/dm^3 urea (Figure 2.24a). However, solutions of YCl$_3$ containing higher urea concentrations yielded, on aging at 115°C for 18 h, rodlike particles somewhat irregular in size, (Figure 2.24b). At temperatures up to 100°C, aqueous solutions of urea yield ammonium and cyanate ions:

$$(NH_2)_2CO \leftrightarrow NH_4^+ + OCN^- \tag{2.43}$$

In acid solutions, cyanate ions react rapidly, according to

$$OCN^- + 2H^+ + H_2O \rightarrow CO_2 + NH_4^+ \tag{2.44}$$

whereas, in neutral and basic solutions, carbonate ions and ammonia are formed:

$$OCN^- + OH^- + H_2O \rightarrow NH_3 + CO_3^{2-} \tag{2.45}$$

Yttrium ions are weakly hydrolyzed in water to $YOH(H_2O)_n^{2+}$. The resulting release of hydronium ions accelerates urea decomposition according to Equation 2.44. The overall reaction for the precipitation of the basic carbonate may be written as:

$$YOH(H_2O)_n^{2+} + CO_2 + H_2O \rightarrow Y(OH)CO_3 \cdot H_2O + 2H^+ + (n-1)H_2O \qquad (2.46)$$

For the reaction at 115°C, the decomposition of excess urea (>2 mol/dm^3) generates a large amount of OH ions which change the medium from acidic to basic (pH 9.7). The reaction of cyanate ions proceeds according to Equation 2.45. The precipitation of rodlike particles may be represented as:

$$2YOH(H_2O)_n^{2+} + NH_3 + 3CO_3^{2-} \rightarrow Y_2(CO_3)_3 \cdot NH_3 \cdot 3H_2O + (2n-3)H_2O + 2OH^- \qquad (2.47)$$

In addition to an excess of urea and a higher aging temperature, longer reaction times (> 12 h) are needed to generate a sufficient amount of free ammonia for the reaction in Equation 2.47 to dominate.

2.5.2.1.2.3   *Precipitation of Complex Oxides*

Complex oxides are oxides such as titanates, ferrites, and aluminates that contain more than one type of metal in the chemical formula. Earlier, we outlined the difficulties of solid-state reaction between a mixture of oxide powders when fine, stoichiometric, high-purity powders are required. Some of these difficulties can be alleviated by the use of *coprecipitation* from a solution of mixed alkoxides, mixed salts, or a combination of salts and alkoxides. A common problem in coprecipitation is that the different reactants in the solution have different hydrolysis rates, resulting in segregation of the precipitated material. Suitable conditions must therefore be found in order to achieve homogeneous precipitation. As an example, consider the preparation of $MgAl_2O_4$ powders [64]. Both Mg and Al are precipitated as hydroxides but the conditions for their precipitation are quite different. $Al(OH)_3$ is precipitated under slightly basic conditions (pH = 6.5 to 7.5), is soluble in the presence of excess ammonia, but is only slightly soluble in the presence of NH_4Cl. However, $Mg(OH)_2$ is completely precipitated only in strongly basic solutions such as NaOH solution. In this case, an intimate mixture of $Al(OH)_3$ and Mg-Al double hydroxide, $2Mg(OH)_2 \cdot Al(OH)_3$ is produced when a solution of $MgCl_2$ and $AlCl_2$ is added to a stirred excess solution of NH_4OH kept at a pH of 9.5 to 10. Calcination of the precipitated mixture above ~400°C yields stoichiometric $MgAl_2O_4$ powder with high purity and fine particle size.

The coprecipitation technique generally produces an intimate mixture of precipitates. In many cases, the mixture has to be calcined to produce the desired chemical composition, thereby requiring a subsequent milling step that can introduce impurities into the powder. An example of this method is the preparation of lead–lanthanum–zirconium–tiatante (PLZT) powders [65]. Generally, it is more desirable to produce a precipitate that does not require the use of elevated temperature calcination and subsequent milling. In a few cases, the precipitated powder may have the same cation composition as the desired product. An example is the preparation of $BaTiO_3$ by the hydrolysis of a solution of barium isopropoxide, $Ba(OC_3H_7)_2$, and titanium tertiary amyloxide, $Ti(OC_5H_{11})_4$ by Mazdiyasni, Doloff, and Brown [66]. The overall reaction can be written as

$$Ba(OC_3H_7)_2 + Ti(OC_5H_{11})_4 + 3H_2O \rightarrow BaTiO_3 + 4C_5H_{11}OH + 2C_3H_7OH \qquad (2.48)$$

The alkoxides are dissolved in a mutual solvent (e.g., isopropanol) and refluxed for 2 h prior to hydrolysis. Water is added slowly while the solution is vigorously stirred, and the reaction is carried out in a CO_2-free atmosphere to prevent the precipitation of $BaCO_3$. After drying the precipitate at 50°C for 12 hr in a helium atmosphere, a stoichiometric BaTiO₃ powder with a purity of more than 99.98% and a particle size of 5 to 15 nm (with a maximum agglomerate size of < 1 µm) is produced. Dopants can be incorporated uniformly into the powders by adding a solution of the metal alkoxide prior to hydrolysis.

The hydrolysis of a mixture of metal alkoxides forms a fairly successful route for the synthesis of complex oxide powders [67], but most metal alkoxides are expensive and their hydrolysis requires carefully controlled conditions because of their sensitivity to moisture. Whereas the controlled hydrolysis of a mixture of salt solutions appears to be more difficult, its use has been demonstrated by Matijevic [61] for a few systems, including BaTiO₃ and SrFeO₄.

2.5.2.1.2.4   *Precipitation under Hydrothermal Conditions*

Precipitation from solution under hydrothermal conditions has been known for decades as a method for synthesizing fine, crystalline oxide particles [68]. Interest in the method has increased in recent years because of a growing need for fine, pure powders for the production of electronic ceramics. The process involves heating reactants, often metal salts, oxide, hydroxide or metal powder, as a solution or a suspension, usually in water, at temperatures between the boiling and critical points of water (100 to 374°C) and pressures up to 22.1 MPa (the vapor pressure of water at its critical point). It is commonly carried out in a hardened steel autoclave, the inner surfaces of which are lined with a plastic (e.g., Teflon) to limit corrosion of the vessel.

Several types of reactions may be employed in hydrothermal synthesis [69]. A common feature is that precipitation of the product generally involves forced hydrolysis under elevated temperature and pressure. Powders prepared by the hydrothermal route have several desirable characteristics, such as fine particle size (10 to 100 nm), narrow size distribution, high purity, good chemical homogeneity, and single-crystal structure. The crystalline phase is commonly produced directly, so a calcination step is not required. As an example, Figure 2.25 shows CeO₂ powders (average particle size ~15 nm) produced from a suspension of amorphous, gelatinous cerium (hydrous) oxide under hydrothermal conditions (~300°C and 10 MPa pressure for 4 h). CeO₂ has a cubic crystal structure and the faceted nature of the particles is an indication that they are crystalline. High-resolution transmission electron microscopy also revealed that the particles are single crystals [70]. The very fine powders have good sintering characteristics but they are very prone to agglomeration, particularly in the dry state, and difficult to consolidate to high packing density. Because of their high surface area, the powders may contain a high concentration of chemically bonded hydroxyl groups on their surfaces which, unless removed prior to sintering, may limit the final density of the fabricated material.

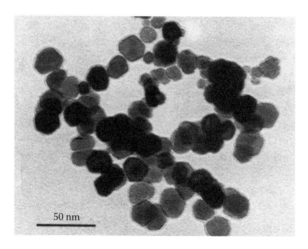

FIGURE 2.25 Powders of CeO₂ prepared by the hydrothermal method.

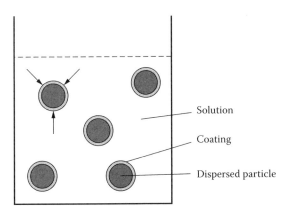

FIGURE 2.26 Schematic of the preparation of coated particles by the precipitation from solution onto dispersed particles.

Hydrothermal synthesis of $BaTiO_3$ powders was reported many years ago [71]. Increased attention has been paid to the method in recent years because of the need for fine powders for the production of thin dielectric layers. One method involves the reaction between TiO_2 gels or fine anatase particles with a strongly alkaline solution (pH > 12 to 13) of $Ba(OH)_2$ at 150 to 200°C, which can be described by the equation:

$$TiO_2 + Ba(OH)_2 \rightarrow BaTiO_3 + H_2O \qquad (2.49)$$

The reaction occurs by a solution–precipitation mechanism [72]. Depending on the reaction time and temperature, particles with an average size in the range of 50 to 200 nm are produced. Another method involves crystallization of an amorphous barium-titanium-acetate gel in a strongly alkaline solution of tetramethylammonium hydroxide for 10 to 15 hr at 150°C [73]. Dissolution of the gel and precipitation of crystalline $BaTiO_3$ particles, coupled with Ostwald ripening of the particles, produced a weakly agglomerated powder with an average particle size in the range of 200 to 300 nm. Hydrothermal $BaTiO_3$ powders, with sizes smaller than ~100 nm, show structural characteristics that are not observed for coarser powders prepared by solid-state reaction at high temperatures [74].

2.5.2.1.2.5 Coated Particles
Coated particles, sometimes referred to as *composite particles*, consist of particles of a given solid coated with a thin or thick layer of another material. Reviews have considered the synthesis and use of coated particles for several applications [75–77]. Coated particles offer some interesting advantages in ceramic powder processing. Thin coatings are particularly useful for modifying the surface characteristics of colloidal dispersions and for uniformly incorporating additives such as sintering aids and dopants into the powder. When compared with mechanically mixed systems, particles or inclusions coated with thick layers generally have improved sintering rates for the fabrication of ceramic composites or complex oxides.

Coated particles can be prepared by several techniques but, here, we shall consider only the method based on precipitation from solution. It can be used to produce thin or thick coatings on particles dispersed in a solution (Figure 2.26). Successful coating of the particles requires control of several variables to produce the desired interaction between a particulate suspension A and the material that is to be precipitated out of solution B [78]. Several types of A–B interactions are possible as follows:

1. B can nucleate *homogeneously* in the solution and grow to form particles that do not interact with A, giving simple mixtures of A and B.

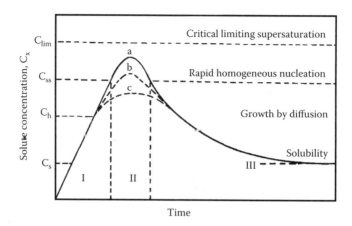

FIGURE 2.27 Modified LaMer diagram for the formation of coated particles by precipitation from solution.

2. Homogeneously nucleated particles of B grow and eventually heterocoagulate with particles of A, producing deposits that are rough and nonuniform, particularly if the B particles are large.
3. The homogeneously nucleated particles of B heterocoagulate with A at an early stage, and the growth of B continues on these aggregates, thereby producing a particulate coating of B on A. The coating would be more homogeneous than that formed in case (2), particularly if the B particles are very small compared to A.
4. B nucleates *heterogeneously* on the surface of A and growth produces a uniform layer of B on A. This may be the most desirable way to deposit smooth coatings on fine particles. To produce coated particles in this way, a number of key requirements must be met. These are described below.

Separation of the nucleation and growth steps: Figure 2.27 is a modified version of the LaMer diagram discussed earlier for the homogeneous precipitation of monodisperse particles. The curve a, it will be recalled, represents the case of a single burst of homogeneous nucleation followed by growth. When the particles of A are present in the solution, heterogeneous nucleation can be initiated on their surfaces when the solute concentration reaches C_h, the critical concentration for heterogeneous nucleation. To produce uniformly coated particles, it is essential to have one rapid burst of heterogeneous nucleation (curve c) without reaching C_{ss}.

Colloidal stability of the dispersion: In order to obtain well-dispersed, coated particles, the dispersion must be stable against flocculation and settling during nucleation and growth. Agglomerates formed during these stages can get bonded by the newly formed surface layers and would be extremely difficult to disperse.

Surface area of the core particles: The surface area of the core particles of A must be sufficient to prevent the solute concentration from reaching C_{ss}, otherwise a system of coated particles and free particles of B will result (curve b). The appropriate surface area of the core particles is linked to the rate of generation of the solute by the reaction, R_g, and the rate of removal of the solute by precipitation, R_r. The minimum surface area of the core particles available for deposition, A_{min}, is associated with the maximum solute concentration, C_{max}, to avoid homogeneous precipitation. For a given R_g and assuming fairly concentrated suspensions such that the interface reaction is rate controlling, A_{min} is defined by:

$$R_g = KA_{min}(C_{ss} - C_s) \tag{2.50}$$

where K is a constant and $C_{max} = C_{ss}$. The maximum surface area for deposition, A_{max}, should be such that the solute concentration exceeds C_h, otherwise only partial coating will result. Therefore, the following equation must also be satisfied.

$$R_g = KA_{max}(C_h - C_s) \qquad (2.51)$$

The maximum production rate and the latitude of the experimental conditions are then associated with the maximum value of the ratio, A_{max}/A_{min}. By equating Equation 2.50 and Equation 2.51 we get:

$$A_{max} / A_{min} = 1 + [(C_{ss} - C_h) / (C_h - C_s)] \qquad (2.52)$$

The optimum conditions, permitting coating in suspensions with high particle concentration, high production rates and ease of processing, depend critically on how close C_h is to C_s and on the separation between C_{ss} and C_h. Conditions should be found such that (1) C_h is close to C_s so that heterogeneous nucleation can begin soon after C_s is passed and (2) C_{ss} is much greater than C_s so that homogeneous precipitation is far removed from the onset of heterogeneous precipitation. In practice, A_{min} can be found by trial and error for a given R_g and for particles with a known size by decreasing the concentration of the particles in the suspension until free precipitates appear. If A_{min} is found to be low, then according to Equation 2.50, $(C_{ss}-C_s)$ is relatively large and coating of the particles in a suspension should be possible. Then, if conditions for avoiding homogeneous precipitation are difficult to achieve, an approach may be to pretreat the particle surface with a nucleation catalyst.

Several examples of the conditions used in the preparation of coated particles by precipitation from solution can be found in the literature, including SiO_2 on Al_2O_3 [79], TiO_2 on Al_2O_3 [80], aluminum (hydrous) oxide on α-Fe_2O_3, chromium (hydrous) oxide, and TiO_2 [81], Al_2O_3 precursor on SiC whisker [82], Y_2O_3 or Y_2O_3/Al_2O_3 precursors on Si_3N_4 [78], yttrium basic carbonate, YOHCO$_3$, or Y_2O_3 on α-Fe_2O_3 [83], and ZnO on ZrO_2 [84]. The degree of crystallinity of the deposited material can have a marked effect on the morphology of the coating. In principle, the coating can be amorphous, polycrystalline, or single crystal. Smooth and uniform coatings are obtained more easily for amorphous coatings [77,81] whereas somewhat rough layers are generated by polycrystalline deposits [85]. Even for amorphous deposits, the morphology can depend on the reaction conditions, as illustrated in Figure 2.28 for SiO_2 coatings deposited on YOHCO$_3$ at room temperature and at 80°C [86].

2.5.2.1.2.6 Industrial Preparation of Powders by Precipitation from Solution

The methods described earlier for the synthesis of mondisperse powders and coated particles have not made significant inroads into industrial production because they are expensive. Coprecipitation and hydrothermal methods are seeing some use, particularly for the synthesis of fine powders for a variety of applications.

The largest use of precipitation is the Bayer process for the industrial production of Al_2O_3 powders. The raw material bauxite is first physically beneficiated then digested in the presence of NaOH at an elevated temperature. During digestion, most of the hydrated alumina goes into solution as sodium aluminate and insoluble impurities are removed by settling and filtration.

$$Al(OH)_3 + NaOH \rightarrow Na^+ + Al(OH)_4^- \qquad (2.53)$$

After cooling, the solution is seeded with fine particles of gibbsite, $Al(OH)_3$, which provide nucleating sites for growth of $Al(OH)_3$. The precipitates are continuously classified, washed to

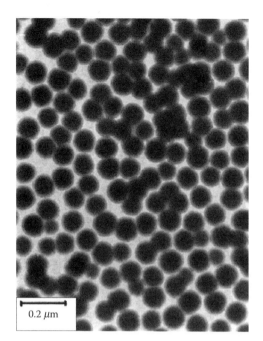
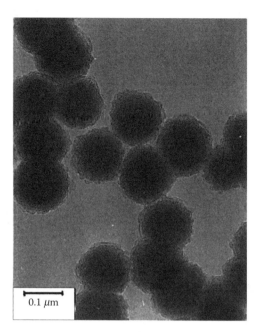

FIGURE 2.28 Transmission electron micrographs of yttrium basic carbonate, YOHCO$_3$, particles coated with SiO$_2$ showing (a) smooth SiO$_2$ coating when the reaction was carried out at room temperature and (b) rough SiO$_2$ coating formed at higher reaction temperature (80°C). (From Giesche, H. and Matijevic, E., Preparation, characterization, and sinterability of well-defined silica/yttria powders, *J. Mater. Res.*, 9, 426, 1994. With permission.)

remove the Na, and calcined. Powders of α-Al$_2$O$_3$ with a range of particle sizes, are produced by calcination at 1100 to 1200°C, followed by grinding and classification, whereas tabular aluminas are produced by calcination at higher temperatures (~1650°C).

2.5.2.2 Evaporation of the Liquid

Evaporation of the liquid provides another method for bringing a solution to supersaturation, thereby causing the nucleation and growth of particles. The simplest case is a solution of a single salt. For the production of fine particles, nucleation must be fast and growth slow. This requires that the solution be brought to a state of supersaturation very rapidly so that a large number of nuclei are formed in a short time. One way of doing this is to break the solution up into very small droplets, in which case the surface area over which evaporation takes place increases enormously. For a solution of two or more salts, a further problem must be considered. Normally the salts will be in different concentrations and will have different solubilities. Evaporation of the liquid will cause different rates of precipitation, leading to segregation of the solids. Here again, the formation of very small droplets will limit the segregation to the droplets, as no mass is transferred between individual droplets. For a given droplet size, the size of the resulting solid particle becomes smaller for more dilute solutions. This means that we can further reduce the scale of segregation by the use of dilute solutions. We now consider some of the practical ways of producing powders by the evaporation of liquid solutions.

2.5.2.2.1 Spray Drying

In spray drying, a solution is broken up into fine droplets by a fluid atomizer and sprayed into a drying chamber (Figure 2.29). Contact between the spray and drying medium (commonly hot air)

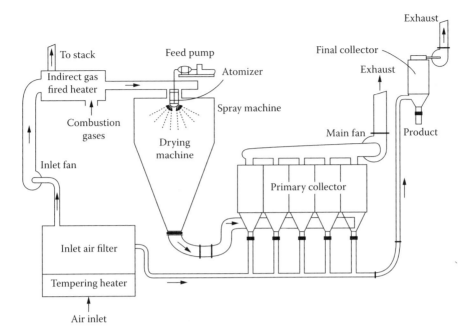

FIGURE 2.29 Schematic of a spray dryer for the production of powders. (Courtesy of L.C. De Jonghe.)

leads to evaporation of moisture. The product, consisting of dry particles of the metal salt, is carried out in the air stream leaving the chamber and collected using a bag collector or a cyclone.

Spray-drying principles, equipment and applications are described in detail by Masters [87]. A variety of atomizers are available and these are usually categorized according to the manner in which energy is supplied to produce the droplets. In rotary atomization (often referred to as centrifugal atomization), the liquid is centrifugally accelerated to high velocity by a spinning disc located at the top of the drying chamber before being discharged into the chamber. In pressure atomization, pressure nozzles atomize the solution by accelerating it through a large pressure difference and injecting it into the chamber. Pneumatic atomization occurs when the solution is impacted by a stream of high-speed gas from a nozzle. Ultrasonic atomization involves passing the solution over a piezoelectric device that is vibrating rapidly. Droplet sizes ranging from less than 10 μm to over 100 μm can be produced by these atomizers.

The solutions in spray drying are commonly aqueous solutions of metal salts. Chlorides and sulfates are often used because of their high solubility. In the drying chamber, the temperature and flow pattern of the hot air as well as the design of the chamber determine the rate of moisture removal from the droplet and the maximum temperature (typically less than ~300°C) that the particles will experience. The key solution parameters are the size of the droplet and the concentration and composition of the metal salt. These parameters control the primary particle size as well as the size and morphology of the agglomerate. The morphology of the agglomerate is not very critical in spray drying of solutions because the particle characteristics are largely determined by subsequent calcination and milling steps. Under suitable conditions, spherical agglomerates with a primary particle size of ~0.1 μm or less can be obtained. The temperature in the drying chamber is commonly insufficient to cause decomposition or solid-state reaction, so the spray-dried powder must be calcined and ground to provide suitable particle characteristics for processing.

Spray drying of solutions has been found to be useful for the preparation of ferrite powders [88]. For Ni-Zn ferrite, a solution of sulfates was broken up into droplets (10 to 20 μm) by a rotary atomizer. The spray-dried powder was in the form of hollow spheres having the same size as the original droplets. Calcination at 800 to 1000°C produced a fully reacted powder consisting of

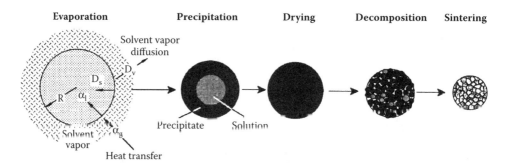

FIGURE 2.30 Schematic of the stages in the spray pyrolysis process. (From Messing, G.L., Zhang, S.-C., and Jayanthi, G.V., Ceramic powder synthesis by spray pyrolysis, *J. Am. Ceram. Soc.*, 76, 2707, 1993. With permission.)

agglomerates with a primary particle size of ~0.2 μm. The ground powder (particle size < 1 μm) was compacted and sintered to almost theoretical density.

2.5.2.2.2 Spray Pyrolysis

By using a higher temperature and a reactive (e.g., oxidizing) atmosphere in the chamber, solutions of metal salts can be dried and decomposed directly in a single step [89]. This technique is referred to as *spray pyrolysis* but other terms, such as *spray roasting, spray reaction, and evaporative decomposition of solutions*, will also be found in the literature. The idealized stages in the formation of a dense particle (agglomerate) from a droplet of solution are shown schematically in Figure 2.30. As the droplet undergoes evaporation, the solute concentration in the outer layer increases above the supersaturation limit, leading to the precipitation of fine primary particles. Precipitation is followed by a drying stage in which the vapor must now diffuse through the pores of the precipitated layer. Decomposition of the precipitated salts produces a porous particle, made up of very fine primary particles, which is heated to produce a dense particle. In practice, a variety of particle morphologies can be produced in the spray pyrolysis process, some of which are shown in Figure 2.31. For the fabrication of advanced ceramics, dense particles are preferred over highly porous or hollow shell-like particles because they give a higher particle packing density in the green article without requiring a grinding step.

Figure 2.32 shows schematically how the conditions leading to precipitation in the droplet and the solution chemistry influence the particle morphology and microstructure. If dense particles are required, we must first achieve homogeneous nucleation and growth in the droplet (referred to as volume precipitation in Figure 2.32a). This is facilitated by a small droplet size and slow drying to reduce gradients in solute concentration and temperature. A large difference between the supersaturation concentration, C_{ss}, and the saturation concentration, C_s, of the solute in solution increases the nucleation rate (see Figure 2.19 and Equation 2.26). It is also important to have a high solute solubility (high C_s), as well as a positive temperature coefficient of solubility so that sufficient solute is available to form agglomerates of touching primary particles. Furthermore, the precipitated solids should not be thermoplastic or melt during the decomposition stage. Figure 2.32b illustrates that there are a number of possibilities for synthesizing multicomponent and composite particles with a variety of microstructures.

The drying of a droplet containing fine precipitates is quite different from that of a liquid droplet. The fine precipitates provide a resistance to the mass transport of the solvent vapor and if the temperature of the drying chamber is too high, boiling of the solution may occur leading to inflation or disintegration of the droplet. If drying is rapid, fracture of the particle can occur, due to the large capillary stresses generated by liquid evaporation from the fine pores between the precipitates. It is important to achieve complete decomposition of the dried salts prior to sintering. For small-scale laboratory equipment in which the decomposition times are small, nitrates and

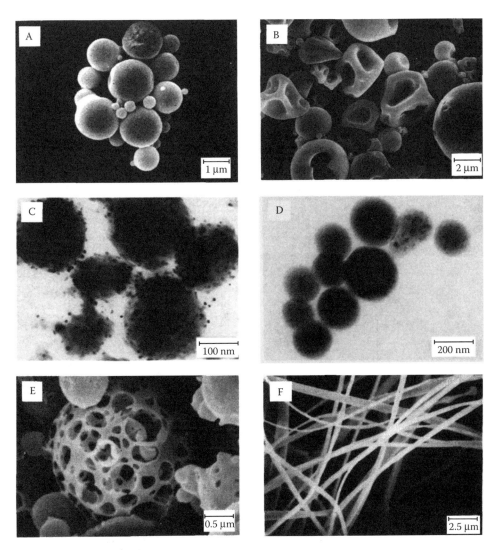

FIGURE 2.31 Examples of the particle morphologies produced in the spray pyrolysis process. (From Messing, G.L., Zhang, S.-C., and Jayanthi, G.V., Ceramic powder synthesis by spray pyrolysis, *J. Am. Ceram. Soc.*, 76, 2707, 1993. With permission.)

acetates are often preferable to sulfates because of their lower decomposition temperatures. Chlorides and oxychlorides are often used industrially because of their high solubilities but the corrosive nature of the gases can be a problem. The particles should be sintered *in situ* to take full advantage of the spray pyrolysis process. The fine pores between the primary particles and the short interparticle collision time in the process favor the formation of dense particles if exposure to a high enough temperature can be achieved.

2.5.2.2.3 Spray Drying of Suspensions

Suspensions of fine particles (slurries) can also be dried by spray drying. In this case, the liquid is removed in such a way as to limit the agglomeration of the dried powder to a scale equal to or less than the size of the droplet. Limiting the scale of the agglomeration provides better homogeneity of the compacted body which, in turn, improves sintering. An example of a powder produced by spray drying of a suspension is shown in Figure 2.33 for fine lead–zirconium titanate particles that were synthesized by precipitation from solution prior to the spray-drying process [90]. Spray drying

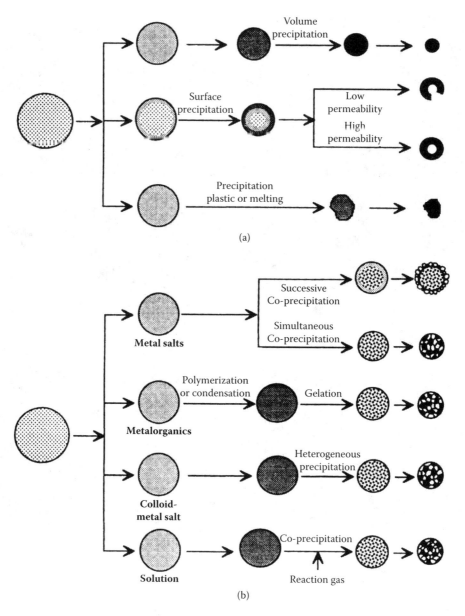

FIGURE 2.32 Effect of precipitation conditions and precursor characteristics on (a) particle morphology and (b) composite particle microstructure in spray pyrolysis. (From Messing, G.L., Zhang, S.C., and Jayanthi, G.V., Ceramic powder synthesis by spray pyrolysis, *J. Am. Ceram. Soc.*, 76, 2707, 1993. With permission.)

of suspensions is used on a large-scale industrially for granulating fine powders to control their flow and compaction characteristics during die pressing (Chapter 7). It is also used for numerous other applications in the food, chemical and pharmaceutical industries [87].

2.5.2.2.4 *Freeze Drying*

In freeze drying, a solution of metal salt is broken up by an atomizer into fine droplets, which are then frozen rapidly by being sprayed into a cold bath of immiscible liquid, such as hexane and dry ice, or directly into liquid nitrogen. The frozen droplets are then placed in a cooled vacuum chamber and the solvent is removed, under the action of a vacuum, by sublimation without any melting.

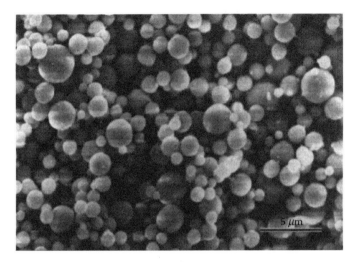

FIGURE 2.33 Scanning electron micrograph of a spray-dried lead-zirconium titanate powder prepared by spray drying of a suspension. (From Schwartz, R.W., Eichorst, D.J., and Payne, D.A., Precipitation of PZT and PLZT powders using a continuous reactor, *Mater. Res. Soc. Symp. Proc.*, 73, 123, 1986. With permission.)

The system may be heated slightly to aid the sublimation. Freeze-drying produces spherical agglomerates of fine primary particles, in which the agglomerate size is the same as that of the frozen droplet. The size of the primary particles (in the range of 10 to 500 nm) depends on the processing parameters such as the rate of freezing, the concentration of metal salt in the solution, and the chemical composition of the salt. After drying, the salt is decomposed at elevated temperatures to produce an oxide.

As we observed for spray drying, the breaking up of the solution into droplets serves to limit the scale of agglomeration or segregation to the size of the droplet. The solubility of most salts decreases with temperature and the rapid cooling of the droplets in freeze-drying produces a state of supersaturation of the droplet solution very rapidly. Particle nucleation is therefore rapid and growth is slow, so the size of the particles in the frozen droplet can be very fine. When compared to the evaporation of the liquid in spray drying, the approach to supersaturation is relatively faster, so freeze-drying produces much finer primary particles with a higher surface area per unit mass. Surface areas as high as 60 m^2/g have been reported for freeze-dried powders.

Freeze drying of solutions has been used on a laboratory scale for the preparation of some oxide powders, in particular ferrite powders. Laboratory equipment and methods are described by Schnettler et al. [91]. Lithium ferrite, $LiFe_5O_8$, powders prepared by freeze drying a solution of oxalates were found to have lower sintering temperature and afforded better control of the grain size when compared to similar powders prepared by spray drying [92]. As outlined for spray drying, the freeze-drying technique is also used for drying slurries [93]. Powders prepared from freeze-dried slurries generally consist of soft agglomerates that could be broken down easily. Pressing of such powders can produce fairly homogeneous green bodies.

2.5.2.3 Gel Route

A few methods utilize the formation of a semirigid gel or a highly viscous resin from liquid precursors as an intermediate step in the synthesis of ceramic powders. They are often used at the laboratory scale, particularly for the synthesis of complex oxides when good chemical homogeneity is required. In the formation of the gel or resin, mixing of the constituents occurs on the atomic scale by a polymerization process. Provided none of the constituents are volatilized during subsequent decomposition and calcination steps, the cation composition of the powder can be identical to that of the original solution. These methods therefore have the ability to achieve good chemical homogeneity. A drawback is that the decomposed gel or resin is commonly not in the form of a powder but consists of charred lumps that must be ground and calcined. Control of the purity and particle size characteristics can therefore be difficult.

2.5.2.3.1 Sol–Gel Processing

The sol–gel route for the production of ceramics was outlined in Chapter 1 and will be discussed in greater detail in Chapter 5. The reader will recall that the process is best applied to the formation of films and fibers and, with careful drying, to a few monolithic ceramics. Here, we would like to point out that the process, though expensive, can also be used for the production of powders. In this method, the gel formed by sol–gel processing is dried and ground to produce a powder. Carefully controlled drying is unnecessary, but dried gels with lower viscosity are easier to grind, lowering the risk of contamination. Gels dried under supercritical conditions have low strength and can be easily ground to a powder using plastic media. Powders of mullite ($3Al_2O_3.2SiO_2$), prepared by supercritical drying of gels, were compacted and sintered to nearly full density below ~1200°C, which is considerably better than mullite prepared by reaction of mixed powders [94]. The high sinterability of the gel-derived powder results from its amorphous structure and high surface area. If crystallization occurs prior to the achievement of full density, the sinterability can be reduced significantly.

2.5.2.3.2 The Pechini Method

The Pechini method, originally developed by Pechini [95] for the preparation of titanates and niobates for the capacitor industry, has since been applied to the preparation of several complex oxides [96,97]. Metal ions from starting materials such as carbonates, nitrates and alkoxides are complexed in an aqueous solution with α-carboxylic acids (e.g., as citric acid). When heated with a polyhydroxy alcohol (e.g., ethylene glycol), polyesterification occurs and, on removal of the excess liquid, a transparent resin is formed. The resin is heated to decompose the organic constituents, ground, and calcined to produce the powder. The steps in the method are outlined in Figure 2.34 for the preparation of $SrTiO_3$ powders [98].

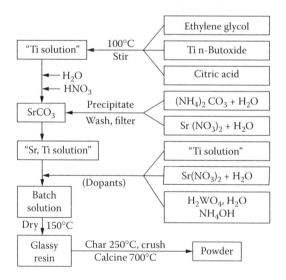

FIGURE 2.34 Flowchart for the preparation of strontium titanate powder by the Pechini method. (From Budd, K.D. and Payne, D.A., Preparation of strontium titanate ceramics and internal boundary layer capacitors by the Pechini method, *Mater. Res. Soc. Symp. Proc.*, 32, 239, 1984. With permission.)

2.5.2.3.3   The Citrate Gel Method

The citrate gel method, developed by Marcilly et al. [99], can be illustrated by the synthesis of the ceramic superconductor $YBa_2Cu_3O_{7-x}$ [100]. Nitrate solutions of Y, Ba, and Cu were added to citric acid solution and the pH was kept at ~6 to prevent precipitation of barium nitrate. Heating the solution at 75°C in air produced a viscous liquid containing polybasic chelates. Further heating at 85°C in a vacuum produced an amorphous solid that was pyrolyzed in air at 900°C to produce a crystalline powder.

2.5.2.3.4   The Glycine Nitrate Process

The glycine nitrate process is one of a general class of combustion methods for the preparation of ceramic powders, either simple oxides or complex oxides (e.g., manganites, chromites, ferrites, and oxide superconductors). A highly viscous mass formed by evaporation of a solution of metal nitrates and glycine is ignited to produce the powder [101]. Glycine, an amino acid, forms complexes with the metal ions in solution which increases the solubility and prevents the precipitation of the metal ions as the water is evaporated, giving good chemical homogeneity. Glycine also serves another important function: it provides a fuel for the ignition step of the process as it is oxidized by the nitrate ions. The reactions occurring during ignition are highly explosive and extreme care must be exercised during this step. Normally, only small quantities should be ignited at a time. The loose mass of fine, crystalline powder (primary particle size smaller than a few tens of nanometers) obtained by this process is believed to result directly from the short exposure to high temperatures in the ignition step. Commonly, no grinding or calcination of the powder is required.

2.5.2.4   Nonaqueous Liquid Reaction

Reactions involving nonaqueous liquids have been used for the synthesis of Si_3N_4 and other nonoxide powders. The powders have higher purity and finer particle size when compared to powders prepared by grinding a solid product. The reaction between liquid $SiCl_4$ and liquid NH_3 has been used on an industrial scale by UBE Industries (Japan), to produce Si_3N_4 powder. Initial products of the reaction are complex and although the reaction can be written as:

$$SiCl_4(l) + 6NH_3 \rightarrow Si(NH)_2 + 4NH_4Cl \qquad (2.54)$$

a more involved sequence of reactions occurs, involving the formation of polymeric silicon diimide and ammonium chloride triammoniate, $NH_4Cl \cdot 3NH_3$. Silicon diimide decomposes to give Si_3N_4 according to the reaction:

$$3Si(NH)_2 \rightarrow Si_3N_4 + N_2 + 3H_2 \qquad (2.55)$$

In the UBE process, the products formed by the interfacial reaction between $SiCl_4$ and NH_3 liquids are collected and washed with liquid NH_3 and calcined at 1000°C to produce an amorphous Si_3N_4 powder. Subsequent calcination at 1550°C in N_2 yields a crystalline powder with a particle size of ~0.2 mm (Figure 2.35 and Table 2.5).

2.5.3   Vapor-Phase Reactions

Reactions involving the vapor phase have been used extensively for the production of oxide powders (e.g., SiO_2 and TiO_2) and nonoxide powders (e.g., Si_3N_4 and SiC). Si_3N_4 and SiC exist in different polymorphic structures and it is useful to know which polymorph is the dominant phase in the powder because this has a strong effect on subsequent processing. Crystalline Si_3N_4 exists in two different hexagonal polymorphs designated α and β, with the α form having the slightly higher

FIGURE 2.35 Scanning electron micrograph of a commercially available Si_3N_4 powder (UBE-SN-E10) produced by the reaction between $SiCl_4$ and NH_3 liquids. (Courtesy UBE, Japan.)

TABLE 2.5
Properties of Commercially Available Silicon Nitride Powders

	Method of Preparation			
	Liquid-Phase Reaction of $SiCl_4/NH_3$	Nitridation of Si in N_2	Carbothermic Reduction of SiO_2 in N_2	Vapor-Phase Reaction of $SiCl_4/NH_3$
Manufacturer	UBE	H. C. Stark	Toshiba	Toya Soda
Grade	SN-E 10	H1	—	TSK TS-7
Metallic impurities (wt%)	0.02	0.1	0.1	0.01
Nonmetallic impurities (wt%)	2.2	1.7	4.1	1.2
α-Si_3N_4 (wt%)	95	92	88	90
β-Si_3N_4 (wt%)	5	4	5	10
SiO_2 (wt%)	2.5	2.4	5.6	—
Surface area (m²/g)	11	9	5	12
Average particle size (μm)	0.2	0.8	1.0	0.5
Tap density (g/cm³)	1.0	0.6	0.4	0.8

free energy at the formation temperature. Particles of α-Si_3N_4 have a more equiaxed particle shape and sinter more readily than β-Si_3N_4 particles which grow in a more elongated shape. The preparation conditions are therefore selected to maximize the amount of the α-Si_3N_4 produced. SiC exists in many polytypes, with the two major forms designated α and β. The β form is more stable at lower temperatures and transforms irreversibly to the α form at ~2000°C. Powders produced above ~2000°C therefore consist of α-SiC (e.g., the Acheson process described earlier). Powders of either the α form or the β form can be used in the production of SiC materials but sintering of β-SiC powders above ~1800 to 1900°C results in transformation to the α-phase which is accompanied by growth of platelike grains and a deterioration of mechanical properties. The use of β-SiC powder requires very fine powders to keep the sintering temperature below ~1800°C.

2.5.3.1 Gas–Solid Reaction

A widely used method for the preparation of Si_3N_4 powders is by direct nitridation, in which Si powder (particle size typically in the range 5 to 20 μm) is reacted with N_2 at temperatures between

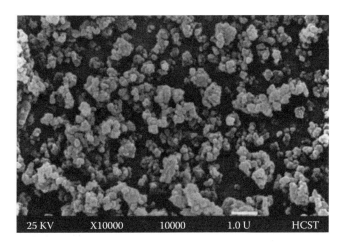

FIGURE 2.36 Scanning electron micrograph of a commercially available Si₃N₄ powder (LC 12) produced by the nitridation of silicon. (Courtesy H. C. Starck, Germany.)

1200 and 1400°C for times in the range 10 to 30 h. The method is used commercially and a powder produced by H. C. Starck (Germany) is shown in Figure 2.36. The nitridation process parameters and mechanisms are similar to those described in Chapter 1 for the production of reaction-bonded Si_3N_4 except that a loose bed of Si powder is used rather than a shaped article. The Si_3N_4 powder consists of a mixture of the α and β phases. Some control of the relative amounts of these two phases is achieved by controlling the reaction temperature, the partial pressure of the N_2 gas in the nitriding atmosphere and the purity of the Si powder.

Silicon nitride powder is also produced by the carbothermic reduction of SiO_2 in a mixture of fine SiO_2 and C powders followed by nitridation between 1200 and 1400°C in N_2. This process is used industrially by Toshiba (Japan). The widespread availability of pure, fine SiO_2 and C makes this method an attractive alternative to the nitridation of Si. Though the overall reaction can be written as

$$3SiO_2 + 6C + 2N_2 \rightarrow Si_3N_4 + 6CO \qquad (2.56)$$

the mechanism is believed to involve the gaseous silicon monoxide, SiO, as follows:

$$3SiO_2(s) + 3C(s) \rightarrow 3SiO(s) + 3CO(g) \qquad (2.57a)$$

$$3SiO(s) \rightarrow 3SiO(g) \qquad (2.57b)$$

$$3SiO(g) + 3C(s) + 2N_2 \rightarrow Si_3N_4(s) + 3CO(g) \qquad (2.57c)$$

Excess carbon is used as an oxygen sink to form gaseous CO and reduce the amount of oxygen on the powder surface. However, any carbon remaining after the reaction has to be burnt out in an oxidizing atmosphere and this may cause some reoxidation of the Si_3N_4 surfaces.

The nitridation and carbothermic reduction methods produce a strongly agglomerated mass of Si_3N_4 that requires milling, washing and classification. The impurities introduced into the powder during these steps can cause a significant reduction in the high-temperature mechanical properties of the fabricated material.

2.5.3.2   Reaction between a Liquid and a Gas

Mazdiyasni and Cooke [102,103] used the reaction between liquid $SiCl_4$, and NH_3 gas in dry hexane at 0°C to prepare a fine Si_3N_4 powder with very low levels of metallic impurities (< 0.03 wt%). The reaction is complex but can be summarized by Equation 2.54 and Equation 2.55. The powder obtained by the reaction is amorphous but crystallizes to α-Si_3N_4 after prolonged heating between 1200 to 1400°C.

2.5.3.3   Reaction between Gases

The types of deposits that can arise from reactions between heated gases are illustrated in Figure 2.37 [104]. Films, whiskers and bulk crystals are produced by heterogeneous nucleation on a solid surface by chemical vapor deposition (Chapter 1). The formation of particles occurs by homogeneous nucleation and growth in the gas phase. It is described by equations similar to those given earlier for the nucleation of liquid droplets from a supersaturated vapor (Equation 2.19 to Equation 2.25). Several gas-phase reactions are used to produce ceramic powders industrially and on a laboratory scale. The methods employ a variety of techniques for heating the reactant gases, including flame, furnace, plasma, and laser heating.

Flame synthesis of TiO_2 and SiO_2 forms two of the largest industrial processes for synthesizing powders by gas-phase reactions [105,106]. The reactions can be written:

$$TiCl_4(g) + 2H_2O(g) \rightarrow TiO_2(s) + 4HCl(g) \tag{2.58}$$

$$SiCl_4(g) + O_2(g) \rightarrow SiO_2(s) + 2Cl_2 \tag{2.59}$$

Particle formation is illustrated in Figure 2.38. In the formation of fumed silica, $SiCl_4$ reacts in a hydrogen flame (~1800°C) to form single spherical droplets of SiO_2. These grow by collision and coalescence to form larger droplets. As the droplets begin to solidify, they stick together on collision but do not coalesce, forming solid aggregates, which in turn continue to collide to form agglomerates. The powder has a fine particle size and good purity (resulting from the use high-

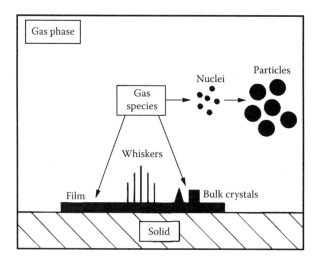

FIGURE 2.37 Schematic diagram illustrating the types of deposits that can form by the reaction between heated gases. (From Kato, A., Hojo, J., and Watari, T., Some common aspects of the formation of nonoxide powders by the vapor reaction method, *Mater. Sci. Res.*, 17, 123, 1984. With permission.)

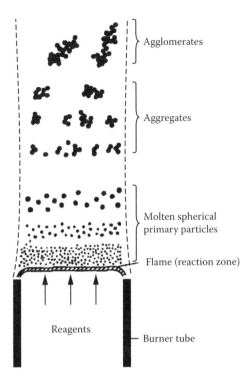

FIGURE 2.38 Schematic diagram illustrating the formation of primary particles, aggregates and agglomerates in gas-phase reactions heated by a flame. (From Ulrich, G.D., Flame synthesis of fine particles, *Chem. Eng. News*, 62, 22, 1984. With permission.)

FIGURE 2.39 Scanning electron micrograph of flame-synthesized SiO_2. (Courtesy of Degussa Company.)

purity gases) but it is agglomerated (Figure 2.39). The simplicity of the reaction system and ease of scale-up provide key benefits for industrial production.

Furnace, plasma and laser heating have been used to produce Si_3N_4 and SiC powders by several gas-phase reactions, including the following:

$$3SiCl_4(g) + 4NH_3(g) \rightarrow Si_3N_4(s) + 12HCl(g) \qquad (2.60)$$

$$3SiH_4(g) + 4NH_3(g) \rightarrow Si_3N_4(s) + 12H_2(g) \qquad (2.61)$$

$$2SiH_4(g) + C_2H_4(g) \rightarrow 2SiC(s) + 6H_2(g) \qquad (2.62)$$

The use of silicon tetrachloride, $SiCl_4$, leads to highly corrosive HCl as a by-product; so, therefore, silane (SiH_4), despite being expensive and flammable in air, is generally preferred as the reactant. For Si_3N_4 production, ammonia is often used as the nitrogen-containing gas because nitrogen is fairly unreactive.

Prochazka and Greskovich [107] used the reaction between SiH_4 and NH_3 at temperatures between 500 and 900°C in an electrically heated silica tube to produce fine amorphous Si_3N_4 powders. Two main parameters were found to control the reaction: the temperature and the NH_3/SiH_4 molar ratio. For a molar ratio >10 and at 500 to 900°C, nearly stoichiometric powders with a cation purity of > 99.99%, surface area of 10 to 20 m^2/g and an oxygen content of < 2 wt% were produced. Subsequent calcination above ~1350°C yielded crystalline α-Si_3N_4 powder. The reaction between $SiCl_4$ and NH_3 is used commercially by Toya Soda (Japan) for the production of Si_3N_4 powder. Table 2.5 summarizes the characteristics of the Toya Soda powder and the other commercial Si_3N_4 powders discussed in this chapter.

Haggerty and coworkers [108,109] used a CO_2 laser as the heat source for the gas-phase synthesis of Si, Si_3N_4 and SiC powders. The frequency of the radiation was chosen to match one of the absorption frequencies of one or more of the reactants, making the laser a very efficient heat source. The reactions are controlled by manipulating the process variables such as the cell pressure, the flow rate of the reactant and dilutant gases, the intensity of the laser beam, and the reaction flame temperature. A laboratory scale reaction cell is shown in Figure 2.40. The range of powder characteristics obtained by this method for Si_3N_4 and SiC using the reactions given in Equation

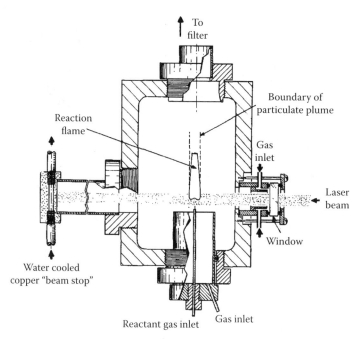

FIGURE 2.40 Laboratory scale reaction cell for the preparation of powders by laser heating of gases. (From Haggerty, J.S., Growth of precisely controlled powders from laser heated gases, in *Ultrastructure Processing of Ceramics, Glasses and Composites*, Hench, L.L. and Ulrich, D.R., Eds., Wiley, New York, 1984, p. 353, chap. 28. With permission.)

TABLE 2.6
Summary of the Range of Characteristics for Si₃N₄ and SiC Prepared by Laser-Heated Gas-Phase Reactions

Powder Characteristic	Si_3N_4	SiC
Mean diameter (nm)	7.5–50	20–50
Standard deviation of diameters (% of mean)	2.3	~2.5
Impurities (wt%)		
Oxygen	0.3	0.33–1.3
Total others	<0.01	—
Major elements	Al, Ca	—
Stoichiometry (%)	0-60 (excess Si)	0–10 (excess C or Si)
Crystallinity	Amorphous-crystalline	Crystalline Si and SiC
Grain size: mean diameter	~0.5	0.5–1.0

Source: From Haggerty, J.S., Growth of precisely controlled powders from laser heated gases, in *Ultrastructure Processing of Ceramics, Glasses and Composites,* Hench, L.L. and Ulrich, D.R., Eds., Wiley, New York, 1984, p. 353, chap. 28. With permission.

2.61 and Equation 2.62 are summarized in Table 2.6. The powders have several desirable characteristics. Although estimates indicate that the production cost can be very competitive with other methods, particularly for the synthesis of submicron powders, the laser heating method has not seen industrial application.

2.6 SPECIAL TOPIC: SYNTHESIS OF NANOSCALE CERAMIC POWDERS

Very fine powders, with particle sizes smaller than 50 to 100 nm, have been the subject of extensive research and development for a variety of potential applications. We shall refer to these powders as *nanoscale* powders, but other terms, such as *nanosize* and *nanocrystalline*, are also used. Whereas large quantities of nanoscale ceramic powders, such as SiO_2, TiO_2, and Al_2O_3, have been produced for decades [105], heightened interest has been generated in recent years as a result of rapid developments in the field often referred to as *nanotechnology*. Methods for the preparation of nanoscale powders have been described in detail in recent texts [110,111], journals, and conference proceedings. They include several of the chemical methods described earlier in this chapter, but particularly the gas-phase reaction methods.

The desirable characteristics for nanoscale powders are the same as those summarized in Table 2.1, except that the particle size is now < 50 to 100 nm. The state of agglomeration is very important because the interesting properties of the powders are much reduced when the particles form hard agglomerates. Solid-state decomposition reactions find little use for the synthesis of nanoscale powders because the powders are often strongly agglomerated (Figure 2.12b). The most widely used methods are based on liquid solution and vapor-phase techniques.

2.6.1 LIQUID SOLUTION TECHNIQUES

Precipitation from solution at or near room temperature often produces amorphous particles or particle precursors, so a calcination step is required to form the crystalline particles. The formation of strong agglomerates of the nanoscale particles during this calcination step is often a problem. Sol–gel processing and the gel routes described earlier often produce solids with strongly agglomerated nanoscale particles. Although the solids can be ground to produce powders in the micrometer range, these methods are limited in their ability to produce unagglomerated or weakly agglomerated

nanoscale particles. One of the most useful solution-based methods for preparing crystalline nanoscale oxide powders industrially and on a laboratory scale is precipitation from solution under
hydrothermal conditions (see Figure 2.25). Industrially, it has been used to prepare oxide powders
such as stabilized ZrO_2, MgO, and $BaTiO_3$.

2.6.2 VAPOR-PHASE TECHNIQUES

Flame synthesis is the primary commercial method for making large quantities of very fine powders,
such as TiO_2 and SiO_2 (see Equation 2.58 and Equation 2.59). Although the method can produce
powders with high purity on an industrial scale, the formation of hard agglomerates of the very
fine particles in the flame is often a problem (see Figure 2.39). The agglomeration problem in flame
synthesis has been addressed in a modified process referred to as the sodium/halide flame process
[112]. The process has been used for the synthesis of a large number of metal and nonoxide ceramic
powders. It is based on the use of sodium reduction of a metal, such as:

$$TiCl_4(g) + 4Na(g) \rightarrow Ti(s) + 4NaCl(s) \qquad (2.63)$$

Or, if the reaction occurs in the presence of a nonmetal, a ceramic is formed, e.g.,

$$TiCl_4(g) + 2BCl_3(g) + 10Na(g) \rightarrow TiB_2(s) + 10NaCl(s) \qquad (2.64)$$

The by-product in the reactions, typically NaCl, has good properties for encapsulating the
nanoscale particles within the flame, providing a good degree of control over the particle size and
the morphology (Figure 2.41). The final powder consists of submicrometer sized salt particles with
nanoscale particles of the desired material embedded in it. The salt can be removed by washing in
water, ammonia, or another appropriate solvent. Alternatively, the salt can be removed by vacuum
sublimation at ~700°C.

A method based on the evaporation of a metal, sometimes referred to as the *inert gas condensation* method, has been used to produce metal or oxide ceramic powders industrially as well as
on a laboratory scale [113]. For the preparation of oxide powders (e.g., TiO_2), a metal (e.g., Ti) is
first evaporated in a chamber containing an inert gas (e.g., He). Collisions between the metal atoms
and the gas molecules in the vicinity of the metal source lead to cooling, resulting in atomic clusters
that collide and grow to form fine particles. These particles are carried along by the gas into a
collection chamber where they are collected on a cold surface. The chamber is then filled with O_2,

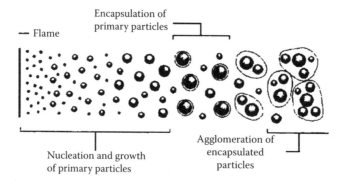

FIGURE 2.41 Illustration of the encapsulation process in the sodium/halide flame. (From Axelbaum, R.L.,
Synthesis of stable metal and nonoxide ceramic particles in sodium/halide flames, *Powder Metall.*, 43, 323,
2000. With permission.)

which promotes rapid oxidation of the particles, after which the powder is scraped off the surface. Spherical particles are produced in this process, and the size of the particles can be controlled by the temperature of the source metal and by the pressure of the inert gas.

Ignition of an aerosol, coupled with rapid quenching of the product, referred to as *liquid-feed flame spray pyrolysis*, has been used to prepare nanoscale powders of several simple oxides and complex oxides [114]. In the process, a solution of inexpensive metal-organic precursors in alcohol is formed into an aerosol with O_2, after which the aerosol is ignited in a reactor. The product consists of single-crystal, nanoscale particles often requiring no further treatment prior to use.

2.7 CONCLUDING REMARKS

In this chapter we have examined a wide range of methods commonly used for the preparation of ceramic powders, and common methods for the synthesis of nanoscale powders were summarized. The methods are based on sound principles of physics and chemistry, which form a framework for understanding how the process variables influence the characteristics of the powder. Practically, the methods vary considerably in the quality of the powder produced and in the cost of production. Generally, higher powder quality is associated with higher production cost. For a given application, we will therefore need to examine whether the higher production cost is justified by the higher quality of the powder produced.

PROBLEMS

2.1

For a powder with the composition of TiO_2, calculate and plot the surface area of 1 g of powder as a function of the particle size. Use a size range of 5 nm to 100 μm and assume that the particles are spherical. Estimate the percentage of TiO_2 molecules at the surface of the particle (relative to the total number of TiO_2 molecules in the volume of the particle) for the same size range and plot the results on a graph.

2.2

Show that the critical speed of rotation for a ball mill, defined as the speed required to take the balls just to the apex of revolution is equal to $(g/a)^{1/2}/2\pi$ rev/s, where a is the radius of the mill and g is the acceleration due to gravity. Determine the rotation speed for a ball mill with a radius of 5 cm which is operating at 75% of the critical rotation speed.

2.3

In an experiment to determine the kinetics of decomposition of $MgCO_3$, a student weighs out 20.00 g of powder and finds that the mass of powder remaining in an isothermal experiment is as follows:

Time (min)	Mass of Powder (g)
0	20.00
10	16.91
20	14.73
30	13.19
50	11.35
80	10.16

Determine whether the following model equations provide a reasonable fit to the reaction kinetics:

    a. The Avrami-Erofe'ev equation.
    b. The geometrical model for a contracting spherical core.
    c. The Jander equation.
    d. A first-order reaction.

2.4

The calcination of $CaCO_3$ described by the equation·

$$CaCO_3(s) \rightarrow CaO(s) + CO_2(g)$$

has a standard free energy given by

$$\Delta G° = 182.50 - 0.16T \ (kJ/mol)$$

where T is the absolute temperature. The partial pressure of CO_2 in air is 3×10^{-3} atm.

    a. If the calcination is carried out in flowing air in an open tube furnace, at what temperature will $CaCO_3$ decompose to CaO?
    b. If the calcination is carried out in a tightly sealed tube furnace, explain qualitatively what will happen.
    c. Will it help to calcine in a tightly sealed tube furnace back-filled with an inert gas such as argon or helium at 1 atm pressure?
    d. If the calcination is carried out in a vacuum furnace with a pressure of 10^{-4} torr, at what temperature will the $CaCO_3$ decompose? What drawbacks, if any, do you see in vacuum calcination?

2.5

Consider the formation of $NiCr_2O_4$ from spherical particles of NiO and Cr_2O_3 when the reaction rate is controlled by diffusion through the product layer.

    a. Sketch an assumed geometry and derive a relation for the rate of formation early in the process.
    b. What governs the particles on which the product layer forms?
    c. At 1300°C, the diffusion coefficients in $NiCr_2O_4$ are as follows: $D_{Cr} > D_{Ni} > D_O$. Which controls the rate of formation of $NiCr_2O_4$? Explain why.

2.6

Outline the derivation of the Jander equation and discuss its limitations. How are the limitations corrected in the Carter equation? One-micron spheres of Al_2O_3 are surrounded by excess ZnO powder to observe the formation of zinc aluminate spinel $ZnAl_2O_4$. It is found that 25% of the Al_2O_3 is reacted to form $ZnAl_2O_4$ during the first 30 min of an isothermal experiment. Determine how long it will take for all the Al_2O_3 to be reacted on the basis of (a) the Jander equation and (b) the Carter equation.

2.7

In the preparation of TiO_2 powder by the Stober process, a student starts out with a solution containing 20 vol% of titanium isopropoxide in isopropanol. Assuming that the reaction is stoichiometric, how much water must be added?

2.8

Discuss the factors involved in the design of a continuous process for the preparation of narrow sized, unagglomerated titania powder. (See Jean, J.H. et al., *Am. Ceram. Soc. Bull.*, 66, 1517, 1987.)

2.9

You wish to prepare approximately 50 g of a Y_2O_3-stabilized ZrO_2 powder containing 8 mol% Y_2O_3 by the process of coprecipitation from solution. You start with a mixed solution of $ZrOCl_2 \cdot 8H_2O$ and $YCl_3 \cdot 6H_2O$ and add it dropwise into a stirred, excess solution of NH_4OH prepared by diluting a concentrated NH_4OH solution containing 30 wt% NH_3. Determine the amount of each starting material and the volume of each solution required for the process, assuming that the reaction goes to completion.

2.10

Compare the key steps in the procedure and the expected particle characteristics for a Y_2O_3-stabilized ZrO_2 powder (8 mol % Y_2O_3) synthesized by the following routes:

   a. Calcination of a ball-milled mixture of submicron ZrO_2 and Y_2O_3 powders
   b. Coprecipitation from solution
   c. Combustion synthesis (e.g., glycine nitrate process).

2.11

Design a process for preparing ZnO powder that is uniformly coated with 0.5 mol% Bi_2O_3. What would the thickness of the coating be if the particle size of the ZnO powder (assumed to be spherical in shape) is 100 nm?

2.12

In the preparation of ZnO particles by spray pyrolysis, a solution containing 0.1 mol/l of zinc acetate is sprayed using a nozzle that produces 40 μm diameter droplets. If the particles are only 50% dense, estimate their diameter.

2.13

Surface oxidation produces an oxide layer (~3 nm thick) on SiC. Assuming the composition of the oxide layer to be that of SiO_2, estimate the wt% of oxygen in SiC powders (assumed to be spherical) with a particle size of (a) 1 μm and (b) 50 nm.

2.14

Discuss the technical and economic factors that will favor the production of Si_3N_4 powders by laser heating of gases over the nitridation of Si. (See Reference 108.)

REFERENCES

1. Snow, R.H., Kaye, B.H., Capes, C.E., and Sresty, G.C., Size reduction and size enlargement, in *Perry's Chemical Engineers' Handbook*, 6th ed., Perry, R.H. and Green, D.W., Eds., McGraw-Hill, New York, 1984, Sec. 8.
2. Lowrinson, G.C., *Crushing and Grinding*, Butterworth, London, 1974.
3. Beddow, J.K., *Particulate Science and Technology*, Chemical Publishing Co., New York, 1980.

4. Polke, R. and Stadler, R., Grinding, in *Concise Encyclopedia of Advanced Ceramic Materials*, Brook, R.J., Ed., MIT Press, Cambridge, MA,1991, p. 187.

5. Somasundaran, P., Theories of grinding, in *Ceramic Processing Before Firing*, Onoda, G.Y., Jr., and Hench, L.L., Eds., John Wiley & Sons, New York, 1978, chap. 11.

6. Gilman, P.S. and Benjamin, J.S., Mechanical alloying, *Annu. Rev. Mater. Sci.*, 13, 279, 1983.

7. Sundaresan, R. and Froes, F.H., Mechanical alloying, *J. Metals*, 39, 22, 1987.

8. McCormick, P.G. and Froes, F.H., The fundamentals of mechanochemical processing, *J. Metals*, 50, 61, 1998.

9. Lin, I.J. and Nadiv, S., Review of the phase transformation and synthesis of inorganic solids obtained by mechanical treatment (mechanochemcial reactions), *Mater. Sci. Eng.*, 39, 193, 1979.

10. Kosmac, T. and Courtney, T.H., Milling and mechanical alloying of inorganic nonmetallics. *J. Mater. Res.*, 7, 1519, 1992.

11. Matteazzi, P., Carbides and silicides by room temperature milling of the elemental powders, *Ceramurgia*, 20, 227, 1990.

12.. El-Eskandarany, M.S. et al., Morphological and structural evolutions of nonequilibrium titanium-nitride alloy powders produced by reactive ball milling, *J. Mater. Res.*, 7, 888, 1992.

13. Patankar, S.N., Xiao, S.-Q., Lewandowski, J.J., and Heuer, A.H., The mechanism of mechanical alloying of $MoSi_2$, *J. Mater. Res.*, 8, 1311, 1993.

14. Yen, B.K., Aizawa, T., and Kihara, T., Reaction synthesis of titanium silicides via self-propagating reaction kinetics, *J. Am. Ceram. Soc.*, 81, 1953, 1998.

15. Ganguli, D. and Chattergee, M., *Ceramic Powder Preparation: A Handbook*, Kluwer Academic Press, Boston, MA, 1997.

16. Segal, D, Chemical preparation of powders, in *Materials Science and Technology*, Vol. 17A, Brook, R.J., Ed., VCH, New York, 1996, p. 69.

17. Riman, R.E., The chemical synthesis of ceramic powders, in *Surface and Colloid Chemistry in Advanced Ceramics Processing*, Pugh, R.J. and Bergström, L., Eds., Marcel Dekker, New York, 1994, p. 29.

18. Segal, D., *Chemical Synthesis of Advanced Ceramic Powders*, Cambridge University Press, New York, 1989.

19. Johnson, D.W., Jr., Innovations in ceramic powder preparation, *Adv. Ceram.*, 21, 3, 1987.

20. Johnson, D.W., Jr., Nonconventional powder preparation techniques, *Am. Ceram. Soc., Bull.*, 60, 221, 1981.

21. Young, D.A., *Decomposition of Solids*, Pergamon Press, Oxford, 1966.

22. Bamford, C.H. and Tipper, C.F.H., Eds., *Comprehensive Chemical Kinetics, Vol. 22: Reactions in the Solid State*, Elsevier, Oxford, 1980.

23. Tompkins, F.C., Decomposition reactions, in *Treatise on Solid State Chemistry, Vol. 4: Reactivity of Solids*, Hannay, N.B., Ed., Plenum Press, New York, 1976, chap. 4.

24. Kubaschewski, O., Evans, E.L., and Alcock, C.B., *Metallurgical Thermochemistry*, 5th ed., Pergamon Press, Oxford, 1979.

25. Kingery, W.D., Bowen, H.K., and Uhlmann, D.R., *Introduction to Ceramics*, 2nd ed., John Wiley & Sons, New York, 1976, p. 415.

26. Hills, A.W.D., The mechanism of the thermal decomposition of calcium carbonate, *Chem. Eng. Sci.*, 23, 297, 1968.

27. Beruto, D. and Searcy, A.W., Use of the Langmuir method for kinetic studies of decomposition reactions: calcite ($CaCO_3$), *J. Chem. Soc., Faraday Trans. I*, 70, 2145, 1974.

28. Hyatt, E.P., Cutler, I.B., and Wadsworth, M.E., Calcium carbonate decomposition in carbon dioxide atmosphere, *J. Am. Ceram. Soc.*, 41, 70, 1958.

29. Ewing, J., Beruto, D., and Searcy, A.W., The nature of CaO produced by calcite powder decomposition in vacuum and in CO_2, *J. Am. Ceram. Soc.*, 62, 580, 1979.

30. Beruto, D. et al., Vapor-phase hydration of submicrometer CaO particles, *J. Am. Ceram. Soc.*, 64, 74, 1981.

31. Beruto, D., Botter, R., and Searcy, A.W., H_2O-catalyzed sintering of $\approx$2-nm-cross-section particles of MgO, *J. Am. Ceram. Soc.*, 70, 155, 1987.

32. Schmalzried, H., *Solid State Reactions*, 2nd ed., Verlag Chemie, Weinheim, Germany, 1981.

33. Linder, J. Studies of solid state reactions with radiotracers, *J. Chem. Phys.*, 23, 410, 1955.

34. Kuczynski, G.C., Formation of ferrites by sintering of component oxides, in *Ferrites: Proceedings of the International Conference*, Hoshino, Y., Iida, S., and Sugimoto, M., Eds., University Park Press, Baltimore, MD, 1971, p. 87.

35. Branson, D.L., Kinetics and mechanism of the reaction between zinc oxide and aluminum oxide, *J. Am. Ceram. Soc.*, 48, 591, 1965.

36. Jander, W., *Z. Anorg. Allg. Chem.*, 163, 1, 1927.

37. (a) Carter, R.E., Kinetic model for solid state reactions, *J. Chem. Phys.*, 34, 2010, 1961.
(b) Carter, R.E., Addendum: kinetic model for solid state reactions, *J. Chem. Phys.*, 35, 1137, 1961.

38. Glatzmier, G. and Koc, R., Method for silicon carbide production by reacting silica with hydrocarbon gas, U.S. Patent 5,324,494, 1994.

39. Koc, R. and Cattamanchi, S.V., Synthesis of beta silicon carbide powders using carbon coated fumed silica, *J. Mater. Sci.*, 33, 2537, 1998.

40. Nielsen, A.E., *Kinetics of Precipitation*, MacMillan, New York, 1964.

41. Overbeek, J.Th.G., Monodisperse colloidal systems, fascinating and useful, *Adv. Colloid Interface Sci.*, 15, 251, 1982.

42. Sugimoto, T., Preparation of monodisperse colloidal particles, *Adv. Colloid Inteface Sci.*, 28, 65, 1987.

43. Haruta, M. and Delmon, B., Preparation of homodisperse solids, *J. Chim. Phys.*, 83, 859, 1986.

44. Pierre, A.C., Sol–gel processing of ceramic powders, *Am. Ceram. Soc. Bull.*, 70, 1281, 1991.

45. Matijevic, E., Production of monodispersed colloidal particles, *Annu. Rev. Mater. Sci.*, 15, 483, 1985.

46. Matijevic, E., Monodispersed colloids: art and science, *Langmuir*, 2, 12, 1986.

47. Christian, J.W., *The Theory of Transformations in Metals and Alloys*, Pergamon Press, Oxford, 1975.

48. (a) McDonald, J.E., Homogeneous nucleation of vapor condensation. I. Thermodynamic aspects, *Am. J. Phys.*, 30, 870, 1962.
(b) McDonald, J.E., Homogeneous nucleation of vapor condensation. II. Kinetic aspects, *Am. J. Phys.*, 31, 31, 1963.

49. Walton, A.G., Nucleation in liquids and solutions, in *Nucleation*, Zettlemoyer, A.C., Ed., Marcel Dekker, New York, 1969, p. 225.

50. Baes, C.F. Jr. and Mesmer, R.E., *The Hydrolysis of Cations*, John Wiley & Sons, New York, 1976.

51. Reiss, H., The growth of uniform colloidal dispersions, *J. Chem. Phys.*, 19, 482, 1951.

52. LaMer, V.K. and Dinegar, R.H., Theory, production and mechanism of formation of monodispersed hydrosols, *J. Am. Chem. Soc.*, 72, 4847, 1950.

53. Hsu, W.P., Ronnquist, L., and Matijevic, E., Preparation and properties of monodispersed colloidal particles of lanthanide compounds. 2. Cerium (IV), *Langmuir*, 4, 31, 1988.

54. Bogush, G.H., and Zukoski, C.F., Studies on the formation of monodisperse silica powders, in *Ultrastructure Processing of Advanced Ceramics*, Mackenzie, J.D. and Ulrich, D.R., Eds., Wiley, New York, 1988, chap. 35.

55. Bogush, G.H., Dickstein, G.L., Lee, K.C., and Zukoski, C.F., Studies of the hydrolysis and polymerization of silicon alkoxides in basic alcohol solutions, *Mater. Res. Soc. Symp. Proc.*, 121, 57, 1988.

56. Stober, W., Fink, A., and Bohn, E., Controlled growth of monodisperse silica spheres in the micron size range, *J. Colloid Interface Sci.*, 26, 62, 1968.

57. Barringer, E.A. and Bowen, H.K., Formation, packing and sintering of monodispersed TiO_2 powders, *J. Am. Ceram. Soc.*, 65, C199, 1982.

58. Barringer, E.A. and Bowen, H.K., High-purity, monodisperse TiO_2 powders by hydrolysis of titanium tetraethoxide. 1. Synthesis and physical properties, *Langmuir*, 1, 414, 1985.

59. Matijevic, E., Monodispersed colloidal metal oxides, sulfides, and phosphates, in *Ultrastructure Processing of Ceramics, Glasses, and Composites*, Hench, L.L. and Ulrich, D.R., Eds., Wiley, New York, 1984, chap. 27.

60. Matijevic, E., Preparation and interactions of colloids of interest in ceramics, in *Ultrastructure Processing of Advanced Ceramics*, Mackenzie, J.D. and Ulrich, D.R., Eds., John Wiley & Sons, New York, 1988, p. 429.

61. Matijevic, E., Control of powder morphology, in *Chemical Processing of Advanced Ceramics*, Hench, L.L. and West, J.K., Eds., John Wiley & Sons, New York, 1992, p. 513.

62. Brace, R. and Matijevic, E., Aluminum hydrous oxide sols. I. Spherical particles of narrow size distribution, *J. Inorg. Nucl. Chem.*, 35, 3691, 1973.

63. Aiken, B., Hsu, W.P., and Matijevic, E., Preparation and properties of monodispersed colloidal particles of lanthanide compounds, *J. Am. Ceram. Soc.*, 71, 845, 1988.

64. Bratton, R.J., Coprecipitates yielding $MgAl_2O_4$ spinel powders, *Am. Ceram. Soc. Bull.*, 48, 759, 1959.

65. Haertling, G.H., Piezoelectric and electrooptic ceramics, in *Ceramic Materials for Electronics*, 2nd ed., Buchanan, R.C., Ed., John Wiley & Sons, New York, 1991, chap. 3.

66. Mazdiyasni, K.S., Doloff, R.T., and Smith, J.S., Jr., Preparation of high-purity submicron barium titanate powders, *J. Am. Ceram. Soc.*, 52, 523, 1969.

67. Mah, T.-I., Hermes, E.E., and Masdiyasni, K.S., Multicomponent ceramic powders, in *Chemical Processing of Ceramics*, Lee, D.I. and Pope, E.J.A., Eds., Marcel Dekker, New York, 1994, chap. 4.

68. Dawson, W.J., Hydrothermal synthesis of advanced ceramic powders, *Am. Ceram. Soc. Bull.*, 67, 1673, 1988.

69. Somiya, S., Hydrothermal preparation and sintering of fine ceramic powders, *Mater. Res. Soc. Symp. Proc.*, 24, 255, 1984.

70. Rahaman, M.N. and Zhou, Y.C., Effect of dopants on the sintering of ultrafine CeO_2 powder, *J. Eur. Ceram. Soc.*, 15, 939, 1995.

71. Christensen, A.N. and Rasmussen, S.E., Hydrothermal preparation of compounds of the type ABO_3 and AB_2O_4, *Acta Chem. Scand.*, 17, 845, 1963.

72. Eckert, J.O. Jr., Hung-Houston, C.C., Gersten, B.L., Lencka, M.M., and Riman, R.E., Kinetics and mechanisms of hydrothermal synthesis of barium titanate, *J. Am. Ceram. Soc.*, 79, 2929, 1996.

73. Hennings, D., Rosenstein, G., and Schreinemacher, H., Hydrothermal preparation of barium titanate from barium-titanate acetate gel precursors, *J. Eur. Ceram. Soc.*, 8, 107, 1991.

74. Frey, M.H. and Payne, D.A., Grain size effect on structure and phase transformation of barium titanate, *Phys. Rev.*, 54, 3158, 1996.

75. Sparks, R.E., Microencapsulation, in *Encyclopedia of Chemical Technology*, Vol. 15, Grayson, M. and Eckroth, D., Eds., John Wiley & Sons, New York, 1981, p. 470.

76. Matijevic, Colloid science of composite systems, in *Science of Ceramic Chemical Processing*, Hench, L.L. and Ulrich, D.R., Eds., John Wiley & Sons, New York, 1986, chap. 50.

77. Garg, A. and Matijevic, E., Preparation and properties of uniformly coated inorganic colloidal particles. 2. Chromium hydrous oxide on hematite, *Langmuir*, 4, 38, 1988.

78. Garg, A. and De Jonghe, L.C., Microencapsulation of silicon nitride particles with yttria and yttria-alumina precursors, *J. Mater. Res.*, 5, 136, 1990.

79. Sacks, M.D., Bozkurt, N., and Scheiffele, G.W., Fabrication of mullite and mullite-matrix composites by transient viscous sintering of composite powders, *J. Am. Ceram. Soc.*, 74, 2428, 1991.

80. Okamura, H., Barringer, E.A., and Bowen, H.K., Preparation and sintering of monosized Al_2O_3-TiO_2 composite powder, *J. Am. Ceram. Soc.*, 69, C22, 1986.

81. Kratohvil, S. and Matijevic, E., Preparation and properties of uniformly coated inorganic colloidal particles. I, aluminum (hydrous) oxide on hematite, chromia, and titania, *Adv. Ceram. Mater.*, 2, 798, 1987.

82. Kapolnek, D. and De Jonghe, L.C., Particulate composites from coated powders, *J. Eur. Ceram. Soc.*, 7, 345, 1991.

83. Aiken, B. and Matijevic, E. Preparation and properties of uniform coated inorganic colloidal particles. IV. Yttrium basic carbonate and yttrium oxide on hematite, *J. Colloid Interface Sci.*, 126, 645, 1988.

84. Hu, C.-L. and Rahaman, M.N., Factors controlling the sintering of particulate composites: II, coated inclusion particles, *J. Am. Ceram. Soc.*, 75, 2066, 1992.

85. Gherardi, P. and Matijevic, E., Interaction of precipitated hematite with preformed colloidal titania dispersions, *J. Colloid Interface Sci.*, 109, 57, 1986.

86. Giesche, H. and Matijevic, E., Preparation, characterization, and sinterability of well-defined silica/yttria powders, *J. Mater. Res.*, 9, 436, 1994.

87. Masters, K., *Spray Drying Handbook*, 5th ed., John Wiley & Sons, New York, 1991.

88. DeLau, J.G.M., Preparation of ceramic powders from sulfate solutions by spray drying and roasting, *Am. Ceram. Soc. Bull.*, 49, 572, 1970.

89. Messing, G.L., Zhang, S.-C., and Jayanthi, G.V., Ceramic powder synthesis by spray pyrolysis, *J. Am. Ceram. Soc.*, 76, 2707, 1993.

90. Schwartz, R.W., Eichorst, D.J., and Payne, D.A., Precipitation of PZT and PLZT powders using a continuous reactor, *Mater. Res. Soc. Symp. Proc.*, 73, 123, 1986.

91. Schnettler, F.J., Montforte, F.R., and Rhodes, W.W., A cryochemical method for preparing ceramic materials, in *Science of Ceramics*, Vol. 4, Stewart, G.H., Ed., The British Ceramic Society, Stoke-on-Trent, U.K., 1968, p. 79.

92. Johnson, D.W. Jr., Gallagher, P.K., Nitti, D.J., and Schrey, F., Effect of preparation technique and calcination temperature on the densification of lithium ferrite, *Am. Ceram. Soc. Bull.*, 53, 163, 1974.

93. Dogan F. and Hausner, H., The role of freeze drying in ceramic powder processing, *Ceram. Trans.*, 1, 127, 1988.

94. Rahaman, M.N., DeJonghe, L.C., Shinde, S.L., and Tewari, P.H., Sintering and microstructure of mullite aerogels, *J. Am. Ceram. Soc.*, 71, C-338, 1988.

95. Pechini, M., Method of preparing lead and alkaline earth titanates to form a capacitor, U.S. Patent 3,330,697, 1967.

96. Eror, N.G. and Anderson, H.U., Polymeric precursor synthesis of ceramic materials, *Mater. Res. Symp. Proc.*, 73, 571, 1986.

97. Lessing, P.A., Mixed-cation oxide powders via polymeric precursors, *Am. Ceram. Soc. Bull.*, 68, 1002, 1989.

98. Budd, K.D. and Payne, D.A., Preparation of strontium titanate ceramics and internal boundary layer capacitors by the Pechini method, *Mater. Res. Soc. Symp. Proc.*, 32, 239, 1984.

99. Marcilly, C., Courty, P., and Delmon, B., Preparation of highly dispersed mixed oxides and oxide solid solutions by pyrolysis of amorphous organic precursors, *J. Am. Ceram. Soc.*, 53, 56, 1970.

100. Chu, C.-T. and Dunn, B., Preparation of high-T_c superconducting oxides by the amorphous citrate process, *J. Am. Ceram. Soc.*, 70, C-375, 1987.

101. Chick, L.A., Pederson, L.R., Maupin, G.D., Bates, J.L., Thomas, L.E., and Exarhos, G.J., Glycine-nitrate combustion synthesis of oxide ceramic powders, *Mater. Lett.*, 10, 6, 1990.

102. Mazdiyasni, K.S. and Cooke, C.M., Synthesis, characterization, and consolidation of Si_3N_4 obtained from ammonolysis of $SiCl_4$, *J. Am. Ceram. Soc.*, 56, 628, 1973.

103. Mazdiyasni, K.S. and Cooke, C.M., Synthesis of high-purity, alpha phase silicon nitride powder, U.S. Patent 3,959,446, 1976.

104. Kato, A., Hojo, J., and Watari, T., Some common aspects of the formation of nonoxide powders by the vapor reaction method, *Mater. Sci. Res.*, 17, 123, 1984.

105. Ulrich, G.D., Flame synthesis of fine particles, *Chem. Eng. News*, 62, 22, 1984.

106. George, A.P., Murley, R.D., and Place, E.R., Formation of TiO_2 aerosol from the combustion supported reaction of $TiCl_4$ and O_2, in *Fogs and Smokes*, Vol. 7, Faraday Symposia of the Chemical Society, London, 1973, p. 63.

107. Prochazka, S. and Greskovich, C., Synthesis and characterization of a pure silicon nitride powder, *Am. Ceram. Soc. Bull.*, 57, 579, 1978.

108. Haggerty, J.S., Growth of precisely controlled powders from laser heated gases, in *Ultrastructure Processing of Ceramics, Glasses and Composites*, Hench, L.L. and Ulrich, D.R., Eds., Wiley, New York, 1984, chap. 28.

109. (a) Cannon, W.R., Danforth, S.C., Flint, J.H., Haggerty, J.S., and Marra, R.A., Sinterable ceramic powders from laser-driven reactions. I. Process description and modeling, *J. Am. Ceram. Soc.*, 65, 324, 1982.
    (b) Cannon, W.R., Danforth, S.C., Haggerty, J.S., and Marra, R.A., Sinterable ceramic powders from laser-driven reactions. II. Powder characteristics and process variables, *J. Am. Ceram. Soc.*, 65, 330, 1982.

110. Moser, W.R., Ed., *Advanced Catalysts and Nanostructured Materials: Modern Synthetic Methods*, Academic Press, San Diego, CA, 1996.

111. Hayashi, C., Uyeda, R., and Tasaki, A., Eds., *Ultra-fine Particles: Exploratory Science and Technology*, Noyes Publications, Westwood, NJ, 1997.

112. Axelbaum, R.L., Synthesis of stable metal and non-oxide ceramic particles in sodium/halide flames, *Powder Metall.*, 43, 323, 2000.

113. Birringer, R., Gleiter, H., Klein, H.P., and Marquard, P., Nanocrystalline materials: an approach to a novel solid structure with gas-like disorder, *Phys. Lett.*, 102A, 365, 1984.

114. Laine, R.M., Marchal, J., Kim, S., Azurdia, J., and Kim, M., Liquid-feed flame spray pyrolysis of single and mixed phase mixed-metal oxide nanopowders, *Ceram. Trans.*, 159, 3, 2005.

3 Powder Characterization

3.1 INTRODUCTION

The powder characteristics, we will recall, have a strong influence on the particle packing homogeneity of the green body and on the microstructural evolution during sintering. Whereas knowledge of the powder characteristics is important in itself for quality control of the starting material in processing, it is also vitally important for providing an improved ability to control the microstructure during sintering.

The extent to which the characterization process is taken depends on the application. In the case of traditional ceramics, which do not have to meet exacting property requirements, a fairly straightforward observation, with a microscope, of the size, size distribution, and shape of the powders may be sufficient. For advanced ceramics, however, detailed knowledge of the powder characteristics is required for adequate control of the microstructure and properties of the fabricated material. Commercial powders are used in most applications. Normally, the manufacturer has carried out most of the characterization experiments and provides the user with the results. The manufacturer's data combined with a straightforward observation of the powder with a microscope are sufficient for many applications.

For a powder prepared in the laboratory, a detailed set of characterization experiments may have to be carried out. Minor variations in the chemical composition and purity of the powder, as we have seen, can have profound effects on the microstructure of advanced ceramics. This realization has led to a growing use of analytical techniques that have the capability of detecting constituents, especially on the surfaces of the particles, at concentrations down to the parts-per-million level.

The important characteristics of a powder can be categorized into four groups: (1) physical characteristics, (2) chemical composition, (3) phase composition, and (4) surface characteristics. For each group, the main powder properties that have a significant influence on processing are summarized in Table 3.1. A vast number of techniques have been developed for the characterization of solids [1]. In this chapter, we concentrate only on those techniques that have broad applications to ceramic powders. The experimental details in performing the characterization will, in general, not be covered as they are usually discussed at length in the manuals supplied by the equipment manufacturers. Instead, we concentrate on the principles of the methods, the range of information that can be obtained with them, and some of their limitations.

3.2 PHYSICAL CHARACTERIZATION

Powders consist of an assemblage of small units with certain distinct physical properties. These small units, loosely referred to as particles, can have a fairly complex structure. A variety of terms have been used to describe them, and this has led to some confusion in the literature. In this book we adopt, with minor modifications, the terminology proposed by Onoda and Hench [2].

TABLE 3.1
Powder Characteristics That Have a
Significant Influence on Ceramic Processing

Physical Characteristics
Particle size and distribution
Particle shape
Degree of agglomeration
Surface area
Density and porosity

Chemical Composition
Major elements
Minor elements
Trace elements

Phases
Structure (crystalline or amorphous)
Crystal structure
Phase composition

Surface Characteristics
Surface structure
Surface composition

3.2.1 Types of Particles

3.2.1.1 Primary Particles

A primary particle is a discrete, low-porosity unit that can be a single crystal, a polycrystalline particle, or a glass. If any pores are present, they are isolated from each other. A primary particle cannot, for example, be broken down into smaller units by ultrasonic agitation in a liquid. It may be defined as the smallest unit in the powder with a clearly defined surface. For a polycrystalline primary particle, the crystals have been referred to variously as crystallites, grains or domains. In this book, we shall use the term *crystal*.

3.2.1.2 Agglomerates

An agglomerate is a cluster of primary particles held together by surface forces, by a solid bridge, or by a liquid. Figure 3.1 is a schematic diagram of an agglomerate consisting of dense, polycrystalline primary particles. Agglomerates are porous, with the pores being generally interconnected. They are classified into two types: soft agglomerates and hard agglomerates. Soft agglomerates are held together by weak surface forces and can be broken down into primary particles by ultrasonic agitation in a liquid. Hard agglomerates consist of primary particles that are chemically bonded by solid bridges, so they cannot be broken down into primary particles by ultrasonic agitation in a liquid. Hard agglomerates, as outlined in Chapter 2, are undesirable in the production of advanced ceramics because they commonly lead to the formation of microstructural flaws.

3.2.1.3 Particles

When no distinction is made between primary particles and agglomerates, the term *particles* is used. Particles can be viewed as small units that move as separate entities when the powder is

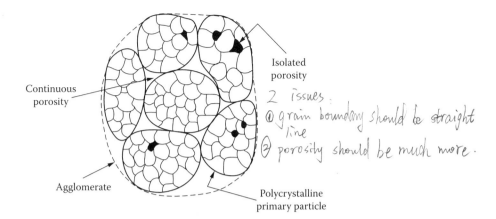

FIGURE 3.1 Schematic diagram of an agglomerate consisting of dense, polycrystalline primary particles.

dispersed by agitation and can consist of primary particles, agglomerates or some combination of the two. Most particle size analysis techniques would refer to such particles.

3.2.1.4 Granules

The term *granules* refers to large agglomerates (~100 to 1000 μm in size) that are deliberately formed by the addition of a granulating agent (e.g., a polymer-based binder) to the powder, followed by tumbling or spray drying. These large, nearly spherical agglomerates improve the flow characteristics of the powder, thereby improving the ease of die filling during compaction.

3.2.1.5 Flocs

Flocs are clusters of particles in a liquid suspension. The particles are held together weakly by electrostatic forces or by organic polymers and can be redispersed by appropriate modification of the interfacial forces through alteration of the solution chemistry. The formation of flocs is normally undesirable because it decreases the packing homogeneity of the green body.

3.2.1.6 Colloids

A *colloid* is any system consisting of a finely divided phase in a fluid. A colloidal suspension (or sol) consists of fine particles dispersed in a liquid. The particles, referred to as *colloidal particles*, undergo Brownian motion and have a slow (often negligible) sedimentation rate under normal gravity. The size range for colloidal particles is approximately 1 nm to 1 μm.

3.2.1.7 Aggregates

An aggregate is a coarse constituent in a mixture, which usually also contains a fine constituent called the *bond*. Pebbles in concrete are an example, with the fine cement particles forming the bond.

The approximate size ranges of the aforementioned types of particles considered are summarized in Table 3.2.

3.2.2 Particle Size and Particle Size Distribution

Ceramic powders generally consist of particles of different sizes distributed over a certain range. Some powders may have a very narrow distribution of sizes (e.g., those prepared by chemical precipitation under controlled conditions), whereas for others, the distribution in sizes may be very broad (e.g., an unclassified powder prepared by milling). Some particles are spherical or equiaxial

TABLE 3.2
Size Range for Particles in Ceramic Processing

Type of Particle	Size Range
Powder	
Colloidal particle	1 nm–1 μm
Coarse particle	1 μm–100 μm
Granule	100 μm–1 mm
Aggregate	> 1 mm

(i.e., with the same length in each direction), but many are irregular in shape. Often, we are required to characterize the particle size and particle size distribution of the powder because these two characteristics, along with the particle shape, have a strong effect on the consolidation and sintering of the powders.

3.2.2.1 Definition of Particle Size

For a spherical particle, the diameter is taken as the size. However, the size of an irregularly shaped particle is a rather uncertain quantity. We therefore need to define what the "particle size" represents. One simple definition of the size of an irregularly shaped particle is the diameter of the sphere having the same volume as the particle. This is not much help because in many cases the volume of the particle is ill-defined or difficult to measure. Usually, the particle size is defined in a fairly arbitrary manner in terms of a number generated by one of the measuring techniques described later. A particle size measured by one technique may therefore be quite different from that measured by another technique, even when the measuring instruments are operating properly.

If an irregularly shaped particle is allowed to settle in a liquid, its terminal velocity may be compared with that for a sphere of the same density settling under similar conditions. For laminar flow, the sphere diameter can be calculated from Stokes law and is commonly referred to as the *Stokes diameter*. Using a microscope, individual particles can be observed and measured. In this case, the particle size is commonly determined from the projected area of the particles (projected area diameter) or a linear dimension measured parallel to some fixed direction (Feret's diameter or Martin's diameter). Some definitions of particle size are given in Table 3.3.

TABLE 3.3
Some Definitions of Particle Size

Symbol	Name	Definition
x_S	Surface diameter	Diameter of a sphere having the same surface area as the particle
x_V	Volume diameter	Diameter of a sphere having the same volume as the particle
x_{SV}	Surface volume diameter	Diameter of a sphere having the same surface area to volume ratio as the particle
x_{STK}	Stokes diameter (or equivalent spherical diameter)	Diameter of a sphere having the same sedimentation rate as the particle for laminar flow in a liquid
x_{PA}	Projected area diameter	Diameter of a circle having the same area as the projected area of the particle
x_C	Perimeter diameter	Diameter of a circle having the same perimeter as the projected outline of the particle
x_A	Sieve diameter	Width of the minimum square aperture through which the particle will pass
x_F	Feret's diameter	Mean value of the distance between pairs of parallel tangents to the projected outline of the particle
x_M	Martin's diameter	Mean chord length of the projected outline of the particle

3.2.2.2 Average Particle Size

Bearing in mind the uncertainty in defining the size of an irregularly shaped particle, we will now attempt to describe the average particle size of the powder. As a start, let us assume that the powder consists of N particles with sizes x_1, x_2, x_3, …, x_N, respectively. We can calculate a mean size, $\bar{x}$, and the standard deviation in the mean, s, according to the equations:

$$\bar{x} = \sum_{i=1}^{N} \frac{x_i}{N} \tag{3.1}$$

$$s = \left(\sum_{i=1}^{N} \frac{(x_i - \bar{x})^2}{N} \right)^{1/2} \tag{3.2}$$

The value of $\bar{x}$ is taken as the particle size of the powder and s gives a measure of the spread in the particle size distribution. In a random (Gaussian) distribution, approximately two out of every three particles will have their sizes in the range of $\bar{x} \pm s$.

Most likely the characterization technique will sort the particles into a small number, n, of size categories, where n is much smaller than N. The technique may also produce a count of the number of particles within each category, so that there would be n_1 particles in a size category centered about x_1, n_2 of size x_2, etc., and n_n of size x_n. Alternatively, the mass or the volume of particles within each size category may be obtained. We will consider the number of particles within a size category giving rise to a number-weighted average. The representation of the data in terms of the mass or volume of particles within a size category will follow along similar lines. The mean size and the standard deviation can be determined from the data according to the equations:

$$\bar{x}_N = \frac{\displaystyle\sum_{i=1}^{n} n_i x_i}{\displaystyle\sum_{i=1}^{n} n_i} \tag{3.3}$$

$$s = \left(\frac{\displaystyle\sum_{i=1}^{n} n_i (x_i - \bar{x})^2}{\displaystyle\sum_{i=1}^{n} n_i} \right)^{1/2} \tag{3.4}$$

For comparison, the volume-weighted average size is

$$\bar{x}_V = \frac{\displaystyle\sum_{i=1}^{n} v_i x_i}{\displaystyle\sum_{i=1}^{n} v_i} = \frac{\displaystyle\sum_{i=1}^{n} n_i x_i^4}{\displaystyle\sum_{i=1}^{n} n_i x_i^3} \tag{3.5}$$

The reader will recognize that the mean particle size determined from Equation 3.3 is the arithmetic mean. This is not the only mean size that can be defined but it is the most significant when the particle size distribution is normal. The geometric mean, $\bar{x}_g$, is the nth root of the product of the diameter of the n particles and is given by:

$$\log \bar{x}_g = \frac{\displaystyle\sum_{i=1}^{n} n_i \log x_i}{\displaystyle\sum_{i=1}^{n} n_i} \tag{3.6}$$

It is of particular value when the distribution is log-normal. The harmonic mean, $\bar{x}_h$, is the number of particles divided by the sum of the reciprocals of the diameters of the individual particles:

$$\bar{x}_h = \frac{\displaystyle\sum_{i=1}^{n} n_i}{\displaystyle\sum_{i=1}^{n} \frac{n_i}{x_i}} \tag{3.7}$$

It is related to the specific surface area and is of importance when the surface area of the sample is concerned.

The mean particle size determined from Equation 3.3 is sometimes referred to as the linear mean diameter and denoted as $\bar{x}_{NL}$ when it is required to distinguish it from the surface mean diameter, $\bar{x}_{NS}$, and the volume mean diameter, $\bar{x}_{NV}$. The surface and volume mean diameters are defined by:

$$\bar{x}_{NS} = \frac{\left(\displaystyle\sum_{i=1}^{n} n_i x_i^2\right)^{1/2}}{\displaystyle\sum_{i=1}^{n} n_i} \tag{3.8}$$

$$\bar{x}_{NV} = \frac{\left(\displaystyle\sum_{i=1}^{n} n_i x_i^3\right)^{1/3}}{\displaystyle\sum_{i=1}^{n} n_i} \tag{3.9}$$

3.2.2.3 Representation of Particle Size Data

A simple and widely used way to describe the data is in terms of a histogram. A rectangle is constructed over each interval, with the height of the rectangle being proportional to the percent or fraction of particles within a size interval. Using the data given in Table 3.4, a histogram is shown in Figure 3.2. A variation of this representation is to construct the rectangles of the histogram

TABLE 3.4
Particle Size Distribution Data

Size Range (µm)	Number in Size Range, n_i	Size, x_i (µm)	Number Fraction	CNPF[a]	$f_N(x)$[b] (1/µm)
<10	35	5	0.005	0.5	0.001
10–12	48	11	0.007	1.2	0.003
12–14	64	13	0.009	2.1	0.004
14–16	84	15	0.012	3.3	0.005
16–18	106	17	0.015	4.8	0.007
18–20	132	19	0.019	6.7	0.009
20–25	468	22.5	0.067	13.4	0.012
25–30	672	27.5	0.096	23.0	0.019
30–35	863	32.5	0.124	35.4	0.025
35–40	981	37.5	0.141	49.5	0.028
40–45	980	42.5	0.141	63.6	0.028
45–50	865	47.5	0.124	76.0	0.025
50–55	675	52.5	0.097	85.7	0.020
55–60	465	57.5	0.067	92.4	0.006
60–70	420	65	0.060	98.4	0.006
70–80	93	75	0.013	99.7	0.001
80–90	13	85	0.002	99.9	0.001
90–100	1	95	0	99.9	—
>100	4	—	0.001	100.0	—

[a] CNPF = cumulative number percent finer

[b] $\bar{x}_N = 40$ µm (Equation 3.3); s = 14 µm (Equation 3.4)

Source: Evans, J.W. and De Jonghe L.C., *The Production of Inorganic Materials*, Macmillan, New York, 1991, 131.

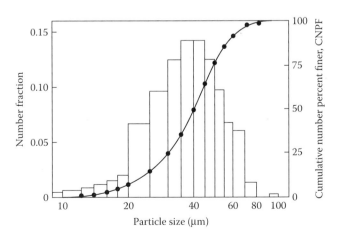

FIGURE 3.2 Particle size distribution data of Table 3.4 represented in terms of a histogram and the cumulative number percent finer (CNPF) than a given size.

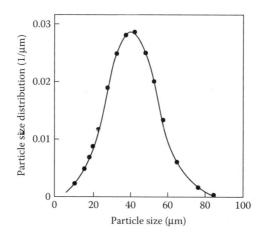

FIGURE 3.3 Particle size distribution data of Figure 3.2 fitted to a smooth curve.

such that their areas are proportional to the number of particles in the intervals. The total area under the histogram is equal to the number of particles counted. To aid the comparison of histograms for different powders, it is useful to reduce this number to 100 by making the areas of the rectangles equal to the percentages of the particles in the intervals.

It is commonly more useful to plot the data as a cumulative size distribution by summing the percent (or fraction) of particles finer than a given size, defined as the *cumulative number percent finer*, CNPF, or larger than a given size, defined as the *cumulative number percent larger*, CNPL. Using the data in Table 3.4, the CNPF is also shown in Figure 3.2, where a smooth curve is drawn through the data points.

In many cases it is necessary to provide a more complete description of the particle size data in terms of a mathematical equation, the parameters of which can be used to compare different powders. If the number of particles were fairly large and the size intervals, Δx, were small enough, then we would be able to fit a reasonably smooth curve through the particle size distribution data. Normally, the cumulative size distribution data is used as the starting point (rather than the histogram) and a smooth curve is fitted through the data. We are now assuming that the cumulative size distribution is some smoothly varying function of x, which we denote as $F_N(x)$. The fractional size distribution function, $f_N(x)$, is then obtained from $F_N(x)$ by taking the derivative:

$$f_N(x) = \frac{d}{dx} F_N(x) \tag{3.10}$$

where $f_N(x)dx$ is the fraction of particles with sizes between x and $x + dx$.

The function $f_N(x)$ represents the measured size distribution and this is used for the description of the data in terms of a mathematical equation. Figure 3.3 shows the measured size distribution determined from the data plotted in Figure 3.2. Usually, the measured size distribution function is fitted in terms of an expected size distribution, such as the normal distribution:

$$f_N(x) = \frac{1}{s\sqrt{2\pi}} \exp\left[-\frac{(x - \bar{x})^2}{2s^2}\right] \tag{3.11}$$

where $\bar{x}$ is the mean particle size and s is the standard deviation in the mean particle size. In practice, the values of $\bar{x}$ and s are adjusted in a curve-fitting routine until the best fit to the data is obtained. As we are assuming the data are smoothly varying functions, the mean size and standard deviation must now be defined in terms of equations appropriate to the normal distribution:

$$\bar{x} = \frac{\int_{-\infty}^{\infty} x f_N(x)dx}{\int_{-\infty}^{\infty} f_N(x)dx} \tag{3.12}$$

$$s = \left[\frac{\int_{-\infty}^{\infty} f_N(x)(x-\bar{x})^2 dx}{\int_{-\infty}^{\infty} f_N(x)dx} \right]^{1/2} \tag{3.13}$$

A serious problem with the normal distribution is that it predicts a finite fraction of particles with sizes less than zero. Furthermore, the largest particle size does not have a limit and is infinite. The normal distribution therefore seldom gives a good fit to real particle size data.

A better equation that has been found to give a good approximation to the size distribution of powders prepared by spray drying or by mechanical milling is the log-normal distribution:

$$f_N(\ln x) = \frac{1}{s\sqrt{2\pi}} \exp\left[-\frac{(\ln x - \bar{x})^2}{2s^2} \right] \tag{3.14}$$

where $\bar{x}$ is now the mean of the natural logarithm of the particle sizes and s is the standard deviation of the natural logarithm of the particle sizes. Unlike the normal distribution, the log-normal distribution accounts for particle sizes greater than zero; however the largest size is still unlimited.

An empirical function that has been observed to describe the particle size distribution of milled powders is the Rosin–Rammler equation. The equation has undergone a number of modifications; one form is

$$F_M(x) = 1 - \exp\left[-\left(\frac{x}{x_{RR}} \right)^n \right] \tag{3.15}$$

where $F_M(x)$ is the cumulative mass fraction of particles with sizes between x and $x + dx$, x_{RR} is the Rosin–Rammler size modulus, equal to the size x for which $F_M(x) = 0.37$, and n is an empirical constant for the powder, typically found to be between 0.4 and 1.0. Another empirical distribution function that has been used to describe the size distribution of milled powders is the Gates–Gaudin–Schuhman equation:

$$F_M(x) = \left(\frac{x}{x_{max}} \right)^n \tag{3.16}$$

where x_{max} is the maximum particle size and n is an empirical constant. Unlike the log-normal distribution, both the Rosin–Rammler and Gates–Gaudin–Schuhman equations have a finite size for the largest particle.

3.2.3 Particle Shape

Particle shape influences the flow properties and packing of powders as well as their interaction with fluids (e.g., viscosity of a suspension). Qualitative terms are sometimes used to give an indication of the nature of the particle shape (e.g., spherical, equiaxial, acicular, angular, fibrous, dendritic, and flaky). However, except for the fairly simple geometries such as a sphere, cube or cylinder, the quantitative characterization of particle shape can be fairly complex. The shape of a particle is commonly described in terms of a shape factor, which provides some measure of the deviation from an idealized geometry, such as a sphere or a cube. For elongated particles, the most common way of representing the shape has been in terms of the *aspect ratio*, defined as the ratio of the longest dimension to the shortest dimension. For powders, the sphere is used as a reference, and the shape factor is defined as:

$$\text{Shape factor} = \frac{1}{\psi} = \frac{\text{surface area of the particle}}{\text{surface area of a sphere with the same volume}} \tag{3.17}$$

where ψ is referred to as the *sphericity*. According to Equation 3.17, the shape factor of a sphere is unity and the shape factor for all other shapes is greater than unity, e.g., $(6/\pi)^{1/3}$ or 1.24 for a cube. Unfortunately, the use of a shape factor (or sphericity) is ambiguous: radically different shapes can have the same shape factor.

Shape coefficients have also been used to characterize particle shape. They relate the measured size, x, determined by one of the techniques described later, and the measured particle surface area, A, or volume, V:

$$A = \alpha_A x^2 \tag{3.18a}$$

$$V = \alpha_V x^3 \tag{3.18b}$$

where α_A is the area shape coefficient and α_V is the volume shape coefficient. According to this definition, the area and volume shape coefficients for the sphere are π and $\pi/6$, respectively, whereas for the cube, the area and volume shape coefficients are 6 and 1, respectively.

Because of its complexity, a detailed quantification of the shapes of irregularly shaped particles may not provide significant practical benefits. The trend in ceramic processing is toward increasing use of spherical or equiaxial particles because they produce better packing homogeneity in the consolidated body. The use of these spherical or equiaxial particles, combined with the direct measurement of the average size and surface area of the particles, provides a far more effective approach.

3.2.4 Measurement of Particle Size and Size Distribution

Excellent accounts of the principles and practice of particle size measurement are given in Reference 3 to Reference 5. Table 3.5 shows the common methods used for the measurement of particle size and particle size distribution, together with their approximate size range of applicability. The main features of these methods are described in the following text.

3.2.4.1 Microscopy

Microscopy is a fairly straightforward technique that offers the advantage of direct measurement of the particle size and simultaneous observation of the individual particles, their shape, and the

TABLE 3.5
Common Methods for the Measurement
of Particle Size and Their Approximate
Range of Applicability

Method	Range (μm)
Microscopy	
Optical	>1
Scanning electron	>0.1
Transmission electron	>0.001
Sieving	20–10000
Sedimentation	0.1–100
Coulter counter	0.5–400
Light scattering	
Scattering intensity	0.1–1000
Brownian motion	0.005–1
X-ray line broadening	<0.1

extent of agglomeration. It usually forms the first step in the characterization of ceramic powders. Optical microscopes can be used for particle sizes down to ~1 μm whereas electron microscopes can extend the range down to ~1 nm. Commonly, the sample is prepared by adding a small amount of the powder to a liquid to produce a dilute suspension. After suitable agitation (e.g., with an ultrasonic probe), a drop of the suspension is placed on a glass slide or a microscope stub. Evaporation of the liquid leaves a deposit that is viewed in the microscope. Good separation of the particles must be achieved. Because of the small amount of powder used in the measurements, care must also be taken to ensure that the deposit is representative of the original powder batch.

Particle size measurements are usually made from micrographs. The method is extremely tedious if it is done manually. Automatic image analysis of micrographs or an electronic display can reduce the amount of work considerably. Normally a large number of particles (a few hundred) need to be measured. As would be apparent, microscopy produces a particle size distribution based on the number of particles within an appropriate size range.

The image of a particle seen in a microscope (or micrograph) is two dimensional and from this image an estimate of the particle size has to be made. Some of the more common measurements of the size are (Figure 3.4):

1. Martin's diameter (x_M) is the length of a line that bisects the area of the particle image. The line can be drawn in any direction but the same direction must be maintained for all of the measurements.
2. Feret's diameter (x_F) is the distance between two tangents on opposite sides of the particle, parallel to some fixed direction.
3. The projected area diameter (x_{PA}) is the diameter of a circle having the same area as the two-dimensional image of the particle.
4. The perimeter diameter (x_C) is the diameter of the circle having the same circumference as the perimeter of the particle.
5. The longest dimension, equal to the maximum value of Feret's diameter.

3.2.4.2 Sieving

The use of sieves for separating particles into fractions with various size ranges is the oldest and one the most widely used classification methods. The particles are classified in terms of their ability

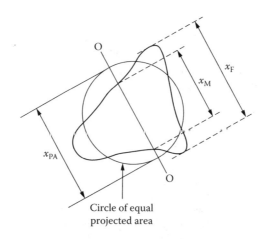

FIGURE 3.4 Representative two-dimensional diameters of particles observed in a microscope.

or inability to pass through an aperture with a controlled size. Sieves with openings between approximately 20 μm and 10 mm are constructed with wire mesh and are identified in terms of a mesh size and a corresponding aperture size. The wire mesh has square apertures, the size of which depends on the number of wires per linear dimension and the diameter of the wire. The mesh size is equal to the number of wires per linear inch of the sieve screen, which is the same as the number of square apertures per inch (Figure 3.5). The mesh number, M, aperture width, a, wire diameter, w, and the open area, A, are related by the following equations:

$$M = 1/(a+w) \tag{3.19a}$$

$$a = 1/M - w \tag{3.19b}$$

$$A = a^2/(a+w)^2 = (Ma)^2 \tag{3.19c}$$

For example, a 400 mesh sieve with an aperture of 38 μm has a wire diameter of 25.5 μm and an open area of 36%. The use of special metal sieves can extend the range of sieving to 5 μm or less whereas punched plate sieves can extend the upper range to ~125 mm.

Summaries of sieve apertures according to the American Standard (ASTM E 11-87) and the British Standard (BS 410) are given in Appendix D. The American Standard apertures are in the

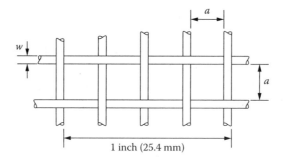

FIGURE 3.5 Dimensions of woven wire cloth in sieves.

progression of the fourth root of two. Originally the apertures were determined taking 75 μm as the reference value but the current International Standard (ISO) uses 45 μm as the reference value.

Sieving may be carried out in the dry or wet state by hand or by a machine. However, most sieving is done in the dry state on a machine designed to impart the necessary shaking, rotating or vibrating motion to the material on the screens. Several sieves are stacked together, with the coarsest mesh aperture at the top and the smallest at the bottom, and placing the powder on the top sieve. A closed pan is placed at the bottom of the stack to collect the fines, and a lid is placed on the top sieve to prevent loss of material. A stack commonly consists of 4 to 6 sieves arranged in a $\sqrt{2}$ progression of sizes. The stack is vibrated for a fixed time and the residual mass of powder on each sieve is measured. For routine classification, sieving is commonly carried out for 20 to 30 min. Reference should be made to the American or British Standard for a complete description of the procedure. Agglomeration of the powder and clogging of the screens during sieving of a dry powder can lead to significant problems below ~40 μm. The use of pulsed jets of air to reduce clogging or wet sieving in which the particles are dispersed in a liquid, can alleviate the problems. Wet sieving is also required when the powder to be tested is already suspended in a liquid.

Adequate separation of the particles into their true size fractions requires a fairly long time. As it is normally too time consuming to sieve for such a long time, the particle size distribution obtained from most sieving operations is only approximate. However, this approximate size characterization can be quite useful for the selection or verification of raw materials in the traditional ceramic industry. The various fractions produced by sieving are weighed so that a particle distribution based on mass is obtained. For elongated particles, the method generally favors measurement of the longer particle dimension. Sieving is not feasible for powders of advanced ceramics, in which the particle size is normally less than a few micrometers. Nylon sieves should be used instead of the common metal sieves for clean powders to avoid contamination with metallic impurities.

3.2.4.3 Sedimentation

A spherical particle falling in a viscous liquid with a sufficiently small velocity soon reaches a constant velocity called the *terminal velocity*, in which the effective weight of the particle is balanced by the frictional force exerted on it by the liquid. The frictional force, F, on the particle is given by Stokes law:

$$F = 6\pi\eta a v \tag{3.20}$$

where η is the viscosity of the liquid, a is the radius of the particle and v is the terminal velocity. Equating F to the effective weight of the particle gives

$$x = \left[\frac{18\eta v}{(d_S - d_L)g} \right]^{1/2} \tag{3.21}$$

where x is the diameter of the sphere, g is the acceleration due to gravity, and d_S and d_L are the densities of the particle and the liquid, respectively. Equation 3.21 is normally referred to as Stokes equation, which is not to be confused with Stokes law, Equation 3.20. Measurement of the sedimentation rate can therefore be used to determine the sphere diameter from Stokes equation. When particle shapes other than spheres are tested, the measured particle size is referred to as the Stokes diameter, x_{STK}, or the equivalent spherical diameter.

Determination of particle size from Equation 3.21 has a limited range of validity. Stokes law is valid for laminar or streamline flow (i.e., in which there is no turbulent flow) and assumes that there are no collisions or interactions between the particles. The transition from laminar to turbulent flow occurs at some critical velocity, v_c, given by:

$$v_c = \frac{N_R \eta}{d_L x} \tag{3.22}$$

where N_R is a dimensionless number called the Reynolds' number. The transition from laminar to turbulent flow occurs when its value is ~0.2. Therefore:

$$\frac{v d_L x}{\eta} < 0.2 \qquad \text{laminar flow} \tag{3.23a}$$

$$\frac{v d_L x}{\eta} > 0.2 \qquad \text{turbulent flow} \tag{3.23b}$$

The maximum particle size for laminar flow can be found by substituting for v from Equation 3.21 into Equation 3.23. For Al_2O_3 particles dispersed in water at room temperature, $x_{max} \approx 100$ μm.

For sufficiently small particles undergoing sedimentation under gravity, Brownian motion resulting from collisions with the molecules of the liquid may displace the particle by a measurable amount. This effect puts a lower limit to the use of gravitational settling in water at ~1 μm. Faster settling can be achieved by centrifuging the suspension, thereby extending the size range down to ~0.1 μm.

In the sedimentation method, the powder may be introduced as a thin layer on top of a column of clear liquid (sometimes referred to as a two-layer or line-start technique) or it may be uniformly dispersed in the liquid (the homogeneous suspension technique). The particle size distribution is determined by measuring the change in concentration (or density) of the suspension as a function of time, height along the suspension, or both. A light beam or a collimated x-ray beam is projected at a known height through a glass cell containing the suspension and the intensity of the transmitted beam is measured by a photocell or an x-ray detector located at the opposite side. According to Equation 3.21, particles with a larger velocity will settle first, followed by successively smaller particles. The intensity, I, of the transmitted beam is given by

$$I = I_o \exp(-KACy) \tag{3.24}$$

where I_o is the intensity of the incident beam (measured as a baseline signal), K is a constant called the extinction coefficient, A is the projected area per unit mass of particles, C is the concentration by mass of the particles, and y is the length of the light path through the suspension. By determining the concentration of particles at various depths from the measured intensity ratio, I/I_o, the settling velocities can be found, from which particle size can be determined using Equation 3.21. The particle size distribution, such as the cumulative mass percent finer, CMPF, is plotted vs. the Stokes diameter.

3.2.4.4 Electrical Sensing Zone Techniques (the Coulter Counter)

In the Coulter counter, the number and size of particles suspended in an electrolyte are measured by causing them to flow through a narrow orifice on either side of which is immersed an electrode (Figure 3.6). As a particle passes through the orifice, it displaces an equivalent volume of the electrolyte and causes a change in the electrical resistance, the magnitude of which is proportional to the volume of the particle. The changes in resistance are converted to voltage pulses, which are

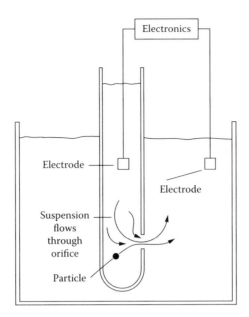

Electronics

Electrode

Electrode

Suspension
flows
through
orifice

Particle

FIGURE 3.6 Electrical sensing zone apparatus (Coulter counter).

amplified, sized, and counted to produce data for the particle size distribution of the suspended particles. As the number and volume of the particles are measured in this technique, the particle size distribution will consist of the CNPF (or CNPL) vs. the volume diameter, x_V.

The effect of particle shape on the data is subject to some doubt. Although it has been claimed that particle shape, roughness and the nature of the material have little effect on the analysis, there is considerable evidence that the measured size parameter corresponds to the overall envelope of the particle [3]. The technique is suitable for dense particles with sizes in the range of 1 to 100 µm but with care and the use of multiple orifices, the range can be extended to 0.5 to 400 µm. It is most sensitive when the particle size is close to the diameter of the orifice. However, blocking of the orifice by larger particles can be a tedious problem. Originally applied to the counting of blood cells, the technique has become fairly popular for particle size analysis of ceramic powders.

3.2.4.5 Light Scattering

A light beam with an intensity I_o entering one end of a column of smoke will emerge from the other end with an intensity I that is less than I_o. However, most of the decrease in the intensity of I in this case is not due to absorption by the smoke particles (which represents an actual disappearance of light) but to the fact that some light is scattered to one side by the smoke particles and thus removed from the direct beam. A considerable intensity, I_s, of scattered light may easily be detected by observing the column from the side in a darkened room.

For a system of fine particles dispersed in a liquid or gas, measurement of I_s provides a versatile and powerful technique for determining particle size data. Large particles tend to reflect light back toward the source in much the same way as this page "backscatters" light toward the reader's eyes. As the particle size is decreased toward the wavelength of the light, there is a greater tendency for the light to be scattered in the forward direction, i.e., at small angles to the direction of the incident beam (Figure 3.7). Many instruments are based on measuring I_s of this forward-scattered light as a function of the scattering angle.

The dependence of I_s on the size of the particles is governed by diffraction theory. Because of the complexity of the general theory, three limiting cases, stated in terms of the ratio of the particle size x to the wavelength λ of the incident light are commonly considered:

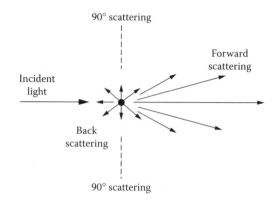

FIGURE 3.7 Illustration of forward scattering of light by small particles.

1. $x \ll \lambda$ Rayleigh scattering theory
2. $x \approx \lambda$ Mie theory
3. $x \gg \lambda$ Fraunhofer diffraction theory

A fourth case, the Rayleigh–Gans theory, is applicable when the x is not significantly smaller than λ. The Rayleigh theory shows that I_s is proportional to the square of the volume of the particle. For larger particles $(x \gg \lambda)$, Fraunhofer diffraction theory applies and the distribution of I_s is similar to that for diffraction from a single slit, where the size of the scattering particle takes the place of the slit width. In the Fraunhofer theory, I_s is proportional to the square to the slit width, so for scattering from particles, I_s is proportional to the projected area of the particle. Based on diffraction theory, the scattering angle, θ, is inversely proportional to the particle size and is given by:

$$\sin \theta \approx \frac{1.22\lambda}{x} \tag{3.25}$$

Smaller particles scatter a small amount of light but according to Equation 3.25, they scatter the light through a larger angle.

With the ready availability of lasers and the reduced cost of computers for data analysis, light scattering has become an important technique for particle size analysis. Instruments consist of a laser source, such as a low power He–Ne laser emitting light with a wavelength of 0.63 μm, a sample cell into which the dry powder or, more commonly, a dilute suspension is placed, optical detectors to convert the scattered light incident upon them into electrical signals, and instrumentation for processing and analyzing the signals. Instruments based on the Fraunhofer diffraction model are very useful because the theory does not involve the refractive index of the particles. The time taken for the analysis is on the order of minutes. A reliable size range for this technique is ~2 to 100 μm but the use of special light collection systems and the Mie theory can extend the range to ~0.1 to 100 μm. Solutions to the Mie, Rayleigh, and Rayleigh–Gans theories include the complex refractive index of the particles, so accurate knowledge of the refractive index is important for these models.

Particles smaller than ~1 μm display Brownian motion when dispersed in a liquid due to collisions with the molecules of the liquid. The motion increases with temperature (increase in momentum of the colliding molecules) and with a decrease in particle size (greater probability of the collision on one side of the particle not matching that on the other size and the enhanced response of a smaller particle to the unbalanced collision). Brownian motion leads to scattering of the incident light and to fluctuations in the average intensity. A light detector (often arranged at right angles to the incident light) records these fluctuations, and the data are used to determine particle size data based on the Stokes–Einstein equation:

$$a = \frac{kT}{3\pi\eta D} \qquad (3.26)$$

where a is the radius of the particle (assumed spherical), k is the Boltzmann constant, T is the absolute temperature, η is the viscosity of the liquid, and D is the diffusion coefficient. This technique, sometimes referred to as dynamic light scattering or photo correlation spectroscopy, can be used for particle size measurements in the range of ~5 nm to ~1 μm. It is expensive compared to the other techniques described so far and is more likely to be encountered in specialized research laboratories.

3.2.4.6 X-Ray Line Broadening

The broadening of x-ray diffraction peaks provides a convenient method for measuring particle sizes below ~0.1 μm [6–8]. As the crystal size decreases, the width of the diffraction peak (or the size of the diffraction spot) increases. An approximate expression for the peak broadening is given by the Scherrer equation:

$$x = \frac{C\lambda}{\beta\cos\theta} \qquad (3.27)$$

where x is the crystal size, C is a numerical constant (≈ 0.9), λ is the wavelength of the x-rays, β is the effective broadening taken as a full width at half maximum (FWHM) in radians, and θ is the diffraction angle for the peak. The effective broadening β is the measured broadening corrected for the broadening due to (1) the instrument and (2) residual stress in the material. Broadening due to the instrument is most accurately determined by using a single crystal with a calibrated broadening. Residual stress arises from cooling or heating the sample, which can trap compressive or tensile stresses in the material due to inhomogeneous or anisotropic thermal expansion. Depending on the magnitude and sign of the stresses, a broadening of the peaks occurs which increases as tan θ. As the broadening due to crystal size (Equation 3.27) varies as $1/(\cos\theta)$, the broadening due to residual stress can be separated. Software available with most modern diffractometers allows for fairly rapid and accurate determination of the crystal size by the line broadening technique.

The reader may have noticed that we have used the term *crystal size* to describe the size obtained by x-ray line broadening. This is because the technique gives a measure of the size of the crystals regardless of whether the particles consist of single crystals, polycrystals, or agglomerates. If the primary particles consist of single crystals, then the crystal size determined by x-ray line broadening will be comparable to the particle size determined by other methods such as electron microscopy. For polycrystalline primary particles or agglomerates, it will be much smaller than the size determined by other methods.

3.2.5 Surface Area

The surface area of powders is important in its own right but can also be used to determine the average particle size when certain assumptions are made concerning the particle shape and the presence or absence of pores. Techniques for measuring the surface area are based on the phenomenon of gas adsorption [3,9]. The term *adsorption* in the present context refers to condensation of gases (adsorbate) on the free surfaces of the solid (adsorbent) and should be distinguished from *absorption* in which the gas molecules penetrate into the mass of the absorbing solid. Adsorption is commonly divided into two categories:

1. Physical adsorption in which the adsorption is brought about by physical forces between the solid and the gas molecules (similar in nature to the weak van der Waals forces that bring about condensation of a vapor to a liquid)
2. Chemical adsorption or chemisorption in which the adsorbed gases will have developed strong chemical bonds with the surface

We are concerned here with physical adsorption, the type of adsorption that forms the basis for the common surface area techniques.

In the determination of the surface area, the amount of gas adsorbed by a fixed mass of solid at a fixed temperature is measured for different gas pressures, p. Commonly, a known volume of gas is contacted with the powder, and the amount of gas adsorbed is determined from the fall in gas pressure by application of the gas laws. A graph of amount of gas adsorbed vs. p (or p/p_o, if the gas is at a pressure below its saturation vapor pressure, p_o), referred to as the *adsorption isotherm*, is plotted. In the literature, there are thousands of measured adsorption isotherms for a wide variety of solids. The majority can be conveniently grouped into five classes (Figure 3.8) — types I to V of the original classification proposed by Brunauer, Deming, Deming, and Teller [10], referred to as the *Brunauer, Emmett, and Teller* (BET) *classification* [11] or simply the *Brunauer classification* [12]. The type VI isotherm, referred to as the *stepped isotherm*, is relatively rare but is of particular theoretical interest. The adsorption of gases by nonporous solids, in the vast majority of cases, gives rise to a type II isotherm. From this isotherm, it is possible to determine a value for the monolayer capacity of the solid, defined as the amount of gas required to form a monolayer coverage on the surface of unit mass of the solid. From this monolayer capacity, the specific surface area of the solid can be calculated.

A kinetic model for adsorption was put forward almost a century ago by Langmuir [13] in which the surface of the solid was assumed to be an array of adsorption sites. He postulated a state of dynamic equilibrium in which the rate of condensation of gas molecules on the unoccupied surface sites is equal to the rate of evaporation of gas molecules from the occupied sites. The analysis leads to the equation:

$$\frac{V}{V_m} = \frac{bp}{1+bp}$$ (3.28)

where V is the volume of gas adsorbed per unit mass of solid at a gas pressure p, V_m is the volume of gas required to form one monolayer, and b is an empirical constant. Equation 3.28 is the familiar Langmuir adsorption equation for the case when adsorption is confined to a monolayer. It is more appropriate to type I isotherms.

By adopting the Langmuir mechanism and introducing a number of simplifying assumptions, Brunauer, Emmett, and Teller [11] were able to derive an equation to describe multilayer adsorption (type II isotherm) that forms the basis for surface area measurement. The equation, commonly referred to as the BET equation, can be expressed in the form:

$$\frac{p/p_o}{V(1-p/p_o)} = \frac{1}{V_m c} + \frac{(c-1)}{V_m c}\frac{p}{p_o}$$ (3.29)

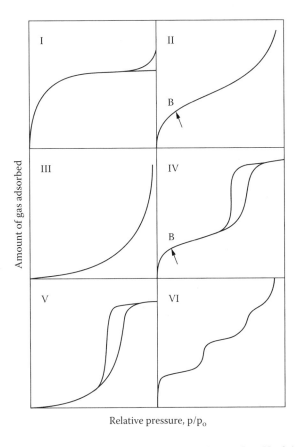

FIGURE 3.8 The five types of adsorption isotherms consisting of types I to V of the original classification proposed by Brunauer, Deming, Deming, and Teller [10], together with type VI, the stepped isotherm.

where c is a constant, and the other terms are as defined earlier. When applied to experimental data, a plot of the left-hand side of Equation 3.29 vs. p/p_o should yield a straight line with a slope s and an intercept i given by

$$s = \frac{c-1}{V_m c} \quad\quad\quad (3.30a)$$

$$i = \frac{1}{V_m c} \quad\quad\quad (3.30b)$$

The monolayer volume, V_m, can, therefore, be obtained from

$$V_m = \frac{1}{s+i} \quad\quad\quad (3.31)$$

We can now use V_m to calculate the specific surface area of the solid from:

$$S_w = \frac{N_A \sigma V_m}{V_o} \qquad (3.32)$$

where the symbols, expressed in their most commonly used units, are defined as follows:

S_w = the specific surface area (m²/g)

N_A = the Avogadro number (6.023×10^{23} mol⁻¹)

σ = area of an adsorbed gas molecule (16.2×10^{-20} m² for nitrogen)

V_m = monolayer volume (in cm³/g)

V_o = the volume of 1 mole of the gas at STP (22,400 cm³/mol)

Substituting these values in Equation 3.27 gives, for nitrogen adsorption,

$$S_w = 4.35 V_m \qquad (3.33)$$

Assuming that the powder is unagglomerated and that the particles are spherical and dense, the particle size can be estimated from the equation

$$x = \frac{6}{S_w d_S} \qquad (3.34)$$

where d_s is the density of the solid.

The validity of the BET equation is commonly stated to extend over the range of relative pressures (p/p_o values) of ~0.05 and ~0.3, but there are numerous examples in which the BET plot departs from linearity at relative pressures below ~0.2. Although several gases have been used as the adsorbate, the standard technique is the adsorption of nitrogen at the boiling point of liquid nitrogen (77K). The molecular area, $\sigma = 16.2 \times 10^{-20}$ m² (16.2 Å²) has gained widespread acceptance for the determination of surface area from type II isotherms. Argon ($\sigma = 16.6 \times 10^{-20}$ m²) is a good alternative because it is chemically inert, is composed of spherically symmetrical, monatomic molecules, and its adsorption is relatively easy to measure at the boiling point of liquid nitrogen. Nitrogen or argon works well for powders with a specific surface area greater than ~1 m²/g. For powders with somewhat lower surface areas, krypton ($\sigma = 19.5 \times 10^{-20}$ m²) at the boiling point of liquid nitrogen may be used.

3.2.6 Porosity of Particles

Depending on the method used to synthesize them, many powders may consist of agglomerates of primary particles which are highly porous. Commonly, it is necessary to characterize, quantitatively, the porosity and pore size distribution of the agglomerates. For pores that are accessible (i.e., pores that are not completely isolated from the external surface), two methods have been used: gas adsorption (sometimes referred to as *capillary condensation*) and mercury intrusion porosimetry (referred to simply as *mercury porosimetry*). The characteristic pore feature of interest is the size, taken as the diameter, radius, or width. A classification of pores according to their size is given in Table 3.6. Although there is considerable overlap in the range of applicability, gas condensation is applicable to pore size measurement in the mesopore range, while mercury porosimetry is better applied to the macropore range.

3.2.6.1 Gas Adsorption

At lower gas pressures, ($p/p_o < $ ~0.3), the adsorbed gas produces multilayer coverage of the solid surface and, as described earlier, this forms the basis for the BET method. At higher pressures, the

TABLE 3.6
Classification of Pores According to Their Size (Diameter or Width)

Category	Pore Size (nm)
Micropores	<2
Mesopores	2–50
Macropores	>50

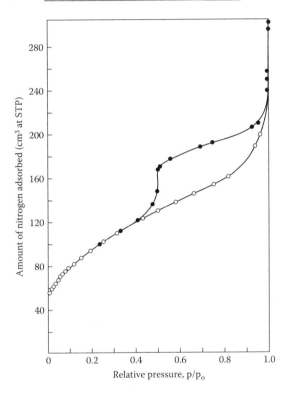

FIGURE 3.9 Nitrogen adsorption data (open circles) and desorption data (filled circles) used in the determination of pore size distribution. (From Allen, T., *Particle Size Measurement*, 4th ed., Chapman and Hall, London, 1990. With permission.)

gas may condense to a liquid in the capillaries of a porous solid, and this condensation can be used to determine the pore size and pore size distribution. When condensation occurs, the isotherm Figure 3.9 is commonly found to possess a hysteresis loop between the adsorption and desorption branches, which is characteristic of type IV isotherms.

Assuming cylindrical capillaries, the relative gas pressure in the capillaries when condensation occurs is governed by the Kelvin equation:

$$\ln \frac{p}{p_o} = \frac{-2\gamma_{LV} V_L \cos\theta}{RTr} \tag{3.35}$$

where p is the gas pressure over a meniscus with radius r, p_o is the saturation gas pressure of the liquid having a plane surface, γ_{LV} is the surface tension of the liquid–vapor interface, V_L is the molar volume of the liquid, θ is the contact angle between the liquid and the pore wall, R is the gas constant, and T is the absolute temperature. From the Kelvin equation it follows that the vapor pressure p over a concave meniscus must be less than p_o. Consequently, capillary condensation of a vapor to a liquid should occur within a pore at some pressure p determined by the radius r for the pore and less than p_o.

The determination of the pore size distribution by application of the Kelvin equation to the capillary condensation part of the type IV isotherms is almost entirely restricted to the use of nitrogen as the adsorbing gas. This is largely a reflection of the widespread use of nitrogen for surface area measurements, so both surface area and pore size distribution can be obtained from the same isotherm. If the volume of gas adsorbed on the external surface of the solid is small compared to that adsorbed in the pores, then the volume of gas adsorbed, V_g, when converted to a liquid (condensed) volume, V_c, gives the pore volume. The relationship between V_g and V_c is

$$V_c = \frac{M_w V_g}{d_L V_o} \tag{3.36}$$

where M_w is the molecular weight of the adsorbate, d_L is the density of the liquefied gas at its saturated vapor pressure, and V_o is the molar gas volume at STP (22.4 l). For nitrogen: $M_w = 28$ g and $d_L = 0.808$ g/cm^3, so that

$$V_c = 1.547 x 10^{-3} V_g \tag{3.37}$$

For nitrogen adsorption at boiling point of liquid nitrogen (77K), $\gamma_{LV} = 8.72 \times 10^{-3}$ N/m, $V_L = 34.68 \times 10^{-6}$ m^3, and θ is assumed to be zero, so Equation 3.35 becomes

$$r = \frac{-0.41}{\log(p/p_o)} nm \tag{3.38}$$

Typical values for p/p_o are in the range of 0.5 to 0.95, so the gas adsorption is appropriate to pore radius values in the range of ~1 to 20 nm.

For any relative pressure p_i/p_o on the isotherm, V_{ci} gives the cumulative volume of the pores having pore radius values up to and equal to r_i, where V_{ci} and r_i are determined from Equation 3.37 and Equation 3.38, respectively. The pore size distribution curve, $v(r)$, vs. r is obtained by differentiating the cumulative pore volume curve with respect to r, i.e., $v(r) = dV_c/dr$. This analysis does not take into account that when capillary condensation occurs, the pore walls are already covered with an adsorbed layer of gas. Thus capillary condensation occurs not directly in the pore itself but in the inner core. The calculated values for the pore size are, therefore, appropriate to the inner core.

A characteristic feature of type IV isotherms, as outlined earlier, is its hysteresis loop. The exact shape of the loop varies from one adsorption system to another, but as seen from Figure 3.9, at any given relative pressure, the amount adsorbed is always greater along the desorption branch than along the adsorption branch. Ink bottle pores, consisting of a cylindrical pore closed at one end and with a narrow neck at the other end (Figure 3.10a), have played a prominent role in the explanation of the hysteresis. It is easily shown that the adsorption branch of the loop corresponds to values of the pore radius of the body and the desorption branch to values of the pore radius of the neck. The representation of a pore as narrow-necked bottle is overspecialized and in practice a series of interconnected pore spaces (Figure 3.10b) rather than discrete bottles is more likely.

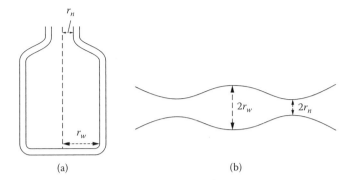

FIGURE 3.10 (a) Ink bottle pores associated with hysteresis in capillary condensation, (b) more practical interconnected pore spaces.

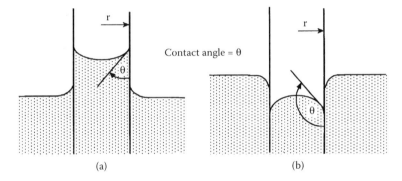

FIGURE 3.11 Capillary rise phenomena for (a) a wetting liquid (contact angle < 90°) and (b) a nonwetting liquid (contact angle > 90°).

3.2.6.2 Mercury Porosimetry

In ceramics, mercury porosimetry is more widely used than gas adsorption because pore sizes in the macropore range are more common. The technique is based on the phenomenon of capillary rise (Figure 3.11). A liquid that wets the walls of a narrow capillary (contact angle, $\theta < 90°$) will climb up the walls of the capillary. However, the level of a liquid that does not wet the walls of a capillary ($\theta > 90°$) will be depressed. For a nonwetting liquid, a pressure must be applied to force the liquid to flow up the capillary to the level of the reservoir. For a capillary with principal radii of curvature r_1 and r_2 in two orthogonal directions, this pressure is given by the equation of Young and Laplace:

$$p = -\gamma_{LV}\left(\frac{1}{r_1} + \frac{1}{r_2}\right)\cos\theta \tag{3.39}$$

where γ_{LV} is the surface tension of the liquid–vapor interface. For a cylindrical capillary of radius r, Equation 3.39 becomes

$$p = \frac{-2\gamma_{LV}\cos\theta}{r} \tag{3.40}$$

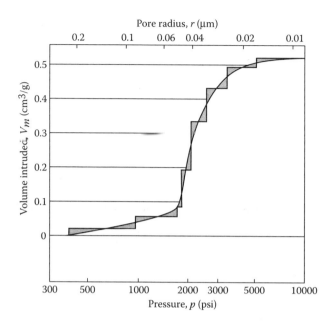

FIGURE 3.12 Mercury porosimetry data showing the intruded volume V_m vs. pressure. (From Allen, T., *Particle Size Measurement*, 4th ed., Chapman and Hall, London, 1990. With permission.)

For mercury, depending on its purity and the type of sample surface, γ_{LV} and θ vary slightly. The average values found in the literature are $\gamma_{LV} = 0.480$ N/m and $\theta = 140°$. Then for p in units of megapascals (MPa) and r in micrometers (μm), Equation 3.40 becomes

$$r = 0.735 / p \qquad (3.41)$$

In mercury porosimetry, a sample holder is partially filled with the powder, evacuated, and then filled with mercury. The volume of mercury intruded into the sample, V_m, is measured as a function of the applied pressure, p. An example of this type of data is shown in Figure 3.12. Assuming that the pores are cylindrical, the value V_{mi} at any value of the applied pressure p_i gives the cumulative volume of all pores having a radius equal to or greater than r_i. This designation is the converse of that in gas adsorption in which the cumulative pore volume is the volume of the pores having a radius less than or equal to r_i. Thus, in mercury porosimetry, the cumulative pore volume decreases as r increases (Figure 3.13), whereas in gas adsorption the cumulative pore volume increases as r increases. In both techniques, however, the pore size distribution $v(r)$ is obtained by differentiating the cumulative pore volume curve with respect to r. The pore size distribution can also be obtained directly from the data for V_m vs. p (Figure 3.12), using

$$v(r) = \frac{dV_m}{dp} \frac{p}{r} \qquad (3.42)$$

The pressures available in mercury porosimeters are such that pore sizes in the range of ~5 nm to 200 μm can be measured but the technique becomes increasingly uncertain as the lower end of the range is approached. It should be remembered that Equation 3.40 assumes that the pores are circular in cross section, which is hardly the case in practice. Furthermore, for ink bottle pores (Figure 3.10) or pores with constricted necks opening to large volumes, the use of Equation 3.40 gives a measure of the neck size, which is not truly indicative of the actual pore size. These types

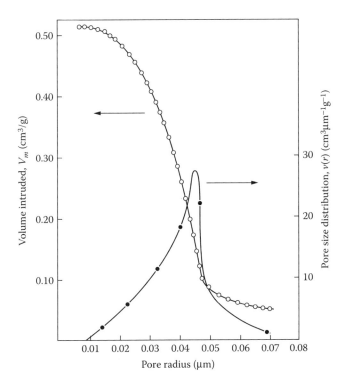

FIGURE 3.13 Mercury porosimetry data of Figure 3.12 replotted to show the intruded volume V_m (open circles) and the pore size distribution $v(r)$ (filled circles) vs. pore radius r. (From Allen, T., *Particle Size Measurement*, 4th ed., Chapman and Hall, London, 1990. With permission.)

of pores can also give rise to hysteresis because they fill at a pressure characteristic of the neck size, whereas they empty at a pressure characteristic of the larger volume of the pore. Finally, the compressibilities of the mercury and of the porous sample are assumed to be independent of the applied pressure, which may not be the case at the lower end of the pore size range involving fairly high pressures.

3.2.6.3 Pycnometry

Pycnometry can be used to determine the isolated porosity in particles. In practice, the apparent density of the particles, d_a, defined as

$$d_a = \frac{\text{mass of solid}}{\text{volume of solid} + \text{volume of isolated pores}} \tag{3.43}$$

is measured. If d_t is the theoretical density of the particles (i.e., the density of the pore-free solid), the amount of isolated porosity is equal to $1 - d_a/d_t$. Commonly, d_t is obtained from a handbook of physical and chemical data. For crystalline materials, it can also be obtained from the crystal structure and the dimensions of the unit cell determined by x-ray diffraction.

For coarser powders (particle size > ~10 μm), a liquid pycnometer is used. A calibrated bottle is weighed (mass m_o) and the powder is added (total mass m_1). A liquid of known density, d_L, is then added (total mass m_2). In a separate run, the pycnometer bottle containing the liquid only (i.e., no powder) is weighed (mass m₃). The apparent density of the particles, d_a, is found from:

$$d_a = \frac{m_1 - m_o}{(m_3 - m_o) - (m_2 - m_1)}(d_L - d_{air}) + d_{air} \tag{3.44}$$

where d_{air} is the density of air. Care must be taken to achieve good wetting of the particle surfaces by the liquid and to remove trapped air (by boiling the liquid).

Helium gas pycnometry is commonly used for powders finer than ~10 μm. The small size of the helium atom enables it to penetrate into very fine pores. The volume occupied by the solid is measured from the volume of gas displaced. The apparent density is then calculated from the mass of the powder used and its measured volume.

3.3 CHEMICAL COMPOSITION

The major, minor, and trace elements present in a powder can all have a significant influence on the subsequent processing and microstructural development. Changes in the concentration of the major elements may result from different powder synthesis methods or from changes in the synthesis conditions. Small concentrations of dopants (~0.1 to 10 at%) are commonly added to improve sintering and properties. Trace impurity elements at concentrations less than a few hundred parts per million are invariably present even in the cleanest powders.

Several techniques are available for the quantitative analysis of chemical composition [14]. For ceramic powders, by far the most widely used technique is optical atomic spectroscopy involving atomic absorption (AA) or atomic emission (AE). High precision of analysis coupled with low detection limits makes the technique very valuable for the determination of major and minor elements as well as for the analysis of trace elements. X-ray fluorescence spectroscopy (XRF) can also be used for the analysis of major, minor, and trace elements, but the sensitivity decreases dramatically for low atomic number elements. XRF cannot, therefore, be used for quantification of elements with atomic number $Z < 9$. Modern AA or AE instruments are simpler to use, less expensive, have similar precision of analysis and have better sensitivity than XRF. They are suited to the analysis of solutions, so the method requires complete dissolution of the powder in a liquid. In comparison, XRF is ideally suited to the analysis of solid samples, and this can be a distinct advantage for ceramic powders that are commonly difficult to dissolve. Table 3.7 includes a summary of the main features of optical atomic spectroscopy and XRF.

3.3.1 OPTICAL ATOMIC SPECTROSCOPY: ATOMIC ABSORPTION AND ATOMIC EMISSION

Optical atomic spectroscopy is based on transitions of electrons between the outer energy levels in atoms. For a multielectron atom, there are many energy levels in the ground state of an atom (i.e., the state of lowest energy). There are also a large number of unoccupied energy levels into which electrons may be excited. In an excited state of an atom, some of the electrons go to the higher energy levels. In the simplest case, we shall consider two energy levels only, one in the ground state of the atom with energy, E_g, and one in the excited state, with energy E_e (Figure 3.14). The frequency of radiation absorbed or emitted during the transition is given by Planck's law,

$$\nu = \frac{E_e - E_g}{h} \tag{3.45}$$

where h is Planck's constant. The transitions that are useful in optical atomic spectroscopy involve fairly low-energy transitions between outer energy levels such that the wavelength of the radiation given by Equation 3.45 occurs in range of ~10 nm to 400 μm.

TABLE 3.7
Common Techniques for the Chemical Analysis of Ceramic Powders

Feature	Flame AA	ICP-AE	Furnace AA	XRF	XRD
Range	Metals and metalloid elements	Metals and metalloid elements	Metals and metalloid elements	$Z > 8$	Crystalline materials
Common state of specimen	Liquid	Liquid	Liquid	Solid	Solid
Detection limit[a] (ppm)	10^{-3}–10^{-1}	10^{-3}–10^{-1}	10^{-5}–10^{-2}	>10	10^{3}–10^{4}
Precision[b] (%)	0.2–1	0.5–2	1–2	0.1–0.5	0.2–0.5
Accuracy[c] (%)	1–2	2–5	5–10	0.1–1	0.5–5

Note: AA = atomic absorption; ICP-AE = inductively coupled plasma atomic emission; XRF = x-ray fluorescence; XRD = x-ray diffraction; Z = atomic number.

[a] Minimum concentration or mass of element that can be detected at a known confidence level.

[b] Relative standard deviation.

[c] Relative error.

Source: From Skoog, D.A., Holler, F.J., and Nieman, T.A., *Principles of Instrumental Analysis*, 5th ed., Harcourt Brace, Philadelphia, PA, 1998; Winefordner, J.D. and Epstein, M.S., *Physical Methods of Chemistry: Vol. IIIA, Determination of Chemical Composition and Molecular Structure*, John Wiley & Sons, New York, 1987, p. 193; Slavin, W., Flames, furnaces, plasmas: how do we choose? *Anal. Chem.*, 58, 589A, 1986; Jenkins, R., *X-Ray Fluorescence Spectrometry*, 2nd ed., John Wiley & Sons, New York, 1999.

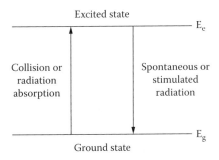

FIGURE 3.14 Schematic showing the transitions in optical atomic spectroscopy.

For atoms at thermal equilibrium at a temperature, T, the number of atoms in the excited state, n_e, is related to the number in the ground state, n_g, by the Boltzmann equation,

$$\frac{n_e}{n_g} = \exp\left[\frac{-(E_g - E_e)}{kT}\right]$$ (3.46)

where k is the Boltzmann constant. The number of atoms in the ground state will be less than the total number n due to the excitation of n_e atoms. However, unless the temperature is very high, n_g will be approximately equal to n. From Equation 3.46, the number of excited atoms, n_e, will be proportional to n. The intensity of the absorbed or emitted radiation, which is proportional to n_e, will therefore be proportional to the total number of atoms, n. As each element has its own characteristic energy levels, the elemental composition can be identified from the wavelength (frequency) of the absorbed or emitted radiation whereas the intensity of the radiation can be used to determine its concentration.

The relationship between intensity of the absorbed or emitted radiation and the concentration of the atoms can be rigorously derived [15], but it is rarely used practically in quantitative analysis. Instead, standard samples with known concentrations are made up, and the intensities are determined for these standards. The concentration of the unknown sample, C_u, is determined from a calibration curve or more simply from the relation

$$C_u = C_s \frac{I_u}{I_s} \tag{3.47}$$

where C_s is the concentration of a standard, and I_u and I_s are the measured intensities of the unknown and standard samples, respectively.

For the analysis of ceramic powders by optical atomic spectroscopy, a portion of the powder has to be converted into individual atoms. In practice, this is achieved by dissolving the powder in a liquid to form a solution, which is then broken up into fine droplets and vaporized into individual atoms by heating. The precision and accuracy of optical atomic spectroscopy are critically dependent on this step. Vaporization is most commonly achieved by introducing droplets into a flame (referred to as flame atomic absorption spectrometry or flame AA). Problems with flame AA include incomplete dissociation of the more refractory elements (e.g., B, V, Ta, and W) in the flame and difficulties in determining elements that have resonance lines in the far ultraviolet region (e.g., P, S, and the halogens). Whereas flame AA is rapid, the instruments are rarely automated to permit simultaneous analysis of several elements. Several types of plasmas have been considered for vaporizing the solution but by far the most widely used is the inductively coupled plasma (ICP). Elements that are only partially dissociated in the flame are usually completely dissociated by the higher temperatures achieved by ICP. A valuable feature of ICP is the capability for simultaneous analysis of several elements. Vaporization in a graphite furnace (furnace AA or electrothermal AA) provides better detection limits than flame AA or ICP for the analysis of trace elements but the analysis is slow. Commonly, it is used only when flame AA or ICP provides inadequate detection limits. Figure 3.15 gives a comparison of the detection limits for flame AA, ICP-AE, and furnace AA for the analysis of trace elements [16].

3.3.2 X-Ray Fluorescence Spectroscopy

In x-ray fluorescence spectroscopy, elemental composition is quantitatively or qualitatively determined by measuring the wavelength and intensity of the electron transitions from the outer to the inner energy levels of the atom [17,18]. The energy associated with these transitions (~0.6 to 60 keV) is significantly greater than that in optical atomic spectroscopy (~a few electron volts). The emitted x-ray spectrum does not depend on the way in which the atomic excitation is produced, but in XRF a beam of energetic x-rays is used.

When a primary beam of monochromatic x-rays falls on a specimen, the x-rays may be absorbed or scattered. Coherent scattering of x-rays from an ordered arrangement of scattering centers (such as the environment present in crystals) leads to diffraction. The use of x-ray diffraction for characterizing powders is described in the next section. The absorption of x-rays leads to electronically excited atoms (ions) due to the ejection of electrons (i.e., the photoelectric effect) or to the transition of electrons to higher energy levels Figure 3.16a. (The ejected electrons, as described later, form the basis for the surface characterization technique of x-ray photoelectron spectroscopy.) As the excited atoms readjust to their ground state, two processes may occur. The first involves a rearrangement of the electrons which leads to the ejection of other electrons (referred to as *Auger electrons*) from higher energy levels Figure 3.16b. The second process leads to the emission of x-rays when the electrons undergo transitions from the outer levels to fill the vacant electron sites in the inner energy levels Figure 3.16c. This secondary beam of x-rays emitted from the specimen

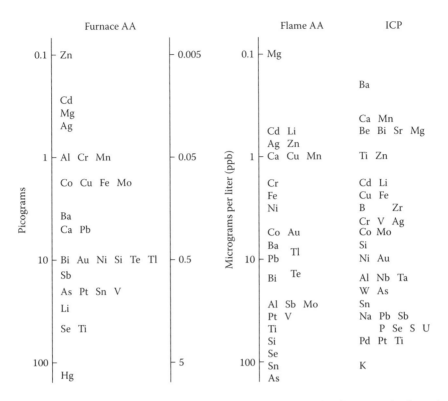

FIGURE 3.15 Comparison of detection limits for trace element analysis by flame atomic absorption, ICP-atomic emission and furnace atomic absorption, plotted on a logarithmic scale of concentration in micrograms per liter (parts per billion). Because furnace detection limits are inherently in units of mass (picograms), they have been converted to concentration by assuming a 20 μl sample. (From Slavin, W., Flames, furnaces, plasmas: how do we choose? *Anal. Chem.*, 58, 589A, 1986. With permission.)

forms the basis of XRF. It is characteristic of the elements in the specimen and can therefore be used to identify the elements. The relationship between the wavelength, λ, of the emitted x-rays, and the atomic number Z of the element was first established by Moseley (1913) and is given by:

$$\frac{1}{\lambda} = A(Z - \alpha)^2 \tag{3.48}$$

where A is a constant for each series of spectral lines and $\alpha \approx 1$.

In XRF, the polychromatic beam of radiation emitted from the specimen is diffracted through a single crystal to isolate narrow wavelength bands (wavelength dispersive spectroscopy) or analyzed with a proportional detector to isolate narrow energy bands (energy dispersive spectroscopy). Because the relationship between wavelength and atomic number is known (Equation 3.48), isolation of individual characteristic lines allows the identification of the element. The elemental concentrations are found from the intensities of the spectral lines (e.g., by comparing with the intensities of standards).

3.4 CRYSTAL STRUCTURE AND PHASE COMPOSITION

Coherent scattering of x-rays from crystalline materials leads to diffraction. Since its discovery in 1912 by von Laue, x-ray diffraction (XRD) has provided a wealth of information about the structure

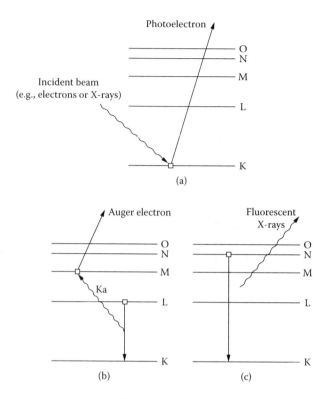

FIGURE 3.16 Interaction of incident beam (X-rays or electrons) with a solid, producing atomic excitation with the emission of a photoelectron (a), followed by de-excitation with the emission of an Auger electron (b) or by the emission of characteristic X-ray photons (c).

and chemical composition of crystalline materials. The technique is described in detail in several texts [6–8,19]. For compositional analysis, XRD is unique in that it is the only analytical method capable of providing qualitative and quantitative information about the crystalline *compounds* (or phases) present in a solid. For example, the technique can determine the percentage of ZnO and Al_2O_3 present in a mixture of the two compounds, whereas other analytical techniques can only give the percentage of the elements Zn, Al, and O in the mixture.

The requirements for x-ray diffraction are (1) the atomic spacing in the solid must be comparable with the wavelength of the x-rays and (2) the scattering centers must be spatially distributed in an ordered way (e.g, the environment present in crystals). The diffraction of x-rays by crystals (Figure 3.17) was treated by Bragg in 1912. The condition for constructive interference, giving rise to intense diffraction maxima, is known as Bragg's law:

$$2d \sin \theta = n\lambda \tag{3.49}$$

where d is the spacing between the lattice planes of the crystal, θ is the angle of incidence (or reflection), n is an integer (sometimes referred to as the *order of diffraction*), and λ is the wavelength of the x-rays (0.15406 nm for Cu K_α radiation). When the reflection angle θ does not satisfy Equation 3.49, destructive interference occurs.

In recent years, computer automation has greatly reduced the amount of work required for x-ray analysis of structure and composition, so that much of the data analysis is now commonly performed by software and computerized search programs. Structural analysis involves the measurement of the lattice parameters of the crystal (the unit cell dimensions) and the structural model

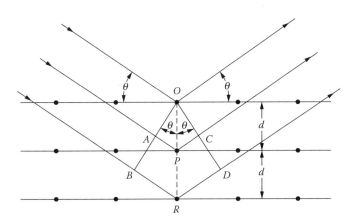

FIGURE 3.17 Diffraction of x-rays by a crystal.

of the crystal (the crystal structure). Whereas the use of single crystals is the preferred method for structural analysis whenever possible, powder diffraction in which the sample consists of a random orientation of a large number of fine particles (polycrystals) can also be used. Determination of the unit cell requires finding a set of unit cell parameters that index all of the observed reflections. Lattice parameter measurements are important for doped powders and solid solutions in which data for compositional effects on the lattice parameters are required (e.g., Y_2O_3-stabilized ZrO_2).

Compositional analysis is based on the fact that an x-ray diffraction pattern is unique for each crystalline material. Thus, if an exact match can be found between the pattern of the unknown material and an authentic sample, chemical identity can be assumed. The International Center for Diffraction Data (ICDD-JCPDS) publishes a database containing powder diffraction patterns for several thousands of materials. Commonly, it is possible to identify an unknown material by searching the ICDD-JCPDS database for a matching pattern.

For a physical mixture, the powder diffraction pattern is the sum of the patterns of the individual materials. The diffraction pattern can therefore be used to identify the crystalline phases in a mixture. The concentrations of the crystalline phases can be determined by methods based on comparing the intensities of the diffraction peaks with standards [6–8]. If the crystal structures of the phases are known, the concentration of each phase can be determined by Rietveld analysis [20,21]. In the Rietveld method, a theoretical diffraction pattern is computed, and the difference between the theoretical and observed patterns is minimized. For quantitative analysis, some care should be taken with specimen preparation if accurate and reliable results are to be obtained. The effects of factors such as preferred orientation, texturing, and particle size broadening must be minimized.

X-ray diffraction, as indicated in Table 3.7, has a detection limit of 0.1 to 1 wt%, so crystalline phases present in concentrations below this limit will not be detected. Furthermore, amorphous phases are not directly measured, but their presence may be quantified by comparing the pattern with a standard that is known to contain no amorphous phase [22].

3.5 SURFACE CHARACTERIZATION

As the particle size of a powder decreases below 1 μm, the surface area of the powder and the volume fraction of the outermost layer of ions on the surface increase quite significantly. We would therefore expect the surface characteristics to play an increasingly important role in the processing of fine powders as the particle size decreases. The surface area (described earlier) and the surface chemistry are the surface characteristics that have the most profound influence on processing. Another characteristic, the surface structure, although having only a limited role in processing, may nevertheless have an important influence on surface phenomena (e.g., vaporization, corrosion, and heterogeneous catalysis).

The characterization of the powder surface area and its role in processing have received considerable attention for decades, but it is only since the 1980s that the importance of surface chemistry in ceramic processing has begun to be recognized. In colloidal processing, the surface chemistry controls the particle–particle interactions in the liquid which, in turn, control the packing homogeneity of the green body. The powder surface chemistry can also have a direct influence on densification and microstructural evolution during sintering, regardless of the powder consolidation method. The surfaces of the particles become interfaces and grain boundaries in the solid which act as sources and sinks for the diffusion of matter. Densification and grain growth processes will therefore be controlled by the structure and composition of the interfaces and grain boundaries. Impurities on the surfaces of the powder, for example, will alter the grain boundary composition and, in turn, the microstructure of the fabricated body. In many ceramics (of which Si_3N_4 is a classic example), controlling the structure and chemistry of the grain boundaries through manipulation of the powder surface chemistry forms one of the most important considerations in processing.

With the enormous and increasing importance of surfaces to modern technological processes, a large number of techniques have been developed to study various surface properties [23–25]. Here we consider only those techniques that have emerged to be most widely applicable to ceramic powders. In general, the surface characterization techniques rely on the interaction of atomic particles (e.g., atoms, ions, neutrons, and electrons) or radiation (e.g., x-rays and ultraviolet rays) with the sample. The interaction produces various emissions that can be used to analyze the sample, as shown in Figure 3.18, for the principal emissions produced by the interaction of an electron beam with a solid. Depending on the thickness of the solid and the energy of the electron beam, a certain fraction of electrons will be scattered in the forward direction, another fraction will be absorbed, and the remaining fraction will be scattered in the backward direction. The forward-scattered electrons consist of electrons that have undergone elastic scattering (i.e., interactions with the atoms of the sample that result in a change in direction but virtually no loss in energy) and electrons that have undergone inelastic scattering (i.e., interactions that result in a change in direction and a reduction in the energy of the incident electrons). The elastically scattered electrons are much greater in number than the inelastically scattered electrons. They are used in the transmission electron microscope to produce diffraction effects in the determination of the structure of the sample. The backscattered electrons are highly energetic electrons that have been scattered in the backward direction. The majority of these will also have undergone elastic collisions with the atoms of the sample.

The incident electron beam can also generate secondary effects in the sample. One of these effects is that the incident electrons can knock electrons out of their orbits around an atom. If the ejected electrons are near the surface of the sample (within ~20 nm) they may have enough energy to escape from the sample and become what are called secondary electrons. These secondary electrons are used in the scanning electron microscope to produce an image. A second type of effect occurs when an electron undergoes a transition from a higher energy level to fill a vacant site in the lower energy level of an excited atom. As we have observed earlier (Figure 3.16), radiation in the form of x-rays or light can be produced in the transition and used to obtain information about the chemical composition of the sample. Some of the x-rays can be absorbed by electrons in the outer orbitals around the atoms. If these electrons are very close to the surface of the sample, they may escape. These electrons, called Auger electrons, can also provide chemical information about the near-surface region.

The depth to which the incident electron beam penetrates into the sample depends on the energy of the beam and the nature of the sample. For electron energies of 1 MeV, the penetration depth is ~1 μm. However, for energies in the range 20 to 1000 eV, the penetration depth is only ~1 nm. An incident electron beam with energy below ~1 keV can therefore be used to probe surface layers

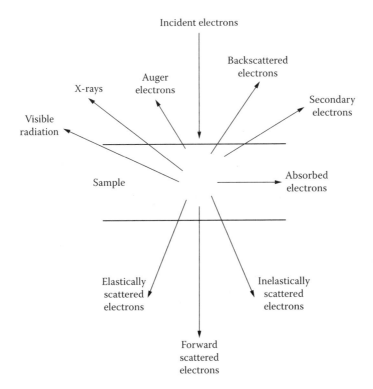

FIGURE 3.18 Emissions produced by the interaction of an electron beam with a solid sample.

one to two monolayers thick. Ions and x-rays can also be used to produce diffraction effects and to eject electrons. They, too, can be used as incident beams to probe the surfaces of materials.

3.5.1 SURFACE STRUCTURE

When viewed on a microscopic or submicroscopic scale, the surface of a particle is not smooth and homogeneous. Various types of irregularities are present. Most ceramic particles are agglomerates of finer primary particles. Features such as porosity and boundaries between the primary particles will be present on the surface. For nonoxide particles, a thin oxidation layer will cover the surfaces of the particles. Even for single-crystal particles, the surface can be fairly complex. Figure 3.19 shows an atomic-scale model of the surface structure of solids which has been con-

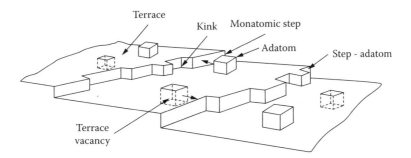

FIGURE 3.19 Model of a heterogeneous solid surface showing different surface sites. (From Somorjai, G.A. and Salmeron, M., Surface characterization of ceramic materials, in *Ceramic Microstructures '76*, Fulrath, R.M. and Pask, J.A., Eds., Westview Press, Boulder, CO, 1977, p. 101.)

structed on the basis of information obtained from a few techniques [26]. The surface has several atomic irregularities that are distinguishable by the number of nearest neighbors (coordination number). The atoms in terraces are surrounded by the largest number of nearest neighbors. Atoms located in steps have fewer neighbors and atoms in kink sites have even fewer. Kink, step, and terrace atoms have large equilibrium concentrations on any real surface. On a rough surface, 10 to 20% of the atoms are in step sites, with ~5% in kink sites. Steps and kinks are also referred to as *line defects*. Atomic vacancies and adatoms, referred to as *point defects*, are also present in most surfaces. Although their concentrations are small (<1% of a monolayer even at temperatures close to the melting point), they are important participants of atomic transport along the surface.

The most widely applicable techniques for characterizing the surface structure of solids include electron diffraction and scanning probe microscopy. The most prominent techniques, low-energy electron diffraction (LEED), scanning tunneling microscopy (STM), and atomic force microscopy (AFM), have limitations for characterizing powder surfaces as they require a fairly flat surface with sufficient coherency over a relatively large area. Readers interested in these techniques are referred to treatments elsewhere on LEED [25,27], STM [28,29], and AFM [29,30].

Modern scanning electron microscopes provide a very simple method for observing the powder structure with a point-to-point resolution of ~2 nm. Finer details of the structure can be observed by TEM. Figure 3.20a shows a high-resolution TEM micrograph of a Si_3N_4 powder having an amorphous layer (3–5 nm thick) on the surface [31]. This surface layer, with the composition of a silicon oxynitride, has a significant effect on the sintering and high-temperature mechanical properties of Si_3N_4. A micrograph of another Si_3N_4 powder observed in a high-voltage TEM (1.5 MeV) shows that the particles are connected by a strong amorphous bridge, presumably a silicon oxynitride (Figure 3.20b). Such hard agglomerates, as we have observed earlier, can have a detrimental effect on the packing of the powder.

3.5.2 SURFACE CHEMISTRY

Electron, ion, and photon emissions from the outermost layers of the surface can be used to provide qualitative or quantitative information about the chemical composition of the surface. The most widely applicable techniques for characterizing the surface chemistry of ceramic powders are Auger electron spectroscopy (AES), x-ray photoelectron spectroscopy (XPS), which is also referred to as electron spectroscopy for chemical analysis (ESCA), and secondary ion mass spectrometry (SIMS). Table 3.8 provides a summary of the main measurement parameters for these three techniques.

All surface analytical techniques (with the exception of Rutherford backscattering) require the use of ultrahigh vacuum environment. The analysis conditions are therefore rather limited and do not correspond to those normally used in ceramic processing in which the powder experiences fairly prolonged exposure to the atmosphere. Although the techniques described here derive their usefulness from being truly surface sensitive, they can also be used to provide information from much deeper layers. This is normally done by sequential (or simultaneous) removal of surface layers by ion beam sputtering and analysis. This mode of analysis in which the composition can be found layer by layer is referred to as *depth profiling*. It is probably one of the most important modes of surface analysis because the composition of the surface is usually different from that of the bulk.

3.5.2.1 Auger Electron Spectroscopy (AES)

Auger electron spectroscopy is based on a two-step process shown schematically in Figure 3.16a and Figure 3.16b. When an electron is emitted from the inner atomic orbital through collision with an incident electron beam, the resulting vacant site is soon filled by another electron from an outer orbital. The energy released in the transition may appear as an x-ray photon (the characteristic x-rays used in electron microprobe techniques for compositional analysis) or may be transferred to another electron in an outer orbital, which is ejected from the atom with a kinetic energy E_K given by

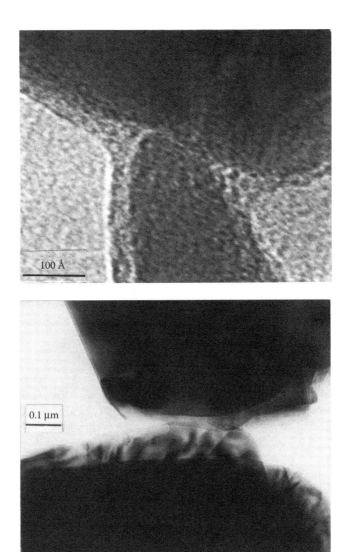

FIGURE 3.20 Surface structure of a Si_3N_4 powder as revealed by (a) high-resolution TEM and (b) high-voltage TEM.

$$E_K = E_1 - E_2 - E_3^*$$ (3.50)

where E_1 and E_2 are the binding energies of the atom in the singly ionized state and E_3^* is the binding energy for the doubly ionized state. The ejected electron (referred to as the Auger electron) moves through the solid and soon loses its energy through inelastic collision with bound electrons. However, if the Auger electron is emitted sufficiently close to the surface, it may escape from the surface and be detected by an electron spectrometer. The number of electrons (or the derivative of the counting rate) is commonly plotted as a function of the kinetic energy of the electron. As each type of atom has its own characteristic electron energy levels, the peaks in the observed Auger spectrum can be used to determine the elemental composition by comparison with standard Auger spectra for the elements.

The technique is fairly rapid and provides high reproducibility of the results. When calibration standards are used, quantitative elemental composition can be obtained to better than ±10%.

TABLE 3.8
Common Methods for the Surface Chemical Analysis of Ceramic Powders

Measurement Parameter	Auger Electron Spectroscopy (AES)	X-ray Photoelectron Spectroscopy (XPS)	Secondary Ion Mass Spectrometry (SIMS)
Incident particle	Electrons (1–20 keV)	X-rays (1254 eV and 1487 eV)	Ions (He$^+$, Ne$^+$, Ar$^+$) (100 eV–30 keV)
Emitted particle	Auger electrons (20–2000 eV)	Photoelectrons (20–2000 eV)	Sputtered ions (~10 eV)
Element range	> Li (Z = 3)	> Li (Z = 3)	> H (Z = 1)
Detection limit	10^{-3}	10^{-3}	10^{-6}–10^{-9}
Depth of analysis	2 nm	2 nm	1 nm
Lateral resolution	> 20 nm	> 150 μm	50 nm – 10 mm

Source: From Walls, J.M., Ed., *Methods of Surface Analysis*, Cambridge University Press, Cambridge, MA, 1989, p. 13. With permission.

Whereas AES in principle can provide information on the chemical composition, it is largely used for elemental analysis only. For ceramic powders, which are mostly insulating, electrostatic charging of the surface may occur and this leads to large shifts in the energies of the Auger electrons, making reliable analysis of the spectra difficult.

3.5.2.2 X-Ray Photoelectron Spectroscopy (XPS)

In XPS, the sample is irradiated by a source of low-energy x-rays that leads to the emission of electrons from the lower energy atomic orbitals by the well-known photoelectric effect (Figure 3.16a). The kinetic energy of the emitted photoelectrons, E_K, is given by

$$E_K = h\nu - E_b - W \tag{3.51}$$

where $h\nu$ is the frequency of the incident x-ray photon, E_b is the binding energy of the photoelectron, and W is the work function of the spectrometer, a factor that corrects for the electrostatic environment in which the electron is produced and measured. By measuring E_K in a spectrometer with a known W, the binding energy can be determined from Equation 3.51. The data are commonly plotted as the number of emitted electrons vs. the binding energy. The binding energy of an electron is characteristic of the atom and orbital from which the electron was emitted. In addition to the valence electrons that take part in chemical bonding, each atom (except hydrogen) possesses core electrons that do not take part in bonding. We are generally concerned with the core electrons in XPS.

XPS is a fairly versatile technique that can be used for qualitative and quantitative analysis of the elemental composition of the surface, as well as for determining the chemical bonding (or oxidation state) of the surface atoms. For qualitative analysis, a low-resolution, wide-scan spectrum (often referred to as a *survey scan*) covering a wide energy range (typically binding energy values of 0 to 1250 eV) serves as the basis for determining the elemental composition of the surface. Every element in the periodic table has one or more energy levels that will result in the appearance of peaks in this range. The position of each peak in the spectrum is compared with standard spectra to determine the elements present. A survey scan for a commercial Si_3N_4 powder prepared by a gas-phase reaction between $SiCl_4$ and NH_3 is shown in Figure 3.21. The presence of residual chlorine and accompanying fluorine, as well as oxygen from the oxide surface layer, is revealed.

For quantitative analysis, the principal peak for each element is selected and its intensity (peak area after removal of the background) is measured. The fractional atomic concentration of an element A is given by:

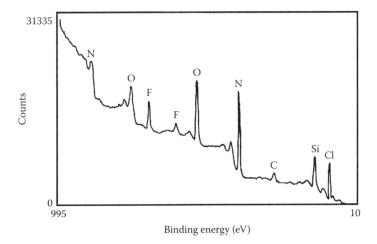

FIGURE 3.21 Survey scan of the surface chemistry of a commercial Si_3N_4 powder by x-ray photoelectron spectroscopy.

$$C_A = \frac{I_A / S_A}{\sum_i I_i / S_i} \tag{3.52}$$

where I_i is the measured peak intensity of the element i and S_i is the atomic sensitivity factor for that peak. The atomic sensitivity factors, which can be calculated theoretically or derived empirically, are usually provided in the reference manuals supplied by the manufacturer of the instrument. The accuracy of the quantitative analysis (less than ±10%) is similar to that for AES.

Information about the chemical bonding and oxidation state of the surface atoms can be determined from the chemical shifts in the peak positions in the XPS spectrum. When one of the peaks of a survey scan is examined at a higher resolution, the binding energy of the maximum is found to depend to a small degree on the chemical environment of the atom responsible for the peak. Variations in the number of valence electrons, and the types of bonds they form, influence the binding energy of the core electrons to produce a measurable shift (of a few eV) in the electron energy. Figure 3.22 illustrates the nature of the chemical shift for the 2p peak in Si and in SiO_2. In the practical determination of the chemical bonding, the measured peak position is compared with reference values for the corresponding peak in different compounds.

Unlike AES in which, especially for insulating samples, electron beam damage may occur, the damage to the sample in XPS is minimal. Furthermore, electrostatic charging of commonly insulating ceramic powders is not a severe problem. AES and XPS provide similar information about the elemental composition, and the methods tend to be complementary rather than competitive. It is, therefore, not uncommon to find instruments that incorporate both techniques, as sketched in Figure 3.23 for a combined AES/XPS instrument.

3.5.2.3 Secondary Ion Mass Spectrometry (SIMS)

SIMS consists of bombarding a surface with a primary beam of ions (He^+, Ne^+ or Ar^+ in the energy range of a few hundred eV to a few keV) which results in the emission of secondary ions, ionized clusters, atoms and atomic clusters. The emitted ions (and ionized clusters) are analyzed directly by a mass spectrometer. As the secondary ions are characteristic of the surface, the technique provides information about the chemical composition of the surface. Some information on the chemical bonding of the atoms can also be extracted by analyzing the emitted ionic clusters.

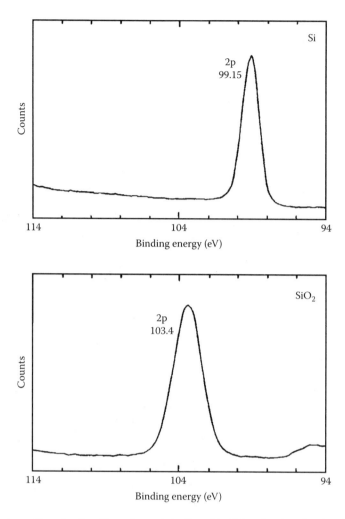

FIGURE 3.22 Peak positions for the silicon 2p peak in Si and in SiO$_2$, showing the chemical shift produced by the formation of SiO$_2$. (Courtesy of Perkin-Elmer Corporation, Eden Prairie, MN.)

There are two distinct modes of analysis. In static SIMS, an ion beam of low current density is used so that the analysis is confined to the outermost layers. In dynamic SIMS, a high current density beam is used to erode successive atomic layers at a relatively fast rate. Whereas the sensitivity of the technique is improved to the parts per billion range, the analytical conditions in dynamic SIMS are not suitable for surface analysis. Here, we limit our attention more specifically to static SIMS.

Ion beam sputtering is the primary process in SIMS but, as remarked earlier, it is also important in depth profiling, a method widely used in AES, XPS, SIMS, and many other techniques to study subsurface composition with quite fine depth resolution. When an energetic ion strikes a surface, it dissipates some of its energy into the surface. In the simplest case when the ion is scattered back out of the surface, the energy transferred to a surface atom after a binary collision is sufficient to cause substantial local damage. However, when the incident ion is scattered into the solid, the result is a collision cascade leading to the emission of sputtered particles (ions, ionized clusters, atoms, and atomic clusters) from the top one or two atomic layers of the surface (Figure 3.24). Analysis of the sputtered particles in SIMS provides a method, albeit intrinsically destructive, of surface composition analysis.

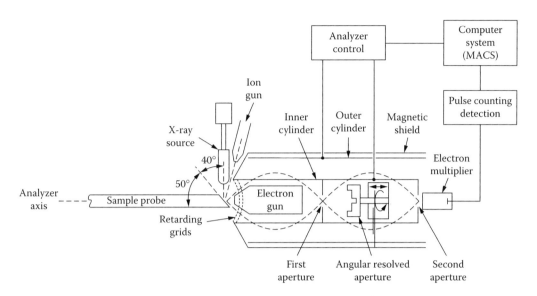

FIGURE 3.23 Schematic of a combined AES/XPS instrument. (Courtesy of Perkin-Elmer Corporation, Eden Prairie, MN.)

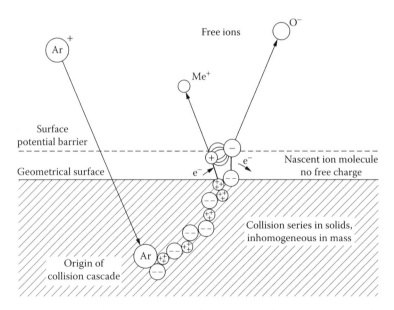

FIGURE 3.24 Model for the emission of secondary ions produced by ion beam sputtering. (From Walls, J.M., Ed., *Methods of Surface Analysis*, Cambridge University Press, Cambridge, MA, 1989, p. 217. With permission.)

SIMS has the advantages of high spatial resolution, high sensitivity for qualitative elemental analysis, and the ability to provide a detailed analysis of the chemical composition of the surface. However, quantitative analysis can be performed only with a certain degree of difficulty. Figure 3.25 shows the positive and negative secondary ion mass spectra of a pressed pellet of the Si_3N_4 powder referred to earlier in Figure 3.21. In addition to oxygen from the native oxide layer, the surface is contaminated with hydroxyl groups as well as small amounts of Cl, F, and Na. The presence of Li and K is also detected.

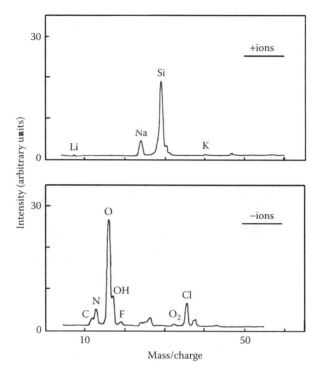

FIGURE 3.25 Positive and negative secondary ion mass spectra of the Si_3N_4 powder referred to in Figure 3.21.

3.6 CONCLUDING REMARKS

In this chapter, we have examined a variety of techniques with broad application to the characterization of ceramic powders. In addition to ceramic processing, powders are also important in several other technologies (e.g., powder metallurgy, catalysis, and pollution control), so the techniques described here have broader applicability. Some of the techniques (e.g., gas adsorption and mercury porosimetry) are also important for characterizing porous compacts whereas others (e.g., techniques for phase analysis, chemical composition, and surface analysis) are very important for solids also. We shall, therefore, encounter the use of these techniques again when we consider powder consolidation and the microstructural characterization of solids produced by sintering. Finally, as stated earlier in the book, proper characterization of the starting powders is essential for understanding their behavior during subsequent processing.

PROBLEMS

3.1

The particle size distribution of a powder is as follows:

Mean size (μm):	4	6	9	12	18	25
Number of particles:	10	16	20	25	15	7

Calculate the arithmetic, geometric, and harmonic mean sizes.

3.2

Discuss what methods you would use to measure the particle sizes of the following powders:

a. A powder with particles less than 50 μm
b. A nanoscale powder with particles less than 20 nm
c. A submicrometer powder
d. A powder with particles in the range 20 to 85 μm which must be analyzed dry
e. A powder with particles less than 50 μm in which the surface area is the important property

3.3

a. Plot the shape factor for a cylinder as a function of the length-to-diameter ratio.
b. Calculate the shape factor for a tetrahedron, an octahedron, and a dodecahedron.
c. An agglomerate consists of n spherical particles of the same size in point contact. Calculate and plot the shape factor as a function of n, for n in the range of 2 to 100.

3.4

In a nitrogen adsorption experiment at the boiling point of liquid N_2, the volume V of gas adsorbed at a pressure p were measured as follows:

p (mm Hg):	80	100	125	140	200
V (cm^3/g):	0.420	0.439	0.464	0.476	0.534

Determine the surface area of the powder.

3.5

The following data were obtained in a liquid pycnometry experiment performed at 20°C: mass of pycnometer = 35.827 g; mass of pycnometer and powder = 46.402 g; mass of pycnometer and water = 81.364 g; mass of pycnometer, powder, and water = 89.894 g. If the theoretical density of the solid is 5.605 g/cm^3, determine the amount of closed porosity in the powder.

3.6

Discuss how you would measure the following:

a. The solid solubility limit of Ca in ZrO_2
b. The effect of Ca concentration on the theoretical density of Ca-doped ZrO_2
c. The Ba/Ti atomic ratio in $BaTiO_3$ powder
d. The presence of the cubic or tetragonal phase in $BaTiO_3$ powder
e. The concentration of SiO_2 impurities in $BaTiO_3$ powder
f. The concentration of the α and β phases in a Si_3N_4 powder
g. The temperature at which ZnO and Al_2O_3 powders react to form $ZnAl_2O_4$
h. The amount of $ZnAl_2O_4$ formed by the reaction between ZnO and Al_2O_3 powders

3.7

In an x-ray photoelectron spectrum, would you expect the C 1s peak to be at a higher or lower binding energy compared to that for the O 1s peak? Would the F 1s peak occur at a higher or lower binding energy compared to that for the O 1s peak? Explain your answer.

3.8

The chemical composition of the oxide layer on Si_3N_4 particles varies as a function of the thickness of the layer. How would you confirm this? What methods would you use to measure the composition as a function of the thickness of the oxide layer?

REFERENCES

1. Wachtman, J.B., *Characterization of Materials*, Butterworth-Heinemann, Boston, MA, 1991.
2. Onoda, G.Y. Jr. and Hench, L.L., Physical characterization terminology, in *Ceramic Processing Before Firing*, Onoda G.Y. Jr. and Hench, L.L., Eds., John Wiley & Sons, New York, 1978, chap. 5.
3. Allen, T., *Particle Size Measurement*, 4th ed., Chapman and Hall, London, 1990.
4. Beddow, J.K., *Particulate Science and Technology*, Chemical Publishing Co., New York, 1980.
5. Jillavenkatesa, A., Dapkunas, S.J., and Lum, L.H., *Particle Size Characterization*, National Institute of Standards and Technology, Special Publication 960-1, Washington, D.C., 2001.
6. Cullity, B.D., *Elements of X-Ray Diffraction*, 2nd ed., Addison-Wesley, Reading, MA, 1978.
7. Klug, H.P. and Alexander, L.F., *X-Ray Diffraction Procedures for Polycrystalline and Amorphous Materials*, John Wiley & Sons, New York, 1974.
8. Jenkins, R. and Snyder, R.L., *Introduction to X-Ray Powder Diffractometry*, John Wiley & Sons, New York, 1996.
9. Gregg, S.J. and Sing, K.S.W., *Adsorption, Surface Area and Porosity*, 2nd ed., Academic Press, New York, 1982.
10. Brunauer, S., Deming, L.S., Deming, W.S., and Teller, E., On a theory of van der Waals adsorption of gases, *J. Am. Chem. Soc.*, 62, 1723, 1940.
11. Brunauer, S., Emmett, P.H., and Teller, E., Adsorption of gases in multimolecular layers, *J. Am. Chem. Soc.*, 60, 309, 1938.
12. Brunauer, S., *The Adsorption of Gases and Vapours*, Oxford University Press, Oxford, 1945.
13. Langmuir, I., The constitution and fundamental properties of solids and liquids. Part 1. Solids, *J. Am. Chem. Soc.*, 38, 2221, 1916.
14. Skoog, D.A., Holler, F.J., and Nieman, T.A., *Principles of Instrumental Analysis*, 5th ed., Harcourt Brace, Philadelphia, PA, 1998.
15. Winefordner, J.D. and Epstein, M.S., *Physical Methods of Chemistry: Vol. IIIA, Determination of Chemical Composition and Molecular Structure*, John Wiley & Sons, New York, 1987, p. 193.
16. Slavin, W., Flames, furnaces, plasmas: how do we choose? *Anal. Chem.*, 58, 589A, 1986.
17. Jenkins, R., *X-Ray Fluorescence Spectrometry*, 2nd ed., John Wiley & Sons, New York, 1999.
18. Jenkins, R., Gould, R.W., and Gedcke, D., *Quantitative X-Ray Spectrometry*, 2nd ed., Marcel Dekker, New York, 1995.
19. Stout, G.H. and Jensen, L.H., *X-Ray Structure Determination*, John Wiley & Sons, New York, 1989.
20. Rietveld, J.M., A profile refinement method for nuclear and magnetic structures, *J. Appl. Cryst.*, 2, 65, 1969.
21. Young, R.A., Ed., *The Rietveld Method*, Oxford University Press, New York, 1993.
22. Von Dreele, R.B. and Kline, J.P., The impact of background function on high accuracy quantitative Rietveld analysis, in *Advances in X-Ray Analysis: Vol. 38*, Predecki, P.K., Bowen, D.K., Gilfrich, J.V. et al., Eds., Plenum Press, New York, 1995, 59.
23. Walls, J.M., Ed., *Methods of Surface Analysis*, Cambridge University Press, Cambridge, MA, 1989.
24. Hudson, J.B., *Surface Science: An Introduction*, Butterworth-Heinemann, Boston, MA, 1992.
25. Woodruff, D.P. and Delchar, T.A., *Modern Techniques of Surface Analysis*, 2nd ed., Cambridge University Press, Cambridge, MA, 1994.

26. Somorjai, G.A. and Salmeron, M., Surface characterization of ceramic materials, in *Ceramic Microstructures '76*, Fulrath, R.M. and Pask, J.A., Eds., Westview Press, Boulder, CO, 1977, p. 101.

27. Van Hove, M.A., Surface crystallography with low-energy electron diffraction, *Proc. R. Soc. Lond.A*, 442, 61, 1993.

28. Van de Leemput, L.E.C. and Van Kempen, H., Scanning tunneling microscopy, *Rep. Prog. Phys.*, 55, 1165, 1992.

29. Hansma, P.K., Elings, V.B., Marti, O., and Bracker, C.E., Scanning tunneling microscopy and atomic force microscopy: applications to biology and technology, *Science*, 242, 209, 1988.

30. Meyer, E., Atomic force microscopy, *Prog. Surf. Sci.*, 41, 1, 1992.

31. Rahaman, M.N., Boiteux, Y., and De Jonghe, L.C., Surface characterization of silicon nitride and silicon carbide powders, *Am. Ceram. Soc. Bull.*, 65, 1171, 1986.

4 Science of Colloidal Processing

4.1 INTRODUCTION

A colloid consists of two distinct phases: a continuous phase (referred to as the *dispersion medium*) and a fine dispersed particulate phase (the *disperse phase*). In general, the two phases may be solids, liquids, or gases, giving rise to various types of colloidal systems as listed in Table 4.1. The dispersed particles generally have dimensions ranging between 1 and 1000 nm, sometimes referred to as the colloidal size range.

In ceramic processing, colloidal suspensions (also referred to as *sols*), consisting of a dispersion of solid particles in a liquid, are being used increasingly in the consolidation of ceramic powders to produce the green body. Compared to powder consolidation by dry or semidry pressing in a die, colloidal methods can lead to better particle packing homogeneity in the green body which, in turn, leads to better control of the microstructure during sintering. Colloidal solutions consist of polymer molecules dissolved in a liquid. The size of the polymer molecules in solution falls in the colloidal size range, so these systems are considered part of colloid science. Polymer solutions are relevant to the fabrication of ceramics by the solution sol–gel route, which we shall consider in the next chapter. In the present chapter, we concentrate on colloidal suspensions.

The principles of colloid science are covered in several excellent texts [1–3]. A basic problem we shall be concerned with is the stability of colloidal suspensions. Clearly the particles must not be too large; otherwise, gravity will produce rapid sedimentation. The other important factor is the attractive force between the particles. Attractive van der Waals forces exist between the particles regardless of whether other forces may be involved. If the attractive force is large enough, the particles will collide and stick together, leading to rapid sedimentation of particle clusters (i.e., to flocculation or coagulation). Although in principle the reduction in the attractive force can also be used to prevent flocculation, the practical methods rely on the introduction of repulsive forces between the particles. Repulsion between electrostatic charges (electrostatic stabilization), between adsorbed polymer molecules (steric stabilization), or between some combination of the two (electrosteric stabilization) are the basic techniques used. The stability of colloidal suspensions controls the packing homogeneity of the consolidated body. A stable colloidal suspension may be consolidated into a densely packed structure whereas an unstable suspension may lead to a loosely packed structure or, under certain conditions, to a particulate gel with a fairly high volume.

Colloidal suspensions exhibit several interesting properties that are important for controlling their stability and consolidation. The motion of the particles when subjected to an electric field (electrophoresis) can be a source of valuable information for determining the surface charge of the particles. Colloidal suspensions, especially concentrated suspensions, also have remarkable rheological properties. A good paint must flow easily when applied to the surface to be painted but then must become rigid enough to prevent it from flowing off a vertical surface. In ceramic processing, the rheological behavior of the suspension can be used as a direct process parameter to control and optimize the structure of the green body.

TABLE 4.1
Types of Colloidal Systems with Some Common Examples

Disperse Phase	Dispersion Medium	Technical Name	Examples
Solid	Gas	Aerosol	Smoke
Liquid	Gas	Aerosol	Mist, fog
Solid	Liquid	Suspension or sol	Paint, printing ink
Liquid	Liquid	Emulsion	Milk, mayonnaise
Gas	Liquid	Foam	Fire extinguisher foam
Solid	Solid	Solid dispersion	Ruby glass (Au in glass), some alloys
Liquid	Solid	Solid emulsion	Bituminous road paving, ice cream
Gas	Solid	Solid foam	Insulating foam

4.2 TYPES OF COLLOIDS

Colloids consisting of particles dispersed in a liquid are generally divided into two broad classes: lyophilic colloids and lyophobic colloids. When the liquid is specifically water, the colloids are described as *hydrophilic* and *hydrophobic*. *Lyophilic* (i.e., liquid-loving) colloids are those in which there is a strong affinity between the dispersed particle and the liquid. The liquid is strongly adsorbed on to the particle surfaces, so the interface between the particle and the liquid is very similar to the interface between the liquid and itself. This system will be intrinsically stable because there is a reduction in the Gibbs free energy when the particles are dispersed. Polymer solutions and soap solutions are good examples of lyophilic colloids.

Lyophobic (liquid-hating) colloids are those in which the liquid does not show affinity for the particle. The Gibbs free energy increases when the particles are distributed through the liquid, and if attractive forces exist between the particles, there will be a strong tendency for the particles to stick together when they come into contact. This system will be unstable and flocculation will result. A lyophobic colloid can, therefore, only be dispersed if the surface is treated in some way to cause a strong repulsion to exist between the particles. Suspensions of insoluble particles in a liquid (e.g., most ceramic particles dispersed in a liquid) are well-known examples of lyophobic colloids. We therefore need to understand the attractive forces that lead to flocculation and how they can be overcome by repulsive forces to produce colloids with the desired stability.

4.3 ATTRACTIVE SURFACE FORCES

Attractive surface forces, generally referred to as *van der Waals* forces, exist between all atoms and molecules regardless of what other forces may be involved. They have been discussed in detail by Israelachvili [4]. We shall first examine the origins of the van der Waals forces between atoms and molecules and later consider the attractive forces between macroscopic bodies such as particles.

4.3.1 VAN DER WAALS FORCES BETWEEN ATOMS AND MOLECULES

The van der Waals forces between atoms and molecules can be divided into three types [5]:

1. Dipole–dipole forces (Keesom forces): A molecule of HCl, because of its largely ionic bonding, has a positive charge and a negative charge separated by a small distance (~0.1 nm), i.e., it consists of a minute electric dipole. A polar HCl molecule will interact with another polar HCl molecule and produce a net attractive force.

FIGURE 4.1 Polarization of a molecule by the field E due to a dipole, giving rise to mutual attraction.

2. Dipole-induced dipole forces (Debye forces): A polar molecule, e.g., HCl, can induce a dipole in a nonpolar atom or molecule, e.g., an argon atom, causing an attractive force between them.
3. Dispersion forces (London forces): This type of force exists between nonpolar atoms or molecules (e.g., between argon atoms).

To understand the origins of dispersion forces, consider an argon atom. Although the argon atom has a symmetrical distribution of electrons around the nucleus so that it has no dipole, this situation is only true as an average over time. At any instant, the electron cloud undergoes local fluctuations in charge density, leading to an instantaneous dipole. This instantaneous dipole produces an electric field that polarizes the electron distribution in a neighboring argon atom, so the neighboring atom itself acquires a dipole. The interaction between the two dipoles leads to an attractive force.

The attractive dispersion force can be calculated as follows [6]. Suppose the first atom has a dipole moment μ at a given instant in time. The electric field, E, at a distance x along the axis of the dipole is

$$E = \frac{1}{4\pi\varepsilon_o}\left(\frac{2\mu}{x^3}\right) \tag{4.1}$$

where ε_o is the permittivity of free space and x is much greater than the length of the dipole. If there is another atom at this point (Figure 4.1), it becomes polarized and acquires a dipole moment, μ' given by:

$$\mu' = \alpha E \tag{4.2}$$

where α is the polarizability of the atom. A dipole μ' in a field E has a potential energy, V, given by:

$$V = -\mu'E \tag{4.3}$$

Substituting for μ and E gives:

$$V = -\left(\frac{1}{4\pi\varepsilon_o}\right)^2\left(\frac{4\alpha\mu^2}{x^6}\right) \tag{4.4}$$

A rigorous derivation by London (1930) using quantum mechanics gives:

$$V = -\frac{3}{4}\left(\frac{1}{4\pi\varepsilon_o}\right)^2\left(\frac{\alpha^2 h\nu}{x^6}\right) \tag{4.5}$$

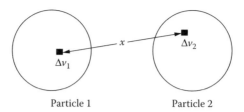

Particle 1   Particle 2

FIGURE 4.2 Additivity of the van der Waals force between macroscopic bodies. The total interaction is taken as the sum of the interactions between infinitesimally small elements in the two bodies.

where h is Planck's constant and v is the frequency of the polarized orbital. The force between the atoms has a magnitude

$$F = \frac{\partial V}{\partial x} = -\frac{B}{x^7} \tag{4.6}$$

where B is a constant. The Keesom and Debye forces are also proportional to $1/x^7$ and lead to a potential energy proportional to $1/x^6$. For dispersion forces, the dependence of V on $1/x^6$ is strictly applicable to separation distances that are small enough (less than a few tens of nanometers) that the interactions between the dipoles can be considered to be instantaneous. In this case, the forces are referred to as *nonretarded* forces. For larger separations, when the forces are described as *retarded* forces, V varies as $1/x^7$. Unless otherwise stated, we shall assume that nonretarded forces operate for the remainder of this chapter.

4.3.2 VAN DER WAALS FORCES BETWEEN MACROSCOPIC BODIES

To determine the van der Waals forces between macroscopic bodies (e.g., two particles), we assume that the interaction between one molecule and a macroscopic body is simply the sum of the interactions with all the molecules in the body. We are, therefore, assuming simple additivity of the van der Waals forces and ignoring how the induced fields are affected by the intervening molecules.

In the computation due to Hamaker (1937), individual atoms are replaced by a smeared-out, uniform density of matter (Figure 4.2). An infinitesimally small volume Δv_1 in the first body exerts an attractive potential on an infinitesimally small volume Δv_2 in the second body according to the equation:

$$V' = -\frac{C}{x^6} \rho_1 \Delta v_1 \rho_2 \Delta v_2 \tag{4.7}$$

where C is a constant, and ρ_1 and ρ_2 are the numbers of molecules per unit volume in the two bodies. Of course, $\rho_1 = \rho_2$ if the two bodies consist of the same material. The additivity of the forces now becomes an integration, with the total potential energy being

$$V_A = \int -\frac{C}{x^6} \rho_1 \rho_2 \, dv_1 dv_2 \tag{4.8}$$

Hamaker showed that Equation 4.8 can be expressed as

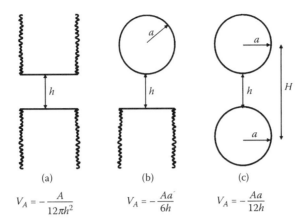

$$V_A = -\frac{A}{12\pi h^2} \qquad V_A = -\frac{Aa'}{6h} \qquad V_A = -\frac{Aa}{12h}$$

FIGURE 4.3 The potential energy of attraction between two bodies with specified geometries: (a) two semiinfinite parallel surfaces, (b) a sphere and a semiinfinite surface, (c) two spheres.

$$V_A = -\frac{A}{\pi^2} \int \frac{dv_1 dv_2}{x^6} \qquad (4.9)$$

where A, called the *Hamaker constant*, is equal to $C\pi^2\rho_1\rho_2$. According to Equation 4.9, V_A is the product of a material-specific constant, A, and a geometrical term.

We will now consider some specific cases of interest in colloid science where analytical solutions to the geometric term in Equation 4.9 may be obtained. For two semiinfinite parallel surfaces separated by a distance h (Figure 4.3a), the potential energy is infinite. However, it is possible to define a finite interaction energy per unit area of one semiinfinite surface as:

$$V_A = -\frac{A}{12\pi h^2} \qquad (4.10)$$

For a sphere of radius a at a distance h from a flat semiinfinite solid (Figure 4.3b), V_A is given by:

$$V_A = -\frac{Aa}{6h}\left[1 + \frac{h}{2a+h} + \frac{h}{a}\ln\left(\frac{h}{2a+h}\right)\right] \qquad (4.11)$$

Finally, the potential energy between two spheres of radius a that are separated by a distance h (Figure 4.3c) is given by:

$$V_A = -\frac{A}{6}\left[\frac{2a^2}{H^2 - 4a^2} + \frac{2a^2}{H^2} + \ln\left(\frac{H^2 - 4a^2}{H^2}\right)\right] \qquad (4.12)$$

where the distance H between the centers of the spheres is equal to $h + 2a$. For small separations, where $h \ll a$, Equation 4.12 gives

$$V_A = -\frac{Aa}{12h} \qquad (4.13)$$

showing that V_A is proportional to $1/h$. In the case of large separations where $h >> a$, the potential energy is

$$V_A = -\frac{16Aa^6}{9h^6} \qquad (4.14)$$

showing that V_A is proportional to $1/h^6$.

4.3.3 THE HAMAKER CONSTANT

We start by making an estimate of A for a simple model. The value of C is given by Equation 4.5 if the x^6 term is removed from the denominator. For identical interacting systems, $A = C\pi^2\rho^2$, so that

$$A = -\frac{3}{4}h\nu\left(\frac{\alpha}{4\pi\varepsilon_o}\right)^2 \pi^2\rho^2 \qquad (4.15)$$

The polarizability of an atom (or molecule) with a spherical shape of radius a is given by:

$$\alpha = 4\pi\varepsilon_o a^2 \qquad (4.16)$$

For a cubic array of atoms, $\rho = (1/2a)^3$. Substituting for α and ρ in Equation 4.15 gives

$$A = -\frac{3}{4}h\nu\pi^2 / 64 \approx h\nu / 8 \qquad (4.17)$$

Taking the quantity $h\nu$ to be comparable with the first ionization potential (of the order 5 to 10 eV),

$$A \approx 1.5 \times 10^{-19} J. \qquad (4.18)$$

For a large number of solids (Table 4.2), A has values in a fairly narrow range between 10^{-20} and 10^{-19} J ($\sim$2.5 to 25 kT, where k is the Boltzmann constant and T is room temperature in Kelvin).

The method of pairwise addition, sometimes referred to as the *microscopic* approach, was originally used by Hamaker [7] to calculate A from the polarizabilities and number densities of the atoms in the two interacting bodies. However, the method suffers from several problems. For example, it neglects many body interactions among the atoms. It is evident that if an atom X in Particle 1 exerts a force on atom Y in Particle 2 (Figure 4.2), then the presence of neighboring atoms in Particle 1 and Particle 2 is bound to influence the interaction between atoms X1 and Y2. It is also not clear how the addition needs to be modified if the intervening medium between Particles 1 and 2 is not vacuum or air but another dielectric medium. A solution to these problems was provided by Lifshitz [8] who treated the bodies and the intervening medium in terms of their bulk properties, specifically their dielectric constants. The Lifshitz theory, sometimes referred to as the *macroscopic* approach, is fairly complex and has been reviewed in a number of textbooks and articles at various levels of sophistication [9–12]. Various methods for calculating A from the Lifshitz theory have been described [13,14] and calculated values for several ceramic materials are given in Table 4.2.

TABLE 4.2
Nonretarded Hamaker Constants at 298K for Ceramic Materials When the Intervening Medium Is Vacuum or Water

Material	Crystal Structure	Hamaker Constant (10^{-20} J)	
		Vacuum	Water
α-Al_2O_3	Hexagonal	15.2	3.67
$BaTiO_3$	Tetragonal	18	8
BeO*	Hexagonal	14.5	3.35
C (diamond)	Cubic	29.6	13.8
$CaCO_3$*	Trigonal	10.1	1.44
CaF_2	Cubic	6.96	0.49
CdS	Hexagonal	11.4	3.40
CsI	Cubic	8.02	1.20
KBr	Cubic	5.61	0.55
KCl	Cubic	5.51	0.41
LiF	Cubic	6.33	0.36
$MgAl_2O_4$	Cubic	12.6	2.44
MgF_2	Tetragonal	5.87	0.37
MgO	Cubic	12.1	2.21
Mica	Monoclinic	9.86	1.34
NaCl	Cubic	6.48	0.52
NaF	Cubic	4.05	0.31
PbS	Cubic	8.17	4.98
6H-SiC	Hexagonal	24.8	10.9
β-SiC	Cubic	24.6	10.7
β-Si_3N_4	Hexagonal	18.0	5.47
Si_3N_4	Amorphous	16.7	4.85
SiO_2 (quartz)	Trigonal	8.86	1.02
SiO_2	Amorphous	6.5	0.46
$SrTiO_3$	Cubic	14.8	4.77
TiO_2*	Tetragonal	15.3	5.35
Y_2O_3	Hexagonal	13.3	3.03
ZnO	Hexagonal	9.21	1.89
ZnS	Cubic	15.2	4.80
ZnS	Hexagonal	17.2	5.74
ZrO_2/3 mol% Y_2O_3	Tetragonal	20.3	7.23

* Average value calculated using a weighted procedure.

Source: From Bergström, L., Hamaker constants of inorganic materials, *Adv. Colloid Interface Sci.*, 70, 125, 1997.

The Hamaker constant can be determined experimentally by direct measurement of the surface forces as a function of separation for bodies separated by vacuum, air, or liquids. Differences between the calculated and measured values depend on the complexity of the theoretical calculations and the accuracy of the physical data used in the models [15]. The first measurements down to separations of 2 nm or less were performed by Tabor and Winterton [16] and by Israelachvili and Tabor [17], using a cantilever spring to measure the force between molecularly smooth surfaces of mica and a piezoelectric crystal to control the separation. By fitting the measured surface force to the theoretical expression for the appropriate geometry, they found that nonretarded forces operate for separations less than ~10 nm and the nonretarded Hamaker constant $A = 10^{-19}$ J. More recently, the relatively new technique of atomic force microscopy [18,19] has seen considerable use [20–23].

4.3.4 EFFECT OF THE INTERVENING MEDIUM

The Lifshitz theory, as outlined earlier, is fairly difficult, and we provide here only approximate relations for the effect of the intervening medium. If we consider two bodies with dielectric constants ε_1 and ε_2 separated by a medium with dielectric constant ε_3, the theory predicts that the Hamaker constant A is proportional to a term given by the equation:

$$A \propto \left(\frac{\varepsilon_1 - \varepsilon_3}{\varepsilon_1 + \varepsilon_3} \right) \times \left(\frac{\varepsilon_2 - \varepsilon_3}{\varepsilon_2 + \varepsilon_3} \right) \tag{4.19}$$

According to Equation 4.19, if $\varepsilon_1 > \varepsilon_3 > \varepsilon_2$ the value of A is negative, that is, there is a repulsive force. It is difficult to explain this in terms of the Hamaker method of pairwise interactions. One might say that the intervening medium with dielectric constant ε_3 "likes" itself more than the two bodies.

In colloidal systems, the bodies 1 and 2 are identical so that

$$A = c \left(\frac{\varepsilon_1 - \varepsilon_3}{\varepsilon_1 + \varepsilon_3} \right)^2 \tag{4.20}$$

where c is a constant. We see that A is always positive; that is, the van der Waals force is always attractive, whether ε_3 is greater or less than ε_1. However, the intervening medium always leads to a reduction in the attractive force when compared to the force when the medium is air or vacuum ($\varepsilon_3 = 1$). Table 4.2 also gives the Hamaker constants calculated from Lifshitz theory for several ceramic materials when the intervening medium is water.

4.4 STABILIZATION OF COLLOIDAL SUSPENSIONS

Consider a suspension of colloidal particles. The particles undergo Brownian motion (similar to the motion of pollen in a liquid) and will eventually collide. The potential energy of attraction, V_A, between two particles with radius a and a distance h apart is given by Equation 4.13. Assuming that this equation is valid down to contact between the particles, then we can calculate V_A by putting $h \approx 0.3$ nm. Taking $A \approx 10^{-20}$ J, for particles with a radius $a = 0.2$ μm, V_A has a value of $\approx 5 \times 10^{-19}$ J. The thermal energy of the particles is kT where k is the Boltzmann constant, and T is the absolute temperature. Near room temperature ($T \approx 300$ K), kT is of the order of 4×10^{-21} J, and this value is much smaller than V_A for the particles in contact. Thus, the particles will stick together on collision because the thermal energy is insufficient to overcome the attractive potential energy. Flocculation will, therefore, occur unless some method is found to produce a repulsion between the particles that is sufficiently strong to overcome the attractive force. There are several ways for achieving this but the most commonly used are (Figure 4.4):

1. *Electrostatic stabilization,* in which the repulsion between the particles is based on electrostatic charges on the particles
2. *Steric stabilization,* in which the repulsion is produced by uncharged polymer chains adsorbed onto to the particle surfaces
3. *Electrosteric stabilization,* consisting of a combination of electrostatic and steric repulsion, achieved by the adsorption of charged polymers (polyelectrolytes) onto the particle surfaces

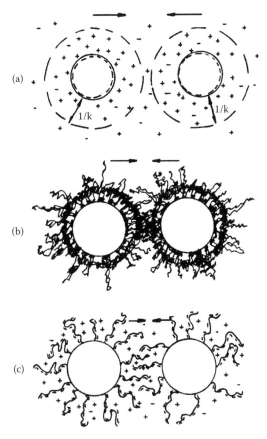

FIGURE 4.4 Schematic representation of (a) electrostatic stabilization for negatively charged particles, (b) steric stabilization, (c) electrosteric stabilization. (From Napper, D.H., *Polymeric Stabilization of Colloidal Dispersions*, Academic Press, New York, 1983.)

4.5 ELECTROSTATIC STABILIZATION

Electrostatic stabilization, as outlined earlier, is said to occur when the repulsion between the particles is achieved by electrostatic charges on the particles. The repulsion is not, however, a simple case of repulsion between charged particles. An electrical double layer of charge is produced around each particle and the repulsion occurs as a result of the interaction of the double layers. By way of an introduction to the analysis of the electrical double layer, we consider how particles acquire an electrostatic charge in an aqueous liquid (water) and the general principle of the double layer.

4.5.1 THE DEVELOPMENT OF CHARGES ON OXIDE PARTICLES IN WATER

The main processes by which particles dispersed in a liquid can acquire a surface charge are (1) preferential adsorption of ions, (2) dissociation of surface groups, (3) isomorphic substitution, and (4) adsorption of charged polymers (polyelectrolytes). Preferential adsorption of ions from solution is the most common process for oxide particles in water, whereas isomorphic substitution is commonly found in clays. The adsorption of polyelectrolytes is the main charging mechanism in electrosteric stabilization and will be discussed later. The dissociation of ionizable surface groups such as sulfate, sulfonate, carboxyl and amino groups is reviewed in Reference 24.

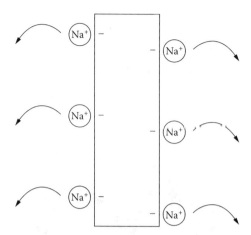

FIGURE 4.5 The production of charges on the basal surfaces of clay particles (e.g., montmorillonite) in aqueous liquids.

4.5.1.1 Isomorphic Substitution

In the crystal lattice of clay minerals, some of the cations are replaced by other cations of lower valence without altering the crystal structure; e.g., Si^{4+} ions are replaced by Al^{3+} or Mg^{2+} ions, and Al^{3+} ions by Mg^{2+} ions. This substitution leads to a deficit of positive charges, which is balanced by other positive ions (e.g., Na^+, K^+, or Ca^{2+}) adsorbed on the basal surfaces (the larger flat surfaces) of the platelike clay particles. The process occurs naturally in the weathering of clays. In the case of the clay mineral pyrophyllite, $Al_2(Si_2O_5)_2(OH)_2$, isomorphic substitution in which Mg^{2+} replaces some of the Al^{3+} ions in the lattice leads to montmorillonite, $Na_{0.33}(Al_{1.67}Mg_{0.33})(Si_2O_5)_2(OH)_2$, another mineral with the same crystal structure in which the charge deficit is balanced by Na^+ ions on the particle surfaces. When the clay mineral montmorillonite is dispersed in water, the Na^+ ions pass freely into solution, leaving negatively charged basal surfaces (Figure 4.5).

The extent of isomorphic substitution is dictated by the nature of the clay, and this is expressed by the cation exchange capacity (CEC). The CEC of a clay is the number of charges on the clay (expressed in coulombs per kilogram) which can be replaced in solution. It is typically in the range of 10^3 to 10^5 C/kg and, for a given clay, is not sensitive to variables such as pH or concentration of the electrolyte in solution.

The edges of clay particles, where imperfections occur due to bond breakage, carry a positive charge in water at low pH, and this decreases to zero as the pH is raised to ~7. The charge is not due to isomorphic substitution but is generally considered to be due to dissociation of OH groups from aluminum octahedra:

$$\text{\textbackslash Al–OH} \longrightarrow \text{\textbackslash Al}^+ + \text{OH}^-$$

The plasticity of clay suspensions results in part from the "house of cards" structure formed by the attraction of the oppositely charged basal surfaces and edges (Figure 4.6).

4.5.1.2 Adsorption of Ions from Solution

In the process of preferential adsorption of ions from solution, an electrolyte such as an acid, a base, or a metal salt is added to the aqueous liquid. Ions preferentially adsorb onto the surface of the dispersed particles, leading to a charge on the particle surface. Because the system consisting of the particle and the electrolyte must be electrically neutral, an equal and opposite countercharge

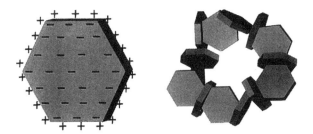

FIGURE 4.6 Schematic illustration of clay particles in water showing (a) surface charges on an individual particle, (b) aggregated particle network formed by the attraction of oppositely charged faces and edges.

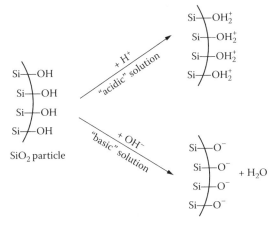

FIGURE 4.7 The production of surface charges on SiO_2 particles by adsorption of ions from acidic or basic solutions.

exists in the solution. Most oxide surfaces are hydrated; for an oxide of a metal M, there will be MOH groups on the surface, as illustrated in Figure 4.7 for SiO_2. In acidic solutions, adsorption of H^+ ions (or hydronium ions, H_3O^+) produces a positively charged surface while in basic solutions, adsorption of OH^- ions (or the dissociation of H^+ ions) leads to a negatively charged surface. Thus, oxide surfaces are positively charged at low pH and negatively charged at high pH. At some intermediate pH, the adsorption of H^+ ions will balance that of the OH^- ions, and the particle surface will be effectively neutral. The pH value where the particle surface has a net charge of zero (denoted as pH_0) is defined as the *point of zero charge* (PZC). The acid/base properties of oxide surfaces are commonly characterized by the PZC. The more "acidic" oxides such as SiO_2 have a low PZC, whereas the more "basic" oxides such as MgO have a high PZC.

The PZC is measured by potentiometric acid–base titrations [25]. As discussed later, it is usually more convenient to measure the ζ potential of the particles, which corresponds to the electrostatic potential at a short distance from the particle surface (at the surface of the Stern layer). The pH where the ζ potential is zero is called the *isoelectric point* (IEP). For oxides, commonly PZC $\approx$ IEP. However, when the ζ-potential measurements are carried out in the presence of surface active species such as multivalent ions and charged dispersant molecules, the PZC may be somewhat different from the IEP. Table 4.3 gives the approximate IEP values for several oxides [26,27].

The PZC of pure oxides can be calculated in two ways. In the Parks theory [27], hydroxyl groups on the surface of the oxide are considered to act as acids or bases, according to the equilibria:

$$MOH \rightleftharpoons MO^- + H^+ \qquad\qquad (4.21a)$$

TABLE 4.3
Nominal Isoelectric Points (IEPs) of Some Oxides

Material	Nominal Composition	IEP
Quartz	SiO_2	2–3
Soda–lime–silica glass	$1.0Na_2O \cdot 0.6CaO \cdot 3.7SiO_2$	2–3
Potassium feldspar	$K_2O \cdot Al_2O_3 \cdot 6SiO_2$	3–5
Zirconia	ZrO_2	4–5
Tin oxide	SnO_2	4–5
Titania	TiO_2	5–6
Barium titanate	$BaTiO_3$	5–6
Fluorapatite	$Ca_{10}(PO)_4 \cdot (F,OH)$	5–6
Kaolinite	$Al_2O_3 \cdot SiO_2 \cdot 2H_2O$	5–6
Mullite	$3Al_2O_3 \cdot 2SiO_2$	6–8
Ceria	CeO_2	6–7
Chromium oxide	Cr_2O_3	6–7
Hydroxyapatite	$Ca_{10}(PO_4)_6(OH)_2$	7–9
Haematite	Fe_2O_3	8–9
Alumina	Al_2O_3	8–9
Zinc oxide	ZnO	9
Calcium carbonate	$CaCO_3$	9–10
Nickel oxide	NiO	10–11
Magnesia	MgO	12

Source: From Hunter, R.J., *Zeta Potential in Colloid Science*, Academic Press, New York, 1981; Parks, G.A., The isoelectric points of solid oxides, solid hydroxides, and aqueous hydroxo complex systems, *Chem. Rev.*, 65, 177, 1965.

and

$$MOH + H \rightleftarrows MOH_2^+ \tag{4.21b}$$

The equilibrium between the positive and negative charges on the surface can be written as:

$$MO^- \ (surface) + 2H^+ \rightleftarrows MOH_2^+ \ (surface) \tag{4.22}$$

with an equilibrium constant, K, given by:

$$K = \frac{[MOH_2^+]}{[MO^-][H^+]^2} \tag{4.23}$$

where the quantities in the square brackets represent the concentrations of the species. At the *PZC*, $[MOH_2^+] = [MO^-]$, so that

$$PZC = \frac{1}{2} \log K \tag{4.24}$$

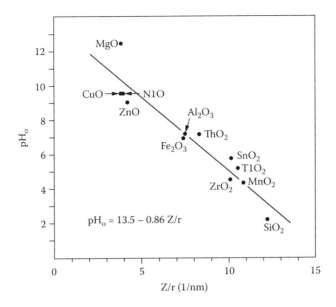

FIGURE 4.8 The Parks relationship between the point of zero charge pH_o and Z/r, where Z is the ionic charge of the cation and r is the sum of the cation radius and the oxygen ion diameter. (From Parks, G.A., The isoelectric points of solid oxides, solid hydroxides, and aqueous hydroxo complex systems, *Chem. Rev.*, 65, 177, 1965.)

By relating K to the free energy change due to the approach of the $2H^+$ ions to the MO^- ion, Parks showed that

$$PZC = A_1 - B_1(Z/r) \qquad (4.25)$$

where A_1 and B_1 are constants, Z is the ionic charge of the cation, and $r = 2r_o + r_+$, where r_o and r_+ are the ionic radii of the oxygen ion and the cation, respectively. A good correlation between the measured PZC values and those calculated from Equation 4.25 is shown in Figure 4.8. Yoon et al. [28] have developed an improved version of the Parks relationship.

The second method considers the surface acidity to result from the electron acceptor character of the oxide surface [29]. This is related to the Lewis acid-base concept where, for an ionic oxide, the acid entity is the cation, with the base being the oxygen anions. For an oxide M_xO_y, the surface acidity has been shown to be related to the ionization potential (*IP*) of the metal M according to

$$PZC = A_2 + B_2(IP) \qquad (4.26)$$

where A_2 and B_2 are constants. The correlation between *PZC* data and Equation 4.26 is also found to be good.

4.5.2 Origins of the Electrical Double Layer

For electrostatically stabilized colloidal suspensions, the charges, as we have seen, consist of a surface charge on the particles and an equal and opposite countercharge in the solution. Suppose the particle has a positive surface charge due to preferential adsorption of positive ions. In the complete absence of thermal motion an equal number of negative ions (counterions) would adsorb onto the positive charge and neutralize it. However, such a compact double layer does not form

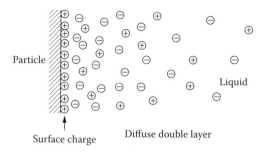

FIGURE 4.9 The distribution of positive and negative charges in the electrical double layer associated with a positively charged surface in a liquid.

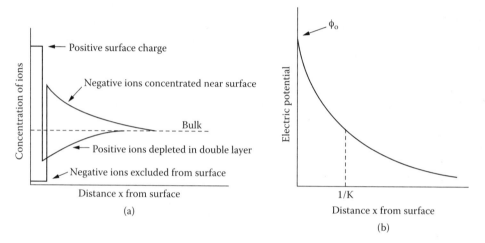

FIGURE 4.10 (a) Concentration of positive and negative ions as a function of distance from the particle surface, (b) the electrical potential ϕ as a function of distance from the particle surface. The distance equal to $1/K$ is the Debye length.

because of thermal motion. Instead, the counterions are spread out in the liquid to form a diffuse double layer as shown in Figure 4.9. There is a fairly rapid change in the concentration of the positive and negative ions as we move away from the surface (Figure 4.10a). As a result, the electrical potential also falls off rapidly with distance from the surface (Figure 4.10b).

As two particles approach one another in the liquid, the diffuse double layers will start to overlap. It is the interaction between the double layers that give rise to the repulsion between the particles. If the repulsion is strong enough, it can overcome the van der Waals attractive force, producing a stable colloidal suspension. As a prelude to examining the interactions between double layers, we start with an isolated double layer associated with a single particle.

4.5.3 ISOLATED DOUBLE LAYER

The electrical properties of an isolated double layer can be described in terms of the variation of the electrical charge or the electrical potential (Figure 4.10). The two are related through the capacitance so that if one is known, the other can be found. We shall consider the potential mainly. We shall also consider the analysis in one dimension (the x direction) and make a number of simplifying assumptions including: (1) the particle surface is taken to be flat and (2) the electrolyte is symmetrical (e.g., HCl), so the valence of the positive ions is equal in magnitude to the valence of the negative ions.

The variation in the electrical potential ϕ with distance x is governed by Poisson's equation:

$$\nabla^2\phi(x) = -\frac{\rho(x)}{\varepsilon\varepsilon_o} \tag{4.27}$$

where $\rho(x)$ is the charge density at a distance x from the surface, ε is the dielectric constant of the liquid medium, and ε_0 is the pemittivity of vacuum. The charge density is obtained from the sum of the contributions of the individual ions:

$$\rho(x) = \sum_i c_i z_i F \tag{4.28}$$

where c_i is the concentration, z_i is the valence of the i ions, F is the Faraday constant, and the summation is over the positive and negative ions in the solution. The concentration c_i is given by the Boltzmann distribution:

$$c_i = c_{\infty i}\exp\left(\frac{-z_i F\phi(x)}{RT}\right) \tag{4.29}$$

where $c_{\infty i}$ is the concentration of the i ions very far from the particle surface, R is the gas constant, and T is the absolute temperature. Equation 4.27 to Equation 4.29 gives:

$$\nabla^2\phi(x) = -\frac{1}{\varepsilon\varepsilon_0}\sum_i c_{\infty i} z_i F\exp\left(\frac{-z_i F\phi(x)}{RT}\right) \tag{4.30}$$

Summing over the positive (+) and negative (−) ions and putting $z_+ = -z = z$ (symmetrical electrolyte) and $c_{\infty+} = c_{\infty-} = c$ (the concentrations of the positive and negative ions far from the surface are equal), Equation 4.30 reduces to

$$\nabla^2\phi(x) = \frac{2czF}{\varepsilon\varepsilon_o}\sinh\left(\frac{zF\phi(x)}{RT}\right) \tag{4.31}$$

For low surface potential, referred to as the *Debye–Hückel approximation*, so that

$$zF\phi \ll RT \quad \text{(Debye–Hückel approximation)} \tag{4.32}$$

and making the substitution

$$K^2 = \frac{2cz^2 F^2}{\varepsilon\varepsilon_o RT} \tag{4.33}$$

Equation 4.31 becomes

$$\nabla^2\phi(x) = K^2\phi(x) \tag{4.34}$$

With the appropriate boundary conditions:

$$\phi(x) = \phi_o \text{ at } x = 0; \quad \phi(x) = 0 \text{ at } x = \infty \tag{4.35}$$

where ϕ_o is the potential at the surface of the particle, the solution of Equation 4.34 is

$$\phi(x) = \phi_o \exp(-Kx) \tag{4.36}$$

The term K occurs frequently in the analysis of the electrical double layer, and 1/K has the dimensions of length. At a distance $x = 1/K$ the potential ϕ has fallen to $1/e$ of its value at the surface of the particle, and beyond this the change in ϕ is small. Thus, 1/K may be considered the *thickness* of the double layer and is usually referred to as the *Debye length*. According to Equation 4.33, the thickness of the double layer depends on a number of experimental parameters. As we shall see later, these parameters can be varied to control the magnitude of the repulsion between two double layers and thereby the stability of the suspension.

Equation 4.31 can be integrated twice to give [30]:

$$\tanh\left(\frac{zF\phi(x)}{4RT}\right) = \tanh\left(\frac{zF\phi_o}{4RT}\right)\exp(-Kx) \tag{4.37}$$

Equation 4.37, referred to as the *Gouy–Chapman equation*, is valid for any value of the surface potential, ϕ_0. However, for ϕ_0 less than ~50 mV, the difference in the ϕ values found from the Debye–Hückel approximation and the Gouy–Chapman equation is insignificant (Figure 4.11).

A more rigorous analysis shows that the electrical double layer consists of a compact layer (about a few molecular diameters thick), referred to as the *Stern layer*, and a more diffuse layer referred to as the *Gouy–Chapman layer*. The electrical potential in the Gouy–Chapman layer decreases exponentially with distance according to Equation 4.36 or Equation 4.37 but decreases less steeply in the Stern layer (Figure 4.12). However, this refinement in the double-layer theory will not be considered any further in this book.

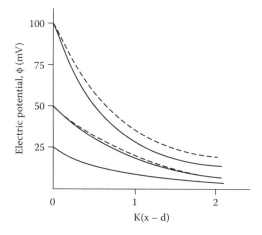

FIGURE 4.11 Comparison of the predictions of the Debye–Hückel equation (dashed line) and Gouy–Chapman equation (full line) for the potential ϕ in the electrical double layer. The Debye–Hückel equation can be used with insignificant error up to ~50 mV. (From Goodwin, J.W., Ed., *Colloidal Dispersions*, The Royal Society of Chemistry, London, p. 61, 1982.)

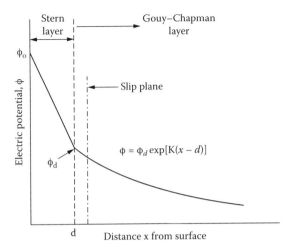

FIGURE 4.12 The electrical double layer consists of a compact Stern layer and a more diffuse Gouy–Chapman layer.

4.5.4 Surface Charge

For electroneutrality, the surface charge density (charge per unit area) must be equal to the integrated charge density in the solution; that is

$$\sigma_o = -\int_0^\infty \rho \, dx \tag{4.38}$$

Substituting for ρ from Equation 4.27, then

$$\sigma_o = \int_0^\infty \varepsilon\varepsilon_o \nabla^2 \phi(x) dx = \left[\varepsilon\varepsilon_o \frac{d\phi}{dx} \right]_0^\infty = -\varepsilon\varepsilon_o \left(\frac{d\phi}{dx} \right)_{x=0} \tag{4.39}$$

Equation 4.31 can be integrated once to give:

$$\left(\frac{d\phi}{dx} \right)_{x=0} = -\left(\frac{8RTc}{\varepsilon\varepsilon_o} \right)^{1/2} \sinh\left(\frac{zF\phi_o}{2RT} \right) \tag{4.40}$$

Substituting in Equation 4.39 gives:

$$\sigma_o = (8\varepsilon\varepsilon_o RTc)^{1/2} \sinh\left(\frac{zF\phi_o}{2RT} \right) \tag{4.41}$$

For low potential (Debye–Huckel approximation), Equation 4.41 becomes:

$$\sigma_o = \varepsilon\varepsilon_o K\phi_o \tag{4.42}$$

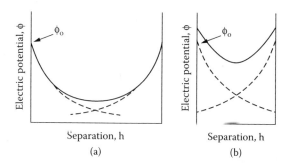

FIGURE 4.13 The electrical potential in the electrolyte solution between two parallel surfaces for (a) large separation and (b) small separation. As the separation is decreased, the potential is increased, implying a repulsion.

The capacitance is defined as the charge divided by the voltage and for a parallel plate capacitor, the capacitance is equal to $\varepsilon\varepsilon_o/t$, where t is the distance between the plates. We can therefore see from Equation 4.42 that the electrical double layer can be treated as a parallel plate capacitor with a thickness of $1/K$.

4.5.5 REPULSION BETWEEN TWO DOUBLE LAYERS

Two colloid particles will begin to interact as soon as their double layers overlap. We consider the simplest case of two parallel surfaces at a distance h apart in an electrolyte. The electrical potential within the double layers consists of two symmetrical curves resembling that shown in Figure 4.10b, and the net effect is roughly additive (Figure 4.13). As the distance h decreases the overlap of the double layers causes the potential to increase, giving rise to a repulsive force that tends to oppose further approach.

The general theory of the interaction between electrical double layers is known as the *DLVO theory* (after Derjaguin, Landau, Verwey, and Overbeek). It applies to the diffuse Gouy–Chapman layer. The theory is beyond the scope of this book, but if we consider the particles to be two parallel surfaces, the repulsion can be approximated by a fairly simple method, called the *Langmuir force method*. The interaction between the particles can be analyzed in terms of a repulsive force or a repulsive potential energy (i.e., the work done in bringing the particles from infinity to the desired distance apart). Normally, the repulsive potential energy is used.

The ionic concentration at a point midway between the two charged surfaces (at A in Figure 4.14) will be greater than that far away from the surfaces (at B). This gives rise to an excess osmotic pressure Π that acts to push the surfaces apart. The excess osmotic pressure at A is given by [1]:

$$\Pi = 2RTc\left[\cosh\left(\frac{zF\phi_m}{RT}\right) - 1\right] \tag{4.43}$$

where ϕ_m is the net electrostatic potential at A arising from the overlap of the double layers. For large separation h, the net potential is roughly additive, so

$$\phi_m = 2\phi \tag{4.44}$$

where ϕ is the potential of an isolated double layer at A. When ϕ_m is small, that is, $zF\phi_m \ll RT$, Equation 4.43 reduces to:

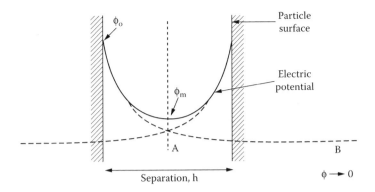

FIGURE 4.14 Overlap of the electrical double layers associated with two parallel surfaces leads to an increase in the ionic concentration. The osmotic pressure that results from the increased concentration acts to push the surfaces apart.

$$\Pi = \frac{cz^2 F^2 \phi_m^2}{RT} \tag{4.45}$$

For low surface potential ϕ_o, the Debye–Hückel approximation can be used:

$$\phi = \phi_o \exp(-Kh/2) \tag{4.46}$$

The repulsive potential energy V_R between the charged surfaces at a separation h is the work done in bringing the surfaces from ∞ to a distance h apart, so that

$$V_R = \int_h^\infty \Pi \, dh \tag{4.47}$$

Using Equation 4.44 to Equation 4.46 and integrating,

$$V_R = 2\varepsilon\varepsilon_o K \phi_o^2 \exp(-Kh) \tag{4.48}$$

Equation 4.48 is valid for low ϕ_o and for large separations when the plates interact under conditions of constant potential or constant charge.

For large ϕ_o, the Gouy–Chapman equation must be used (Equation 4.37). Using Equation 4.45 (valid for low ϕ_m) and integrating [1], we obtain:

$$V_R = 2\varepsilon\varepsilon_o K \left(\frac{4RT\gamma}{zF} \right)^2 \exp(-Kh) \tag{4.49}$$

where γ is defined by

$$\gamma = \tanh\left(\frac{zF\phi_o}{4RT} \right) \tag{4.50}$$

For two spherical particles of radius a at a distance h apart (see Figure 4.3c), when the double layer around the particles is very extensive such that $\mathrm{K}a < 5$, V_R is given approximately by [2].

$$V_R \approx 2\pi a \varepsilon \varepsilon_o \phi_o^2 \exp(-\mathrm{K}h) \qquad (4.51)$$

Equation 4.51 is valid for low ϕ_o and for constant potential or constant charge. For large values of $\mathrm{K}a$ (> 10) and for low ϕ_o, V_R is given by:

$$V_R = 2\pi a \varepsilon \varepsilon_o \phi_o^2 \ln[1 + \exp(-\mathrm{K}h)] \qquad (4.52)$$

4.5.6 Stability of Electrostatically Stabilized Colloids

We now consider what happens when two colloid particles approach one another. As the double layers start to overlap, the resulting repulsion opposes the attraction from van der Waals interactions. The total potential energy V_T is given by:

$$V_T = V_A + V_R \qquad (4.53)$$

Using the convention that repulsive potentials are positive and attractive potentials are negative, we have shown in Figure 4.15 an example of V_A for the van der Waals attraction and V_R for the double layer repulsion. The resultant curve for V_T shows a deep minimum at M_1 corresponding approximately to contact between the particles, a secondary minimum at M_2, and a maximum between M_1 and M_2. For two particles initially far apart approaching one another to a separation M_2, if the thermal energy (kT) of the particles is small compared to the depth of M_2, the particles will not be able to escape from one another. Flocculation will result, leading to a sediment of loosely packed particles. Restabilization of the colloid can be achieved by heating (increasing kT), by changing the electrolyte concentration to increase the double layer repulsion, or a combination of the two. This process of restabilizing a flocculated suspension is referred to as *peptization*.

In practice, the resultant curve for V_T depends on the relative magnitudes of V_A and V_R. For a given system (e.g., a suspension of Al_2O_3 particles in water), V_A is approximately constant but V_R

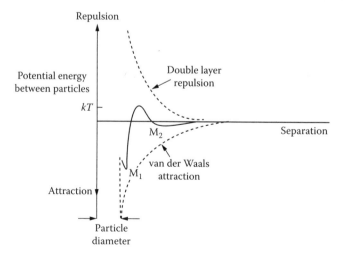

FIGURE 4.15 The potential energy between two particles in a liquid resulting from the effects of the van der Waals attraction and the double layer repulsion.

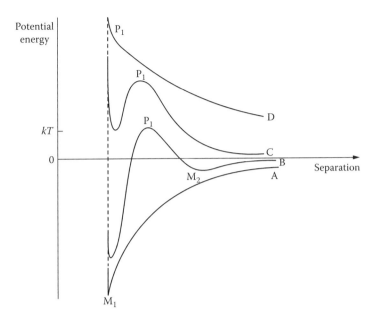

FIGURE 4.16 Four examples of the resultant potential energy between two particles. The repulsion between the particles increases in going from curve A to curve D.

can be changed significantly by changing the concentration and the valence of the ions in solution (see Equation 4.51). These two experimental parameters provide a useful way for controlling the stability of the colloid. Figure 4.16 summarizes four main types of curves for V_T. In curve A, the repulsion is weak and V_T is not significantly different from the curve for the van der Waals curve. The particles attract one another and reach equilibrium at the primary minimum M_1. A sediment consisting of clusters of particles almost in contact is produced. This is known as coagulation, in contrast to flocculation in which, as we have outlined above, the particles are loosely held together at larger separations. In curve B the repulsion is increased and this leads to a shallow secondary minimum at M_2 which is shallow compared to kT. There is therefore little chance for flocculation. However, the height of the maximum at P_1 is comparable to kT so that particles may be able to surmount the energy barrier at P_1 and fall into the primary minimum at M_1. Irreversible coagulation will result because the depth of M_1 is much greater than kT.

The double layer repulsion has become so large in curve C that there is no secondary minimum at M_2. Furthermore, the height of the maximum P_1 is much greater than kT so that the particles have almost no chance of surmounting the energy barrier. Curve C, therefore, represents the situation for a stable colloid. Finally, in curve D, the attraction is much smaller than the repulsion over all separations so that there is no minimum in the curve. An extremely stable colloid will be produced under these conditions, but the strong repulsion limits the concentration of particles in suspension. Addition of a salt can lead to compression of the double layer and a reduction in the viscosity.

Some general features of the DLVO theory have been confirmed [31] by direct measurement of the surface forces between two sapphire platelets immersed in an aqueous solution of NaCl at pH values from 6.7 to 11 using a surface force apparatus [32]. Measurements of the electrical double layer and van der Waals forces were found to fit the theoretical predictions.

4.5.7 Kinetics of Flocculation

Flocculation is a kinetic process, and the rate at which a colloidal suspension flocculates forms one of its most important characteristics. Smoluchowski (1917) distinguished between rapid flocculation and slow flocculation, and developed a theory based on the rate of collision between the particles [2]. Rapid flocculation is considered to take place in the absence of a potential barrier and is limited only by the rate of diffusion of the particles toward one another. The flocculation time, defined as the time $t_{1/2}$ required for the number of particles to be reduced by one half of the initial value, is given by:

$$t_{1/2} = \frac{3\eta}{4kTn_o} \qquad (4.54)$$

where η is the viscosity of the liquid, k is the Boltzmann constant, T is the absolute temperature, and n_o is the particle concentration (number per unit volume). For particles of radius a in water, Equation 4.54 becomes:

$$t_{1/2} \approx \frac{2 \times 10^{11}}{n_o} \text{ sec} \qquad (4.55)$$

where n_o is expressed in units of number per cubic centimeter. The particle volume fraction f is related to n_o by the relation $n_o = f/(4\pi a^3/3)$, so Equation 4.55 can also be written as:

$$t_{1/2} \approx \frac{a^3}{f} \text{ sec} \qquad (4.56)$$

where a is expressed in microns (μm).

For slow flocculation in the presence of a potential barrier, analysis leads to

$$R_s = R_f / W \qquad (4.57)$$

where R_s and R_f are the rates of slow and rapid flocculation, respectively, and W is a factor known as the *stability ratio*. For a repulsive potential energy barrier with a maximum V_{max}, the stability ratio is given by:

$$W = \frac{1}{2Ka} \exp\left(\frac{V_{max}}{kT}\right) \qquad (4.58)$$

showing that W depends exponentially on V_{max} and linearly as the normalized double layer thickness $1/(Ka)$.

4.5.8 ELECTROKINETIC PHENOMENA

Electrokinetic phenomena involve the combined effects of motion and an electric field. When an electric field is applied to a colloidal suspension, the particles move with a velocity that is proportional to the applied field strength. The motion is called *electrophoresis*. It is a valuable source of information on the sign and magnitude of the charge and on the potential associated with the double layer. The measured potential, called the ζ potential, is an important guide to the stability of lyophobic colloids [25]. A widely used method for measuring the ζ potential is the microelectrophoretic technique, in which the motion of individual particles is followed in a microscope. The technique is used with very dilute suspensions. Modern instrumentation provides for automated, rapid measurements and for the use of concentrated suspensions.

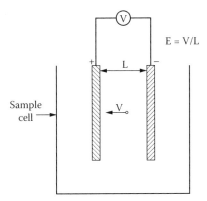

FIGURE 4.17 Schematic of a particle electrophoresis cell.

4.5.8.1 Microelectrophoretic Technique

A schematic of a particle electrophoresis apparatus is shown in Figure 4.17. The suspension is placed in a cell, and a dc voltage V is applied to two electrodes at a fixed distance l apart. The sign of the particle charge is obtained directly, as it is opposite to that of the electrode toward which the particle is migrating. The particle velocity is measured by using a microscope and the velocity per unit field strength (the electrophoretic mobility) is used to determine the ζ potential and the surface charge.

A particle with a charge q in an electric field E experiences a force directed toward the oppositely charged electrode given by:

$$F = qE \tag{4.59}$$

In a viscous medium, the terminal velocity v is reached rapidly, so Stokes law can be used to account for the force on the particle:

$$F = 6\pi\eta av \tag{4.60}$$

where η is the viscosity of the medium and a is the radius of the particle. From these two equations, the electrophoretic mobility is obtained:

$$u = \frac{v}{E} = \frac{q}{6\pi\eta a} \tag{4.61}$$

For particles in a dilute electrolyte solution (i.e., for $Ka < 0.1$), the potential on the surface of the particle can be taken as that of an isolated particle, that is,

$$\zeta = \frac{q}{4\pi\varepsilon\varepsilon_o a} \tag{4.62}$$

Substituting for q in Equation 4.61 and Equation 4.62 gives:

$$u = \frac{2\varepsilon\varepsilon_o\zeta}{3\eta} \tag{4.63}$$

For concentrated electrolyte solutions ($Ka > 200$), the Helmholtz–Smoluchowski equation can be used:

$$u = \frac{\varepsilon \varepsilon_o \zeta}{\eta} \qquad (4.64)$$

For values of Ka between 0.1 and 200, Henry's equation applies.

$$u = \frac{2\varepsilon \varepsilon_o \zeta}{3\eta}[1 + f(Ka)] \qquad (4.65)$$

where $f(Ka)$ is a function of Ka, having the values:

Ka	0	0.1	1	5	10	50	100	∞
$f(Ka)$	0	0.0001	0.027	0.16	0.239	0.424	0.458	0.5

4.5.8.2 Significance of the ζ Potential

The ζ potential determined from the electrophoretic mobility represents the potential at the surface of the electrokinetic unit moving through the solution. It is not the potential at the surface of the particle but must correspond to a surface removed from the particle surface by at least one hydrated radius of the counterion (sometimes referred to as the shear plane). The ζ potential is close to the Stern potential (ϕ_d in Figure 4.12). It is the appropriate potential to be used in effects that depend on the diffuse double layer (e.g., the double layer repulsion discussed earlier) provided that a reasonable guess can be made as to the distance of the ζ potential from the surface of the particle.

The surface charge calculated from the ζ potential is the charge within the shear surface. If the particle surface is assumed to be planar, the surface charge density, σ_d, may be determined approximately from Equation 4.41 or Equation 4.42. As the ζ potential and σ_d depend only on the surface properties and the charge distribution in the electrical double layer, they are independent of the particle size.

Typical data are shown in Figure 4.18 for the electrophoretic mobility and surface charge density of TiO_2 (rutile) as a function of pH in aqueous solutions of potassium nitrate [33]. The ζ potential will show the same variation with pH as it is proportional to the mobility. For many oxides, ζ-potential values determined from the mobility generally fall within the range of +100 to –100 mV.

The ζ potential provides a measure of the stability of colloidal suspensions. Suspensions prepared at pH values close to the isoelectric point (IEP) may flocculate fairly rapidly because the repulsion may not be sufficient to overcome the van der Waals attraction. Farther away from the IEP, we should expect the rate of flocculation to be slower. In practice, for good stability, suspensions are often prepared at pH values comparable to those of the plateau regions of the ζ potential or electrophoretic mobility curve. For the data shown in Figure 4.18, this corresponds to pH values of <5 or >7.

4.6 STERIC STABILIZATION

Steric stabilization is the term used to describe the stabilization of colloidal particles which results from the interaction between uncharged polymer chains adsorbed onto the particle surfaces (Figure 4.3b). The interactions among the polymer chains are fundamentally different from those among the charged ions in electrostatic stabilization and are dominated by the configurational entropy of the chains. Steric stabilization is commonly associated with suspensions in organic liquids, but it

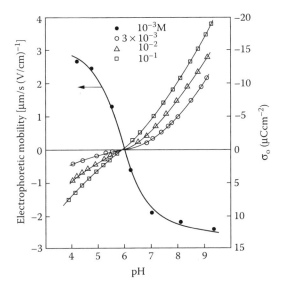

FIGURE 4.18 Surface charge density and electrophoretic mobility as functions of pH for rutile in aqueous solutions of potassium nitrate. (From Jang, H.M. and Fuerstenau, D.W., The specific adsorption of alkaline-earth cations at the rutile/water interface, *Colloids Surf.* 21, 235, 1986.)

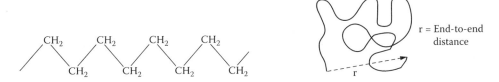

FIGURE 4.19 Two configurations of a polymer chain: (a) an extended chain in a highly extended polymer; (b) a coiled chain in a solution. The end-to-end distance of the coil is r.

is also effective for aqueous solvents. Exploited for over 4500 years since the Egyptians applied the principle empirically in the production of ink for writing on papyrus, steric stabilization is today utilized in a wide range of industrial products such as paints, inks, coatings, and pharmaceuticals. In ceramic processing, it is widely used for the production of stable suspensions in the consolidation of ceramic powders by casting methods such as slip casting and tape casting.

Steric stabilization is treated in detail by Napper [34,35]. For effective stabilization, the adsorbed polymer chains must be well anchored to the particle surfaces to reduce the risk of desorption, and the adsorbed layer must be reasonably thick to provide sufficient repulsion when the particles approach one another.

4.6.1 ADSORPTION OF POLYMERS FROM SOLUTION

Polymer chains existing freely in solution can adopt a large number of different configurations and, in thermal equilibrium, this leads to a large configurational entropy contribution to their free energy [36]. Because of the mutual van der Waals attraction between the monomer units, the chains have a tendency to collapse to a small solid ball, but this is resisted by the reduction in configurational entropy that it would entail. Polymer chains in solution therefore tend to adopt an open, random coil structure (Figure 4.19). The diameter of the coil is a difficult parameter to calculate. It can be taken as approximately the root mean square (rms) end-to-end distance $\langle r^2 \rangle^{1/2}$ of the polymer chain, given by:

$$\left\langle r^2 \right\rangle^{1/2} = l\sqrt{N} \qquad (4.66)$$

where l is the length of a monomer unit and N is the number units in the chain (i.e., the degree of polymerization). We see that for $l \approx 0.1$ nm and $N \approx 10^4$, then $\left\langle r^2 \right\rangle^{1/2} \approx 10$ nm.

Homopolymers, such as polystyrene and polyethylene oxide, consisting of a single type of monomer in the polymer chain, adsorb physically on the particle surfaces by weak van der Waals forces. Individual monomer segments are weakly attracted to the particle surface, but a large number of segments may contact the surface at any given time, so the sticking energy of each chain is still much greater than its thermal energy ($\approx kT$, where k is the Boltzmann constant, and T is the absolute temperature). The result is irreversible adsorption. Under these conditions, the adsorbed polymer chain adopts a different conformation from the coil conformation in free solution but has a similar characteristic size when "trains" of adsorbed segments act as anchor points between "loops" and "tails" of segments extending into solution (Figure 4.20a). Strong segmental attraction to the particle surface is undesirable because it leads to a flattened conformation.

Steric stabilization by adsorbed homopolymers suffers from the conflicting requirements that the liquid be a poor solvent to ensure strong adsorption but a good solvent to impart a strong repulsion when the adsorbed polymer chains overlap. At low polymer concentrations, an individual polymer chain can become simultaneously adsorbed on two (or more) surfaces, resulting in an attractive interaction known as *bridging flocculation* (see Section 4.6.4).

A more effective approach is to use polymer chains consisting of two parts (graft or block copolymers). Figure 4.20b and Figure 4.20c show these two types of copolymers in which one part is nominally insoluble and anchors (chemically or physically) onto the particle surface, and the other is soluble in the liquid and extends into the liquid. Examples of commonly chosen block copolymers in organic solvents (e.g., toluene) are poly(vinyl pyrrolidone)/polystyrene (PVP/PS) or poly(ethylene oxide)/polystyrene (PEO/PS).

The incorporation of a single polar group to the end of the polymer chain can help to strengthen the anchoring to the particle surface by the formation of hydrogen bonds (in aqueous solvents) or coordinate bonds (in aqueous or nonaqueous solvents). A useful way to describe the polar interactions is by the Lewis acid/base concept or the more general donor-acceptor concept [38]. In the original form of the concept, a Lewis acid is defined as a substance capable of accepting a pair of electrons from another species, whereas a Lewis base is a substance capable of donating a pair of

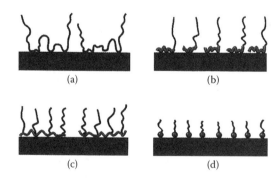

(a)   (b)

(c)   (d)

FIGURE 4.20 Schematic illustration of the adsorbed polymer conformation on a ceramic surface as a function of the chemical structure and composition: (a) homopolymer consisting of tails, loops, and trains, (b) block copolymer consisting of a short anchoring block and an extended chain block, (c) copolymer consisting of extended segments attached to anchored backbone, (d) short-chain polymer (surfactant) consisting of an anchoring functional head group and an extended tail. (From Lewis, J.A., Colloidal processing of ceramics, *J. Am. Ceram. Soc.*, 83, 2341, 2000.)

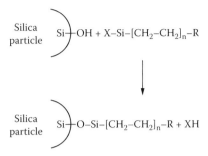

FIGURE 4.21 Schematic illustration of the formation of a chemically bonded polymer chain on the surface of an oxide particle (e.g., SiO_2).

electrons. A basic functional group, for example, attaches itself to acidic functional sites on the particle surface, whereas an acidic functional attaches itself to basic functional group. The resulting polymer-coated surface has a brush-like structure, with the chains attached at one end while the nonadsorbing chain protrudes into the solvent (Figure 4.20d). An example is polyisobutene succinamide (OLOA-1200), one of the most widely used dispersants in nonaqueous solvents, in which the nitrogen atom in the basic succinamide group can form a coordinated bond with a Lewis acid site (e.g., an electron deficient metal atom) on the particle surface (see Chapter 7).

Another way of achieving strong anchoring is to chemically attach the polymer chain to the particle surface via functional groups that react with specific sites [39,40]. This eliminates the need for anchoring segments on the chain. One type of approach involves the reaction of surface hydroxyl groups with chlorosilane or alkoxysilane groups of the polymers. This is illustrated in Figure 4.21 by a silane terminated polymer chain with silica.

4.6.2 ORIGINS OF STERIC STABILIZATION

Let us consider what happens when two particles covered with polymer molecules come within range of one another (Figure 4.22a). The polymer chains will start to interact when the distance between the particles is equal to 2L, where L is the thickness of the adsorbed polymer layer (see Figure 4.20). For a long-chain homopolymer, $L \approx \langle r^2 \rangle^{1/2}$. On further approach, such that the interparticle distance is between L and $2L$ (Figure 4.22b), the polymer chains may interpenetrate. The concentration of the polymer is increased in the interpenetration region, and, in some solvents, this can lead to a repulsion. (In other solvents, as discussed below, the interpenetration of the coils can lead to an attraction, leading to flocculation.) The repulsion is said to arise from a mixing (or

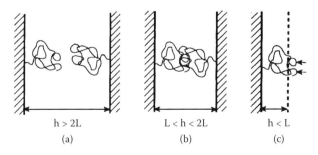

h > 2L	L < h < 2L	h < L
(a)	(b)	(c)

FIGURE 4.22 The range of steric repulsion in polymeric stabilization. (a) There is no interpenetration for large separations. (b) When the particles become close enough, interpenetration of the coils leads to a repulsion in good solvents or an attraction in poor solvents. (c) For small separations, compression of the coils always leads to a repulsion.

TABLE 4.4
Interactions between Adsorbed Polymer Chains in Steric Stabilization

Interaction	Nature of Force
Mixing effect (osmotic effect)	Repulsive (good solvent)
	Attractive (poor solvent)
Elastic effect (entropic effect of volume restriction effect)	Repulsive (always)

rather a demixing) effect due to the need of the polymer chains to avoid other chains in the interpenetration region of increased concentration. The repulsion can also be thought of as corresponding to an osmotic pressure (i.e., due to the increased concentration in the interpenetration region) and is sometimes described in these terms.

As the distance of approach between the particles decreases further to less than L (Figure 4.22c), not only does interpenetration occur but the polymer chain on one particle may be compressed by the rigid surface of the other particle. This compression generates an elastic contribution to the stabilization that always opposes flocculation. The elastic contribution to the repulsion is sometimes described as an entropic effect or a volume restriction effect. At these separations, there are regions of space which are no longer accessible to a given chain. Conformations that would otherwise have been accessible are excluded so that there is a loss of configurational entropy. The loss in entropy is given by the Boltzmann equation:

$$\Delta S = k \ln \Omega \tag{4.67}$$

where k is the Boltzmann constant and Ω is the loss in the number of configurations of the polymer chain. The free energy change due to the loss in entropy is

$$\Delta G_{en} = -T\Delta S \tag{4.68}$$

where T is the absolute temperature. The term ΔG_{en} is always positive (because ΔS is negative), and this is equivalent to a repulsion.

To summarize at this stage, the interactions (Table 4.4) between the adsorbed polymer chains may be separated into (1) a mixing effect that produces either a repulsion or an attraction and (2) an elastic effect that is always repulsive. The mixing effect is also described as an osmotic effect, whereas the elastic effect is described as an entropic effect or a volume restriction effect. The free energy of the polymeric interaction can be written as:

$$\Delta G_{steric} = \Delta G_{mix} + \Delta G_{elastic} \tag{4.69}$$

where ΔG_{mix} and $\Delta G_{elastic}$ are the free energy changes due to the mixing and elastic effects, respectively.

4.6.3 EFFECTS OF SOLVENT AND TEMPERATURE

A difference between electrostatic and steric stabilization is the significant influence that the solvent and the temperature can have in steric stabilization. As we saw above, interpenetration of the polymer chains gives rise to a mixing effect. At a certain temperature, referred to as the Θ (theta)

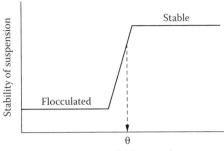

FIGURE 4.23 Schematic representation of the stability of a suspension as a function of the temperature of the polymer solution. The stability changes dramatically within a few degrees of the Θ temperature of the solution.

temperature, the interpenetration of the polymer chains does not lead to a change in the free energy of mixing ($\Delta G_{mix} = 0$) and the polymer solution behaves like an ideal solution. The solvent is referred to as a Θ solvent.

At temperatures greater than the Θ temperature, the interpenetration of the polymer chains leads to an increase in the free energy of mixing ($\Delta G_{mix} > 0$), that is to say that the polymer chains prefer contact with the solvent to contact with one another. The solvent is referred to now as a *good solvent* (or a better than Θ solvent). The repulsion between the chains leads to stability of the suspension. Below the Θ temperature, the polymer chains prefer contact with one another to contact with the solvent ($\Delta G_{mix} < 0$) and in this case, the solvent is referred to as a *poor solvent* (or a worse than Θ solvent). The attraction between the chains may lead to flocculation of the suspension. As shown schematically in Figure 4.23, we would, therefore, expect the stability of the suspension to change very markedly within a few degrees of the Θ temperature of the polymer solution. In practice, for many sterically stabilized suspensions, the temperature at which the stability changes dramatically (referred to as the *critical flocculation temperature*, CFT) correlates very well with the Θ temperature of the solution (Table 4.5).

TABLE 4.5
Comparison of the Theta (Θ) Temperature of Solutions of Steric Stabilizers with the Critical Flocculation Temperatures (CFT) of Suspensions

Stabilizer	Molecular Weight	Dispersion Medium	Θ (K)	CFT (K)
Poly(ethylene oxide)	10,000	0.39 M MgSO$_4$	315 ± 3	318 ± 2
Poly(acrylic acid)	51,900	0.2 M HCl	287 ± 5	283 ± 2
Poly(vinyl alcohol)	57,000	2 M NaCl	300 ± 3	310 ± 3
Polyisobutylene	760,000	2-Methylbutane	318	327
Polyisobutylene	760,000	Cyclopentane	461	455
Polystyrene	110,000	Cyclopentane	410	427
Polystyrene	110,000	Cyclopentane	280	293

Source: From Napper, D.H., Polymeric stabilization, in *Colloidal Dispersions*, Goodwin, J.W., Ed., The Royal Society of Chemistry, London, 1982, p. 99, chap. 5.

4.6.4 STABILITY OF STERICALLY STABILIZED SUSPENSIONS

The total energy of interaction V_T for sterically stabilized suspensions is now the sum of the steric interaction, ΔG_{steric}, given by Equation 4.69, and the van der Waals energy of attraction V_A; that is,

$$V_T = \Delta G_{steric} + V_A \tag{4.70}$$

Napper [34] gives theoretical expressions for ΔG_{mix} and $\Delta G_{elastic}$. For interactions between adsorbed homopolymers in the region $L < h < 2L$ (see Figure 4.22), where chain conformations from only trains and loops are taken into consideration, ΔG_{mix} is given by [41,42]:

$$\Delta G_{mix} = \frac{32\pi akT}{5v_1} \frac{(\bar{\phi}_2^a)^2}{L^4} \left(\frac{1}{2} - \chi\right) \left(L - \frac{h}{2}\right)^6 \tag{4.71}$$

where a is the radius of the particle, k is the Boltzmann constant, T is the absolute temperature, $\bar{\phi}_2^a$ is the average volume fraction of segments in the adsorbed layer (measured as ≈ 0.37), v_1 is the molar volume of the solvent, and χ is the Flory–Huggins parameter: $\chi > 0.5$ for a poor solvent and $\chi < 0.5$ for a good solvent. At smaller interparticle separations, $h < L$, the polymer segment density is assumed to be uniform, and the contributions from the elastic and mixing interactions are given by:

$$\Delta G_{mix} = \frac{4\pi aL^2 kT}{v_1} (\bar{\phi}_2^a)^2 \left(\frac{1}{2} - \chi\right) \left(\frac{h}{2L} - \frac{1}{4} - \ln\frac{h}{L}\right) \tag{4.72}$$

$$\Delta G_{elastic} = \frac{2\pi aL^2 kT \rho_2 \bar{\phi}_2^a}{M_2^a} \left\{ \frac{h}{L} \ln\left[\frac{h}{L}\left(\frac{3 - h/L}{2}\right)^2\right] - 6\ln\left(\frac{3 - h/L}{2}\right) + 3(1 - h/L)\right\} \tag{4.73}$$

where ρ_2 is the density and M_2^a the molecular weight of the adsorbed polymer.

The interaction energy in sterically stabilized systems has been measured [43,44] using the surface force apparatus described earlier, and the variation in V_T is can be illustrated with the results shown in Figure 4.24 for the interaction between mica surfaces in the presence of a solution of polystyrene in cyclopentane. The degree of polymerization of the polymer (i.e., the number of segments in the chain) is 2×10^4 which gives $\langle r^2 \rangle^{1/2} \approx 15$ nm. In the absence of polymer molecules there is simply a short-range van der Waals attraction (curve a). This is replaced by a deep, long-ranged attractive well at low coverage of the surface by the polymer chains (curve b). At progressively increasing surface coverage, the attractive well becomes shallower and the range of the final repulsive potential increases (curves b to c). Finally, when equilibrium coverage of the mica surfaces is attained, the interaction is repulsive at all separations (curve d).

The results shown in Figure 4.24 serve to illustrate the influence of two important effects in steric stabilization: the relative unimportance of van der Waals attraction in steric stabilization, especially at fairly high coverage, and the significance of the surface coverage by the adsorbed polymer chains. In certain cases, low coverage may actually lead to flocculation. This occurs when the particle has usable anchoring sites that are not occupied, and for homopolymers having a large number of segments that can contact the particle surface. Polymers form a neighboring particle can then attach themselves and form bridges, leading to flocculation (see Figure 4.26). This type of flocculation is referred to as *bridging flocculation*. As outlined earlier, the use of block or graft copolymers or polymers with functional end groups can reduce this type of flocculation.

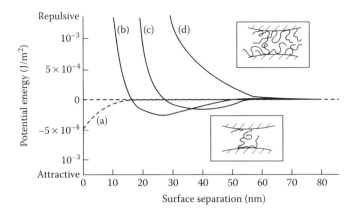

FIGURE 4.24 Interaction energy between mica surfaces in the presence of a solution of polystyrene in cyclopentane. In the absence of polymer molecules (a), there is a simple van der Waals attraction. At low surface coverage bridging (lower inset) dominates. As more polymer adsorbs (b) and (c), the steric repulsion counteracts the attraction, until at high surface coverage (d), the interaction is repulsive at all separations. (From Klein, J., Entropic interactions, *Phys. World*, 2, 35, 1989.)

4.6.5 STABILIZATION BY POLYMERS IN FREE SOLUTION

It has been suggested that polymers in free solution can also impart stability to colloidal suspensions. To understand how this type of stabilization might occur, let us consider the approach of two particles from a large separation (Figure 4.25a). Closer approach of the particles must be accompanied by demixing of the polymer molecules and the solvent in the interparticle region. Consequently, work must be done to make the polymer molecules leave the interparticle region. This

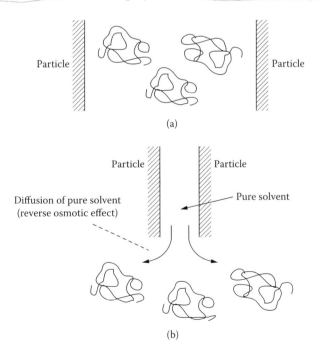

FIGURE 4.25 Polymeric stabilization by free polymer molecules in solution. (a) Depletion stabilization at separations greater than the diameter of the polymer coil, and (b) depletion flocculation at small separations.

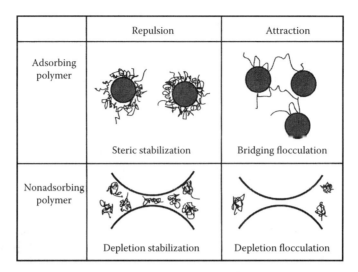

FIGURE 4.26 Summary of the most important effects occurring in the stabilization of suspensions with uncharged polymer chains. (From Horn, R.G., Surface forces and their action in ceramic materials, *J. Am. Ceram. Soc.*, 73, 1117, 1990.)

corresponds to a repulsion between the particles that, if high enough, can lead to stabilization of the suspension. This type of stabilization is referred to as *depletion stabilization*.

If the repulsion is not high enough, the particles can approach closer, and when the separation becomes smaller than the size of the polymer chain in solution (i.e., approximately $\langle r^2 \rangle^{1/2}$), all of the polymer chains would have been excluded from the interparticle region (Figure 4.25b). Closer approach of the particles will be favored because we have now created a type of reverse osmotic effect. The pure solvent between the particles will diffuse into the surrounding region in an attempt to lower the concentration of the polymer. This reverse osmotic effect is equivalent to an attraction between the particles and flocculation occurs. This type of flocculation is referred to as *depletion flocculation*.

The important effects in the stabilization of suspensions with uncharged polymer chains are summarized in Figure 4.26.

4.7 ELECTROSTERIC STABILIZATION

Suspensions can also be stabilized by electrosteric repulsion, involving a combination of electrostatic repulsion and steric repulsion (Figure 4.4c). Electrosteric stabilization requires the presence of adsorbed polymers and a significant double layer repulsion. It is commonly associated with suspensions in aqueous liquids but several studies have indicated that electrostatic effects can be important in some nonaqueous systems [46]. A common way of achieving electrosteric stabilization in aqueous liquids is through the use of polyelectrolytes, i.e., polymers that have at least one type of ionizable group (e.g., carboxylic or sulfonic acid group) that dissociates to produce charged polymers. The polymers can be either homopolymers such as poly(acrylic acid), block copolymers, or graft copolymers. Polyelectrolytes are widely used industrially in the preparation of highly concentrated ceramic suspensions (>50 vol% particles) which are subsequently consolidated and sintered to produce dense articles.

4.7.1 DISSOCIATION OF POLYELECTROLYTES IN SOLUTION

In aqueous solvents, polymers with ionizable groups develop electrostatic charges by dissociation. The dissociation of the polymers as well as their adsorption is strongly influenced by the properties

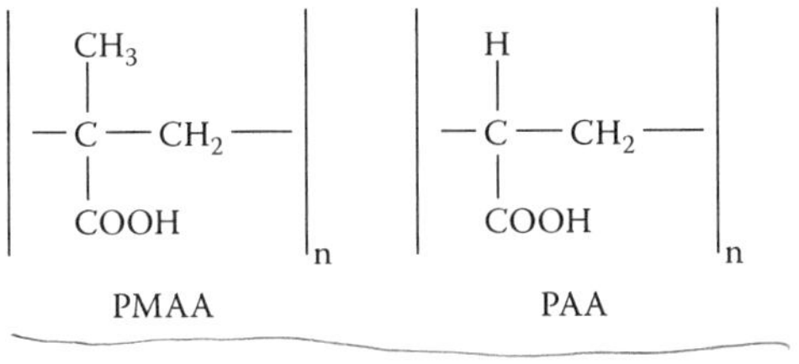

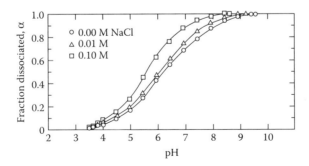

FIGURE 4.27 Schematic diagram showing the polymer segments of poly(methacrylic acid), PMAA, and poly(acrylic acid), PAA.

of the solvent and the particle surfaces [47–50]. Figure 4.27 shows schematically the structure of two common homopolymers used for electrosteric stabilization: poly(methacrylic acid), denoted as PMAA, and poly(acrylic acid), PAA, having a functional carboxylic acid (COOH) group. The sodium or ammonium salts of these polymers, where the H in the COOH group is replaced by Na or NH_4, are also commonly used in electrosteric stabilization. The molecular weight of the polymers can vary between ~1,000 and ~50,000. These polymers are examples of *anionic polyelectrolytes*, in that they dissociate to give negatively charged species. *Cationic polyelectrolytes* are positively charged on dissociation.

The functional groups of PMAA and PAA can exist as COOH or dissociated to COO^-. The dissociation reaction can be written in a general form as:

$$A - COOH + H_2O \rightleftarrows A - COO^- + H_3O^+ \tag{4.74}$$

Depending on the pH and the ionic concentration of the solution, the fraction of the functional groups that is dissociated (i.e., COO) and that is not dissociated (i.e., $COOH^-$) will vary. As the fraction of COOH groups dissociated, α, increases from 0 to 1, the charge on the polymer varies from neutral to highly negative. Figure 4.28 shows how α varies as a function of the pH and NaCl salt concentration for the sodium salt of PMAA with molecular weight of ~15,000. As the pH and salt concentration increases, the extent of dissociation and the negative charge of the polymer increase. At a pH greater than ~8.5, the polymer is highly negative with $\alpha \approx 1$. Under these conditions, the free polymer is in the form of relatively large expanded random coils (diameter ~10 nm) because of the repulsion between the charges segments. As the pH decreases, the number of negative charges decreases, with the polymer being effectively neutral at a pH of ~3.5 and forming small coils (diameter ~3 nm) or clumps.

FIGURE 4.28 Fraction of acid groups dissociated vs. pH for the sodium salt of poly(methacrylic acid), PMAA-Na, in water containing three different salt concentrations. (The molecular weight of the polyelectrolyte is 15,000.) (From Cesarano, J., III, Aksay, I.A., and Bleier, A., Stability of aqueous α-Al_2O_3 suspensions with poly(methacrylic acid) polyelectrolytes, *J. Am. Ceram. Soc.*, 71, 250, 1988.)

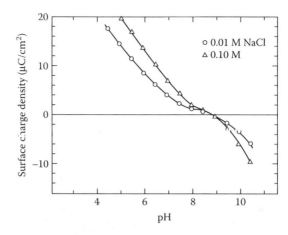

FIGURE 4.29 Surface charge density vs. pH for α-Al$_2$O$_3$ powder (Sumitomo AKP-30) in water containing two different salt concentrations. (From Cesarano, J., III, Aksay, I.A., and Bleier, A., Stability of aqueous α-Al$_2$O$_3$ suspensions with poly(methacrylic acid) polyelectrolytes, *J. Am. Ceram. Soc.*, 71, 250, 1988.)

4.7.2 ADSORPTION OF POLYELECTROLYTES FROM SOLUTION

Adsorption of polyelectrolyte onto the particle surface is commonly dominated by electrostatic interactions and is often strongly favored if the particle surface and the polyelectrolyte have opposite charges. As discussed earlier, the surface charge of oxide particles dispersed in aqueous solvents depends on the pH of the suspension. For example, Figure 4.29 shows the surface charge density, σ_o, of Al$_2$O$_3$ particles (Sumitomo AKP-30) as a function of pH for two NaCl salt concentrations. By comparing the data in Figure 4.28 and Figure 4.29, we see that considerable electrostatic attraction should occur between the negatively charged polyelectrolyte and the positively charged Al$_2$O$_3$, particularly in the pH range of ~3.5 to 8.5.

Figure 4.30 shows the measured adsorption isotherms for the sodium salt of PMAA (PMAA-Na) on the Al$_2$O$_3$ particles at various pH values. The results are plotted in the form of milligrams

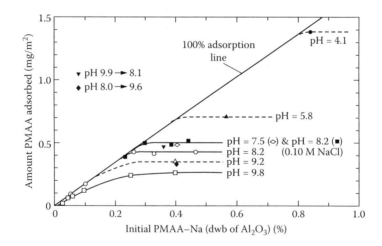

FIGURE 4.30 Amount of PMAA salt adsorbed on α-Al$_2$O$_3$ powder as a function of the initial amount of PMAA-Na added to the suspension. (From Cesarano, J., III, Aksay, I.A., and Bleier, A., Stability of aqueous α-Al$_2$O$_3$ suspensions with poly(methacrylic acid) polyelectrolytes, *J. Am. Ceram. Soc.*, 71, 250, 1988.)

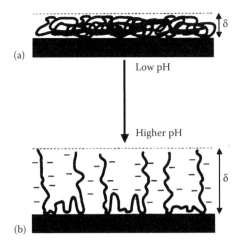

FIGURE 4.31 Schematic illustration of the conformation of an anionic polyelectrolyte on a ceramic surface as a function of pH. (δ is the thickness of the adsorbed polymer layer.) (From Lewis, J.A., Colloidal processing of ceramics, *J. Am. Ceram. Soc.*, 83, 2341, 2000.)

of PMAA adsorbed per square meter of the Al_2O_3 surface vs. the initial amount of PMAA-Na added (as a percent of the dry weight of Al_2O_3, i.e., on a dry weight basis, dwb). The solid diagonal line represents the adsorption behavior that would occur if 100% of the PMAA added were to adsorb. It is clear that the amount of PMAA adsorbed increases with decreasing pH. At low pH values ($\alpha \rightarrow 0$), the polymer chain adopts a fairly compact conformation and adsorbs in a dense layer with a small thickness (Figure 4.31a). The projected area per adsorbed chain is relatively small and more adsorbed chains are required to establish a monolayer. When fully ionized ($\alpha \rightarrow 1$) at higher pH values, the polyelectrolyte adsorbs in an open layer with a large thickness and fewer chains are needed to form a monolayer (Figure 4.31b).

4.7.3 STABILITY OF ELECTROSTERICALLY STABILIZED SUSPENSIONS

At low concentration of adsorbed polyelectrolyte, charge neutralization between the oppositely charged polyelectrolyte and the particle surface can occur, leading to weak interparticle repulsion and to flocculation. Bridging flocculation can also occur when a charged homopolymer adsorbs on two or more neighboring particles. At higher concentration of adsorbed polyelectrolyte, the stability of the suspension improves because of the long-range repulsion resulting from electrostatic and steric interactions [51]. The electrostatic repulsion dominates at larger particle separation, whereas the steric repulsion becomes more pronounced at shorter separation.

By studying the sedimentation behavior of Al_2O_3 suspensions stabilized with PMAA-Na, the conditions for stability have been determined as a function of pH and polyelectrolyte concentration. The observations can be conveniently plotted on a map, called a *colloid stability map* (Figure 4.32). The map shows, for a given concentration of Al_2O_3 particles (20 vol%), the regions of stability and instability (tendency toward flocculation). Regions below the curve are unstable. Regions near and slightly above the curve are stable and dispersed as a result of electrosteric repulsion due to the adsorbed polyelectrolyte. For a given pH value, the concentration of adsorbed polyelectrolyte at the stability boundary corresponds to the saturation concentration of the adsorption isotherm (Figure 4.30). Regions further above the curve have appreciable amounts of free polymer in solution. At pH values below ~3.5, the adsorbed polymer is essentially neutral, so the electrostatic repulsion results from the surface charge of the particles.

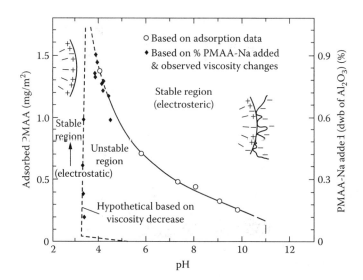

FIGURE 4.32 Colloid stability map for suspensions of α-Al$_2$O$_3$ powder (20 vol%) as a function of adsorbed PMAA and pH. (From Cesarano, J., III, Aksay, I.A., and Bleier, A., Stability of aqueous α-Al$_2$O$_3$ suspensions with poly(methacrylic acid) polyelectrolytes, *J. Am. Ceram. Soc.*, 71, 250, 1988.)

4.8   STRUCTURE OF CONSOLIDATED COLLOIDS

So far in this chapter we have been concerned mainly with the principles of colloid stability. The key factors and how they can be manipulated to influence the stability of the suspensions have been discussed. The next step, which is of great practical importance to ceramic processing, is to relate how the colloidal properties of the suspension can be manipulated to produce the desired structure in the consolidated body.

An initial step in the colloidal processing of many ceramic powders may involve the removal of hard agglomerates and large particles, if present, from the suspension by sedimentation. After this, consolidation of the suspension is achieved by a variety of methods. For suspensions with a low concentration of particles, methods such as gravitational settling, centrifuging, and filtration can be used, whereas the casting methods such as pressure casting, slip casting, tape casting, and gel casting are suitable for concentrated suspensions. Stable suspensions of submicron particles are known to settle very slowly under normal gravity, so the settling rate is normally increased by centrifuging or by filtration. The consolidated colloid has a densely packed structure (Figure 4.33a). Unstable suspensions settle faster because the particles flocculate while settling, poducing a loosely packed structure (Figure 4.33b).

The structure of the consolidated colloid formed is actually more complex than the arrangements illustrated in Figure 4.33. The microstructures of spherical, nearly monodisperse SiO$_2$ particles

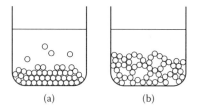

(a)   (b)

FIGURE 4.33 Schematic of a consolidated colloid formed from (a) a stable suspension and (b) a flocculated suspension.

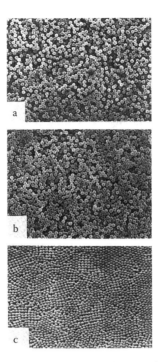

FIGURE 4.34 Microstructures of particle domains formed by centrifugal consolidation of SiO$_2$ colloidal suspensions at (a) $\zeta = 0$ mV, (b) $\zeta = 68$ mV, and (c) $\zeta = 110$ mV. The average particle diameter is 0.7 μm. (From Aksay, I.A., Microstructure control through colloidal consolidation, *Adv. Ceram.*, 9, 94, 1984.)

(Figure 4.34) indicate that the difference in structure is not controlled by the way in which the individual particles arrange themselves in the sediment but rather the way they group themselves to form densely packed multiparticle units called domains [52]. Figure 4.34C shows that the domains in a consolidated colloid formed from a stable suspension are arranged in a periodic pattern not unlike that found in crystalline grains. Domains are also formed if the particles are irregular in shape but the periodic arrangement of the domains is lost.

The microstructures of Figure 4.34 also show that the size of the domains can be altered through manipulation of the interparticle forces. The domain size increases with higher double layer repulsion. The efficiency of the domain packing increases, leading to high packing density in the consolidated colloid (Figure 4.34c). As the repulsion decreases (or equivalently as the attraction increases), the domain size decreases and the interdomain pore size increases. This interdomain porosity is the main cause of the low packing density of flocculated colloids.

We can attempt to describe the structure of the consolidated colloid in terms of the particle interactions. For highly attractive forces, the particles will "stick" on contact with another particle, leading to the formation of a highly disordered system (Figure 4.34a). As the repulsion increases, the particle will be able to undergo a certain amount of rearrangement into low-energy positions leading to an increase in the domain size and a reduction of the interdomain pore size (Figure 4.34b). Finally, when the repulsion becomes dominant, each additional particle can interact with the others to produce larger domains with small interdomain pores (Figure 4.34c). A problem, however, is that the domain boundaries constitute flaws that interfere with sintering. Sintering of the domains leads to differential stresses that cause an enlargement of the flaws at the domain boundaries, limiting the final density and the properties of the sintered material (see Figure 6.8).

We have discussed the structure of consolidated colloids formed from electrostatically stabilized suspensions but the same principles apply to sterically stabilized suspensions. For example, studies with polystyrene spheres stabilized with polymers have yielded structures similar to those shown in Figure 4.34.

4.9 RHEOLOGY OF COLLOIDAL SUSPENSIONS

The term *rheology* refers to the deformation and flow characteristics of matter. Rheological measurements monitor changes in flow behavior in response to an applied stress (or strain). In ceramic processing, the viscosity of suspensions is the key rheological parameter of interest but other parameters, such as the yield stress in compression (or shear) and the viscoelastic properties (e.g., storage modulus and the loss modulus) under an oscillatory stress, are also important in many systems.

Rheological measurements are widely used to characterize the properties of colloidal suspensions [53,54]. They are used as a method of analysis, for example, in determining the optimum concentration of dispersant required to stabilize a suspension by measuring the viscosity of the suspension as a function of the concentration of dispersant added. More often, they are used as a quality control technique to minimize the batch-to-batch variation of suspension properties prior to consolidation (e.g., by spray drying, slip casting or tape casting). Rheological measurements are also used as a direct processing parameter. For example, in slip casting or tape casting, we require, on the one hand, that the suspension contains the highest possible fraction of particles. This is because a concentrated suspension serves to reduce the shrinkage during drying of the cast and to produce a green body with high packing density. On the other hand, we also want the suspension to have a low enough viscosity so that it can be cast into the desired shape. Rheological measurements provide an important means of optimizing these requirements.

4.9.1 RHEOLOGICAL PROPERTIES

In simple shear, which is the most common way of determining the rheological properties, the shear stress, τ, is related to the shear strain rate, $\dot{\gamma}$, by:

$$\tau = \eta\dot{\gamma} \tag{4.75}$$

where η is the viscosity. In general, the stress and the stain are tensors, with each having nine components. If η is independent of the shear rate (or the shear stress), the material is said to be Newtonian. Many simple liquids (e.g., water, alcohols, and some oils), as well as many molten glasses, show Newtonian behavior. For more complex systems such as polymer solutions and colloidal suspensions, η is not independent of the shear rate and the behavior is said to be non-Newtonian. We must now write

$$\eta(\dot{\gamma}) = \frac{d\tau}{d\dot{\gamma}} \tag{4.76}$$

where $\eta(\dot{\gamma})$, the viscosity at a given strain rate, referred to as the *instantaneous viscosity*, is found from the gradient of the shear stress vs. strain rate curve. The *apparent viscosity*, at any given strain rate, is the shear stress divided by the strain rate.

The viscosity of colloidal suspensions often shows a dramatic dependence on the shear rate, and this dependence is used as a means to classify the rheological behavior. Figure 4.35 shows the curves for Newtonian behavior and types of non-Newtonian behavior found with colloidal suspensions. When the viscosity increases with increasing shear rate, the behavior is described as *shear thickening* (or *dilatant*). Suspensions containing a high concentration of nearly equiaxial

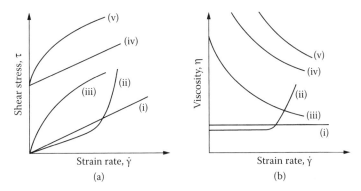

FIGURE 4.35 Typical rheological behavior of colloidal suspensions: (i) Newtonian, (ii) shear thickening or dilatant, (iii) shear thinning or pseudoplastic, (iv) Bingham plastic, (v) pseudoplastic with a yield stress.

particles may show shear thickening. Moderately concentrated suspensions of small, elongated particles may show *shear thinning or pseudoplasticity*, where the viscosity decreases with increasing shear rate. Shear thickening or shear thinning behavior is sometimes described by an empirical relation of the form:

$$\tau = K\dot{\gamma}^n \tag{4.77}$$

where K is called the consistency index, and n is an exponent that indicates the deviation from Newtonian behavior. If $n = 1$, the suspension is Newtonian; when $n < 1$, the suspension shows shear thinning or pseudoplastic flow; and when n is > 1, the flow is described as shear thickening or dilatant.

For systems where the viscosity decreases with increasing shear rate after an initial threshold stress called the yield stress, τ_y, the flow behavior is described as plastic. Concentrated suspensions of clay particles commonly exhibit plastic behavior. A material obeying the equation:

$$\tau - \tau_y = \eta\dot{\gamma} \tag{4.78}$$

where η is independent of the shear rate, is said to show Bingham-type behavior. The viscosity of most plastic materials, however, is dependent on the shear rate and the flow behavior is often described by a power law relation:

$$\tau - \tau_y = K\dot{\gamma}^n \tag{4.79}$$

where K and n were defined earlier in Equation 4.77.

The rheological properties of concentrated suspensions often depend not only on the shear rate but also on the time. When the viscosity decreases with time under shear but recovers to its original value after flow ceases, the behavior is known as *thixotropic* (Figure 4.36). This type of behavior is more often observed in flocculated suspensions and colloidal gels. When the suspension is sheared, the flocs are broken down leading a distribution of floc sizes. Often the regeneration of the flocs is slow which causes the resistance to flow to decrease. The opposite behavior, when the viscosity increases with shear rate and is also time dependent, is known as *rheopectic*.

In practice, dilatant and rheopectic behavior are often undesirable because at high shear rates the suspension becomes too stiff to flow smoothly. Plastic behavior is desirable for many ceramic-forming methods because the suspension will flow under high stress but will retain its shape

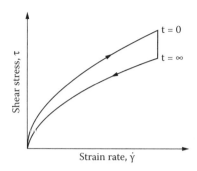

FIGURE 4.36 Thixotropic behavior in which the rheology depends not only on the strain rate but also on time.

when the stress is removed after forming. Pseudoplastic (shear thinning) behavior is often an acceptable compromise.

For the range of values encountered with colloidal suspensions, a concentric cylinder rotating viscometer is often used to measure the viscosity. Less often used are the cone and plate viscometer and the parallel plate viscometer. For liquids and solutions, a capillary viscometer is often used. Information on several types of viscometers can be found in Reference 55.

4.9.2 Factors Influencing the Rheology of Colloidal Suspensions

The interparticle forces discussed earlier in this chapter have a significant influence on the rheological behavior, and the stability of the suspension can often be inferred from the observed behavior. The particle characteristics, particularly the particle concentration, can also have a significant effect on the suspension rheology. Commonly, the influence of the liquid medium is small, unless an excess concentration of dissolved polymer is present.

4.9.2.1 Influence of Interparticle Forces

4.9.2.1.1 Hard-Sphere Systems

Hard-sphere colloidal systems do not experience interparticle interactions until they come into contact, at which point the interaction is infinitely repulsive. Such systems represent the simplest case, where the flow is affected only by hydrodynamic (viscous) interactions and Brownian motion. Hard-sphere systems are not often encountered in practice, but model systems consisting of SiO_2 spheres stabilized by adsorbed stearyl alcohol layers in cyclohexane [56,57] and polymer lattices [58,59] have been shown to approach this behavior. They serve as a useful starting point for considering the more complicated effects when interparticle forces are present.

In a quiescent suspension, the particles are moving continuously in a random manner with Brownian motion. The time t_B taken for a particle to diffuse a distance equal to its radius a is given by the relation:

$$a^2 \approx Dt_B \tag{4.80}$$

with the diffusion coefficient D given by the Stokes–Einstein equation:

$$D = \frac{kT}{6\pi\eta_L a} \tag{4.81}$$

where k is the Boltzmann constant, T is the absolute temperature, and η_L is the viscosity of the liquid medium. From Equation 4.80 and Equation 4.81,

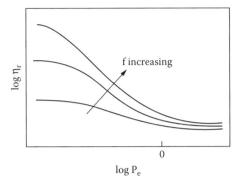

FIGURE 4.37 Schematic representation of the relative viscosity vs. the Peclet number for suspensions with different particle concentration f. (From Goodwin, J.W., Some uses of rheology in colloid science, in *Colloidal Dispersions*, Goodwin, J.W., Ed., The Royal Society of Chemistry, London, 1982, p. 165, chap. 8.)

$$t_B = \frac{6\pi\eta_L a^3}{kT} \tag{4.82}$$

We can take t_B to represent a characteristic time required to restore the structure of the suspension from a disturbance caused by Brownian motion. The time scale t_S for viscous flow due to a shear stress can be taken as the reciprocal of the strain rate $\dot\gamma$. The dimensionless parameter equal to the ratio t_B/t_S, which is referred to as the *Peclet number* (or the reduced strain rate),

$$P_e = \frac{6\pi\eta_L a^3 \dot\gamma}{kT} \tag{4.83}$$

gives the relative importance of Brownian motion and viscous forces.

For hard-sphere systems, experimental data for the relative viscosity η_r (equal to the viscosity of the suspension η divided by the viscosity of the liquid medium η_L) vs. P_e lie on curves of the form illustrated in Figure 4.37. At low strain rates when P_e is $\ll 1$, the structure of the suspension is not significantly altered by the shear because Brownian motion dominates over the viscous forces. At higher strain rates, the viscous forces start to affect the suspension structure and shear thinning occurs. At very high strain rates when P_e is $\gg 1$, the viscous forces dominate and the plateau region represents the viscosity of a suspension with a hydrodynamically controlled structure.

In the limit of infinite dilution when all interactions between the particles are neglected, the relative viscosity of hard-sphere suspensions is given by the Einstein equation:

$$\eta_r = 1 + 2.5f \tag{4.84}$$

where f is the volume fraction of the particles. Equation 4.84 can be used with negligible error for $f < 0.01$. At low strain rates ($P_e \ll 1$), when two-particle interactions are taken into account:

$$\eta_r = 1 + 2.5f + 6.2f^2 \tag{4.85}$$

Equation 4.85 has been shown to provide an adequate fit to experimental data for $f < 0.15$.

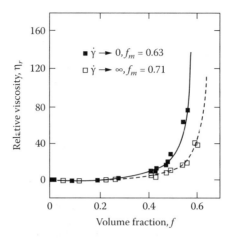

FIGURE 4.38 Relative viscosity vs. volume fraction of particles for sterically stabilized suspensions of silica spheres (radius = 110 nm) in cyclohexane. The data are for low and high strain rates, $\dot{\gamma}$. (From Van der Werff, J.C. and De Kruif, C.G., Hard-sphere colloidal dispersions: the scaling of rheological properties with particle size, volume fraction, and shear rate, *J. Rheol.*, 33, 421, 1989.)

As f is increased above 0.15, interactions involving multiple particles become important and no rigorous hydrodynamic theory exists. The problem is treated empirically and several equations are available in the literature. One of the most useful is the Krieger–Dougherty equation [60]:

$$\eta_r = \left(1 - \frac{f}{f_m} \right)^{-[\eta]f_m} \tag{4.86}$$

where $[\eta]$ is the intrinsic viscosity of the suspension, equal to 2.5 for a suspension of hard spheres, and f_m is the volume fraction of particles at which the viscosity becomes practically infinite. For sterically stabilized SiO_2 spheres in cyclohexane, Figure 4.38 shows that the relative viscosity at low and high strain rates can be well fitted by Equation 4.86, with $f_m = 0.63$ at low strain rate and $f_m = 0.71$ at high strain rate [61].

4.9.2.1.2 Soft-Sphere Systems

Colloidal suspensions can be classified as soft-sphere systems because the repulsive interactions occur at some characteristic distance from the particle surface. For electrostatic and steric stabilization, this distance is the Debye length (1/K) and the thickness of the adsorbed polymer layer, respectively. For sterically stabilized suspensions, the adsorbed polymer layer leads to an increase in the hydrodynamic radius of the particle. When the adsorbed layer is densely packed, the principles described above for hard-sphere systems are applicable, provided that the volume fraction of particles f is replaced by an effective volume fraction f_{eff} given by:

$$f_{eff} = f \left(1 + \frac{\delta}{a} \right)^3 \tag{4.87}$$

where δ is the thickness of the polymer layer and a is the particle radius [62–64]. The use of Equation 4.87 with hard-sphere scaling principles is successful only when the ratio δ/a is small (less than ~0.1). With increasing thickness of the adsorbed polymer layer, the experimental data deviate from the predicted behavior, particularly at high volume fraction of particles. In principle, it should also be possible to scale electrostatically stabilized suspensions in a similar manner. For these systems, the Debye length $1/K$ is a reasonable estimate for the parameter δ in Equation 4.87.

4.9.2.1.3 Flocculated Systems

Flocculated suspensions are dominated by attractive interparticle interactions and form disordered, metastable structures. Because of these nonequilibrium structures, the rheological behavior is difficult to characterize. It is common to distinguish between two types of flocculated systems which are dependent on the magnitude of the interparticle attraction: weakly flocculated and strongly flocculated. Weakly flocculated suspensions are characterized by a shallow minimum having a total potential energy of interaction in the range $1 < -V_T^{min}/kT < 20$. This state can be achieved by adding a nonadsorbing polymer to an otherwise stable suspension [65] or by controlling the repulsion between the particles [66] to achieve flocculation into a secondary minimum of the DLVO potential (see Figure 4.15). Weakly flocculated suspensions exhibit reversible flocculation, which facilitates deformation of the structure during shear and its reformation in a short time after removal of the stress. With increasing shear stress, the links between individual particles or particle clusters are broken down, so significant shear thinning can occur at low stresses and low particle concentration [54]. Strongly flocculated suspensions are characterized by a deep minimum in the total potential energy $-V_T^{min}/kT\ (> 20)$ and an irreversible rheological behavior during shear. They exhibit significant shear thinning and a significant yield stress.

4.9.2.2 Influence of Particle Interactions on the Viscosity

The interactions between the particles have a dramatic effect on the viscosity of the suspension, as illustrated in Figure 4.39 for Al_2O_3 suspensions containing 50 vol% particles which are stabilized with the polyelectrolyte poly(acrylic acid), PAA (see Section 4.7). The viscosity decreases dramatically as the concentration of PAA is increased and, at some critical PAA concentration corresponding to the amount required to form a complete monolayer on the particle surface, it reaches a low plateau region that is almost independent of the PAA concentration. This dramatic change in viscosity reflects the change from a flocculated suspension at low PAA concentration to an elec-

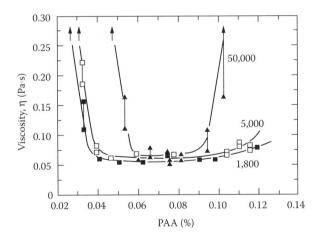

FIGURE 4.39 Viscosity vs. the amount of poly(acrylic acid), PAA, added (as a percent of the dry weight of Al_2O_3) for various PAA molecular weights. The suspensions, prepared at a pH of 9, contained 50 vol% α-Al_2O_3 particles (Sumitomo AKP-20). The unfilled data points are the initial values and the filled points are the values after 10 min. (From Cesarano, J., III and Aksay, I.A., Processing of highly concentrated aqueous α-alumina suspensions stabilized with polyelectrolytes, *J. Am. Ceram. Soc.*, 71, 1062, 1988.)

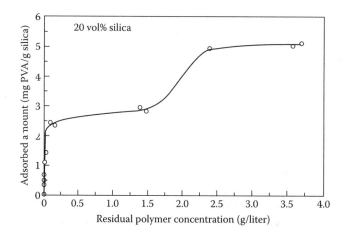

FIGURE 4.40 Adsorption isotherm for suspensions containing 20 volume percent SiO_2 particles in solutions of poly(vinyl alcohol). (From Sacks, M.D. et al., Dispersion and rheology in ceramic processing, *Adv. Ceram.*, 21, 495, 1987.)

trosterically stabilized system above the critical PAA concentration. Figure 4.39 also shows that the critical PAA concentration required to achieve monolayer coverage of the particle surface is dependent on the molecular weight. After the critical concentration is reached, further additions of PAA serve mainly to produce excess polymer in the solution, and the viscosity of the suspension starts to increase again at high concentration.

A significant decrease in the viscosity is also found for systems stabilized by electrostatic or steric mechanisms by moving from conditions where the suspension is flocculated to those where it is stable. Data for the viscosity of aqueous suspensions of nearly monodisperse SiO_2 particles (20 vol%) stabilized electrostatically by varying the pH or sterically by adsorbed layers of poly(vinyl alcohol) (PVA) serve to illustrate some of the effects [67]. The adsorption behavior of PVA onto the SiO_2 particle surfaces is shown in Figure 4.40 for a suspension prepared at a pH value (3.7) that is close to the isoelectric point. Initially, all of the PVA added to the suspension is adsorbed on to the particle surfaces. As the PVA concentration increases, the particle surfaces become saturated and a plateau region occurs in the adsorption isotherm. This first plateau is believed to be due to the development of monolayer coverage of PVA on the particle surfaces. Further increases in the PVA concentration lead to an increase in the adsorption and the occurrence of a second plateau region, believed to be due to the development of a denser packing of the adsorbed polymer molecules or to the occurrence of multilayer adsorption.

Figure 4.41 shows the effect of the adsorbed PVA on the relative viscosity of the suspension for suspensions prepared at a pH of 3.7. The data also show the viscosity for an electrostatically stabilized suspension prepared at a pH of 7.0 without any PVA. For the suspensions containing no PVA, we see that the stabilized suspension (pH = 7.0) has a fairly low viscosity and shows Newtonian behavior while the unstabilized suspension (pH = 3.7) has a much higher viscosity and shows a high degree of shear thinning.

Considering now the effect of the PVA, we see that initially, the viscosity increases as the amount of adsorbed PVA increases from 0 to 1.1 mg per gram of SiO_2 and a high degree of shear thinning occurs. The increase in the viscosity is most likely due to bridging flocculation discussed earlier (see Section 4.6.4). When the amount of adsorbed PVA becomes greater than 1.1 mg/g SiO_2, the trend is reversed, and the viscosity decreases with increasing amount of adsorbed PVA. For an amount of adsorbed PVA equivalent to that in the middle of the second plateau of the adsorption isotherm of Figure 4.40 (~2.9 mg/g SiO_2), the viscosity is fairly low and the behavior is almost

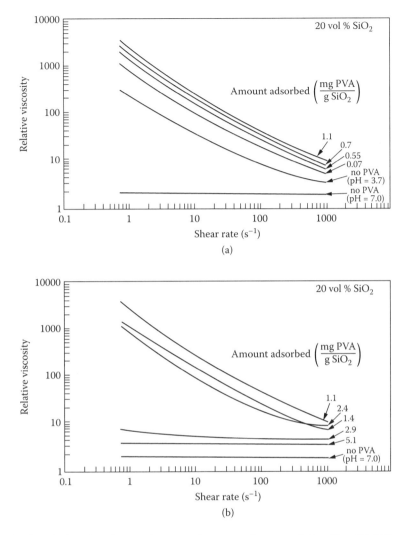

FIGURE 4.41 The relative viscosity vs. shear rate for suspensions containing 20 vol% SiO$_2$ particles. The suspensions prepared at pH = 3.7 (near the isoelectric point of the SiO$_2$) and pH = 7.0 contained no poly(vinyl alcohol). All of the suspensions containing poly(vinyl alcohol) were prepared at pH = 3.7. (From Sacks, M.D. et al., Dispersion and rheology in ceramic processing, *Adv. Ceram.*, 21, 495, 1987.)

Newtonian. For higher PVA concentrations (5.1 mg/g SiO$_2$), the viscosity shows a small further decrease but it is still higher than that for the electrostatically stabilized suspension (pH = 7.0).

4.9.2.3 Influence of Particle Characteristics

In addition to the interparticle forces discussed above, the particle concentration also has a significant effect on the rheological properties of the suspension. At the same level of stability, the viscosity of the suspension increases with increasing volume fraction of particles, as illustrated by the results (Figure 4.38). The viscosity increases rapidly as the particle concentration approaches f_m, the maximum volume fraction of particles that can be accommodated in the suspension before flow ceases.

For electrostatically and sterically stabilized suspensions (soft-sphere systems), the presence of an electrical double layer or an adsorbed polymer layer, as discussed earlier, leads to an increase

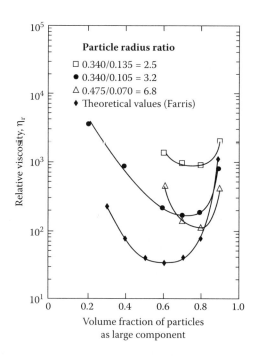

FIGURE 4.42 Effect of particle size ratio on the relative viscosity of stable suspensions of bimodal polymer particles. (The particle sizes are in units of μm.) The data were obtained at a shear stress = 1 dyn/cm² and for suspensions with a particle concentration $f = 0.65$. (From Hoffmann, R.L., Factors affecting the viscosity of unimodal and multimodal colloidal dispersions, *J. Rheol.*, 36, 947, 1992.)

in the hydrodynamic radius of the particle and a consequent reduction in f_m with decreasing particle size (Equation 4.87).

As discussed later (Chapter 6), the packing of particles can be significantly improved by mixing spheres of two different sizes (bimodal distribution) or by using a broad, continuous particle size distribution. These particle characteristics can also have a significant effect on the rheological properties of a suspension. Several studies have shown how the use of a bimodal or a broad, continuous particle size distribution can lower the viscosity and increase f_m, but in many cases large particles (> 5 to 10 μm) have been used where colloidal effects play no significant role [68]. For electrostatically stabilized suspensions of bimodal polymer particles, Figure 4.42 shows how the relative viscosity varies with the composition of the mixture and the ratio of the particle sizes [69]. The data are also compared with a theoretical model proposed by Farris [70]. For even a small size ratio of 2.5, the viscosity of the suspension decreases significantly and increasing the size ratio to 3.2 produces a further decrease. However, the decrease in the viscosity due to a further increase in the size ratio from 3.2 to 6.8 is far smaller than expected, and this has been attributed to the enhanced importance of the Debye length with decreasing particle size. For a constant Debye length, δ/a increases with decreasing particle size and, as can be seen from Equation 4.87, f_{eff} increases.

The shape of the particles also has an effect on the viscosity. Most advanced ceramics are fabricated using nearly equiaxial particles, but the effect of shape would be important, for example, in the colloidal processing of ceramic composites reinforced with whiskers (short single-crystal fibers) or platelets. Figure 4.43 shows the results of theoretical calculations for the intrinsic viscosity [η] as a function of the axial ratio for particles with the shape of prolate ellipsoids. For an axial ratio of 15 to 20, which may be relevant to the use of whiskers or platelets in ceramic composites, the results show that [η] has a value of ~4.5 compared to 2.5 for spherical particles.

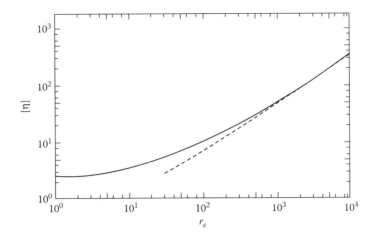

FIGURE 4.43 Intrinsic viscosity as a function of axial ratio for prolate ellipsoids. (From Goodwin, J.W., Some uses of rheology in colloid science, in *Colloidal Dispersions*, Goodwin, J.W., Ed., The Royal Society of Chemistry, London, 1982, p. 165, chap. 8.)

4.10 CONCLUDING REMARKS

In this chapter we outlined the basic principles of electrostatic, steric, and electrosteric stabilization of colloidal suspensions. A considerable gap still exists in our understanding of many aspects of colloid stability, particularly in the case of steric stabilization. Nevertheless, the principles outlined in this chapter provide a basis for manipulating the experimental parameters in order to produce a suspension of the desired stability. Colloidal methods have been used for a long time in the forming of traditional, clay-based ceramics, and they are now being used increasingly in the advanced ceramics sector. The use of colloidal methods for improving the packing uniformity of the green body has yielded clear benefits in the control of the fired microstructure. Many of the principles outlined in this chapter will be encountered again in Chapter 7 when we consider the methods used to consolidate ceramic powders.

PROBLEMS

4.1

Calculate the van der Waals potential energy V_A, the electrostatic repulsive potential energy V_R, the gravitational potential energy V_G, and the total potential energy V_T between two colloidal particles of radius $a = 0.2$ μm separated by a distance $h = 10$ nm in a dilute hydrochloric acid solution at 27°C, given that the Hamaker constant $A = 10^{-20}$ J, the surface potential of the particles $\phi_o = 25$ mV, the dielectric constant of the solution $\varepsilon = 80$, the concentration of hydrochloric acid $c = 5$ millimoles/liter, and the density of the particles $d = 4.0$ g/cm^3.

4.2

Compare the mechanism by which clay mineral particles (e.g., kaolinite) acquire a surface charge in water with the mechanism by which oxide particles (e.g., SiO_2) acquire a surface charge in water.

4.3

The clay mineral montmorillonite has the formula:

$$Na_{0.33}(Al_{1.67}Mg_{0.33})(Si_2O_5)_2(OH)_2$$

Assuming that all of the Na^+ ions pass into solution, determine the ion exchange capacity (in C/kg) of montmorillonite.

4.4

Stable suspensions of Al_2O_3 particles and SiO_2 particles are prepared separately at a pH = 6. If the two suspensions are mixed, discuss the colloidal stability of the resulting suspension.

4.5

Derive Equation 4.37. Hence, show that for a high surface potential the double layer potential at a large distance from a flat surface decays as:

$$\phi = \frac{4RT}{zF}\exp(-Kx)$$

4.6

Calculate the free energy of repulsion V_R between two double layers, each having an area of 1 cm^2 at a surface potential $\phi_o = 4RT/F$, in a monovalent electrolyte giving a Debye length 1/K equal to 10^{-6} cm and a distance of separation equal to 10 nm.

4.7

The surface potential of a flat surface in dilute HCl solution at 27°C is 50 mV. If the acid concentration is 5 mM/l, calculate the surface charge density. The dielectric constant of the solution can be assumed to be equal to that of water (i.e., 78).

4.8

a. How does the double layer repulsion change if the medium water is replaced by methanol, assuming that the surface potential and the ionic concentration remain the same?
b. If a monovalent electrolyte is replaced by a bivalent electrolyte, but the Debye length 1/K remains the same, how would the concentration of the electrolyte have to change?

4.9

In a particle electrophoresis experiment, the particles in a dilute suspension are observed to travel an average distance of 1 mm in 25 sec when a potential difference of 100 V is applied to the electrodes that are 5 cm apart. Determine the ζ potential of the particles.

4.10

Derive Equation 4.66 for the root mean square end-to-end distance $\langle r^2 \rangle^{1/2}$ of a polymer chain. Derive expressions for the average end-to-end distance $\langle r \rangle$ and the most probable end-to-end distance r_{mp} in terms of $\langle r^2 \rangle^{1/2}$.
Determine $\langle r^2 \rangle^{1/2}$ for polystyrene molecules in cyclopentane given that the degree of polymerization of polystyrene is 2×10^4.

4.11

Determine the van der Waals potential energy V_A, the free energy due to the mixing effect ΔG_{mix}, and the free energy due to the elastic effect $\Delta G_{elastic}$, between two mica particles of radius a = 0.2 μm separated by a distance $h = 2L/3$ in a solution of polystyrene in cyclopentane at 27°C, given that the Hamaker constant $A = 5 \times 10^{-20}$ J, the Flory-Huggins parameter $\chi = 0.25$, the degree of polymerization of the polystyrene molecules = 2×10^4, and L = $\langle r^2 \rangle^{1/2}$, the root mean square end-to-end distance of the polystyrene molecules in the solvent. State any assumptions that you make.

4.12

Using Equation 4.87, plot the effective volume fraction of particles f_{eff} vs. the volume fraction of solid particles f, for particles with radii of 10, 1, 0.1, and 0.01 μm and assuming an adsorbed layer with a constant thickness of 10 nm.

4.13

Assuming that the suspension viscosity can be described by the Krieger–Dougherty equation, determine the relative viscosity for suspensions containing 40 vol% of solid spherical particles when:

  a. The particles are hard spheres with a radius of 1 μm. Assume $[\eta] = 2.5$ and $f_m = 0.63$.
  b. The spherical particles with a radius of 1 μm have an adsorbed polymer layer with a thickness of 5 nm.
  c. The spherical particles with a radius of 0.1 μm have an adsorbed polymer layer with a thickness of 5 nm.
  d. The particles are elongated. Assume $[\eta] = 4.5$ and $f_m = 0.5$.

REFERENCES

1. Overbeek, J.Th.G., *Colloid and Surface Chemistry: A Self-Study Course*, Parts 1–4, Massachusetts Institute of Technology, Cambridge, MA, 1972.
2. (a) Hunter, R.J., *Foundations of Colloid Science*, Vol. 1, Oxford University Press, Oxford, 1986. (b) Hunter, R.J., *Foundations of Colloid Science*, Vol. 2, Oxford University Press, Oxford, 1989.
3. Goodwin, J.W., Ed., *Colloidal Dispersions*, The Royal Society of Chemistry, London, 1982.
4. Israelachvili, J.N., *Intermolecular and Surface Forces*, 2nd ed., Academic Press, London, 1992.
5. Tabor, D., Attractive surface forces, in *Colloidal Dispersions*, Goodwin, J.W., Ed., The Royal Society of Chemistry, London, 1982, chap. 2.
6. Tabor, D., *Gases, Liquids and Solids*, 3rd ed., Cambridge University Press, Cambridge, 1991, chap. 12.
7. Hamaker, H.C., The London-van der Waals attraction between spherical particles, *Physica*, 4, 1058, 1937.
8. Lifshitz, E.M., The theory of molecular attractive forces between solids, *Sov. Phys. JETP*, 2, 73, 1956.
9. Mahanty, J. and Ninham, B.W., *Dispersion Forces*, Academic Press, New York, 1976.
10. Margenau, H. and Kestner, N.R., *Theory of Intermolecular Forces*, 2nd ed., Pergamon Press, Oxford, 1971.
11. Israelachvili, J.N., and Tabor, D., Van der Waals forces: theory and experiment, *Prog. Surf. Membrane Sci.* 7, 1, 1973.
12. Israelachvili, J.N., Van der Waals forces in biological systems, *Q. Rev. Biophys.*, 6, 341, 1974.
13. French, R.H., Origins and applications of London dispersion forces and Hamaker constants in ceramics, *J. Am. Ceram. Soc.*, 83, 2117, 2000.
14. Bergström, L., Hamaker constants of inorganic materials, *Adv. Colloid Interface Sci.*, 70, 125, 1997.
15. Ackler, H.D., French, R.H., and Chiang, Y.-M., Comparisons of Hamaker constants for ceramic systems with intervening vacuum or water: from force laws and physical properties, *J. Colloid Interface Sci.*, 179, 460, 1996.

16. Tabor, D. and Winterton, R.H.S., The direct measurement of normal and retarded van der Waals forces, *Proc. R. Soc. Lond.*, A312, 435, 1969.

17. Israelachvili, J.N. and Tabor, D., The measurement of van der Waals dispersion forces in the range 1.5 to 130 nm, *Proc. Roy. Soc. Lond.*, A331, 19, 1972.

18. Hansma, P.K., Elings, V.B., Marti, O., and Bracker, C.E., Scanning tunneling microscopy and atomic force microscopy: application to biology and technology, *Science*, 242, 209, 1988.

19. Meyer, E., Atomic force microscopy, *Prog. Surf. Sci.* 41, 3, 1992.

20. Ducker, W.A, Senden, T.J., and Pashley, R.M., Measurement of forces in liquids using a force microscope, *Langmuir*, 8, 1831, 1992.

21. Larson, I., Drummond, C.J., Chan, D.Y.C., and Grieser, F., Direct force measurements between TiO_2 surfaces, *J. Am. Chem. Soc.*, 115, 11885, 1993.

22. Biggs, S. and Mulvaney, P., Measurement of the forces between gold surfaces in water by atomic force microscopy, *J. Chem. Phys.*, 100, 8501, 1994.

23. Senden, T.J., Drummond, C.J., and Kékicheff, P., Atomic force microscopy: imaging with electrical double layer interactions, *Langmuir*, 10, 358, 1994.

24. Healey, T.W. and White, L.R., Ionizable surface group models of aqueous interfaces, *Adv. Colloid Interface Sci.*, 9, 303, 1978.

25. Hunter, R.J., *Zeta Potential in Colloid Science*, Academic Press, New York, 1981.

26. Reed J.S., *Introduction to the Principles of Ceramic Processing*, Wiley, New York, 1988, p. 134.

27. Parks, G.A., The isoelectric points of solid oxides, solid hydroxides, and aqueous hydroxo complex systems, *Chem. Rev.*, 65, 177, 1965.

28. Yoon, R.H., Salman, T., and Donnay, G., Predicting points of zero charge of oxides and hydroxides, *J. Colloid Interface Sci.*, 70, 483, 1979.

29. Carre, A., Roger, F., and Varinot, C., Study of acid/base properties of oxide, oxide glass, and glass-ceramic surfaces, *J. Colloid Interface Sci.*, 154, 174, 1992.

30. Shaw D.J., *Introduction to Colloid and Surface Chemistry*, 3rd ed., Butterworths, London, 1980.

31. Horn, R.G., Clarke, D.R., and Clarkson, M.T., Direct measurement of surface forces between sapphire crystals in aqueous solution, *J. Mater. Res.*, 3, 413, 1988.

32. Israelachvili, J.N. and Adams, G.E., Measurement of forces between two micro surfaces in aqueous electrolyte solutions in the range 0–100 nm, *J. Chem. Soc.,* 74, 975, 1978.

33. Jang, H.M. and Fuerstenau, D.W., The specific adsorption of alkaline-earth cations at the rutile/water interface, *Colloids Surf.* 21, 235, 1986.

34. Napper, D.H., *Polymeric Stabilization of Colloidal Dispersions*, Academic Press, New York, 1983.

35. Napper, D.H., Polymeric stabilization, in *Colloidal Dispersions*, Goodwin, J.W., Ed., The Royal Society of Chemistry, London, 1982, chap. 5.

36. Flory, P.J., *Statistical Mechanics of Chain Molecules*, Hanser Publishers, New York, 1969.

37. Lewis, J.A., Colloidal processing of ceramics, *J. Am. Ceram. Soc.*, 83, 2341, 2000.

38. Jensen, W.B., *The Lewis Acid-Base Concepts*, John Wiley & Sons, New York, 1980.

39. Laible, R. and Hamann, K., Formation of chemically bound polymer layers on oxide surfaces and their role in colloidal stability, *Adv. Colloid Interface Sci.*, 13, 65, 1980.

40. Green, M., Kramer, T., Parish, M., Fox, J., Lalanandham, R., Rhine, W., Barclay, S., Calvert, P., and Bowen, H.K., Chemically bonded organic dispersants, *Adv. Ceram.*, 21, 449, 1987.

41. Vincent, B., Luckham, P.F., and Waite, F.A., The effect of free polymer on the stability of sterically stabilized dispersions, *J. Colloid Interface Sci.* 73, 508, 1980.

42. Vincent, B., Edwards, J., Emmett, S., and Jones, A., Depletion flocculation in dispersions of sterically-stabilized particles (soft spheres), *Colloids Surf.*, 18, 261, 1986.

43. Klein, J., Entropic interactions, *Phys. World*, 2, 35, 1989.

44. Patel, S.S. and Tirrell, M., Measurement of forces between surfaces in polymer fluids, *Annu. Rev. Phys. Chem.*, 40, 597, 1989.

45. Horn, R.G., Surface forces and their action in ceramic materials, *J. Am. Ceram. Soc.*, 73, 1117, 1990.

46. Fowkes, F.M., Dispersions of ceramic powders in organic media, *Adv. Ceram.*, 21, 412, 1987.

47. Cesarano, J., III, Aksay, I.A., and Bleier, A., Stability of aqueous α-Al_2O_3 suspensions with poly(meth-acrylic acid) polyelectrolytes, *J. Am. Ceram. Soc.*, 71, 250, 1988.

48. Cesarano, J., III and Aksay, I.A., Processing of highly concentrated aqueous α-alumina suspensions stabilized with polyelectrolytes, *J. Am. Ceram. Soc.*, 71, 1062, 1988.

49. Biggs, S. and Healy, T.W., Electrosteric stabilization of colloidal zirconia with low molecular weight poly(acrylic acid), *J. Chem. Soc. Faraday Trans.*, 90, 3415, 1994.

50. Rojas, D.J., Claesson, P.M., Muller, D., and Neuman, R.D., The effect of salt concentration on adsorption of low-charge-density polyelectrolytes and interactions between polyelectrolyte-coated surfaces, *J. Colloid Interface Sci.*, 205, 77, 1998.

51. Marra, J. and Hair, M.L., Forces between two poly(2-vinyl pyridine)-covered surfaces as a function of ionic strength and polymer charge, *J. Phys. Chem.*, 92, 6044, 1988.

52. Aksay, I.A., Microstructure control through colloidal consolidation, *Adv. Ceram.*, 9, 94, 1984.

53. Goodwin, J.W., Some uses of rheology in colloid science, in *Colloidal Dispersions*, Goodwin, J.W., Ed., The Royal Society of Chemistry, London, 1982, chap. 8.

54. Bergström, L., Rheology of concentrated suspensions, in *Surface and Colloid Chemistry in Advanced Ceramic Processing*, Pugh, R.J. and Bergström, L., Eds., Marcel Dekker, New York, 1994, chap. 5.

55. Barnes, H.A., Hutton, J.F., and Walters, K., *An Introduction to Rheology*, Elsevier Science, Amsterdam, 1989.

56. Jones, D.A.R., Leary, B., and Boger, B.V., The rheology of a concentrated suspension of hard spheres, *J. Colloid Interface Sci.*, 147, 479, 1991.

57. De Hek, H. and Vrij, A., Interactions in mixtures of colloidal silica spheres and polystyrene molecules in cyclohexane, *J. Colloid Interface Sci.*, 84, 409, 1981.

58. Krieger, I.M., Rheology of monodisperse lattices, *Adv. Colloid Interface Sci.*, 3, 111, 1972.

59. Frith, W.J., Strivens, T.A., and Russel, W.B., The rheology of suspensions containing polymerically stabilized particles, *J. Colloid Interface Sci.*, 139, 55, 1990.

60. Krieger, I.M. and Dougherty, T.J., A mechanism for non-Newtonian flow in suspensions of rigid spheres, *Trans. Soc. Rheol.*, 3, 137, 1959.

61. Van der Werff, J.C. and De Kruif, C.G., Hard-sphere colloidal dispersions: the scaling of rheological properties with particle size, volume fraction, and shear rate, *J. Rheol.*, 33, 421, 1989.

62. Choi, G.N. and Krieger, I.M., Rheological studies of sterically stabilized model dispersions of uniform colloidal spheres. II. Steady-shear viscosity, *J. Colloid Interface Sci.*, 113, 101, 1986.

63. Jones, D.A.R., Leary, B., and Boger, D.V., The rheology of a sterically stabilized suspension at high concentration, *J. Colloid Interface Sci.*, 150, 84, 1991.

64. Mewis, J., Frith, W.J., Strivens, T.A., and Russell, W.B., The rheology of suspensions containing polymerically stabilized particles, *AIChE J.*, 35, 415, 1989.

65. Reynolds, P.A. and Reid, C.A., Effect of nonadsorbing polymers on the rheology of a concentrated nonaqueous dispersion, *Langmuir*, 7, 89, 1991.

66. Velamakanni, B.V., Chang, J.C., Lange, F.F., and Pearson, D.S., New method for efficient colloidal particle packing via modulation of repulsive lubricating hydration forces, *Langmuir*, 6, 1323, 1990.

67. Sacks, M.D., Khadilkar, C.S., Scheiffele, G.W., Shenoy, A.V., Dow, J.H., and Sheu, R.S., Dispersion and rheology in ceramic processing, *Adv. Ceram.*, 21, 495, 1987.

68. Smith, P.L. and Haber, R.A., Reformulation of an aqueous alumina slip based on modification of particle-size distribution and particle packing, *J. Am. Ceram. Soc.*, 75, 290, 1992.

69. Hoffmann, R.L., Factors affecting the viscosity of unimodal and multimodal colloidal dispersions, *J. Rheol.*, 36, 947, 1992.

70. Farris, R.J., Prediction of the viscosity of multimodal suspensions from unimodal viscosity data, *Trans. Soc. Rheol.*, 12, 281, 1968.

5 Sol–Gel Processing

5.1 INTRODUCTION

A *sol* is a suspension of colloidal particles in a liquid or a solution of polymer molecules. A *gel* is a semirigid mass formed when the colloidal particles are linked to form a network or when the polymer molecules are cross-linked or interlinked. The term *sol–gel processing* is used broadly to describe ceramic processing methods that involve the preparation and gelation of a sol. Two different sol–gel routes are commonly distinguished: (1) the *particulate* (or *colloidal*) gel route, in which the sol consists of dense colloidal particles (1 to 1000 nm) and (2) the *polymeric* gel route, in which the sol consists of polymer chains but has no dense particles > 1 nm. In many cases, particularly when the particle size approaches the lower limit of the colloidal size range, it may be difficult to distinguish between a particulate gel and a polymeric gel.

The subject of sol–gel science is covered in great depth in Reference 1. The basic steps in sol–gel processing were outlined in Chapter 1 when we surveyed the common methods used for the production of ceramics. This chapter provides a more detailed examination of the science and practice of the process for the fabrication of ceramics and glasses. Particular attention will be paid to the sol–gel processing of SiO_2 glass not only because of its practical interest but also because of the heightened understanding of the process mechanisms and structural evolution developed from numerous studies.

Figure 5.1 illustrates the routes that can be followed in sol–gel processing. The starting compounds (precursors) for the preparation of the sol consist of inorganic salts or metal–organic compounds, but we shall focus mainly on the *metal alkoxides*, a class of metal–organic precursors that are most widely used in sol–gel research. The chemical reactions that occur during the conversion of the precursor solution to the gel have a significant influence on the structure and chemical homogeneity of the gel. Therefore, a basic problem is to understand how the chemical reaction rates are controlled by the processing variables such as precursor chemical composition, concentration of reactants, solution pH, and temperature. The problem becomes more complex when two or more alkoxides are used in the fabrication of multicomponent gels (i.e., gels containing more than one metal cation). There will be a loss of chemical homogeneity if steps are not taken to control the reaction.

After preparation, the gel contains a large amount of liquid existing in fine interconnected channels, and it must be dried prior to conversion to a useful material. Conventional drying by evaporation of the liquid gives rise to capillary pressure that causes shrinkage of the gel. The resulting dried gel is called a *xerogel*. The capillary pressure can be quite large in polymeric gels because the pores are normally much finer than those in colloidal gels, so polymeric gels are prone to warping and cracking. Two general approaches have been used to circumvent these problems. The use of chemicals added to the precursor solution prior to gelation, referred to as *drying control chemical additives* (DCCAs), permit relatively rapid drying but the mechanism by which they operate is not very clear. The removal of the liquid under supercritical conditions eliminates the liquid–vapor interface, preventing the development of capillary stresses, so the gel undergoes relatively little shrinkage. The dried gel, called an *aerogel*, is fragile and may shrink considerably during sintering.

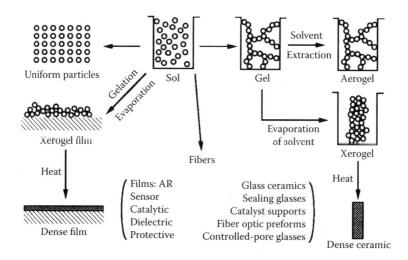

FIGURE 5.1 Schematic illustration of the routes that could be followed in sol–gel processing. (From Brinker, C.J. and Scherer, G.W., *Sol-Gel Science*, Academic Press, New York, 1990. With permission.)

Most gels have an amorphous structure even after drying and contain fine pores, which, in many cases, can be fairly uniform in size. If a dense ceramic is required as the end product, these characteristics are very favorable for good densification during sintering. When compared to the production of similar ceramics by traditional processing routes involving the compaction and sintering of crystalline particles, a reduction in the sintering temperature, particularly in the case of polymeric gels, forms a characteristic advantage of the sol–gel route. In some cases, crystallization of the gel prior to full densification may limit the sintering rate for compositions that are crystallizable.

Applications sol–gel processing can provide substantial benefits, such as the various special shapes that can be obtained directly from the gel state (e.g., monoliths, films, fibers and particles), control of the chemical composition and microstructure, and low processing temperatures. However, the disadvantages are also real. Many metal alkoxides are fairly expensive and most are very sensitive to moisture, so they must be handled in a dry environment (e.g., an inert atmosphere glove box). The large shrinkage of the gel during the drying and sintering steps makes dimensional control of large articles difficult. It is often difficult to dry monolithic gels thicker than a few mm or films thicker than ~1 μm without cracking. The sol–gel process is, therefore, seldom used for the production of thick articles. Instead, it has seen considerable use for the production of small or thin articles such as films, fibers, and powders, and its use in this area is expected to grow in the future.

5.2 TYPES OF GELS

5.2.1 PARTICULATE GELS

Particulate gels consist of a skeletal network of essentially anhydrous particles held together by surface forces (Figure 5.2). The structure of the particles normally corresponds to that of the bulk solid with the same composition. For example, colloidal SiO_2 particles have the same structure as bulk silica glass produced by melting. Hydroxyl groups are present only on the surfaces of the particles. The pores in particulate gels are much larger than in polymeric gels and the capillary stress developed during drying is therefore lower, so less shrinkage occurs. Because of the larger pores, the permeability of colloidal gels is higher, and this, combined with the lower capillary pressure, means that colloidal gels are less likely to crack during drying. The structure of the dried gel is characterized by a relatively high porosity (~70–80%) and pores that are larger than the particles (e.g., the average pore size is typically one to five times the particle size).

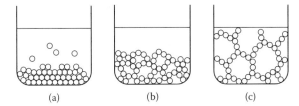

(a)   (b)   (c)

FIGURE 5.2 Sediments formed from (a) a stable colloid and (b) a flocculated colloid compared with (c) a particulate gel.

Particulate gels are commonly prepared by taking a stable colloidal suspension (see Chapter 4) and reducing its volume by evaporating some of the liquid or reducing the colloid stability by the addition of an electrolyte such as an acid, a base, or a metal salt. The attractive van der Waals forces dominate and the particles stick on contact, forming a skeletal network. Methods for the preparation of SiO_2 gels from colloidal particles have been described by Rabinovitch et al. [2,3] and by Scherer and Luong [4]. This class of gels provides an interesting contrast with the properties of polymeric gels prepared from silicon alkoxides.

The structure of the dried gel has important consequences for its sintering and conversion to the final article. Sintering of amorphous materials occurs by viscous flow. Because of the larger pores (hence lower driving force for densification), particulate gels sinter at temperatures well above the glass transition temperature, T_g. In contrast, polymeric gels with their much finer pores sinter near T_g. For particulate SiO_2 gels (particle size $\approx$ 50 nm), sintering is performed in the range of 1200–1500°C.

While polymeric gels are the subject of considerable research and publication, the particulate systems include some of the most successful industrial applications of sol–gel technology. An early application was in the nuclear fuels industry in the late 1950s [5]. The goal of the work was to prepare small spheres (several tens of micrometers) of radioactive oxides, such as $(U,Pu)O_2$ and $(Th,U)O_2$, as fuels for nuclear reactors. The sol–gel process avoided the generation of harmful dust, as would be produced in conventional ceramic processing methods, and allowed the formation of spherical particles for efficient packing of fuel rods. In the last few decades, the particulate sol–gel route has been applied to the production of a range of materials for nonnuclear applications, such as porous oxides for catalyst supports and chromatographic columns [1].

For the formation of multicomponent gels that contain more than one metal cation (e.g., SiO_2–TiO_2 glass), the particulate gel route typically involves the mixing of two or more sols (e.g., sols of SiO_2 and TiO_2), or mixing a metal (e.g., Ti) salt solution with a sol (e.g., SiO_2), followed by gelling. The range of chemical homogeneity in the gelled material can, at best, be of the order of the particle size containing between 10^3 to 10^9 molecules. When the concentration of one component is small (i.e., less than a few percent), the production of good chemical homogeneity becomes difficult. The polymeric gel route commonly provides exceptional chemical homogeneity because the mixing can be achieved at a molecular level rather than at the colloidal level.

5.2.2 POLYMERIC GELS

Polymeric gels consist of a skeletal network of polymer chains that form by the entanglement and cross-linking of growing polymer chains or polymer clusters resulting from the hydrolysis, condensation, and polymerization of precursors in solution. Our main focus will be on polymeric gels formed from metal alkoxide precursors. Depending on the conditions used in the preparation, the structure of the polymer chains can vary considerably. The polymerization of silicon alkoxide, for example, can lead to complex branching of the polymer as shown in Figure 5.3 [6], but under certain conditions (e.g., low water concentration), little branching will occur.

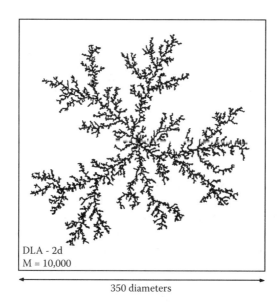

350 diameters

FIGURE 5.3 Fractal polymer made by branching of polyfunctional monomer. (Computer simulation of two-dimensional aggregation.) (From Meakin, P., Models for colloidal aggregation, *Annu. Rev. Phys. Chem.*, 39, 237, 1988. With permission.)

Gelation is accompanied by a sharp increase in the viscosity that essentially freezes-in the polymer structure at the point of gelling. At this stage, the gel consists of a weak amorphous solid structure and an interconnected network of very fine pores filled with liquid. The volume of the liquid-filled pores is very high, typically 90 to 95% of the total volume, and the diameters of the pore channels are typically of the order of 2 to 10 nm. The frozen-in structure can change appreciably during subsequent aging and drying of the gel. Under certain conditions, the aging gel can shrink considerably while expelling liquid. Because of the very fine pores, large capillary stresses are developed during conventional drying, so polymeric gels are very prone to cracking. The capillary stresses are equivalent to the application of external compressive stresses on the gel. Conventional drying therefore collapses the weak polymer network and results in additional cross-linking of the polymer structure. Cross-linking and collapse of the gel continue until the structure can withstand the compressive action of the capillary stresses. Depending on the structure of the gel, the porosity of the dried gel can be anywhere between ~30 and 70%.

Compared to the particulate gel case, structural effects of the dried gel on sintering are more pronounced as well as more complex. The average pore size of polymeric gels is usually much finer than that of colloidal gels, so the driving force for sintering is usually much higher. The result is a lowering of the sintering temperature for polymeric gels. For example, in the case of SiO_2, viscous sintering of polymeric gels occurs generally between 800 and 1000°C. The local chemical structure also has an important influence on the sintering of polymeric gels. While there is some collapse of the gel structure during drying, the solid skeletal phase which makes up the dried gel is not identical to the corresponding bulk glass produced, for example, by melting. The gel structure contains fewer cross-links and additional free volume compared to the melt-prepared glass. This means that during sintering, the gel structure will change to become more highly cross-linked with a corresponding reduction in its free volume and its surface area.

Although there are many potential applications of the polymeric gel route, the preparation of thin films is by far the most important use. Thin films benefit from most of the advantages of the method while avoiding most of the disadvantages. However, even films suffer from problems (e.g., cracking) if attempts are made to prepare films thicker than ~1 μm.

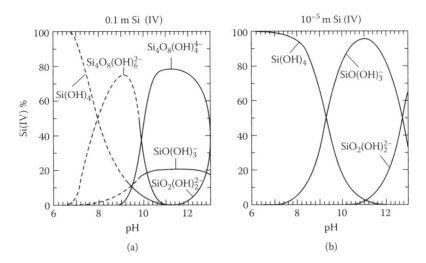

FIGURE 5.4 Distribution of aqueous silicate species at 25°C in (a) 0.1-m Si(IV) and (b) 10^{-5}-m Si (IV). Ionic strength I = 3 m. (From Baes, C.F. and Messmer, R.E., *The Hydrolysis of Cations*, John Wiley & Sons, New York, 1976. With permission.)

5.3 SOL–GEL PROCESSING OF AQUEOUS SILICATES

The formation of silica gel is usually studied in solutions of aqueous silicates (e.g., silicic acid) or silicon alkoxides (e.g., tetraethoxysilane, TEOS). The chemistry of silica in aqueous systems is discussed in detail by Iler [7]. Silicon is hydrolyzed even in dilute acid, and as shown in Figure 5.4. silicic acid, $Si(OH)_4$, often referred to as monosilicic acid, orthosilicic acid, or soluble silica, is the dominant mononuclear species in solution below pH values of ~7 [8]. The Si-OH group is called a *silanol* group, so that $Si(OH)_4$ contains four silanol groups. Above pH ≈ 7, further hydrolysis produces anionic species:

$$Si(OH)_4(aq) \rightarrow SiO_x(OH)_{4-x}^{x-} + xH^+ \tag{5.1}$$

where $SiO(OH)_3^-$ (x = 1 in Equation 5.1) is the dominant mononuclear species. Because $SiO(OH)_3^-$ is a very weak acid, $SiO_2(OH)_2^{2-}$ is observed in appreciable quantities only above pH values of ~12 (Figure 5.4).

Solutions of silicic acid thicken slowly and finally form a gel. Because the gel appears outwardly similar to organic gels, it was generally thought that $Si(OH)_4$ polymerized into siloxane chains (i.e., chains with Si—O—Si bonds) that branched and cross-linked similar to many organic polymers. However, Iler [7] clearly states that silicic acid polymerizes into discrete particles that in turn aggregate into chains and networks. Polymerization occurs in three stages:

1. Polymerization of monomers to form particles
2. Growth of particles
3. Linking of particles into chains, then networks that extend throughout the liquid medium, thickening into a gel

In aqueous silicate chemistry, the term *polymerization* is commonly used in its broadest sense to include reactions that result in an increase in molecular weight of silica. It includes the condensation of silanol groups

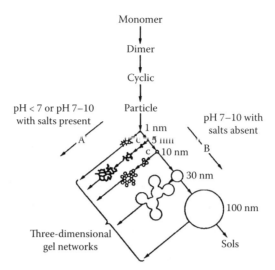

FIGURE 5.5 Polymerization behavior of silica. In basic solution (B), particles in sol grow in size with decrease in numbers; in acid solution or in presence of flocculating salts (A), particles aggregate into three-dimensional networks and form gels. (From Iler, R.K., *The Chemistry of Silica*, John Wiley & Sons, New York, 1979. With permission.)

$$\equiv Si - OH + HO - Si \equiv \ \rightarrow \ \equiv Si - O - Si \equiv + H_2O \tag{5.2}$$

to give molecularly coherent units of increasing size, irrespective of whether these are polymeric species, particles, or particle networks. Condensation reactions take place in such a way as to maximize the number of Si—O—Si bonds and minimize the number of terminal hydroxyl groups through internal condensation. Thus, rings are formed quickly to which monomers add, giving particles that in turn condense to the more compact state, leaving OH groups on the outside. The particles serve as nuclei for further growth, which occurs by an Ostwald ripening process [9]. The smaller, more soluble particles dissolve and precipitate on the larger, less soluble particles, so the number of particles decreases but the average particle size increases. Because of the greater solubility, particle growth is enhanced at higher temperatures, particularly at pH values > 7. After the gel network has formed, the structure becomes stronger as the necks between the particles become thicker because of solution of silica and its reprecipitation at the necks. A schematic of the polymerization process for aqueous silica is shown in Figure 5.5.

5.3.1 Effect of pH

The polymerization of aqueous silica is divided into three approximate pH domains [1,7]: pH < ~2, pH ≈ 2–7, and pH > ~7 (Figure 5.6). The pH value of 2 occurs as a boundary because the point of zero charge (PZC) and the isoelectric point (IEP) both occur close to this pH value (Chapter 4). The pH of 7 forms another boundary because the solubility and dissolution rates of silica are maximized at or above this pH value and because the appreciable surface charge creates a significant repulsion, so particle growth occurs without aggregation or gelation. Silica sols have a maximum stability at pH ≈ 2 and a minimum stability with rapid gelling at pH ≈ 6.

5.3.1.1 Polymerization in the pH range of 2 to 7

Because the gelation times decrease with increasing pH values between 2 and 6, it is generally assumed that above the IEP the gelation rate is proportional to the concentration of OH⁻. The

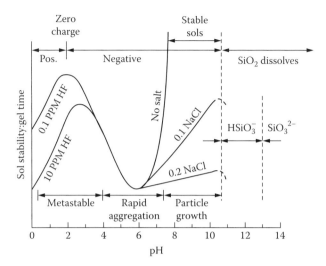

FIGURE 5.6 Effect of pH in the colloidal silica–water system. (From Iler, R.K., *The Chemistry of Silica*, John Wiley & Sons, New York, 1979. With permission.)

base-catalyzed polymerization occurs by a nucleophilic mechanism according to the following reaction sequence:

$$\equiv Si - OH + OH^- \xrightarrow[\text{fast}]{} \equiv Si - O^- + H_2O \tag{5.3}$$

$$\equiv Si - O^- + HO - Si \equiv \xrightarrow[\text{slow}]{} \equiv Si - O - Si \equiv + OH^- \tag{5.4}$$

The surface silanols will be deprotonated depending on their acidity, which depends on the other substituents on the silicon atom. When basic OR and OH groups are replaced by O–Si, the reduced electron density on Si increases the acidity of the protons on the remaining silanols. In any given distribution of silicate species, the most acidic silanols are the ones contained in the most highly condensed species and these will be most likely to be deprotonated according to Equation 5.3. Therefore, condensation according to Equation 5.4 occurs preferentially between the more highly condensed species and the less highly condensed, neutral species. The formation of dimers is slow (Equation 5.4), but once formed, dimers react with monomers to form trimers, which, in turn, react with monomers to form tetramers. Cyclization (to form rings) is rapid because of the proximity of the chain ends and the rapid depletion of monomers. Cyclic trimers may also form, but the strain resulting from the reduced Si—O—Si bond angles makes them much less stable in this pH range. (As described later, they are, however, quite stable at pH values above ~12.)

Further growth occurs by continued addition of lower-molecular-weight species to more highly condensed species (by conventional polymerization or by Ostwald ripening) and by aggregation of the condensed species to form chains and networks. Near the IEP, where there is no electrostatic repulsion, growth and aggregation occur together and may be indistinguishable. However, particle growth becomes negligible when the particles reach a size of 2 to 4 nm because of low solubility of silica in this pH range and the greatly reduced size-dependent solubility (Figure 5.7). The gels consist of chains and networks of very fine particles. The presence of a salt (e.g., NaCl) has little effect on the gelation in this pH range.

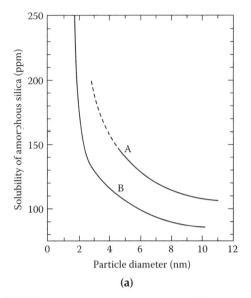
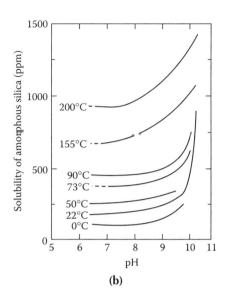

(a)   **(b)**

FIGURE 5.7 (a) Relation between solubility of amorphous silica in water at 25°C and particle diameter. Particles A made at 80 to 100°C at pH 8. Particles B made at 25 to 50°C at pH 2.2. (From Iler, R.K., *The Chemistry of Silica*, John Wiley & Sons, New York, 1979. With permission.) (b) Solubility of amorphous silica vs. pH at different temperatures. (From Iler, R.K., *The Chemistry of Silica*, John Wiley & Sons, New York, 1979. With permission.)

5.3.1.2 Polymerization above pH ≈ 7

Polymerization occurs by the same nucleophilic mechanism described by Equation 5.3 and Equation 5.4. In this pH range, however, all the condensed species are likely to be ionized and therefore mutually repulsive. Growth occurs by the addition of monomers to the highly condensed particles rather than by aggregation. Particles with diameters of 1 to 2 nm are formed in a few minutes. Above pH values of ~12, where most of the silanols are deprotonated, the condensed species are cyclic trimers and tetramers. Cyclic trimers are stable in this pH range because the planar, cyclic configuration permits the greatest separation of charge between the deprotonated sites.

Because of the greater solubility of silica in this pH range and the greater size-dependence of solubility (Figure 5.7), growth of the primary particles occurs by Ostwald ripening. Particles grow rapidly to a size (5 to 10 nm) that depends on the temperature. Because the ripening process occurs by the dissolution of the smaller particles and reprecipitation on the larger particles, the growth rate depends on the particle size distribution. In the absence of any salt, the electrostatic repulsion between the particles leads to stable sols without the formation of chains or aggregates. The addition of salt (e.g., NaCl) leads to a reduction in the thickness of the electrical double layer (Chapter 4), thereby reducing the stability of the sol and producing a dramatic reduction of the gelation time (Figure 5.6).

5.3.1.3 Polymerization below pH ≈ 2

Because the gelation times decrease with decreasing pH below ~2, the polymerization rate is proportional to the concentration of H⁺. The acid-catalyzed polymerization mechanism is not clear. Iler and others propose a mechanism that involves a siliconium ion intermediate ($\equiv Si^+$):

$$\equiv Si - OH + H_3O^+ \rightarrow \; \equiv Si^+ + 2H_2O \tag{5.5}$$

$$\equiv Si^+ + HO - Si \rightarrow \, \equiv Si - O - Si \equiv + H^+ \qquad (5.6)$$

However, in the polymerization of silicon alkoxides discussed in the following section, Brinker and Scherer [1] suggest that the condensation process more likely proceeds via an associative $\equiv SiOHR(OH_2)^+$ intermediate.

In the absence of fluoride ion, the solubility of silica in this pH range is very low, and for moderate acidities (pH = 0-2), the silicate species should not be highly ionized. Therefore, it is likely that the formation and aggregation of the primary particles occur together and that Ostwald ripening contributes little to the growth after the particles exceed a diameter of ~2 nm. As a result, the gel networks are composed of exceedingly fine particles. Traces of F^- or additions of HF lead to a reduction of the gelation times and produce gels similar to those formed above pH $\approx$ 2. Because F^- and OH^- have a similar size, they have the same influence on the polymerization behavior.

5.4 METAL ALKOXIDES

Metal alkoxides are the most common precursors used in sol–gel processing because they react readily with water. We recall that these compounds are also used in the synthesis of fine oxide particles (e.g., the Stober process discussed in Chapter 2). Metal alkoxides are a class of metal–organic compounds having the general formula $M(OR)_z$, where M is a metal of valence z and R is an alkyl group. They can be considered as derivatives of either an alcohol in which the hydroxylic hydrogen is replaced by the metal or of a metal hydroxide in which the hydrogen is replaced by an alkyl group. Accordingly, the chemistry of the metal alkoxides involves the metal–oxygen–carbon bond system. Metal alkoxides have also been referred to in the literature as *organometallic* compounds, but it is not correct to call alkoxides organometallic. An organometallic has a metal–carbon (M—C) bond (unlike most alkoxides), whereas a metal–organic need not have M and C directly bonded.

The synthesis and properties of metal alkoxides have been considered in detail in a book by Bradley et al. [10], and several important synthesis reactions are described in a review article [11]. Here we consider only a few of the main features relevant to the sol–gel processing of ceramics and glasses.

5.4.1 Preparation of Metal Alkoxides

The method used for the preparation of a metal alkoxide depends, in general, on the electronegativity of the metal. The main methods may be divided into two groups: (1) reactions between metals and alcohols for the more electropositive metals (those with relatively low electronegativity values) and (2) reactions involving metal chlorides for the less electropositive metals (those with higher electronegativity values). In addition, there are miscellaneous methods, such as alcohol interchange, transesterification, and esterification reactions, which are useful for the synthesis of some alkoxides.

5.4.1.1 Reactions between Metals and Alcohols

The alkoxides of the more electropositive metals with valences up to 3 can be prepared by the simple reaction:

$$M + zROH \rightarrow M(OR)_z + (z/2)H_2 \qquad (5.7)$$

The nature of the alcohol also has a significant effect on these reactions. For example, sodium reacts vigorously with methanol and ethanol, but the reaction rate is considerably slower with isopropanol and extremely slow in *tert*-butanol. For the alkali metals (e.g., Li, Na, K) and the alkali earth metals (e.g., Ca, Sr, Ba), the reaction proceeds without the use of a catalyst. Metals such as Be, Mg, Al, the lanthanides, and Y require catalysts to cause them to react with alcohols. Depending on the metal, various catalysts have been used, including I_2, $HgCl_2$, and $BeCl_2$. Although the mechanism of these catalytic reactions is not well understood, one suggestion is that the catalyst aids the removal of the surface oxide layer that acts as a passivating layer on the metal.

5.4.1.2 Reactions Involving Metal Chlorides

For less electropositive metals, the alkoxides are obtained via the anhydrous metal chloride. For some highly electronegative metals, such B, Si, and P, the direct reaction between metal chlorides and alcohols is effective:

$$MCl_z + zROH \rightarrow M(OR)_z + zHCl \tag{5.8}$$

However, for most metals the reaction must be forced to completion by using a base such as ammonia:

$$MCl_z + zROH + NH_3 \rightarrow M(OR)_z + zNH_4Cl \downarrow \tag{5.9}$$

The reaction described by Equation 5.9 forms the most useful procedure for the preparation of many alkoxides (including Zr, Hf, Si, Ti, Fe, Nb, Ge, V, Ta, Th, Sb, U, and Pu) and is used widely for commercial production. The reaction between anhydrous metal chlorides and sodium alkoxide in the presence of excess alcohol and an inert solvent such as benzene or toluene is also a useful method:

$$MCl_z + zNaOR \rightarrow M(OR)_z + zNaCl \downarrow \tag{5.10}$$

The nature of the alcohol has a significant influence on the preparation of alkoxides by the reaction involving metal chlorides and alcohols. For the lower straight-chain alcohols such as methanol and ethanol, the alcohol undergoes a relatively straightforward reaction with the metal chloride and the base. For other alcohols, side reactions may assume a dominant role, so the yield of the alkoxide product is usually low.

5.4.1.3 Miscellaneous Methods

The alkoxides of the alkali metals may also be prepared by dissolving the metal hydroxide in the alcohol. For example, sodium ethoxide may be produced from sodium hydroxide and ethanol:

$$NaOH + C_2H_5OH \rightleftharpoons NaOC_2H_5 + H_2O \tag{5.11}$$

Alkoxides of some highly electronegative metals (e.g., B, Si, Ge, Sn, Pb, As, Se, V, and Hg) can be prepared by an *esterification* reaction involving an oxide and an alcohol:

$$MO_{z/2} + zROH \rightleftharpoons M(OR)_z + (z/2)H_2O \tag{5.12}$$

The reactions described by Equation 5.11 and Equation 5.12 are reversible so that the water produced must be removed continually. In practice, this is usually done by using solvents such as benzene or xylene, which form azeotropes with the water. The azeotropic mixture behaves like a single substance in that the vapor produced by distillation has the same compositions as the liquid, and it can be easily fractionated out by distillation.

Metal alkoxides have the ability to exchange alkoxide groups with alcohols, and this has been used in the preparation of new alkoxides for a variety of metals, including Zn, Be, B, Al, Si, Sn, Ti, Zr, Ce, Nb, Nd, Y, and Yb. The reaction is called *alcoholic interchange* or *alcoholysis*. The general reaction can be written:

$$M(OR)_z + zR'OH \rightleftharpoons M(OR')_z + zROH \tag{5.13}$$

To complete the reaction, the alcohol ROH produced in the reaction is removed by fractional distillation. Benzene or xylene, which forms azeotropes with alcohol, is used to aid the removal of the alcohol by fractional distillation. As an example, the alcoholysis of aluminum isopropoxide with *n*-butanol can be used to prepare aluminum *n*-butoxide:

$$Al(O-{}^iC_3H_7)_3 + 3\,{}^nC_4H_9OH \rightleftharpoons Al(O-{}^nC_4H_9)_3 + 3\,{}^iC_3H_7OH \tag{5.14}$$

where the superscripts *i* and *n* refer to the secondary (or iso) and normal alkyl chains, respectively.

Metal alkoxides undergo transesterification with carboxylic esters, and this affords a method of conversion from one alkoxide to another. The reaction is reversible and can be written as:

$$M(OR)_z + zCH_3COOR' \rightleftharpoons M(OR')_z + zCH_3COOR \tag{5.15}$$

Fractional distillation of the more volatile ester CH_3COOR in Equation 5.15 is required to complete the reaction. Transesterification reactions have been used for the preparation of alkoxides of various metals including Zr, Ti, Ta, Nb, Al, La, Fe, Ga, and V. For example, zirconium *tert*-butoxide has been prepared from the reaction between zirconium isopropoxide and *tert*-butyl ester followed by distillation of isopropyl acetate:

$$Zr(O-{}^iC_3H_7)_4 + 4CH_3COO-{}^tC_4H_9 \rightarrow Zr(O-{}^tC_4H_9)_4 + 4CH_3COO-{}^iC_3H_7 \tag{5.16}$$

Double alkoxides are metal alkoxides with two different metal cations chemically combined within each molecule or molecular species. They have an advantage over a mixture of individual alkoxides for the preparation of multicomponent gels with good chemical homogeneity. However, they are difficult to prepare in a ceramic laboratory and are expensive when available commercially. Their stability can vary considerably, depending on the nature of the two metals and the alkoxide group. Bradley et al. [10] provide a summary of the methods used to synthesize double alkoxides. One of these involves dissolving each alkoxide in a mutual solvent, mixing the solutions, and refluxing at elevated temperatures. An example is the synthesis of the double alkoxide, $NaAl(OC_2H_5)_4$, by the reaction between sodium ethoxide and aluminum ethoxide:

$$NaOC_2H_5 + Al(OC_2H_5)_3 \rightarrow NaAl(OC_2H_5)_4 \tag{5.17}$$

An overview of several double alkoxides that can be used for the preparation of multicomponent ceramics is given by Mah et al. [12].

TABLE 5.1
Physical State of the Alkoxides of Some
Metals with Different Electronegativities

Alkoxide	State
$Na(OC_2H_5)$	Solid (decomposes above ~530K)
$Ba(O-^iC_3H_7)_2$	Solid (decomposes above ~400K)
$Al(O-^iC_3H_7)_3$	Liquid (bp 408K at 1.3 kPa)
$Si(OC_2H_5)_4$	Liquid (bp 442K at atmospheric pressure)
$Ti(O-^iC_3H_7)_4$	Liquid (bp 364.3K at 0.65 kPa)
$Zr(O-^iC_3H_7)_4$	Liquid (bp 476K at 0.65 kPa)
$Sb(OC_2H_5)_3$	Liquid (bp 367K at 1.3 kPa)
$Te(OC_2H_5)_4$	Liquid (bp 363K at 0.26 kPa)
$Y(O-^iC_3H_7)_3$	Solid (sublimes at ~475K)

5.4.2 BASIC PROPERTIES

5.4.2.1 Physical Properties

The physical properties of metal alkoxides depend primarily on the characteristics of the metal (e.g., the electronegativity, valence, atomic radius, and coordination number) and secondarily on the characteristics of the alkyl group (e.g., the size and shape). There is a change from the solid, nonvolatile ionic alkoxides of some of the alkali metals to the volatile covalent liquids of elements with valence of 3, 4, 5, or 6 (e.g., Al, Si, Ti, Zr, Sb, and Te), whereas alkoxides of metals with intermediate electronegativities, such as La and Y, are mainly solids (Table 5.1).

The alkyl group has a striking effect on the volatility of metal alkoxides. Many metal methoxides are solid, nonvolatile compounds (e.g., sodium methoxide). As the number of methyl groups increases and the size of the metal atom decreases, methoxides become sublimable solids or fairly volatile liquids (e.g., silicon tetramethoxide). Many metal alkoxides are strongly associated by intermolecular forces that depend on the size and shape of the alkyl group. The degree of association of metal alkoxides is sometimes described by the term *molecular complexity*, which refers to the average number of empirical units in a complex. Figure 5.8 shows a schematic of a coordination complex of aluminum isopropoxide with a molecular complexity of 3.

The physical properties of the alkoxides are determined by two opposing tendencies. One tendency is for the metal to increase its coordination number by utilizing the bridging property of the alkoxo groups. The opposite tendency is the screening or steric effect of the alkyl groups, which interfere with the coordination process. The degree of the screening depends on the size and shape of the alkyl group. The result is that alkoxides with a wide range of properties, ranging from nonvolatile polymeric solids to volatile monomeric liquids, can be achieved depending on

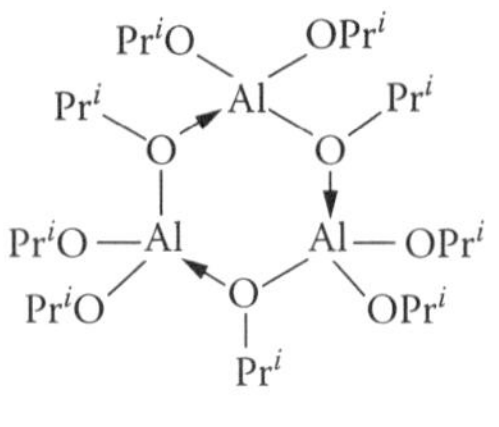

FIGURE 5.8 Coordination complex of aluminum isopropoxide consisting of three molecules.

TABLE 5.2
Boiling Point and Molecular Complexity of Some Titanium Alkoxides

Alkoxide	Molecular Weight	Boiling point (K) at 0.65 kPa	Molecular Complexity
$Ti(OCH_3)_4$	172	(Solid)	—
$Ti(OC_2H_5)_4$	228	411.3	2.4
$Ti(O-^nC_3H_7)_4$	284.3	410	(unknown)
$Ti(O-^iC_3H_7)_4$	284.3	364.3	1.4
$Ti(O-^tC_4H_9)_4$	340.3	366.8	1.0

the nature of the alkyl group. Table 5.2 shows the boiling point and molecular complexity of some titanium alkoxides. In spite of the large increase in molecular weight, the boiling point decreased from 138.3°C for the ethoxide to 93.8°C for the *tert*-butoxide. For the same molecular weight, there is also a reduction in the boiling point for the branched isopropoxide compared to the linear *n*-propoxide.

5.4.2.2 Chemical Properties

Metal alkoxides are characterized by the ease with which they undergo hydrolysis. In many cases, the alkoxides are so sensitive to traces of moisture that special precautions must be taken in their handling and storage. The use of an inert, dry atmosphere (e.g., normally available in a glove box) and dehydrated solvents is essential in most experiments. Hydrolysis in excess water commonly leads to the formation of insoluble hydroxides or hydrated oxides. However, when restricted amounts of water are added, metal alkoxides undergo partial hydrolysis reactions, yielding soluble species that can take part in polymerization reactions. The initial step involves a hydrolysis reaction in which alkoxide groups (OR) are replaced by hydroxyl groups (OH):

$$M(OR)_z + H_2O \rightarrow M(OH)(OR)_{z-1} + ROH \qquad (5.18)$$

Subsequent condensation reactions involving the hydroxy metal alkoxide produces polymerizable species with M–O–M bonds plus alcohol (ROH) or water as a by-product:

$$M(OH)(OR)_{z-1} + M(OR)_z \rightarrow (RO)_{z-1}M - O - M(OR)_{z-1} + ROH \qquad (5.19)$$

$$2M(OH)(OR)_{z-1} \rightarrow (RO)_{z-1}M - O - M(OR)_{z-1} + H_2O \qquad (5.20)$$

Because most metal alkoxides containing lower aliphatic alkyl groups are actually coordinated complexes and not single molecules (Figure 5.8), the reactions described by Equation 5.18 to Equation 5.20 must be considered to be somewhat simplified. The rate of hydrolysis of metal alkoxides depends on the characteristics of the metal and those of the alkyl group. In general, silicon alkoxides are among the slowest to hydrolyze and, for a given metal alkoxide, the hydrolysis rate increases as the length of the alkyl group decreases.

In excess water, aluminum alkoxides initially form the monohydroxide (boehmite), which later may convert to the trihydroxide (bayerite) [13]:

$$Al(OR)_3 + 2H_2O \rightarrow AlO(OH)\downarrow + ROH \qquad (5.21)$$

$$AlO(OH) + H_2O \rightarrow Al(OH)_3 \downarrow \qquad (5.22)$$

Boron alkoxides form the oxide, B_2O_3, or boric acid, $B(OH)_3$, when reacted with excess water. Oxide formation can be written as:

$$2B(OR)_3 + 3H_2O \rightarrow B_2O_3 + 6ROH \qquad (5.23)$$

As discussed later, the formation of insoluble precipitates as represented by Equation 5.21 to Equation 5.23 makes it impossible for polymerization reactions to occur and must be prevented if good chemical homogeneity is to be achieved by the polymeric gel route.

Silicon alkoxides show a different type of reaction. They form soluble silanols in excess water rather than an insoluble oxide or hydroxide. The hydrolysis reaction may be written as:

$$Si(OR)_4 + xH_2O \rightarrow Si(OR)_{4-x}(OH)_x + xROH \qquad (5.24)$$

Complete hydrolysis leading to the formation of the silicic acid monomer $Si(OH)_4$ generally does not occur except at low pH and high water concentration. Under most conditions, condensation of the silanol groups occurs prior to the replacement of all the OR groups by hydroxyl groups, leading to the formation of polymeric species, as in Equation 5.19 and Equation 5.20, for example.

Metal alkoxides are soluble in their corresponding alcohols. In practice, dissolution of solid alkoxides or dilution of liquid alkoxides is normally performed in the corresponding alcohol. As outlined in the following text for the case of silicon alkoxides, the alcohol may also serve an additional function as a mutual solvent for the alkoxide and water when they are immiscible. Metal alkoxides also have the ability to exchange alkoxide groups with alcohols, as in Equation 5.13, for example.

5.5 SOL–GEL PROCESSING OF SILICON ALKOXIDES

5.5.1 PRECURSORS

The most commonly used precursors for the sol–gel processing of silica are *tetraethyoxysilane*, $Si(OC_2H_5)_4$, abbreviated TEOS, which is also referred to as tetraethylorthosilicate or silicon tetra-ethoxide, and *tetramethoxysilane*, $Si(OCH_3)_4$, abbreviated TMOS, which is also referred to as tetramethyorthosilicate or silicon tetramethoxide. The physical properties of these two precursors are given in Table 5.3.

TABLE 5.3
Physical Properties of the Commonly Used Silicon Alkoxides

Name	Molecular Weight	Boiling Point (K)	Specific Gravity (at 293K)	Refractive Index (at 293K)	Solubility
$Si(OC_2H_5)_4$ Tetraethoxysilane	208.33	442	0.934	1.3818	Alcohols
$Si(OCH_3)_4$ Tetramethoxysilane	152.22	394	1.032	1.3688	Alcohols

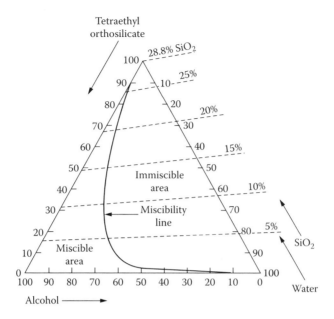

FIGURE 5.9 TEOS, H_2O and Synasol (95% EtOH, 5% water) ternary phase diagram at 25°C. For pure ethanol, the miscibility line is shifted slightly to the right. (From Cogan, H.D. and Setterstrom, C.A., Properties of ethyl silicate, *Chem. Eng. News*, 24, 2499, 1946. With permission.)

5.5.2 HYDROLYSIS AND CONDENSATION

Following Equation 5.18 to Equation 5.20, the hydrolysis and condensation reactions for silicon alkoxides can be written as:

$$\text{hydrolysis}$$
$$\equiv Si - OR + H_2O \quad \rightleftharpoons \quad \equiv Si - OH + ROH \tag{5.25}$$
$$\text{esterification}$$

$$\text{alcohol condensation}$$
$$\equiv Si - OR + HO - Si \equiv \quad \rightleftharpoons \quad \equiv Si - O - Si \equiv + ROH \tag{5.26}$$
$$\text{alcoholysis}$$

$$\text{water condensation}$$
$$\equiv Si - OH + HO - Si \equiv \quad \rightleftharpoons \quad \equiv Si - O - Si \equiv + H_2O \tag{5.27}$$
$$\text{hydrolysis}$$

where R represents the alkyl group and the reverse reactions have also been included. The alkoxysilanes and water are immiscible, so a mutual solvent, commonly the corresponding alcohol of the alkoxy group, is used to homogenize the mixture, as shown in Figure 5.9 for TEOS [16]. The alcohol can also take part in esterification and alcoholysis reactions (Equation 5.25 and Equation 5.26).

Silica gels are often synthesized by employing an acid (e.g., HCl) or a base (e.g., NH_3) as a catalyst. Gels have been prepared under a wide range of catalyst concentration (0.01 to 7 M) and

TABLE 5.4
Sol–Gel Silicate Compositions for Bulk Gels, Fibers, Films, and Powders

	Mole %					
SiO$_2$ Gel Types	TEOS	EtOH	H$_2$O	HCl	NH$_3$	H$_2$O/Si (w)
Bulk						
One-step acid	6.7	25.8	67.3	0.2	—	10
One-step base	6.7	25.8	67.3	—	0.2	10
Two-step acid-base						
First-step acid	19.6	59.4	21.0	0.01	—	1.1
Second-step acid (A2)	10.9	32.8	55.7	0.6	—	5.1
Second-step base (B2)	12.9	39.2	47.9	0.01	0.016	3.7
Fibers	11.31	77.26	11.31	0.11	—	1.0
Films	5.32	36.23	58.09	0.35	—	10.9
Monodisperse spheres	0.83	33.9	44.5	—	20.75	53.61

Source: From Brinker, C.J. and Scherer, G.W., *Sol-Gel Science*, Academic Press, New York, 1990. With permission.

reactant concentration (H$_2$O:Si ratio w of < 1 to > 50). Table 5.4 gives examples of the conditions used to synthesize gels in the form of monoliths, fibers, films, and particles.

It is found that the synthesis conditions have a significant influence on the structural evolution of the sol–gel silicate, which in turn affects the behavior of the gel during subsequent drying and sintering. The synthesis conditions are often divided into acid-catalyzed conditions and base-catalyzed conditions. Three broad categories of synthesis conditions can be identified. Under conditions of acid catalysis and low w, the sols consist of weakly branched polymers. With base catalysis and high w, highly condensed particulate sols are formed. In conditions that are intermediate between these two extremes (acid catalysis/high w or base catalysis/low w), the sols have intermediate structures (branched to highly branched polymers).

The structure of the sol–gel silicate depends on the relative rates of the hydrolysis and condensation reactions, as well as on the reverse reactions (Equation 5.25 to Equation 5.27). The influence of the reaction conditions on the mechanism and kinetics of the reactions forms a key issue in understanding the structural evolution.

5.5.2.1 Acid-Catalyzed Conditions

Hydrolysis and condensation occur by bimolecular nucleophilic displacement reactions involving protonated alkoxide groups [1]. In the first step, it is likely that an OR group bonded to Si is rapidly protonated. Electron density is withdrawn from Si, making it more electrophilic and thus more susceptible to attack by water. The H$_2$O molecule acquires a small positive charge while the positive charge on the protonated alkoxide is correspondingly reduced, making alcohol a better leaving group. The transition state decays by displacement of alcohol accompanied by inversion of the Si tetrahedron:

$$\text{(5.28)}$$

Condensation to form siloxane bonds occurs by either an alcohol-producing reaction (Equation 5.26) or a water-producing reaction (Equation 5.27). The most basic silanol species, namely silanols contained in monomers or weakly branched oligomers (very-short-chain polymers), are most likely to be protonated so that condensation reactions are likely to occur between neutral species and protonated silanols situated on monomers and the end groups of chains.

5.5.2.2 Base-Catalyzed Conditions

Under basic conditions, hydrolysis and condensation also occurs by a bimolecular nucleophilic substitution reaction. It is likely that water dissociates to produce hydroxyl anions in a rapid first step. The hydroxyl anion then attacks Si directly, displacing OR, followed by inversion of the Si tetrahedron:

$$
\begin{array}{c}
\text{RO} \\
\text{RO} \diagdown \\
\text{HO}^- + \ \text{Si} - \text{OR} \rightleftharpoons \\
\diagup \\
\text{RO}
\end{array}
\quad
\begin{array}{c}
\text{RO} \qquad \text{OR} \\
\delta^- \diagdown \ \diagup \ \delta^- \\
\text{HO} \text{-----} \text{Si} \text{-----} \text{OR} \rightleftharpoons \\
| \\
\text{OR}
\end{array}
\quad
\begin{array}{c}
\text{OR} \\
\diagup \text{OR} \\
\text{HO} - \text{Si} \ + \text{OR}^- \\
\diagdown \\
\text{OR}
\end{array}
\qquad (5.29)
$$

The most widely accepted condensation mechanism involves the attack of a nucleophilic deprotonated silanol on a neutral silicate species, as outlined earlier for the condensation in aqueous silicates (Equation 5.4). This condensation mechanism pertains above the PZC (or IEP) of silica (pH > ~2) because the surface silanols are deprotonated (i.e., they are negatively charged) and the mechanism changes with the charge on the silanol. Condensation between larger, more highly condensed species, which contain more acidic silanols, and smaller, less weakly branched species is favored. The condensation rate is maximized near neutral pH, where significant concentrations of both protonated and deprotonated silanols exist. A minimum rate is observed near the PZC (or IEP).

5.5.3 Polymer Growth

Structural evolution of sol–gel silicates in solution has been investigated on several length scales using a variety of techniques, such as nuclear magnetic resonance (NMR), Raman and infrared spectroscopy, and x-ray, neutron, and light scattering. Reference 1 provides an excellent account of the information obtained with these techniques and the models proposed to describe polymer growth and gelation in silicate systems. On a length scale of 1 to 200 nm, the evolving structures are not uniform objects described by Euclidean geometry but tenuous structures called *fractal structures* (an example of which is given in Figure 5.3). Fractal geometry is discussed in a classic text by Mandelbrot [15], and the connection of fractals to problems in materials science is discussed by Feder [16].

A *mass fractal* is defined as an object whose mass m increases with its radius r according to

$$ m \propto r^{d_m} \qquad (5.30) $$

where d_m is called *mass fractal dimension* of the object. For a Euclidean object, $m \propto r^3$, whereas for a fractal $d_m < 3$, so the density of the fractal ($\rho \propto m/r^3$) decreases as the object gets bigger. A surface fractal has a surface area S that increases faster than r^2:

$$ S \propto r^{d_s} \qquad (5.31) $$

where d_s is called the *surface fractal dimension*.

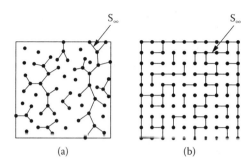

FIGURE 5.10 Gelation theories (schematic): (a) Flory's model (molecular functionality $f = 3$), (b) Percolation on a square lattice. In each case, a spanning S_∞ gel molecule is embedded in the sol. (From Zarzycki, J. Physical-chemical factors in sol–gel processes, in *Science of Ceramic Chemical Processing*, Hench, L.L. and Ulrich, D.R., Eds., John Wiley & Sons, New York, 1986, p. 21, chap. 2. With permission.)

Euclidean objects (e.g., dense spherical particles) are most likely to form in systems (e.g., aqueous silicates) in which the particle is slightly soluble in the solvent. In this case, monomers can dissolve and reprecipitate until the equilibrium structure (having a minimum surface area) is obtained. In nonaqueous systems (e.g., silicon alkoxide–alcohol–water solutions), the solubility of the solid phase is so limited that condensation reactions are virtually irreversible. Bonds form at random and cannot convert to the equilibrium configuration, thereby leading to fractal polymeric clusters.

5.5.3.1 Polymer Growth Models

The models put forward to describe polymer growth in silicate systems are divided into two types: classical (or equilibrium) growth models and kinetic growth models [1,17,18].

5.5.3.1.1 Classical Growth Models

The first theory of polymer growth accompanying gelation was put forward more than 50 years ago by Flory [19a and 19b] and by Stockmayer [20]. Bonds are formed at random between adjacent nodes on an infinite Cayley tree or Bethe lattice (Figure. 5.10). The model is qualitatively successful in correctly describing the emergence of an infinite cluster at some critical extent of reaction and in providing good predictions of the gel point. However, it has come under some criticism because it does not provide an entirely realistic picture of polymer growth. Because of the unphysical nature of the Cayley tree, cyclic configurations of the growing polymers are avoided in the model. The purely branched clusters formed on the Cayley tree have a mass fractal dimension $d_m = 4$. Because the volume of the cluster increases as r^3, where r is the radius of the cluster, the theory predicts that the density ρ increases in proportion to r. This result is physically unrealistic because the density cannot increase indefinitely as the cluster grows.

5.5.3.1.2 Kinetic Growth Models

The kinetic models avoid the unrealistic assumptions of the classical theory and makes predictions about the growth and fractal structure of the simulated structures that are in good agreement with experimental observations. They are based on Smoluchowski's equation describing the rate at which the number of clusters of a given size changes with time during an aggregation process [6]. Most kinetic growth models produce objects with self-similar fractal properties, i.e., they look self-similar under transformation of scale, as would be achieved, for example, by changing the magnification of a microscope. Depending on the conditions, growth in silicate systems may occur predominantly by the condensation of monomers with growing clusters (referred to as *monomer–cluster growth*) or by condensation reactions of clusters with other clusters (*cluster–cluster* growth). The simulated structures resulting from these two types of growth processes are illustrated in Figure. 5.11.

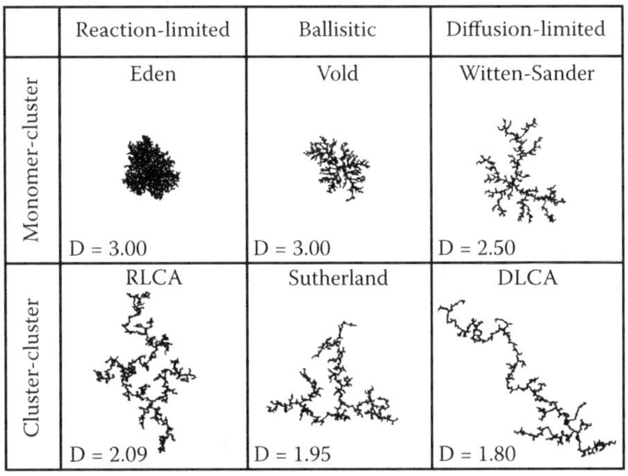

	Reaction-limited	Ballisitic	Diffusion-limited
Monomer-cluster	Eden D = 3.00	Vold D = 3.00	Witten-Sander D = 2.50
Cluster-cluster	RLCA D = 2.09	Sutherland D = 1.95	DLCA D = 1.80

FIGURE 5.11 Simulated structures resulting from various kinetic growth models. Fractal dimensions are listed for three-dimensional clusters even though their two-dimensional analogs are displayed. Each cluster contains 1000 primary particles. Simulations by Meakin [22]. (From Martin, J.E., Molecular engineering of ceramics, in *Atomic and Molecular Processing of Electronic and Ceramic Materials: Preparation, Characterization and Properties*, Aksay, I.A. et al., Eds., Materials Research Society, Pittsburg, PA, 1987, p. 79. With permission.)

Monomer–cluster or cluster–cluster growth can be limited by diffusion or by reaction. In diffusion-limited monomer–cluster aggregation (DLMCA), simulated by the Witten and Sander model [23] in Figure 5.11, it is assumed that monomers are released one by one from sites arbitrarily far from a central cluster. The monomers travel by a random walk diffusion mechanism and stick irreversibly at first contact with the growing cluster. Because of this trajectory, the monomers cannot penetrate deeply into a cluster without intercepting a cluster arm, and the arms effectively screen the interior of the cluster from incoming monomers. Growth occurs preferentially at exterior sites, resulting in objects in which the density decreases radially from the center of mass ($d_m = 2.45$ in three dimensions).

Reaction-limited monomer–cluster aggregation (RLMCA) is distinguished from DLMCA in that there is an energy barrier to bond formation. The effect of this barrier is to reduce the condensation rate because many collisions can occur before the monomer and cluster form a bond. In this process, all potential growth sites are sampled by the monomers. The probability of attachment to a particular site per encounter is dictated by the local structure rather than the large-scale structure, which governs the probability that a monomer will encounter a given site. RLMCA is simulated by the Eden model [24], originally developed to simulate cell colonies. In the model, unoccupied perimeter sites are selected randomly and occupied with equal probability. Because all sites are accessible and filled with equal probability, the Eden model leads to compact, smooth clusters ($d_m = 3$). The "poisoned" Eden growth model [24] represents a modification of the Eden model whereby a certain fraction of sites is prohibited from being occupied, i.e., these sites do not undergo polymerization. Depending on the number of poisoned sites and their distribution, the poisoned Eden growth model generates structures that vary from being uniformly nonporous (nonfractal) to surface fractals to mass fractals.

Cluster–cluster aggregation models [25] describe growth that results when a "sea" of monomers undergoes random walks, forming a collection of clusters that continue to grow by condensation reactions with each other and with remaining monomers. Under diffusion-limited conditions (DLCA), clusters stick irreversibly on contact, whereas under reaction-limited conditions (RLCA), the sticking probability is less than unity. Compared to monomer–cluster growth, the strong mutual screening of colliding clusters creates very open structures even under reaction-limited conditions. It is also evident from Figure 5.11 that in contrast to monomer–cluster growth, cluster–cluster growth produces objects with no obvious centers.

5.5.3.2 Structural Evolution of Sol–Gel Silicates

In silicates, the condensation rate is sufficiently small that reaction-limited aggregation is assumed to occur under both acid and base catalysis. The cross-linking between the polymer chains is much higher at high pH and high H_2O:Si ratio (w), so highly branched clusters are formed under these conditions, whereas more weakly branched clusters are formed at low pH. Based on our description of the hydrolysis and condensation mechanisms and the polymer growth models, we can now summarize the structural evolution of sol–gel silicates and its description by the kinetic models.

5.5.3.2.1 pH < ~2

Both hydrolysis and condensation occur by a bimolecular nucleophilic displacement mechanism involving protonated alkoxide groups. The rate of hydrolysis is large compared to the rate of condensation. For w greater than ~4, hydrolysis is expected to be completed at an early stage of the reaction prior to any significant condensation. After monomers are depleted, condensation between completely hydrolyzed species occurs by reaction-limited cluster–cluster aggregation, leading to weakly branched structures with a mass fractal dimension $d_m \approx 2$.

Under low-water conditions ($w < 4$), condensation occurs prior to the completion of hydrolysis. Condensation between the incompletely hydrolyzed species is also expected to occur by reaction-limited cluster–cluster aggregation, but because the OR groups effectively reduce the functionality of the condensing species, the structures will be more weakly branched when compared to the high-water conditions.

5.5.3.2.2 pH > ~7

Hydrolysis and condensation occur by bimolecular nucleophilic displacement reactions involving OH^- and Si–O^- anions. For $w > 4$, the hydrolysis of all polymeric species is expected to be complete. Dissolution reactions provide a continual source of monomers. Because condensation occurs preferentially between weakly acidic species that tend to be protonated and strongly acidic species that are deprotonated, growth occurs primarily by reaction-limited monomer–cluster aggregation (equivalent to nucleation and growth), leading to compact, nonfractal structures.

For $w << 4$, unhydrolyzed sites are incorporated into the growing cluster. The probability of condensation at these sites is less than at hydrolyzed sites. Under these conditions, growth is described by a "poisoned" Eden model. Depending on the number and distribution of the poisoned sites, mass fractals, surface fractals, or uniformly porous objects can result. The addition of more water in a second hydrolysis step is expected to completely hydrolyze the clusters and further growth should be described by the Eden model.

5.5.3.2.3 pH ≈ 2–7

Hydrolysis occurs by an acid-catalyzed mechanism involving a basic, protonated alkoxy substituent. Condensation occurs by a base-catalyzed mechanism involving a deprotonated silanol. Hydrolysis will therefore occur on monomers and weakly branched oligomers that subsequently condense preferentially with clusters (monomer-cluster growth). However, the availability of monomers at later stages in the reaction decreases with decreasing pH. The predominant growth mechanism therefore changes from monomer–cluster to cluster–cluster with decreasing pH and increasing time of reaction. A wide range of structures, from weakly branched to highly branched, might be expected in this pH range of ~2 to 7.

5.5.3.3 Rheological Measurements

Rheological measurements are most often used to characterize the bulk properties of a solution (e.g., the viscosity), but the dependence of the rheological properties on the concentration, molecular weight, and shear rate can be used to infer structural information in sol–gel systems.

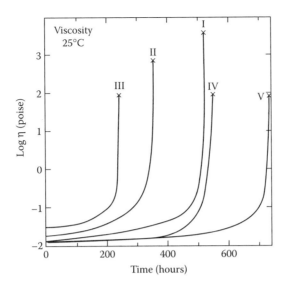

FIGURE 5.12 Temporal changes in solution viscosity for acid- and base-catalyzed TEOS systems. Crosses indicate gel points. Samples I-V are identified in Table 5.5. (From Sakka, S. and Kamiya, Y., The sol–gel transition in the hydrolysis of metal alkoxides in relation to the formation of glass fibers and films, *J. Non-Cryst. Solids*, 48, 31, 1982. With permission.)

TABLE 5.5
Compositions and Behavior of Sol–Gel Systems Investigated by Sakka et al. [26–28]

Solution	Si(OC₂H₅)₄ (g)	H₂O (g)	C₂H₅OH (g)	Mole Ratio (w) of H₂O to Si(OC₂H₅)₄	Catalyst[a]	Time for gelling (h)	Spinnability
I	169.5	14.7	239.7	1	HCl	525	Yes
II	382.0	33.0	83.4	1	HCl	360	Yes
III	169.5	292.8	37.5	20	HCl	248	No
IV	50	3.8	47.6	1	NH₄OH	565	No
V	50	7.6	47.6	2	NH₄OH	742	No

[a] Mole ratio of HCl or NH₄OH to the Si(OC₂H₅)₄ is 0.01.

Source: From Brinker, C.J. and Scherer, G.W., *Sol-Gel Science*, Academic Press, New York, 1990. With permission.

Figure 5.12 shows the changes in the solution viscosity for acid-catalyzed and base-catalyzed TEOS systems described in Table 5.5. The sudden increase in viscosity is generally used to identify the gel point in a crude way. The gel point is often defined as the time at which the viscosity is observed to increase rapidly or the time to reach a given viscosity (e.g., 1000 Pa·sec).

The rheology of silicate systems prepared from TEOS (Table 5.5) was studied by Sakka and coworkers [26–28] and correlated with the observed ability of certain compositions for fiber formation (spinnability). They measured the dependence of the reduced viscosity, η_{sp}, [equal to $(\eta_r - 1)$ in Equation 4.84] on silica concentration C and the dependence of the intrinsic viscosity $[\eta]$ on the number-averaged molecular weight M_n for acid-catalyzed and base-catalyzed systems prepared with w values in the range of 1 to 20. Figure 5.13 compares the concentration dependence of η_{sp}/C for composition 1 (Table 5.5) after various periods of aging (t/t_{gel}) with that for Ludox® (sol of silica particles) and for sodium metasilicate (chainlike silicates). For a sol of noninteracting

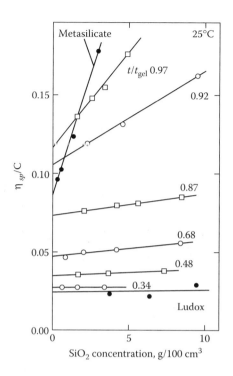

FIGURE 5.13 Dependence of the reduced viscosity of solution I (see Table 5.5) on concentration for various values of t/t_{gel}. Data for LUDOX® (sol of silica particles) and for sodium metasilicate (chain-like silicates) are shown for comparison. (From Sakka, S. and Kamiya, Y., The sol–gel transition in the hydrolysis of metal alkoxides in relation to the formation of glass fibers and films, *J. Non-Cryst. Solids*, 48, 31, 1982. With permission.)

spherical particles (e.g., Ludox®), η_{sp} is given by the Einstein relation (Equation 4.84), which can be expressed as:

$$\eta_{sp} / C = K_1 / \rho \tag{5.32}$$

where K_1 is a constant equal to 2.5 and ρ is the density of the particles. The silicate species present at $t/t_{gel} = 0.34$ can therefore be inferred to be compact and noninteracting.

As t/t_{gel} increases, there is a progressively larger dependence of η_{sp}/C on C. According to the Huggins equation [29], the reduced viscosity of solutions of chainlike or linear polymers (e.g., metasilicate) is given by:

$$\eta_{sp} / C = [\eta] + K_2 [\eta]^2 C \tag{5.33}$$

where K_2 is a constant. The larger dependence of η_{sp}/C on C with aging time may be explained in terms of a change in the silicate structure from small noninteracting species to extended, weakly branched polymers.

Figure 5.14 shows plots of log $[\eta]$ vs. log M_n, where M_n is the number averaged molecular weight for acid-catalyzed TEOS systems in which w was varied from 1 to 20. For organic polymer solutions, $[\eta]$ is related to M_n by the expression:

$$[\eta] = K_3 M_n^\alpha \tag{5.34}$$

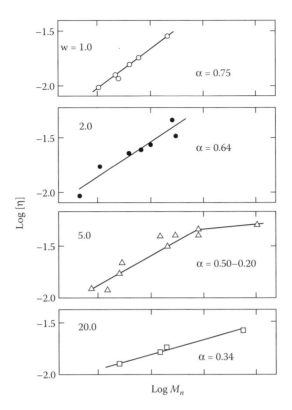

FIGURE 5.14 Log intrinsic viscosity [η] vs. log number-averaged molecular weight M_n for acid-catalyzed hydrolysis of TEOS and w = 1-20. The parameter α is defined in Equation. 5.34. (From Sakka, S., Formation of glass and amorphous oxide fibers from solution, *Mater. Res. Symp. Proc.*, 32, 91, 1984. With permission.)

where K_3 is a constant that depends on the type of polymer, the solvent and the temperature. The exponent α depends on the structure of the polymer: α = 0 for rigid spherical particles, α = 0.5 to 1.0 for flexible, chainlike, or linear polymers, and α = 1.0 to 2.0 for rigid, rodlike polymers [30]. The results of Figure 5.14 indicate that spinnable systems (w = 1 or 2) are composed of flexible chain-like or linear polymers (α = 0.64–0.75), whereas the nonspinnable systems are composed of more highly branched structures.

5.5.4 GELATION

The clusters grow until they begin to impinge on one another and gelation occurs by a linking of these clusters by a percolation process. Near the gel point, bonds form at random between the nearby clusters, linking them into a network. The gel point corresponds to the percolation threshold, when a single cluster (called the *spanning cluster*) appears that extends throughout the sol (Figure 5.10). The spanning cluster coexists with the sol containing many smaller clusters, which gradually become attached to the network. It reaches across the vessel that contains it, so the sol does not pour when the vessel is tipped. By creating a continuous solid network, the spanning cluster is responsible for an abrupt rise in the viscosity and the appearance of an elastic response to stress.

The theories put forward to explain gelation can be divided into three main classes: classical theory, percolation theory, and kinetic models [1]. Classical theory is based on the theory of Flory and Stockmeyer, mentioned earlier. It provides a good description of the gel point, but as we outlined earlier, it gives an unrealistic description of polymer growth in sol–gel silicates. Percolation theory avoids the unrealistic assumptions of the classical theory and makes predictions of the gelling systems that are in good accord with experimental observations [31,32]. A disadvantage is that

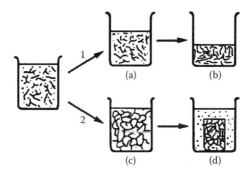

FIGURE 5.15 Two modes of shrinkage of a wet gel: shrinkage produced when the liquid is allowed to evaporate (route 1) and expulsion of liquid from the gel giving rise to shrinkage without evaporation (syneresis) (route 2). (From Partlow, D.P. and Yoldas, B.E., Colloidal vs. polymer gels and monolithic transformation in glass-forming systems, *J. Non-Cryst. Solids*, 46, 153, 1981. With permission.)

only a few results can be obtained analytically, so these models must be studied by computer simulations. Kinetic models are based on Smoluchowski's analysis of cluster growth [6]. The geometry of the clusters is not considered specifically in the theory, but the size distribution and shape of the clusters determined by computer simulations are in good agreement with experiment. The predictions of polymer growth and fractal structure are also in good agreement with experimental observations. The growing clusters eventually overlap and become immobile, so further bonding involves a percolation process. The evolution of the properties in the vicinity of the gel point is generally in agreement with the critical behavior predicted by percolation theory.

5.5.5 AGING OF GELS

The condensation reactions that cause gelation continue long after the gel point, leading to strengthening, stiffening, and shrinkage of the network. These changes have a significant effect on subsequent drying and sintering. If the gel is aged in the original pore liquid, small clusters continue to diffuse and attach to the main network. As these new links form, the network becomes stiffer and stronger. Many gels exhibit the phenomenon of syneresis, shown in Figure 5.15, where the gel network shrinks and expels the liquid from the pores [33]. Most likely, shrinkage results from condensation between neighboring groups on the surface of the solid network, as illustrated in Figure 5.16 [1].

Studies of syneresis in silica gels show that the rate depends on the processing conditions in much the same way as the condensation reactions leading to gelation. The shear modulus of the gel increases with aging time and the rate of increase of the modulus is larger at higher temperatures. However, when compared to drying where evaporation of the liquid is allowed to take place, the shrinkage and the shear modulus of the gel increase much less rapidly with time.

5.5.6 DRYING OF GELS

After preparation, polymeric gels typically consist of a weak amorphous solid structure containing an interconnected network of very fine pores filled with liquid. Often an excess volume of alcohol is used as a common solvent, so the liquid composition is predominantly an alcohol. This gel is sometimes referred to as an *alcogel*. Particulate gels consist of a network in which the pores are filled with an aqueous liquid, and in this case the gel is sometimes referred to as an *aquagel* (or a *hydrogel*). The gel must be dried prior to its conversion to the final article by sintering. The simplest method, referred to as *conventional* drying, is to remove the liquid by evaporation in air or in a drying chamber such as an oven. Drying must be carried out slowly and under carefully controlled conditions, especially in the case of polymeric gels, if monolithic crack-free bodies are to be

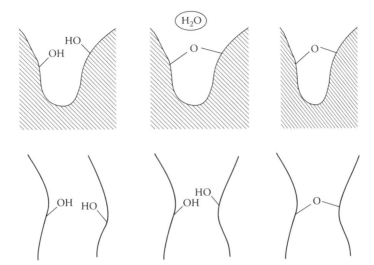

FIGURE 5.16 (Top) Shrinkage results from condensation between neighboring groups on a surface as the strain in the new bond relaxes. (Bottom) Movement of flexible chains may permit new bonds to form that prevent the chains from returning to their original position; this permits extensive shrinkage as long as the network remains flexible. (From Brinker, C.J. and Scherer, G.W., *Sol-Gel Science*, Academic Press, New York, 1990. With permission.)

obtained. The gel produced by conventional drying is referred to as a *xerogel*. Alternatively, the gel may be dried by removal of the liquid under supercritical conditions. In the ideal case, no shrinkage occurs during supercritical drying, so the dried gel is highly porous (e.g., typically ~90 to 95% porosity in polymeric gels). The gel produced by supercritical drying is referred to as an *aerogel*.

In practice, the drying stage often presents one of the major difficulties in sol–gel processing. It is often difficult to dry polymeric gels thicker than 1 mm or films thicker than 1 µm. We will examine the main factors that control the drying process, outline the key problems, and indicate how the problems can be alleviated.

5.5.6.1 Conventional Drying

Drying is a complex process involving the interaction of three independent processes: (1) evaporation, (2) shrinkage, and (3) fluid flow in the pores. As in most complex physical phenomena, an insight into the problem is best achieved through a combination of theoretical modeling and experimental investigations. A detailed analysis of the drying process has been carried out by Scherer [34a,b]. Although special attention is given to the drying of gels, the theory is fairly general that it can also be applied to the drying of ceramics formed by other methods (e.g., slip casting and extrusion) described in Chapter 7.

5.5.6.1.1 Stages in Drying

The drying process can be divided into two major stages: (1) a constant-rate period (CRP) where the evaporation rate is nearly constant, and (2) a falling-rate period (FRP), where the evaporation rate decreases with time or the amount of liquid remaining in the body (Figure 5.17). In some materials it is possible to further separate the FRP into two parts: the first falling-rate period (FRP1), in which the evaporation rate decreases approximately linearly with time, and the second falling-rate period (FRP2), in which the rate decreases in a curvilinear manner.

The drying process is illustrated schematically in Figure 5.18. During the CRP, the liquid–vapor meniscus remains at the surface of the gel. Evaporation occurs at a rate close to that of a free liquid surface (e.g., an open dish of liquid). For every unit volume of liquid that evaporates, the volume

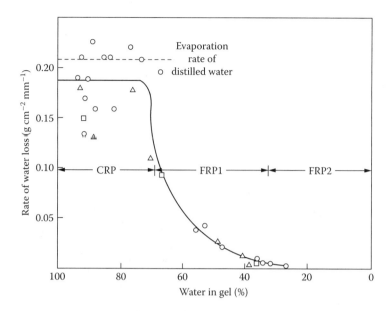

FIGURE 5.17 Rate of water loss from alumina gel vs. water content of the gel for various initial thicknesses (◯ = 7.5; ○ = 3.0; □ = 1.8; △ = 0.8 mm). During the first stage (CRP) the evaporation rate from the gel is about the same as from a dish of water. (From Dwivedi, R.K., Drying behavior of alumina gels, *J. Mater. Sci. Lett.*, 5, 373, 1986. With permission.)

of the gel decreases by one unit volume. This stage of constant evaporation rate accompanied by shrinkage lasts until the end of the CRP, when shrinkage stops and the FRP begins. At the end of the CRP, the gel may shrink to as little as one tenth of its original volume. During the FRP, the liquid recedes into the gel.

5.5.6.1.2 Driving Force for Shrinkage: Decrease in Interfacial Energy

Consider a tube of radius a held vertically in a reservoir of liquid that wets it (Figure 5.19). If the contact angle is θ, the negative pressure under the liquid–vapor meniscus in the capillary is

$$p = -\frac{2\gamma_{LV} \cos\theta}{a} \tag{5.35}$$

where γ_{LV} is the specific surface energy (surface tension) of the liquid–vapor interface. Assuming for simplicity that $\theta = 0$, the liquid is drawn up the tube to a height h, given by:

$$h = \frac{2\gamma_{LV}}{a\rho_L g} \tag{5.36}$$

where ρ_L is the density of the liquid and g is the acceleration due to gravity. The potential energy (*PE*) gained by the liquid is equivalent to raising a mass of liquid $\pi a^2 h \rho_L$ through a height $h/2$. Hence,

$$PE = \pi a^2 h^2 \rho_L g / 2 \tag{5.37}$$

This energy comes from the wetting of the walls of the tube by the liquid. We may describe the process in the following way. The surface energy of a solid arises from the asymmetric forces at

Stages of drying

(a) Initial condition

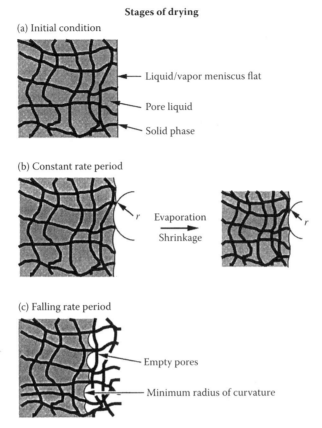

— Liquid/vapor meniscus flat

— Pore liquid

— Solid phase

(b) Constant rate period

r Evaporation

Shrinkage

r

(c) Falling rate period

— Empty pores

— Minimum radius of curvature

FIGURE 5.18 Schematic illustration of the drying process: (a) Before evaporation occurs, the meniscus is flat, (b) capillary tension develops in liquid as it "stretches" to prevent exposure of the solid phase, and network is drawn back into liquid. The network is initially so compliant that little stress is needed to keep it submerged, so the tension in the liquid is low and the radius of the meniscus is large. As the network stiffens, the tension rises and, at the critical point (end of the constant-rate period), the radius of the meniscus drops to equal the pore radius, (c) during the falling-rate period, the liquid recedes into the gel. (From Scherer, G.W., Theory of drying, *J. Am. Ceram. Soc.*, 73, 3, 1990. With permission.)

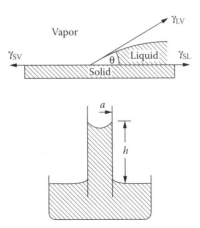

γ_{LV}

Vapor

γ_{SV} θ Liquid γ_{SL}

Solid

a

h

FIGURE 5.19 Capillary rise phenomenon for a wetting liquid with a contact angle θ.

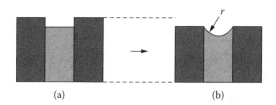

(a)   (b)

FIGURE 5.20 To prevent exposure of the solid phase (a), the liquid must adopt a curved liquid–vapor interface (b). Compressive forces on the solid phase cause shrinkage. (From Scherer, G.W., Theory of drying, *J. Am. Ceram. Soc.*, 73, 3, 1990. With permission.)

the free surface. If we cover the surface with another material, for example, a liquid, we reduce this asymmetry and hence reduce the surface energy. The energy liberated in the process is available for pulling the liquid up the tube. The energy liberated can also be thought of as giving rise to a capillary pressure or capillary force that acts on the liquid to pull it up the tube. Quantitatively, if the liquid rises up the tube to cover 1 m^2 of the surface, then we destroy 1 m^2 of solid–vapor interface and create 1 m^2 of solid–liquid interface. The energy given up by the system is then:

$$\Delta E = \gamma_{SV} - \gamma_{SL} \tag{5.38}$$

where γ_{SV} and γ_{SL} are the specific surface energies of the solid–vapor and solid–liquid interfaces, respectively.

Consider now a wet gel with pores that are assumed, for simplicity, to be cylindrical. If evaporation occurs to expose the solid phase, a solid–liquid interface is replaced by a solid–vapor interface. If the liquid wets the solid (i.e., the contact angle $\theta < 90°$), then, as seen from Figure 5.19, $\gamma_{SV} > \gamma_{SL}$. The exposure of the solid phase would lead to an increase in the energy of the system. To prevent this, liquid tends to spread from the interior of the gel to cover the solid–vapor interface. (This is analogous to the example of liquid flow up a capillary tube discussed earlier.) Because the volume of the liquid has been reduced by evaporation, the meniscus must become curved, as indicated in Figure 5.20. The hydrostatic tension in the liquid is related to the radius of curvature, r, of the meniscus by:

$$p = -\frac{2\gamma_{LV}}{r} \tag{5.39}$$

The negative sign in this equation arises from the sign convention for stress and pressure. The stress in the liquid is positive when the liquid is in tension; the pressure follows the opposite sign convention, so tension is negative pressure.

The maximum capillary pressure, p_R, in the liquid occurs when the radius of the meniscus is small enough to fit into the pore. For liquid in a cylindrical pore of radius a, the minimum radius of the meniscus is

$$r = -\frac{a}{\cos\theta} \tag{5.40}$$

The maximum tension is

$$p_R = -\frac{2(\gamma_{SV} - \gamma_{SL})}{a} = -\frac{2\gamma_{LV} \cos\theta}{a} \tag{5.41}$$

It is equal to the tension under the liquid-vapor meniscus (Equation 5.35). The capillary tension in the liquid imposes a compressive stress on the solid phase, causing contraction of the gel. It is smaller than the maximum value during most of the drying process.

For silica gels, shrinkage occurs faster when evaporation is allowed to take place, indicating that capillary pressure is the dominant factor driving the shrinkage. However, other factors can make a contribution to the driving force, and these can be significant in other systems.

Osmotic pressure (Π) is produced by a concentration gradient. A common example is the diffusion of pure water through a semipermeable membrane to dilute a salt solution on the other side. A pressure Π would have to be exerted on the salt solution or a tension $-\Pi$ must be exerted on the pure water to prevent it from diffusing into the salt solution. Gels prepared by the hydrolysis of metal alkoxides contain a solution of liquids (e.g., water and alcohol) that differ in volatility. Evaporation creates a composition gradient and liquid diffuses from the interior to reduce the gradient. If the pores are large, a counterflow of liquid to the interior occurs and no stress is developed. On the other hand, if the pores are small enough to inhibit flow, diffusion away from the interior can produce a tension in the liquid. The balancing compression on the solid phase (which, in principle, can approach the value of Π) can cause shrinkage of the gel.

Disjoining forces are short-range forces resulting from the presence of solid–liquid interfaces. An example of such forces is the repulsion between electrostatically charged double layers discussed in Chapter 4. Short-range forces in the vicinity of a solid surface can also induce some degree of structure in the adjacent liquid. The molecules in the more ordered regions adjacent to the solid surface have reduced mobility compared to those in the bulk of the liquid. Disjoining forces are important in layers that are within ~1 nm of the solid surface. They would be expected to be important in gels with very fine pores. For gels with relatively large pore sizes, disjoining forces will be important only in the later stages of drying.

5.5.6.1.3 Transport of Liquid

Transport of liquid during drying can occur by (1) flow if a pressure gradient exists in the liquid and (2) diffusion if a concentration gradient exists. According to Fick's first law, the flux, J, caused by a concentration gradient, ∇C, is given by:

$$J = -D\nabla C \tag{5.42}$$

where D is the diffusion coefficient. The flux J is defined as the number of atoms (molecules or ions) diffusing across unit area per second down the concentration gradient, and in one dimension (e.g., the x direction), $\nabla C = dC/dx$. Diffusion may be important during drying if the liquid in the pores consists of a solution and a concentration gradient develops by preferential evaporation of one component of the solution. In general, however, diffusion is expected to be less important than flow.

Liquid (or fluid) flow in most porous media obeys Darcy's law:

$$J = -\frac{K\nabla p}{\eta_L} \tag{5.43}$$

where J is the flux of liquid (the volume flowing across unit area per unit time down the pressure gradient), ∇p is the pressure gradient (equal to dp/dx in one dimension), η_L is the viscosity of the liquid, and K is the permeability of the porous medium. As in Fick's law, in which the physics of

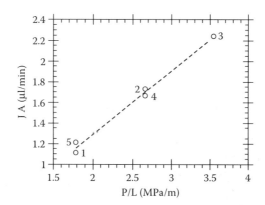

FIGURE 5.21 Flow rate JA (where J is the flux and A is the area of sheet of gel) vs. pressure gradient in a silica gel is linear, in accordance with Darcy's law, Equatio 5.43; numbers next to points indicate order of experiments. (From Scherer, G.W., Drying of ceramics made by sol–gel processing, in *Drying '92*, Mujumdar, A.S., Ed., Elsevier, New York, 1992, p. 92. With permission.)

the diffusion process is subsumed in the diffusion coefficient, in Darcy's law the parameters of the porous medium that control the liquid flow are accounted for in terms of the permeability. A variety of models, based on the representation of pores by an array of tubes, has been put forward to account for the permeability of porous media. One of the most popular, based on its simplicity and accuracy, is the *Carman–Kozeny* equation:

$$K = \frac{P^3}{5(1-P)^2 S^2 \rho_S^2} \tag{5.44}$$

where P is the porosity, S is the specific surface area (the surface area per unit mass of the solid phase), and ρ_S is the density of the solid phase. For a silica gel prepared from an alkoxide, $S \approx$ 400–800 m^2/g, $P \approx 0.9$, and $\rho_S \approx$ 1.5–1.8 g/cm^3. Substituting these values in Equation 5.44 gives K in the range of 10^{-13} to 10^{-14} cm^2 (i.e., very low). Although Equation 5.44 is fairly successful for many types of porous materials, it also fails often, and so it must be used with caution.

Darcy's law is obeyed by many materials, including some with very fine pores ($\leq$ 10 nm in size). Figure 5.21 shows the data for silica gel where the flux is indeed proportional to the pressure gradient [36]. The experiment was performed by casting a sheet of gel (~2-mm thick) onto a Teflon® filter and then imposing a pressure drop across it and measuring the rate of flow though it.

5.5.6.1.4 The Physical Process of Drying

Particulate gels and polymeric gels show the same general behavior in drying: a constant-rate period followed by a falling-rate period (Figure 5.17), but because of the profound difference in structure, the gels respond differently to the compressive stresses imposed by capillary forces. Initially, colloidal gels show a very small elastic contraction, but this is insignificant compared to the total drying shrinkage (typically 15 to 30 vol%) that occurs predominantly by rearrangement of the particles. Sliding of the particles over one another leads to denser packing and an increase in the stiffness of the gel. Eventually, the network is stiff enough to resist the compressive stresses and shrinkage stops. Rearrangement processes are difficult to analyze, so theoretical equations for the strain rate and shrinkage during drying have not been developed for particulate gels.

Polymeric gels are highly deformable. As for other materials with a polymeric structure, the response to a stress is viscoelastic, i.e., a combination of an instantaneous elastic deformation and a time-dependent viscous deformation. The elastic deformation arises from stretching and bending of the polymer chains, whereas the viscous deformation arises from reorientation and relaxation of polymer chains (or clusters) into lower energy configurations. Assuming the gel to be a continuum, constitutive equations relating the strain to the imposed stress have been developed. As a result of such analysis, considerable insight has been gained into the stress development in the gel and the shrinkage during drying. In the following discussion, particular attention will therefore be paid to the drying of polymeric gels.

5.5.6.1.4.1 The Constant-Rate Period (CRP)

When evaporation starts, the temperature at the surface of the gel drops because of a loss of heat due to the latent heat of vaporization of the liquid. However, heat flow to the surface from the atmosphere quickly establishes thermal equilibrium where transfer of heat to the surface balances the heat loss due to the latent heat of vaporization. The temperature at the surface becomes steady and is called the wet-bulb temperature (T_w). The surface of the gel is therefore at the wet-bulb temperature during the CRP. The rate of evaporation, $\dot{V}_E$, is proportional to the difference between the vapor pressure of the liquid at the surface, p_V, and the ambient vapor pressure, p_A:

$$\dot{V}_E = H(p_V - p_A) \tag{5.45}$$

where H is a factor that depends on the temperature, the velocity of the drying atmosphere, and the geometry of the system. Since $\dot{V}_E$ increases as p_A decreases, T_W decreases with a decrease in ambient humidity. For polymeric gels, the ambient vapor pressure must be kept high to avoid rapid drying, so the temperature of the sample remains near the ambient.

Let us consider a wet gel in which some liquid suddenly evaporates. The liquid in the pores stretches to cover the dry region and a tension develops in the liquid. The tension is balanced by compressive stresses on the solid phase of the gel. Because the network is compliant, the compressive forces cause it to contract into the liquid and the liquid surface remains at the exterior surface of the gel (Figure 5.18b). In a polymeric gel, it does not take much force to submerge the solid phase, so initially the capillary tension of the liquid is low and the radius of the meniscus is large. As drying proceeds, the network becomes stiffer because new bonds are forming (e.g., by condensation reactions) and the porosity is decreasing. The meniscus deepens (i.e., the radius decreases) and the tension in the liquid increases (Equation 5.39). When the radius of the meniscus becomes equal to the pore radius in the gel, the liquid exerts the maximum possible stress (Equation 5.41). This point marks the end of the CRP. Beyond this, the tension in the liquid cannot overcome the further stiffening of the network. The liquid meniscus recedes into the pores, marking the start of the FRP (Figure 5.18c). Thus, the characteristic features of the CRP are (1) the shrinkage of the gel is equal to the rate of evaporation, (2) the liquid meniscus remains at the surface, and (3) the radius of the liquid meniscus decreases.

At the end of the CRP, shrinkage virtually stops. According to Equation 5.41, for an alkoxide gel with $\gamma_{LV} \cos \theta \approx 0.02-0.07$ J/m^2 and $a \approx 1-10$ nm, the capillary tension, p_R, at the critical point is $\approx 4-150$ MPa. This shows that the gel can be subjected to enormous pressures at the critical point. The amount of shrinkage that precedes the critical point depends on p_R, which, according to Equation 5.41 increases with the interfacial energy, γ_{LV}, and with decreasing pore size, a. If surfactants are added to the liquid to reduce γ_{LV}, then p_R decreases. As a result, less shrinkage occurs and the porosity of the dried gel increases.

5.5.6.1.4.2 The Falling-Rate Period (FRP)

When shrinkage stops, further evaporation forces the liquid meniscus into the pores and the evaporation rate decreases (Figure 5.18c). This stage is called the falling-rate period (FRP). As outlined earlier, the FRP can be divided into two parts. In the first falling-rate period (FRP1), most of the evaporation is still occurring at the exterior surface. The liquid in the pores near the surface exists in channels that are continuous with the rest of the liquid. (The liquid is said to be in the funicular state.) These continuous channels provide pathways for liquid flow to the surface (Figure

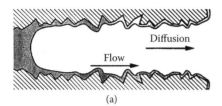 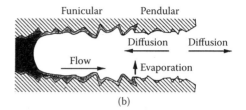

(a)   (b)

FIGURE 5.22 Schematic diagram illustrating fluid transport during the falling-rate period. After the critical point, the liquid–vapor meniscus retreats into the pores of the body: (a) In the first falling-rate period, liquid is in the funicular state, so transport by liquid flow is possible. There is also some diffusion in the vapor phase. (b) During the second falling-rate period evaporation occurs inside the body, at the boundary between the funicular (continuous liquid) and pendular (isolated pockets of liquid) regions. Transport in the pendular region occurs by diffusion of vapor. (From Scherer, G.W., Theory of drying, *J. Am. Ceram. Soc.*, 73, 3, 1990. With permission.)

5.22a). At the same time, some liquid evaporates in the pores and the vapor diffuses to the surface. In this stage of drying, as air enters the pores, the surface of the gel may lose its transparency.

As the distance between the liquid–vapor interface (the drying front) and the surface increases, the pressure gradient decreases and the flux of liquid also decreases. If the gel is thick enough, eventually a stage is reached where the flux becomes so slow that the liquid near the surface is in isolated pockets. (The liquid is now said to be in the pendular state.) Flow to the surface stops and the liquid is removed from the gel by diffusion of the vapor. This marks the start of the second falling-rate period (FRP2), where evaporation occurs inside the gel (Figure 5.22b).

5.5.6.1.4.3   Drying from One Surface

In many cases the wet gel is supported so that liquid evaporates from one surface only (Figure 5.23). As evaporation occurs, capillary tension develops first on the drying surface. This tension draws liquid from the interior to produce a uniform hydrostatic pressure. If the permeability of the gel is high, liquid flow is produced by only a small pressure gradient. On the other hand, if the permeability is low (or the gel is fairly thick), a significant pressure gradient is developed. The solid network is therefore subjected to a greater compression on the drying surface. This causes the gel to warp upwards (Figure 5.23a). Later in the drying process, the liquid–vapor interface moves into the interior of the gel and the pores are filled with air. The gel network surrounding the air-filled pores is relieved of any compressive stress, but the lower part of the gel still contains liquid so that it is subjected to compression due to capillary forces. This causes the gel to warp in the opposite direction (Figure 5.23b).

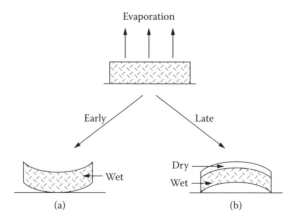

FIGURE 5.23 Warping of plate of gel dried by evaporation from upper surface. Plate warps upward initially (a), then reverses curvature after top surface becomes dry (b).

5.5.6.1.5 Drying Stresses

During the CRP, the pores remain full of incompressible liquid. The change in liquid content must be equal to the change in pore volume, which is related to the volumetric strain rate, $\dot{V}$. Equating these changes, we obtain:

$$\dot{V} = -\nabla \cdot J = -\nabla \cdot \left(\frac{K}{\eta_L} \nabla p \right) \tag{5.46}$$

The flux of liquid to the surface also matches the evaporation rate (i.e., $J = \dot{V}_E$), so Equation 5.43 requires that

$$\dot{V}_E = \frac{K}{\eta_L} \nabla p \big|_{surface} \tag{5.47}$$

To calculate the drying stresses, Equation 5.46 must be solved using Equation 5.47 as a boundary condition. The methods for solving the equation are discussed in Reference 34. For a viscoelastic flat plate (including one that is purely elastic or viscous), the stress in the solid phase of the gel in the plane of the plate (*x-y* plane) is given by:

$$\sigma_x = \sigma_y = \langle p \rangle - p \tag{5.48}$$

where p is the negative pressure (tension) in the liquid and $\langle p \rangle$ is the average pressure in the liquid. According to this equation, if the tension in the liquid is uniform, $p = \langle p \rangle$ and there is no stress on the solid phase. However, when p varies through the thickness, the network tends to contract more where p is high and this differential strain causes warping or cracking. The situation is analogous to the stress produced by a temperature gradient: cooler regions contract relative to warmer regions and the differential shrinkage causes the development of stresses.

If the evaporation rate is high, p can approach its maximum value, given by Equation 5.41, while $\langle p \rangle$ is still small, so that the total stress at the surface of the plate is

$$\sigma_x \approx \frac{2\gamma_{LV} \cos \theta}{a} \tag{5.49}$$

For slow evaporation, the stress at the drying surface of the plate is

$$\sigma_x \approx \frac{L\eta_L \dot{V}_E}{3K} \tag{5.50}$$

where the half thickness of the plate is L and evaporation occurs from both faces of the plate.

5.5.6.1.6 Cracking during Drying

It is commonly observed that cracking is more likely to occur if the drying rate is high or the gel is thick. It is also observed that cracks generally appear at the critical point (the end of the CRP) when shrinkage stops and the liquid–vapor meniscus moves into the body of the gel. In the theoretical analysis developed by Scherer [34], cracking is attributed to stresses produced by a pressure gradient in the liquid. The stress that causes failure is not the macroscopic stress σ_x acting

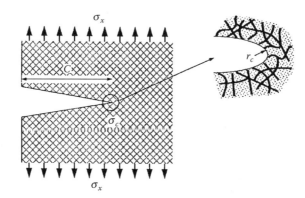

FIGURE 5.24 Surface flaw of length c and tip radius r_c acted upon by far-field stress σ_x creates amplified stress σ_c at the crack tip. Insert indicates that r_c is on the scale of the mesh size of the network. (From Scherer, G.W., Drying of ceramics made by sol–gel processing, in *Drying '92*, Mujumdar, A.S., Ed., Elsevier, New York, 1992, p. 92. With permission.)

on the network. If the drying surface contains flaws, such as that sketched in Figure 5.24, the stress is concentrated at the tip of the flaw. The stress at the tip of a flaw of length c is proportional to

$$\sigma_c = \sigma_x (\pi c)^{1/2} \tag{5.51}$$

Fracture occurs when $\sigma_c > K_{IC}$, where K_{IC} is the critical stress intensity of the gel. Assuming that the flaw size distribution is independent of the size and drying rate of the gel, the tendency to fracture would be expected to increase with σ_x. According to Equation 5.50, σ_x increases with the thickness, L, of the sample and the drying rate, $\dot{V}_E$. Scherer's analysis therefore provides a qualitative explanation for the dependence of cracking on L and $\dot{V}_E$.

In another explanation, it has been suggested that cracking in gels occurs as a result of local stresses produced by a distribution of pores sizes [37]. As sketched in Figure 5.25, after the critical point, liquid is removed first from the largest pores. It has been suggested that the tension in the neighboring small pores deforms the pore wall and causes cracking. This explanation does not appear to be valid [34]. The drying stresses are macroscopic because the pressure gradient extends

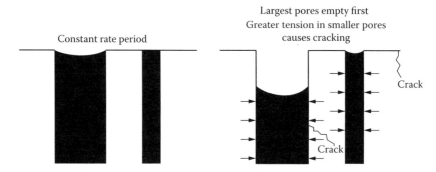

FIGURE 5.25 Illustration of microscopic model: during the constant-rate period, meniscus has same radius of curvature for pores of all sizes; after the critical point, the largest pores are emptied first. The capillary tension compressing the smaller pores causes local stresses that crack the network. (After Zarzycki, J., Prassas, M., and Phalippou, J., Synthesis of glasses from gels: the problem of monolithic gels, *J. Mater. Sci.*, 17, 3371, 1982. With permission.)

through the thickness of the gel (i.e., the stresses are not localized). If the stresses were localized on the scale of the pores, the gel would be expected to crumble to dust as the drying front advanced. Instead, gels crack into only a few pieces. However, flaws that lead to failure may be created by local stresses (resulting, for example, from nonuniform pore sizes) and then propagated by the macroscopic stresses.

5.5.6.1.7 Avoidance of Cracking

Scherer's theory of drying provides useful guidelines for controlling stress development and therefore for avoiding fracture. Cracking is attributed to macroscopic stresses produced by a pressure gradient, ∇p, in the liquid. Fast evaporation leads to a high ∇p and hence to large differential strain. To avoid cracking, the gel must be dried slowly. However, the "safe" drying rates are so slow that gels thicker than ~1 cm require uneconomically long drying times. A number of procedures may be used to increase the safe drying rates. The boundary condition at the surface of the gel (Equation 5.47) shows that high $\dot{V}_E$ leads to high ∇p. Furthermore, low permeability (low K) leads to high ∇p for a given $\dot{V}_E$. From Equation 5.44, it can be shown that K increases roughly as the square of the pore size. An obvious strategy is to increase the pore size of the gel. One approach that has been used for silica gels involves mixing colloidal silica particles with the alkoxide (TEOS). In addition to producing gels with large pores, the silica particles also strengthen the gel. Another approach is to use colloidal particles to produce gels that are easier to dry (e.g., particulate silica gel). However, gels with larger pores require higher sintering temperatures for densification, so some compromise between ease of drying and ease of sintering must be made.

The capillary pressure sets a limit on the magnitude of the drying stresses, i.e., $\sigma_x \leq p_R$, and is probably responsible for the creation of critical flaws. The probability of fracture can therefore be reduced by reducing the capillary pressure through (1) increasing the pore size (discussed earlier) and (2) decreasing the liquid-vapor interfacial tension (see Equation 5.41). The interfacial tension can be reduced by using a solvent with a lower volatility than water that, in addition, has a low γ_{LV}. The surface tension of a liquid can also be reduced by raising the temperature. Beyond the critical temperature and pressure, there is no tension. Because $p = 0$, ∇p must also be zero so that no drying stresses can be produced. This approach forms the basis of supercritical drying described in the next section.

Another approach is to strengthen the gel so that it is better able to withstand the drying stresses. For example, aging the gel in the pore liquid at slightly elevated temperatures stiffens the gel network and also reduces the amount of shrinkage in the drying stage. For gels prepared from metal alkoxides, certain organic compounds added to the alkoxide solution have been claimed to speed up the drying process considerably while avoiding cracking of the gel [38]. It has been reported that, with the use of these compounds, gels thicker than 1 cm can be dried in ~1 day. These compounds, referred to as *drying control chemical additives* (DCCAs), include formamide (NH_2CHO), glycerol ($C_3H_8O_3$), and oxalic acid ($C_2H_2O_4$). Although the role of these compounds during drying is not clear, it is known that they increase the hardness (and presumably the strength) of the wet gel. However, they also cause serious problems during sintering because they are difficult to burn off. Decomposition leading to bloating of the gel and chemical reaction (e.g., the formation of carbonates) severely limits the effectiveness of these compounds.

5.5.6.2 Supercritical Drying

In *supercritical* drying (sometimes referred to as *hypercritical* drying), the liquid in the pores is removed above the critical temperature, T_c, and the critical pressure, p_c, of the liquid. Under these conditions, there is no distinction between the liquid and the vapor states. The densities of the liquid and vapor are the same, there is no liquid–vapor meniscus and no capillary pressure, and so there are no drying stresses.

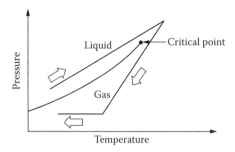

FIGURE 5.26 Schematic illustration of the pressure–temperature schedule during supercritical drying.

TABLE 5.6
Critical Points of Some Fluids

Fluid	Formula	Critical Temperature (K)	Critical Pressure (MPa)
Carbon dioxide	CO_2	304.1	7.36
Nitrous oxide	N_2O	309.8	7.24
Freon-13	$CClF_3$	301.9	3.86
Freon-23	CHF_3	298.9	4.82
Freon-113	$CCl_2F-CClF_2$	487	3.40
Freon-116	CF_3-CF_3	292.7	2.97
Methanol	CH_3OH	513	7.93
Ethanol	C_2H_5OH	516	6.36
Water	H_2O	647	22.0

In supercritical drying, the wet gel is placed in an autoclave and heated fairly slowly (less than ~0.5°C /min) to a temperature and pressure above the critical point. The temperature–pressure path during the drying process is sketched in Figure 5.26. The temperature and pressure are increased in such a way that the liquid–vapor phase boundary is not crossed. After equilibration above the critical point, the fluid is released slowly.

Table 5.6 shows the critical points of some fluids. The critical temperature and pressure for ethanol are 243°C and 6.4 MPa (~63 atm). The need to cycle an autoclave through such high temperature and pressure makes the process time consuming, expensive, and somewhat dangerous. The total drying time can be as long as 2 to 3 days for a large gel. An alternative is to replace the liquid in the pores with a fluid having a much lower critical point. Carbon dioxide has a T_c of 31°C and a p_c of 7.4 MPa. The low T_c means that the drying process can be carried out near ambient temperatures. Carbon dioxide is also inexpensive. In supercritical drying with CO_2, the liquid (e.g., alcohol) in the pores of the gel must first be replaced. This is accomplished by placing the gel in the autoclave and flowing liquid CO_2 through the system until no trace of alcohol can be detected. Following this, the temperature and pressure of the autoclave are raised slowly to ~40°C and ~8.5 MPa. After maintaining these conditions for ~30 min, CO_2 is slowly released.

With supercritical drying, monolithic gels as large as the autoclave can be produced [39]. Very little shrinkage (< 1% linear shrinkage) occurs during supercritical drying, so the aerogel is very porous and fragile. The solid phase occupies less than 10 vol% of the aerogel. Because of the low density, the sintering of aerogels to produce dense polycrystalline articles is impractical. Considerable shrinkage (> 50% linear shrinkage) would have to take place, and this makes dimensional control of the article difficult. A common problem is that crystallization of the gel prior to the attainment of high density can severely limit the final density [9]. Because of these problems, supercritical drying is better suited to the fabrication of porous materials and powders. Monolithic aerogels are of interest because of their exceptionally low thermal conductivity. Powders with good chemical homogeneity can be easily produced by grinding the fragile aerogel.

5.5.6.3 Structural Changes during Drying

We saw earlier that there are fundamental differences in structure between particulate gels and polymeric gels, and within polymeric gels, between gels prepared by acid-catalyzed and base-catalyzed reactions. During conventional drying, we would expect the structure of these gels to evolve differently under the compressive action of the capillary stresses. Polymeric gels will gradually collapse and cross-link as unreacted OH and OR groups come into contact with each other (see Figure 5.16). When the structure becomes stiff enough to resist the capillary stresses, residual porosity will be formed. For gels produced by acid-catalyzed reactions, in which the polymer chains are weakly cross-linked, the structure can be highly compacted before it is sufficiently cross-linked to produce residual porosity (Figure 5.27a). The xerogel will be characterized by a relatively high density and very fine pores. In contrast, after drying, base-catalyzed gels will have a relatively low density and larger pores because the structure of the polymeric clusters will be more difficult to collapse (Figure 5.27b). Silica gels prepared by acid-catalyzed reactions can lead to xerogels with porosities as low as 35 to 40%, compared to values as high as 60 to 70% for similar xerogels produced by the base-catalyzed route.

For particulate gels, the structure will fold or crumple under the action of the capillary stresses. Neck formation or an increase in the coordination number of the gel will continue until the structure is strong enough to resist the capillary stresses, at which point residual porosity is formed. Because the particulate structure of the gel can better withstand the smaller capillary stresses arising from the larger pores, the shrinkage will be much smaller than that for polymer gels. The structure of the xerogel will be a somewhat contracted and distorted version of the original structure (Figure 5.27c and Figure 5.27d). The porosity is typically in the range of 70 to 80%.

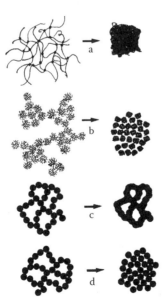

FIGURE 5.27 Schematic representation of the structural evolution during drying for (a) acid-catalyzed gels, (b) base-catalyzed gels, (c) colloidal gel aged under conditions of high silica solubility, and (d) colloidal gel composed of weakly bonded particles. (From Brinker, C.J. and Scherer, G.W., Sol → gel → glass: I, gelation and gel structure, *J. Non-Cryst. Solids*, 70, 301, 1985. With permission.)

5.5.7 Gel Densification during Sintering

Particulate gels, as discussed earlier, contract slightly during drying. Because the solid phase is composed of a fully cross-linked network that is similar in structure to that of the corresponding melted glass, the drying process does not alter the structure of the solid skeletal phase of the gel. Polymeric gels undergo significant contraction and further polymerization by condensation reactions during drying. Even so, the resulting polymer structure of the solid skeletal phase of the xerogel is still less highly cross-linked than the corresponding melted glass. For example, the number of nonbridging oxygen atoms (i.e., those ending in OH and OR groups) has been estimated in the range of 0.33 to 1.48 for every silicon atom in polymeric silica xerogels, compared to a value of ~0.003 for a melted silica glass with 0.05 wt% water. One effect of the lower cross-link density is that the solid skeletal phase of the gel has a lower density than that of the corresponding melted glass. This is equivalent to saying that the skeletal phase has extra free volume when compared to the corresponding glass produced by melting.

The structure of the xerogel has important implications for densification. Compared to the corresponding dense glass prepared by melting, xerogels have a high free energy that will act as a powerful driving force for sintering. Three characteristics contribute to this high free energy [40]. The surface area of the solid–vapor interface makes the largest contribution, estimated at 30 to 300 J/g (which corresponds to 100 to 1000 m^2/g of interfacial area). The reduction of the surface area provides a high driving force for densification by viscous flow. The two other contributions to the high free energy result from the reduced cross-link density of the polymer chains compared to the corresponding melted glass. Polymerization reactions can occur according to

$$Si(OH)_4 \rightarrow SiO_2 + 2H_2O \qquad \Delta G_{f(298K)} = -14.9 \, kJ \, / \, mol \qquad (5.52)$$

More weakly cross-linked polymers containing more nonbridging oxygen atoms will therefore make a greater contribution to the free energy. For silica gels containing 0.33 to 1.48 OH groups per silicon atom, the contribution to the free energy resulting from the reaction described by Equation 5.52 is estimated to range from ~20 to 100 J/g. The free volume of the solid skeletal phase is also expected to provide a contribution to the free energy. Figure 5.28 summarizes the free energy vs. temperature relations for polymeric and particulate gels, glass, and an ideal supercooled liquid of the same oxide composition.

The densification of gels during sintering has been studied extensively by Scherer and coworkers [41–43]. Figure 5.29 shows data for the linear shrinkage of three silica gels during sintering at a constant heating rate of 2°C/min. The particulate gel (curve C) shrinks only at elevated temperatures (~1200°C) and its densification behavior can be accurately described by models for viscous sintering of porous melted glass [9]. The shrinkage of the polymeric gels differs markedly from that of the particulate gel. Starting from fairly low temperatures, continuous shrinkage occurs and the extent of the shrinkage depends on the method used to prepare the gel. At any temperature, the shrinkage of the acid-catalyzed gel (curve A) is significantly greater than that of the base-catalyzed gel (curve B). The sintering of polymeric gels is not as simple as that of particulate gels and it cannot be explained on the basis of viscous sintering alone.

For polymeric gels, four mechanisms can operate during the conversion of the dried gel to a dense glass: (1) capillary contraction of the gel, (2) condensation polymerization leading to an increase in the cross-link density, (3) structural relaxation by which the structure approaches that of a supercooled liquid, and (4) viscous sintering. The temperature range in which each of these mechanisms contribute to the densification depends on the structure of both the porous and solid phases of the gel (e.g., pore size and skeletal density) as well as the rate of heating and the previous thermal history. A good illustration of the temperature range in which each of these mechanisms operate is provided in Figure 5.30, which shows the shrinkage and weight loss of a borosilicate

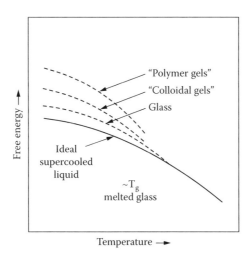

FIGURE 5.28 Schematic representation of free energy vs. temperature relations between dried polymeric gels, dried colloidal gels, glass, and an ideal supercooled liquid of the same oxide composition. (From Brinker, C.J. and Scherer, G.W., Sol → gel → glass: I, gelation and gel structure, *J. Non-Cryst. Solids*, 70, 301, 1985. With permission.)

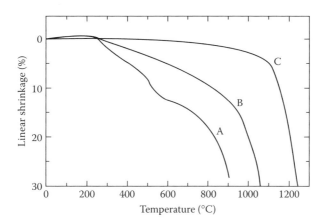

FIGURE 5.29 Linear shrinkage vs. temperature during constant heating rate sintering at 2°C/min for an acid-catalyzed silica gel (A), a base-catalyzed silica gel (B), and a particulate silica gel (C). The acid- and base-catalyzed gels were prepared with a water concentration of 4 moles H_2O per mole of TEOS. (From Brinker, C.J. et al., Structure of gels during densification, in *Science of Ceramic Chemical Processing*, Hench, L.L. and Ulrich, D.R., Eds., John Wiley & Sons, New York, 1986, p. 37, chap. 3. With permission.)

gel during heating at a constant rate of 0.5°C/min. Three regions have been identified corresponding to the following trends:

Region I: weight loss without shrinkage (25–150°C)
Region II: weight loss with concurrent shrinkage (150–525°C)
Region III: shrinkage without weight loss (> 525°C)

The weight loss in region I is because of the desorption of physically adsorbed water and alcohol. A very small shrinkage is observed and this occurs as a result of the increase in the surface energy due to the desorption process. The surface energy increases from ~0.03 J/m^2 for a surface

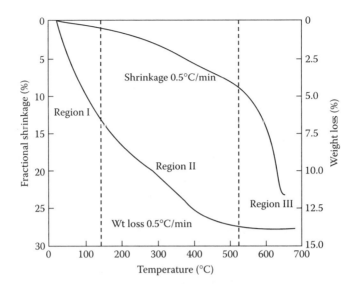

FIGURE 5.30 Linear shrinkage and weight loss for a borosilicate gel during heating at a constant rate of 0.5°C/min. (From Brinker, C.J., Scherer, G.W., and Roth, E.P., Sol → gel → glass: II, physical and structural evolution during constant heating rate experiments, *J. Non-Cryst. Solids*, 72, 345, 1985. With permission.)

saturated with water and alcohol to ~0.15 J/m² for a pure silanol surface (i.e., produced after desorption). This increase in surface energy is equivalent to an increase in the capillary pressure.

Weight loss in region II is attributed to two processes. Water is removed as a by-product of polycondensation reactions:

$$\equiv Si - OH + HO - Si \equiv \quad \rightarrow \quad \equiv Si - O - Si \equiv + H_2O \tag{5.53}$$

The reaction described by Equation 5.53 may occur within the skeletal phase or on the surface of the skeletal phase. The second process is the oxidation of carbonaceous residues present originally as unhydrolyzed alkoxy groups. Experiments indicate that the removal of carbon is essentially complete by 400°C but that hydrogen is continuously removed over the whole temperature range (150–525°C).

The shrinkage in region II can be attributed solely to densification of the solid skeletal phase of the gel. Alternatively, the removal of very fine pores by viscous sintering may be postulated but the pore size would have to be less than ~0.4 nm. Pores of this size would, however, be indistinguishable from free volume in the skeletal phase. Two mechanisms have been identified for the densification of the skeletal phase: (1) polymerization reactions (e.g., Equation 5.53), which lead to a higher cross-link density of the polymer chain, and (2) structural relaxation of the polymeric network as the structure approaches the configuration of the melt-prepared glass. Structural relaxation occurs by diffusive motion of the polymer network without the expulsion of water and other products. The relative contribution of each mechanism to the overall densification in region II is difficult to quantify.

We would expect skeletal densification to be insignificant for particulate gels in which the solid network has the same structure as the corresponding melt-prepared glass and to make the greatest contribution to shrinkage for weakly cross-linked gels. This is fully borne out by the results of Figure 5.29.

The large, fairly fast shrinkage in region III is consistent with a mechanism of densification by viscous flow. The pores in the gel that were formed after the removal of the liquid in the drying stage are removed in this process. The driving force for densification is the reduction in the

TABLE 5.7

Dominant Mechanisms of Shrinkage during the Sintering of a Borosilicate Gel and their Relative Contributions to the Total Shrinkage

Region	Temperature Range (°C)	Mechanism	Relative Contribution (%)
I	25–150	Capillary contraction	3
II	150–525	Condensation polymerization and structural relaxation	33
III	>525	Viscous flow	63

Source: From Scherer, G.W., Brinker, C.J., and Roth, E.P., Sol → gel → glass: III, viscous sintering, *J. Non-Cryst. Solids*, 72, 369, 1985. With permission.

solid–vapor interfacial area associated with the pores. While the weight loss in this region is low, studies indicate that polymerization reactions and structural relaxation can have a strong influence on the kinetics of densification.

Table 5.7 summarizes the dominant shrinkage mechanisms and their relative contribution to the overall shrinkage for the borosilicate gel described in Figure 5.30. The significant contribution due to skeletal densification by condensation polymerization and structural relaxation should not go unnoticed.

So far, we have discussed the sintering of gels under the influence of the available driving forces. Other factors must also be considered because they can have a significant influence on the sintering process. Densification involves transport of matter, so it is limited by the kinetics of the process. The kinetics determine how long the process will take and depend not only on the physical and chemical structure of the gel but also on the time-temperature schedule of heating. The rate of heating, for example, may have a significant effect on the densification. Increased heating rate reduces the time spent at each increment of temperature and therefore reduces the amount of viscous flow that can occur in a given temperature interval. On the other hand, increased heating rate also reduces the amount of cross-linking and structural relaxation that can occur over the same temperature interval, so the viscosity of the gel at each temperature is reduced. Under certain conditions, the rate of viscous sintering can increase at sufficiently high heating rates [41].

Even after drying, the gel structure normally contains a large amount of unhydrolyzed alkoxy groups. These are oxidized and removed as gases below ~400°C. Two major problems may arise if the sintering process is not carried out carefully. The first problem is that trapped gases in the pores can lead to bloating of the gel at higher temperatures. This may occur, for example, if the heating rate is so fast that the pores become isolated prior to complete oxidation of the alkoxy groups. Further heating to higher temperatures causes an increase in the pressure of the gas in the pores, resulting in enlargement of the pores and bloating. The second problem is the presence of carbon residues in the gel due to incomplete oxidation. In some cases, the gel may turn black around 400°C. Generally, for this part of the sintering process, the temperature must be high enough to oxidize the alkoxy groups but at the same time low enough to prevent the pores from sealing off.

We have assumed that a dense glass is produced at the end of the firing process. This is not always so. For crystallizable compositions such as ceramics, crystallization may occur prior to full densification. A crystalline body is generally more difficult to densify than the corresponding amorphous body, and so crystallization prior to full densification will, in general, hinder densification [9]. If full densification is achieved prior to crystallization, a controlled heating step is necessary to nucleate and grow the crystalline grains.

In the production of a glass by the polymeric gel route, the network structure evolves very differently from that of the corresponding glass prepared by melting. An important question is whether the structure and properties of the fabricated sol–gel-derived glass is any different from that for the corresponding melt-prepared glass. Current indications are that for gels densified above the glass transition temperature, the structure and properties can be indistinguishable from those for the corresponding melt-prepared glass.

5.6 SOL–GEL PREPARATION TECHNIQUES

Having considered some of the basic physics and chemistry of the sol–gel method, we come now to some of the practical issues involved in the preparation of gels. For the production of simple oxides (e.g., SiO_2), the preparation techniques are fairly straightforward. Further considerations must be taken into account for the production of complex oxides to achieve the required chemical composition and uniformity of mixing. Gel compositions with one type of metal cation (such as SiO_2) yield simple oxides on pyrolysis and are referred to as single-component gels. Multicomponent gels have compositions with more than one type of metal cation and yield complex oxides on pyrolysis.

A wide range of ceramic and glass compositions have been prepared by sol–gel processing and details of the experimental procedure can be found in the literature (see, for example, Reference 44 for a partial list). We will not repeat the procedures for the preparation of specific gel compositions. Instead, the common techniques will be outlined with the aid of a few specific examples.

5.6.1 PREPARATION OF PARTICULATE GELS

5.6.1.1 Single-Component Gels

For single-component gels, colloidal particles are dispersed in water and peptized with acid or base to produce a sol. Gelation can be achieved by: (1) removal of water from the sol by evaporation to reduce its volume or (2) changing the pH to slightly reduce the stability of the sol. Earlier in this chapter we discussed the sol–gel processing of aqueous silicates and mentioned the preparation of SiO_2 gels from fine particles made by flame oxidation [2–4]. The preparation of alumina sols from alkoxides has been described by Yoldas [45a,b]. Aluminum alkoxides such as aluminum *sec*-butoxide and aluminum isopropoxide are readily hydrolyzed by water to form hydroxides. Which hydroxide is formed depends on the conditions used in the hydrolysis. The initial hydrolysis reaction of aluminum alkoxides can be written as:

$$Al(OR)_3 + H_2O \rightarrow Al(OR)_2(OH) \tag{5.54}$$

The reaction proceeds rapidly with further hydrolysis and condensation:

$$2Al(OR)_2(OH) + H_2O \rightarrow (RO)(OH)Al-O-Al(OH)(OR) + 2ROH \tag{5.55}$$

Assuming the formation of polymers that are not too highly cross-linked, the incorporation of n aluminum ions into the chain is given by the formula $Al_nO_{n-1}(OH)_{(n+2)-x}(OR)_x$. As the reaction proceeds, the number of OR groups (i.e., x) relative to n should decrease to a value that depends on the hydrolysis temperature and the concentration of OR groups in the solvent. It should be kept in mind, as we mentioned earlier, that Equation 5.54 and Equation 5.55 are not exact formulas but merely represent simplifications.

Hydrolysis by cold water ($20°C$) results in the formation of a monohydroxide that is predominantly amorphous. The structure contains a relatively high concentration of OR groups. It is believed that the presence of the OR groups is directly related to the structural disorder in the amorphous phase since their removal (e.g., by aging in the solvent) inevitably leads to conversion

TABLE 5.8
Peptizing Effect of Various Acids on the Precipitate Formed by the Hydrolysis of Aluminum sec-Butoxide

Acid	Formula	Condition of Precipitate[a]
Nitric	HNO_3	Clear sol
Hydrochloric	HCl	Clear sol
Perchloric	$HClO_4$	Clear sol
Hydrofluoric	HF	Unpeptized
Iodic	HIO_4	Unpeptized
Sulfuric	H_2SO_4	Unpeptized
Phosphoric	H_3PO_4	Unpeptized
Boric	H_3BO_3	Unpeptized
Acetic	CH_3COOH	Clear sol
Trichloroacetic	CCl_3COOH	Clear sol
Monochloroacetic	$CH_2ClCOOH$	Clear to cloudy
Formic	$HCOOH$	Clear to cloudy
Oxalic	$H_2C_2O_4.2H_2O$	Unpeptized
Phthalic	$C_8H_4O_3$	Unpeptized
Citric	$H_3C_6H_5O_7.H_2O$	Unpeptized
Carbolic	C_6H_5OH	Unpeptized

[a] After 7 d at 95°C.

Source: Yoldas, B.E., Alumina gels that form porous transparent Al_2O_3, *J. Mater. Sci.*, 10, 1856, 1975. With permission.

of the amorphous hydroxide to a crystalline hydroxide, boehmite ($AlO(OH)$), or bayerite ($Al(OH)_3$). Aging at room temperature leads to the formation of bayerite by a process involving the solution of the amorphous hydroxide and subsequent precipitation as the crystalline phase. Aging of the amorphous hydroxide above 80°C leads to rapid conversion to boehmite. Because the conversion of the amorphous hydroxide to boehmite or bayerite is accompanied by the liberation of OR groups, the rate of conversion is inhibited by the presence of alcohol in the solvent during the aging process. Hydrolysis of aluminum alkoxides by hot water (80°C) results in the formation of boehmite, which is relatively unaffected by aging.

Using aluminum alkoxides as the starting material, the production of Al_2O_3 by the colloidal gel route involves the following main steps: (1) hydrolysis of the alkoxide to precipitate a hydroxide, (2) peptization of the precipitated hydroxide (e.g., by the addition of acids) to form a clear sol, (3) gelation (e.g., by evaporation of solvent), (4) drying of the gel, and (5) sintering of the dried gel. The formation of the sol can be a critical part of the process. Although boehmite and the amorphous hydroxide prepared by cold water hydrolysis can be peptized to a clear sol, bayerite will not form a sol, and its formation during hydrolysis should therefore be avoided. In addition, the nature of the acid has a significant effect on the peptization step. Table 5.8 shows the peptizing effect of various acids on the precipitate formed by hydrolysis of aluminum *sec*-butoxide. The results are similar when aluminum isopropoxide is used. It appears that only strong or fairly strong acids that do not form chemical complexes (or form only very weak complexes) with aluminum ions are effective for achieving peptization. For these acids, the concentration of the acid also has an effect. Peptization requires the addition of at least 0.03 mole of acid per mole of alkoxide (followed by heating at ~80°C for a sufficient time).

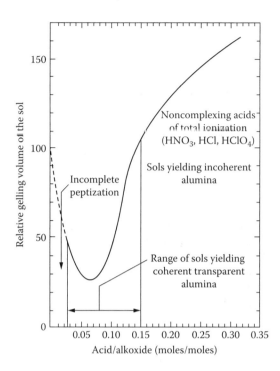

FIGURE 5.31 Effect of acid concentration on the volume of the gel formed from a peptized boehmite sol. (From Partlow, D.P. and Yoldas, B.E., Colloidal vs. polymer gels and monolithic transformation in glass-forming systems, *J. Non-Cryst. Solids*, 46, 153, 1981. With permission.)

The amount of acid used in the peptization step also has a significant influence on the gelation of the sol and on the properties of the fabricated aluminum oxide. There is a critical acid concentration at which the volume of the gel is a minimum. For nitric acid, this critical concentration is ~0.07 mole per mole of alkoxide. At this minimum volume, the gel contains an equivalent of 25 wt% of Al_2O_3. Deviation from the critical acid concentration, to higher or lower values, causes a sharp increase in the volume of the gel (Figure 5.31). At higher acid concentration, the gels may contain an equivalent of only 2 to 3 wt% Al_2O_3. Because of the large shrinkages that occur, gels containing an equivalent of less than ~4 wt% of Al_2O_3 do not retain their integrity after drying and sintering.

5.6.1.2 Multicomponent Gels

In the case of multicomponent particulate gels, a primary concern is preventing segregation of the individual components in order to achieve uniform mixing. Various routes have been used for their preparation, including (1) coprecipitation of mixed oxides or hydroxides, (2) mixing sols of different oxides or hydroxides, and (3) mixing of sols and solutions. In the coprecipitation technique, the general approach is to mix different salt solutions or alkoxide solutions to give the required composition, followed by hydrolysis with water. The precipitated material is usually referred to as a gel but, unlike the gels produced from dispersions of colloidal particles, it is not normally dispersible in water. The success of the method depends on controlling the concentration of the reactants, as well as the pH and temperature of the solution, to produce mixed products with the desired chemical homogeneity (see Chapter 2).

When gels are formed by mixing sols of different oxides or hydroxides, the uniformity of mixing is, at best, of the order of the colloidal particle size. The chemical homogeneity will therefore be worse than that obtained in the coprecipitation method assuming ideal coprecipitation (i.e., without aggregation). An example of this technique is the preparation of alumino-silicate gels with the mullite composition, $(3Al_2O_3.2SiO_2)$, by the mixing of boehmite sol and silica sol [46,47]. Although these sols can be prepared in the laboratory (by, for example, the hydrolysis of alkoxides), they are also available commercially. In the pH range of ~2.5 to 8, the surfaces of the boehmite particles are positively charged while those of the silica particles are negatively charged. If the mixture of the two sols is gelled within this pH range, then a homogeneous colloidal gel can be obtained because of the attraction and intimate contact between the oppositely charged boehmite and silica particles.

The alumino-silicate system can also be used to illustrate the third method of mixing sols and solutions. In one case, boehmite sol is mixed with a solution of TEOS in ethanol [46]. Gelling is achieved by heating the mixture to evaporate some solvent. Alternatively, silica sol is mixed with a solution of aluminum nitrate and the mixture is gelled by heating.

5.6.2 PREPARATION OF POLYMERIC GELS

For single-component gels such as SiO_2 gel, we have considered in detail the conditions that lead to the formation of polymeric gels. Turning now to the preparation of multicomponent gels, further considerations must be taken into account. As an example, consider the formation of SiO_2–TiO_2 glasses. A convenient starting point for the preparation of the gel is the hydrolysis and condensation of a mixed solution of a silicon alkoxide (e.g., TEOS) and a titanium alkoxide (e.g., titanium tetraethoxide). From our earlier discussion of the properties of alkoxides, we would expect the hydrolysis of the titanium alkoxide to be much faster than that of the silicon alkoxide. Uncontrolled additions of water to the mixture of the two alkoxides will lead to vigorous hydrolysis of the titanium alkoxide and the formation of precipitates that are useless for polymerization. The problem of mismatched hydrolysis rates must therefore be considered when gels with good chemical homogeneity are required.

In general, five different approaches can be used to prepare multicomponent gels: (1) use of double alkoxides, (2) partial hydrolysis of the slowest reacting alkoxide, (3) use of a mixture of alkoxides and metal salts, (4) slow addition of small amounts of water, and (5) matching the hydrolysis rates of the individual alkoxides. Of these, methods (2), (3), and (4) are more commonly used.

5.6.2.1 Use of Double Alkoxides

The use of double alkoxides as the starting material for sol–gel processing seeks to eliminate the problem of mismatched hydrolysis rates by forming a molecule or molecular species in which the metal cations are mixed in the desired ratio. The double alkoxide is hydrolyzed and polymerized in the same way described earlier for simple alkoxides containing one metal cation, and the gel has the same ratio of the two metals as the double alkoxide. The homogeneity of mixing therefore extends to the atomic level. The mechanisms of hydrolysis and condensation of double alkoxides are, however, unclear because the topic has been studied thoroughly in only a few cases [48,49].

The high degree of homogeneity obtained by this route is illustrated by the work of Dislich [50], who prepared magnesium aluminum spinel, $MgAl_2O_4$, from a double alkoxide formed by reacting magnesium methoxide with aluminum sec-butoxide (Figure 5.32). On heating the dried gel, the major x-ray reflections of spinel started to appear at ~250°C and crystallization was completed by ~400°C. In comparison, particulate gels of the same material prepared by Bratton [51] using a coprecipitation technique still contained gibbsite, $Al(OH)_3$, at temperatures of ~400°C.

Despite the high degree of chemical homogeneity obtainable by the double alkoxide route, the method is used only on a limited basis because many double alkoxides are difficult to prepare or are unstable. Furthermore, for a given two-component gel, changing the chemical composition of the gel by changing the atomic ratios of the two metal cations becomes a tedious task because a new double alkoxide must be synthesized in each case.

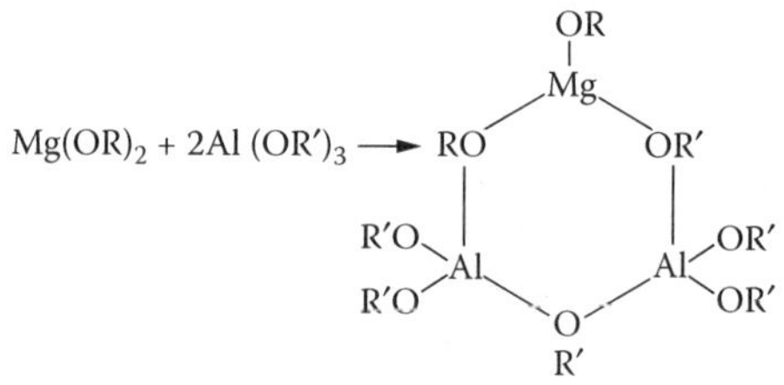

$$Mg(OR)_2 + 2Al(OR')_3 \longrightarrow$$

$$R = CH_3, R' = CH(CH_3)_2$$

FIGURE 5.32 Magnesium aluminum double alkoxide formed by the reaction of a solution containing 1 mol magnesium methoxide to 2 mol aluminum *sec*-butoxide in alcohol. (After Dislich, H., New routes to multi-component oxide glasses, *Angew. Chem., Int. Ed.*, 10, 363, 1971.)

5.6.2.2 Partial Hydrolysis of the Slowest Reacting Alkoxide

This method, involving partial hydrolysis of the slowest reacting alkoxide prior to adding the other alkoxide was invented by Thomas [52] and popularized by the work of Yoldas [15,53]. The idea is that the newly added, unhydrolyzed alkoxide will condense with the partially hydrolyzed sites formed by the previous hydrolysis (heterocondensation) rather than with itself (homocondensation). Unlike the method involving the use of double alkoxides, in which molecular-level homogeneity is predicted, the chemical homogeneity of the gel will depend on the size of the polymeric species to which the last component is added.

As an example, consider the SiO_2–TiO_2 system, for which the starting materials are commonly TEOS and titanium tetraethoxide, $Ti(OEt)_4$. The hydrolysis rate of TEOS, we recall, is much slower than that of the titanium ethoxide. If the TEOS is diluted with ethanol (e.g., 1 mol of ethanol to 1 mol of TEOS) and partial hydrolysis of the solution is carried out with the addition of 1 mol of water per mole of TEOS, the majority of the silanol species will contain one OH group:

$$Si(OR)_4 + H_2O \rightarrow (RO)_3Si-OH + ROH \tag{5.56}$$

As noted earlier, these soluble silanols will undergo condensation reactions on aging. For the present situation in which the number of OH groups in the silanol is limited to about 1 per molecule and the system is diluted, the condensation rate is slow during the first few hours if the temperature is reasonably low (e.g., room temperature):

$$(RO)_3Si-OH + RO-Si(OR)_2(OH) \xrightarrow{\text{slow}} (RO)_3Si-O-Si(OR)_2(OH) + ROH \tag{5.57}$$

If titanium ethoxide is introduced into the solution with vigorous stirring, heterocondensation occurs:

$$(RO)_3Si-OH + RO-Ti(OR)_3 \xrightarrow{\text{fast}} (RO)_3Si-O-Ti(OR)_3 + ROH \tag{5.58}$$

Further condensation with, for example, other silanols occurs with aging:

$$(RO)_3Si-O-Ti(OR)_3 + HO-Si(OR)_3 \rightarrow (RO)_3Si-O-\overset{\displaystyle OR}{\underset{\displaystyle OR}{Ti}}-O-Si(OR)_3 + ROH \tag{5.59}$$

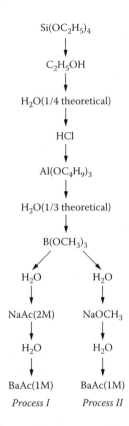

FIGURE 5.33 Synthesis of a five-component oxide by sequential additions of alkoxides (or salts) to partially hydrolyzed condensates. (From Brinker, C.J. and Mukherjee, S.P., Conversion of monolithic gels to glasses in a multicomponent silicate glass system, *J. Mater. Sci.*, 16, 1980, 1981. With permission.)

The heterocondesation reactions described by Equation 5.58 and Equation 5.59 occur at a faster rate than the self-condensation reaction described by Equation 5.57, so dissimilar constituents (i.e., molecular species of silicon and titanium rather than those of silicon alone) tend to become neighbors. Therefore, a gel with good chemical homogeneity can be expected. Furthermore, the product remains in solution because there are too few hydroxyl groups to cause precipitation. After the silicon and titanium have been incorporated into the polymeric network, further addition of water completes the hydrolysis and condensation reactions to produce a single-phase gel. Figure 5.33 summarizes the application of the method to the preparation of a five-component borosilicate glass [54].

5.6.2.3 Use of a Mixture of Alkoxides and Metal Salts

For some metals it is inconvenient to use alkoxides because they are not available, difficult to prepare or use, or too expensive. This is particularly the case with group 1 and group II elements whose alkoxides are solid and nonvolatile, and, in many cases, have low solubility. Metal salts provide a viable alternative, provided they are readily converted to the oxide by thermal or oxidative decomposition [55,56]. The salt should preferably be soluble in alcohol so that it can be mixed with alkoxide solutions without premature hydrolysis of the alkoxide. Salts of organic acids, in particular acetates, but also citrates, formates, and tartrates are potential candidates. Nitrates are possibly the only suitable inorganic salts because others, such as chlorides or sulfates, are more thermally stable and their anions are difficult to remove. However, nitrates are highly oxidizing and care should be taken during heating of the gel to reduce the risk of explosions. A problem that

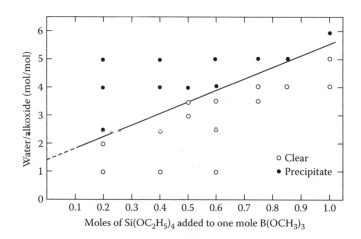

FIGURE 5.34 Regions of precipitate and clear solution formation for the hydrolysis of a solution containing various ratios of TEOS and boron methoxide. (From Yoldas, B.E., Monolithic glass formation by chemical polymerization, *J. Mater. Sci.*, 14, 1843, 1979. With permission.)

is often encountered with this method is that the salt often crystallizes during the drying of the gel, leading to a loss of chemical homogeneity.

Little is known about the reaction mechanism by which the polymerizing alkoxide incorporates the ions of the metal salt into the gel structure. The normal method of preparation [54–57] is first to form a solution of all components that are to be added as alkoxides, as described in the previous method, and then add one or more salts as solutions in alcohol or, if this is not possible, in the water to be used for further hydrolysis (Figure 5.33). All components are then uniformly dispersed and subsequent gelation should then incorporate the elements into a gel network.

5.6.2.4 Slow Addition of Small Amounts of Water

For most metal alkoxides, hydrolysis with excess water yields insoluble oxide or hydroxide precipitates that are useless for further polymerization reactions (Equation 5.21 to Equation 5.23). However, if small amounts of water are added slowly to a sufficiently dilute solution, it is possible to form polymerizable molecular species from these alkoxides also. For example, when a dilute solution of boron alkoxide in alcohol is exposed to water, soluble transient molecular species such as $B(OR)_2(OH)$ and $B(OR)(OH)_2$, representing various degrees of hydrolysis, form initially, e.g.,

$$B(OR)_3 + H_2O \rightarrow B(OR)_2(OH) + ROH \tag{5.60}$$

These species can undergo condensation reactions leading to the formation of a polymer network, as discussed earlier for the silanols.

For the SiO_2–B_2O_3 system, Yoldas [13] investigated the effect of water additions on the homogeneity of a solution of TEOS and boron methoxide, $B(OCH_3)_3$. His results are shown in Figure 5.34. For a given solution, if the water content exceeds a certain value, solution homogeneity is lost because of precipitation. Boron methoxide hydrolyzes much faster than TEOS and precipitates as B_2O_3. For lower water content, the partially hydrolyzed molecular species are soluble, so a clear solution is obtained. Condensation reactions between the partially hydrolyzed species lead to the production of a homogeneous gel. The line separating the clear solution from that containing precipitates can be represented by:

$$M_{water} = A(M_{ME}) + B(M_{TEOS}) \qquad (5.61)$$

where M_{water}, M_{ME}, and M_{TEOS} are the number of moles of water, boron methoxide, and TEOS, respectively. For $M_{ME} = 1$, the results of Figure 5.34 give $A \approx 1.5$ and $B \approx 4$. These values indicate that, to cause precipitation by the addition of water slowly, ~1.5 mol of water is required for 1 mole of boron methoxide (in accordance with Equation 5.23) and an additional 4 mol of water is required for each mole of TEOS added to the system.

5.6.2.5 Matched Hydrolysis Rates

We noted earlier that for a given metal M, the hydrolysis rate of the alkoxide $M(OR)_x$ depends on the length of the alkyl group, R. This suggests that it may be possible to match the hydrolysis rates of the alkoxides of different metals by careful choice of the alkoxide group. As an example, consider the SiO_2–TiO_2 system. Silicon alkoxides are among the slowest to hydrolyze, but of these tetramethoxysilane (TMOS) has the fastest hydrolysis rate. By selecting a titanium alkoxide with a sufficiently long alkyl group, it may be possible to match its hydrolysis rate with that of TMOS. Yamane et al. [58] found that the hydrolysis of a mixed solution of TMOS and titanium *tert*-amyloxide, $Ti[OC(CH_3)_2C_2H_5]_4$, produced a homogeneous gel.

In general, this method is rarely used because of the difficulty in matching the hydrolysis rates closely or in obtaining alkoxides with the desired hydrolysis rates. A more promising approach has been described by Livage and coworkers [59,60] who showed that chelating agents such as acetylacetone can react with alkoxides at a molecular level, giving rise to new molecular precursors. The whole hydrolysis–condensation process can therefore be modified. For example, with the use of chelating agents, the hydrolysis of transition metal alkoxides can be slowed, so better chemical homogeneity can be achieved in multicomponent gels.

5.6.2.5.1 *Comparison of the Preparation Methods*

To compare the homogeneity of multicomponent gels prepared by different methods, Yamane et al. [58] produced gels by methods (2), (4), and (5) outlined above for a composition corresponding to 93.75 mol% SiO_2 and 6.25 mol% TiO_2. Each gel was melted to form a glass and the optical transmission of the glass was used as a measure of the homogeneity of the gel. In this way, it was found that the homogeneity of the three glasses did not differ significantly. The homogeneity of the glasses was also compared with that of a glass produced by melting a mechanically mixed gel. To produce the mixed gel, SiO_2 and TiO_2 gels were prepared separately by hydrolysis of the individual alkoxides, after which they were mixed mechanically. The three gels prepared from mixtures of alkoxides gave glasses with a far better homogeneity than the glass produced from the mechanically mixed gel.

5.7 APPLICATIONS OF SOL–GEL PROCESSING

Applications of sol–gel processing are discussed in a book [61] and a review of the topic with an extensive listing of references to the technology is given in Reference 1. The applications of sol–gel processing derive from the specialty shapes obtained directly from the gel state (e.g., films, fibers, monoliths, and particles), coupled with compositional and microstructural control, and low temperature processing. Sol–gel processing also has its problems. Many alkoxides are expensive and most require special handling in a dry inert atmosphere (e.g., in a glove box). Drying is often a limiting step for articles thicker than ~1 mm, and the shrinkages during drying and sintering are large. Some advantages and disadvantages of the process are summarized in Table 5.9.

Potential applications of sol–gel processing are numerous but the actual number of successful applications is rather few. Thin films and coatings benefit from most of the advantages of sol–gel

TABLE 5.9
Advantages and Disadvantages of the Polymeric Gel
Route Compared with Conventional Fabrication
Methods for Ceramics and Glasses

Advantages:
  1. High purity
  2. Good chemical homogeneity with multicomponent systems
  3. Low temperature of preparation
  4. Preparation of ceramics and glasses with novel compositions
  5. Ease of fabrication for special products such as films and fibers

Disadvantages:
  1. Expensive raw materials
  2. Large shrinkage during fabrication
  3. Drying step leads to long fabrication times
  4. Limited to the fabrication of small articles
  5. Special handling of raw materials usually required

processing while avoiding the disadvantages. They represent one of the few successful commercial applications. (However, even films suffer from cracking problems when attempts are made to prepare films thicker than ~1 μm). Sol–gel processing allows fibers to be drawn from viscous sols and is used to prepare continuous, refractory fibers with high strength, stiffness and chemical durability for the reinforcing phase in composites. For the production of monolithic ceramics and glasses, it is unlikely that the process will be a successful alternative to conventional fabrication methods (glass melting or powder processing) unless novel compositions with unique properties are the result.

5.7.1 THIN FILMS AND COATINGS

Prior to gelation, a sol or solution can be used for preparing thin films by common methods such as dipping, spinning, and spraying. We shall consider the techniques of (1) dip coating, where the object to be coated is lowered into the solution and withdrawn at a suitable speed, and (2) spin coating, where the solution is dropped onto the object, which is spinning at a high speed. In practice, dip coating is currently the more widely used.

One requirement is that the contact angle between the solution and the surface of the object be low so that the solution wets and spreads over the surface. Usually water is present in the sol but moisture from the atmosphere is also sufficient to cause hydrolysis and condensation reactions. The final film is obtained after sintering. There are various advantages and disadvantages inherent in each coating technique. Dip coating does not require any specialized apparatus. Both internal and external surfaces wetted by the solution are coated. The solution must not be too sensitive to moisture because it commonly undergoes some exposure to the atmosphere during the coating process. Spin coating leads to coating of one side of the object only and the coating solution can be kept away from moisture prior to being dropped onto the object. Edge effects may occur for objects that are not axisymmetric. Because the spinning of large objects is impractical, the objects to be coated are commonly in the shape of small, flat disks.

5.7.1.1 Dip Coating

The physics of dip coating has been reviewed by Scriven [62], who divided the batch dip-coating process into five stages: immersion, start-up, deposition, drainage, and evaporation (Figure 5.35). With volatile solvents such as alcohols, evaporation normally accompanies start-up, deposition, and drainage. The continuous dip-coating process (Figure 5.35f) is simpler because it separates

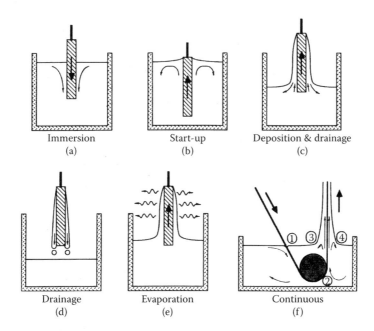

FIGURE 5.35 Stages of the dip-coating process: (a–e) batch and (f) continuous. (From Scriven, L.E., Physics and applications of dip coating and spin coating, *Mater. Res. Soc. Symp. Proc.*, 121, 717, 1988. With permission.)

immersion from the other stages, essentially eliminates start-up, hides drainage in the deposited film, and restricts evaporation to the deposition stage and afterward.

The thickness of the film in the deposition region (position 3 in Figure 5.35f) is controlled by a competition between as many as six forces: (1) viscous drag upward on the liquid by the moving substrate, (2) force of gravity, (3) resultant force of surface tension in the concavely curved meniscus, (4) inertial force of the boundary layer liquid arriving at the deposition region, (5) surface tension gradient, and (6) the disjoining or conjoining pressure, important for films less than ~1 μm thick.

When the liquid viscosity η_L and substrate speed U are high enough to hold the meniscus curvature down, then the deposited film thickness h is governed by a balance between the viscous drag (proportional to $\eta U/h$) and gravity force (proportional to $\rho_L gh$, where ρ_L is the density of the liquid and g is the acceleration due to gravity). Thus,

$$h = c_1 \left(\frac{\eta_L U}{\rho_L g} \right)^{1/2} \tag{5.62}$$

where the constant of proportionality c_1 is ~0.8 for Newtonian liquids [63]. When the liquid viscosity and substrate speed are not high enough, as in sol–gel processing, this balance of forces is modulated by the ratio of viscous drag to the liquid–vapor surface tension γ_{LV} according to the following relationship derived by Landau and Levich [64]:

$$h = 0.944 \left(\frac{\eta_L U}{\gamma_{LV}} \right)^{1/6} \left(\frac{\eta_L U}{\rho_L g} \right)^{1/2} \tag{5.63}$$

Rearranging the terms in Equation 5.63 gives:

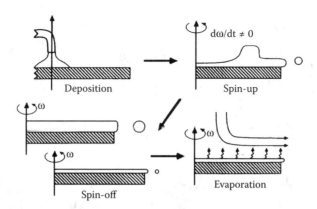

FIGURE 5.36 Stages of the spin-coating process. (From Scriven, L.E., Physics and applications of dip coating and spin coating, *Mater. Res. Soc. Symp. Proc.*, 121, 717, 1988. With permission.)

$$h = 0.944 \frac{\left(\eta_L U\right)^{2/3}}{\gamma_{LV}^{1/6}\left(\rho_L g\right)^{1/2}} \tag{5.64}$$

so the thickness of the film is predicted to vary as $(\eta_L U)^{2/3}$.

The applicability of Equation 5.62 to Equation 5.64 to film deposition in sol–gel processing has been tested in only a few studies. Strawbridge and James [65] investigated the relationship between film thickness and viscosity for an acid-catalyzed TEOS solution (H_2O: TEOS ratio = 1.74) deposited on glass substrates at substrate withdrawal speeds U in the range of 5 to 15 cm/min. Significant deviations from a modified form of Equation 5.62 were observed as the speed was increased. For silica sols in which the precursor structures varied from weakly branched polymers to highly condensed particles, the results of Brinker and Ashley reported in Reference 1 indicate that for polymeric sols, h varies approximately as $U^{2/3}$, in accordance with Equation 5.64, and with previous results of Dislich and Hussmann [66].

5.7.1.2 Spin Coating

As illustrated in Figure 5.36, spin coating can be divided into four stages: deposition, spin-up, spin-off, and evaporation [62,67]. As in dip coating, evaporation may occur throughout the process. Initially, an excess of liquid is deposited on the surface, which is at rest or rotating slowly. In the spin-up stage, liquid flows radially outward, driven by the centrifugal force generated by the rotating substrate. In the spin-off stage, excess liquid flows to the perimeter of the substrate, leaving as droplets. As the film gets thinner, the rate of removal of excess liquid by spin-off becomes slower because of the greater resistance to flow and the greater effect of evaporation in raising the viscosity by concentrating the nonvolatile component of the solution. In the fourth stage, evaporation becomes the primary mechanism of thinning.

The key stage in spin coating is spin-off. Apart from edge effects, a film of liquid tends to become uniform in thickness as it gets thinner and, once uniform, tends to remain so with further thinning, provided the viscosity is insensitive to shear and does not vary over the substrate. The reason for the strong tendency towards uniformity is the way the two main forces balance, i.e., the rotation-induced centrifugal force, which drives radially outward flow, and the resisting viscous force, which acts radially inward.

The thickness of an initially uniform film during spin-off is given by [62]:

$$h(t) = \frac{h_o}{(1 + 4\rho_L\omega^2 h_o^2 t / 3\eta_L)^{1/2}} \tag{5.65}$$

where h_o is the initial thickness, t is the time, and ω is the angular velocity. In the theory, both ω and ρ_L are assumed to be constant. Even films that are initially nonuniform tend strongly towards uniformity following Equation 5.65.

The spinning substrate creates a steady forced convection in the vapor above the substrate such that the mass-transfer coefficient, κ, is uniform. The evaporation rate all over the substrate is therefore quite uniform also. Assuming a crude model in which evaporation occurs only after the rate of thinning by flow falls to a certain rate $e\rho_A / \rho_A^o$, where ρ_A is the mass of volatile solvent per unit volume, ρ_A^o is the initial value, and e is the evaporation rate that depends on the gas-phase mass-transfer coefficient, Meyerhofer [68] found that the final thickness and the total elapsed time to achieve this thickness are given by:

$$h_{final} = \left(1 - \frac{\rho_A^o}{\rho_A}\right)\left(\frac{3\eta_L e}{2\rho_A^o\omega^2}\right)^{1/3} \tag{5.66}$$

$$t_{final} = t_{spin-off} + h_{spin-off}\left(\frac{\rho_A^o}{e\rho_A}\right) \tag{5.67}$$

According to Equation 5.66, the final thickness of the film varies inversely as $\omega^{2/3}$ and as $\eta_L^{1/3}$.

5.7.1.3 Structural Evolution in Films

The physics and chemistry governing polymer growth and gelation in films are essentially the same as those discussed earlier for bulk gels but certain factors distinguish the structural evolution in films [1]. The overlap between the deposition and evaporation stages establishes a competition between evaporation (which compacts the structure) and condensation reactions (which stiffen the structure, thereby increasing the resistance to compaction). We recall that the gelation and drying stages are normally separated in bulk systems. The aggregation, gelation, and drying in films deposited by dipping or spinning occur in seconds to minutes compared with days or weeks for bulk gels. Films therefore undergo considerably less aging, so the result is a more compact dried structure. Fluid flow by draining, evaporation, or spin-off coupled with the attachment of the precursors to the substrate imposes a shear stress within the film during deposition. After gelation, drying and sintering lead to *tensile* stresses in the plane of the film [9]. Except near the edges, shrinkage of the adherent film occurs in the thickness direction only. Bulk gels are generally not constrained in any dimension.

5.7.1.4 Applications of Films

Current and potential applications of films prepared by sol–gel processing are numerous [1,61], and they are expected to grow in the future. Table 5.10 summarizes some possible applications of films and coatings prepared from metal alkoxide solutions [69].

5.7.2 Fibers

Fibers can be produced by two sol–gel routes. In one route, a dense monolithic preform is produced by sol–gel processing, drying and sintering, followed by conventional fiber drawing above the glass

TABLE 5.10

Applications of Films Prepared from Metal Alkoxide Solutions

Application	Example	Composition
Mechanical	Protection	SiO_2
Chemical	Protection	SiO_2
Optical	Absorbing	TiO_2–SiO_2; SiO_2–R_mO_n oxides of Fe, Cr and Co
	Reflecting	In_2O_3–SnO_2
	Anti-reflecting	Na_2O–B_2O_3–SiO_2
Electrical	Ferroelectric	$BaTiO_3$; $KTaO_3$; PLZT
	Electronic conductor	In_2O_3–SnO_2; SnO_2–CdO
	Ionic conductor	β-alumina
Catalytic	Photocatalyst	TiO_2
	Catalyst carrier	SiO_2; TiO_2; Al_2O_3

Source: Sakka, S., Sol–gel synthesis of glasses: present and future, *Am. Ceram. Soc. Bull.*, 64, 1463, 1985. With permission.

softening temperature. The preparation of monolithic SiO_2 gels from fine particles made by flame oxidation is described by Rabinovitch et al. [2,3] and by Scherer and Luong [4]. Compared to fibers drawn from the melt, the monolithic gel route does not require a melting step, thereby avoiding the introduction of impurities from the crucible. However, the route suffers from long processing times and the fiber quality for optical applications is inferior to that prepared from CVD performs. Therefore the monolithic gel route does not appear to have any technological or commercial advantage over current CVD processes.

The other sol–gel route is to draw the fibers directly from a viscous sol near room temperature and then covert them into dense glass or ceramic fibers by sintering. The advantage of this route is that very refractory and chemically durable fibers can be produced at relatively low temperatures which would be difficult to prepare by conventional drawing from a melt. The preparation of SiO_2 fibers from TEOS precursors is reviewed by Sakka [70]. The solution conditions for fiber formation (spinnability) are acid catalysis and low water content, which means that the polymeric species in the solution are more weakly branched polymers. Under these conditions, the drawn fibers are microporous and contain a significant amount of residual organics that are difficult to remove completely prior to sintering. Many compositions also produce fibers with nonuniform cross sections. The influence of the starting composition on spinnability and fiber cross section are summarized in Figure 5.37. The fibers produced by this sol–gel route are not suitable for optical applications but have current or potential applications as the reinforcing phase in composites and as refractory textiles. In addition to SiO_2, examples of high-strength, refractory fibers produced commercially by the sol–gel route include compositions in the SiO_2–Al_2O_3, SiO_2–ZrO_2, and SiO_2–B_2O_3–Al_2O_3 systems [1,70].

5.7.3 MONOLITHS

Monoliths are defined as bulk gels (smallest dimension $\geq$ 1 mm) cast into shape and processed without cracking. The ability to form complex shapes by casting the gel coupled with relatively low temperature sintering are attractive features of the route, but cracking during drying and the large shrinkage during sintering can present significant challenges in achieving the final shape. The good applications of monolithic gels take advantage of the process advantages, such as purity, chemical homogeneity, porosity, low temperature fabrication, and the ability to produce novel chemical compositions. The main applications are optical components, graded refractive index (GRIN) glasses, and transparent aerogels used as Cherenkov detectors and as insulation [1].

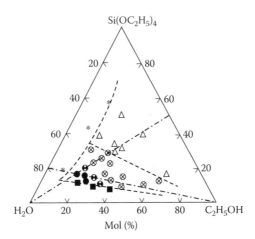

FIGURE 5.37 Relation between fiber drawing behavior and composition of TEOS–water–ethanol solutions after hydrolysis at 80°C with an acid concentration of 0.01 mol HCl per mole of TEOS: (∗) immiscible, (■) not spinnable, (△) no gel formation, (●) circular cross section, (⊗) noncircular cross section, (⊖) circular and noncircular cross section. (From Sakka, S., Sol–gel synthesis of glasses: present and future, *Am. Ceram. Soc. Bull.*, 64, 1463, 1985. With permission.)

5.7.4 POROUS MATERIALS

Sol–gel processing is an attractive route for the production of porous ceramics and glasses because the characteristic properties of gels such as high surface area, high porosity, and small pore size are not easily attainable by more conventional methods. In addition to the transparent aerogel applications mentioned above, these unique properties are exploited in a variety of applications, which include filtration, separation, catalysis, and chromatography [1,61,71].

5.8 CONCLUDING REMARKS

In this chapter, we have examined the physics and chemistry of sol–gel processing and the ways in which these relate to the practical fabrication of ceramics and glasses. It is apparent that sol–gel processing offers considerable opportunities in both science and technology. An understanding of the scientific issues provides the most useful basis for successful application of the method. Although sol–gel processing offers considerable advantages as a fabrication method, its disadvantages are also very real. Potential applications of sol–gel processing are numerous but the successful applications (e.g., films, fibers, and porous membranes) benefit from most of the advantages of the process while avoiding the disadvantages. Although the successful applications are expected to grow substantially, the method cannot compete with the more conventional fabrication methods for the mass production of monolithic ceramics and glasses.

PROBLEMS

5.1

Summarize the polymerization and growth mechanisms of aqueous silicates in the pH range of (a) <2; (b) 2–7; and (c) >7.

5.2

Summarize the structural evolution of sol–gel silicates in the pH range of (a) <2; (b) 2–7; and (c) >7.

5.3

Estimate the surface free energy of 1 g of a silica gel that is 50% porous with pores (assumed spherical) of 5 nm in diameter. Compare your answer with the surface free energy of 1 g of silica glass spheres with a diameter of 10 μm. The specific surface energy of both materials can be assumed to be 0.25 J/m^2.

5.4

You wish to prepare approximately 50 g of SiO_2 by the sol–gel process under the following conditions: You start with a solution of 50 mol % TEOS in ethanol and you add 10 mol of H_2O per mole of alkoxide. Determine the volume of each starting material required for the process, assuming that the reaction goes to completion.

5.5

A silica gel prepared from an alkoxide has a surface area $S \approx 500$ m^2/g, a porosity $P \approx 0.9$, and a density $\rho \approx 1.5$ g/cm^3. Estimate the permeability K of the gel.
Compare the estimated value of K with the value determined from the data in Fig. 5.21.

5.6

An alcogel contains 5 vol % of solid. Careful drying leads to the production of a xerogel that is approximately 50% porous. Estimate the linear shrinkage during drying. On sintering, a fully dense solid is obtained but there is a weight loss of 20%. Estimate the linear shrinkage during sintering.

5.7

In another experiment, the alcogel in Problem 5.6 is dried supercritically with negligible shrinkage. Sintering in this case also produces a fully dense solid with a 20% weight loss. Estimate the linear shrinkage during sintering.

5.8

Explain the meaning of the term *drying control chemical additives* (DCCAs). Briefly describe the role of DCCAs in drying. What drawbacks do you see with the use of DCCAs?

5.9

Compare the drying and sintering characteristics of SiO_2 gels prepared by the following two methods:

   a. Particulate gel route by gelling of fumed SiO_2 particles
   b. Polymeric gel route by hydrolysis of silicon ethoxide.

5.10

Describe three methods for the preparation of $BaTiO_3$ by the sol–gel process and comment on the chemical homogeneity of the gel produced by each method.

5.11

In a dip-coating experiment, a student uses a solution of TEOS in ethanol with a viscosity $\eta = 2$ mPa.s and a density $\rho = 1.0$ g/cm^3, a withdrawal speed U = 10 cm/min. Assuming a surface tension of the solution $\gamma_{LV} = 0.04$ J/m^2, use Eq. (5.64) to estimate the thickness of the liquid film deposited on the substrate.

If the film is dried and sintered to full density ($\rho_f = 2.2$ g/cm^3), determine the thickness of the final film. If required, the density of the solvent $\rho_s = 0.8$ g/cm^3.

5.12

Compare the key steps in the production of 1-μm thick TiO$_2$ films on an Al$_2$O$_3$ substrate by the following two methods:

a. Spin coating of a solution of titanium alkoxide
b. Spin coating of a suspension of flame-synthesized TiO$_2$ particles.

REFERENCES

1. Brinker, C.J. and Scherer, G.W., *Sol-Gel Science*, Academic Press, New York, 1990.
2. Rabinovich, E.M., Johnson, D.W., MacChesney, J.B., and Vogel, E.M., Preparation of high-silica glasses from colloidal gels: I, preparation for sintering and properties of sintered glasses, *J. Am. Ceram. Soc.*, 66, 683, 1983.
3. Johnson, D.W., Rabinovich, E.M., MacChesney, J.B., and Vogel, E.M., Preparation of high-silica glasses from colloidal gels: II, sintering, *J. Am. Ceram. Soc.*, 66, 688, 1983.
4. Scherer, G.W. and Luong, J.C., Glasses from colloids, *J. Non-Cryst. Solids*, 63, 163, 1984.
5. Segal, D., *Chemical Synthesis of Advanced Ceramic Materials*, Cambridge University Press, Cambridge, 1989, chap. 4.
6. Meakin, P., Models for colloidal aggregation, *Annu. Rev. Phys. Chem.*, 39, 237, 1988.
7. Iler, R.K., *The Chemistry of Silica*, John Wiley & Sons, New York, 1979.
8. Baes, C.F. and Messmer, R.E., *The Hydrolysis of Cations*, John Wiley & Sons, New York, 1976.
9. Rahaman, M.N., *Ceramic Processing and Sintering*, 2nd ed., Marcel Dekker, New York, 2003, chap. 11.
10. Bradley, D.C., Mehrotra, R.C., and Gaur, D.P., *Metal Alkoxides*, Academic Press, London, 1978.
11. Okamura, H. and Bowen, H.K., Preparation of alkoxides for the synthesis of ceramics, *Ceram. Int.*, 12, 161, 1986.
12. Mah, T.-I., Hermes, E.E., and Masdiyasni, K.S., Multicomponent ceramic powders, in *Chemical Processing of Ceramics*, Lee, B.I. and Pope, E.J.A., Eds., Marcel Dekker, New York, 1994, chap. 4.
13. Yoldas, B.E., Monolithic glass formation by chemical polymerization, *J. Mater. Sci.*, 14, 1843, 1979.
14. Cogan, H.D. and Setterstrom, C.A., Properties of ethyl silicate, *Chem. Eng. News*, 24, 2499, 1946.
15. Mandelbrot, B.B., *The Fractal Geometry of Nature*, W.H. Freeman, New York, 1983.
16. Feder, J., *Fractals*, Plenum Press, New York, 1988.
17. Schaefer, D.W., Fractal models and the structure of materials, *Mater. Res. Soc. Bull.*, 13(2), 22, 1988.
18. Martin, J.E., Molecular engineering of ceramics, in *Atomic and Molecular Processing of Electronic and Ceramic Materials: Preparation, Characterization and Properties*, Aksay, I.A., McVay, G.L., Stoebe, T.G., and Wager, J.F., Eds., Materials Research Society, Pittsburg, PA, 1987, p. 79.
19. (a) Flory, P.J., Molecular size distribution in three-dimensional polymers. I. Gelation, *J. Am. Chem. Soc.*, 63, 3083, 1941.
    (b) Flory, P.J., Constitution of three-dimensional polymers and the theory of gelation, *J. Phys. Chem.*, 46, 132, 1942.
20. Stockmeyer, W.H., Theory of molecular size distribution and gel formation in branched-chain polymers, *J. Chem. Phys.*, 11, 45, 1943.
21. Zarzycki, J. Physical-chemical factors in sol-gel processes, in *Science of Ceramic Chemical Processing*, Hench, L.L. and Ulrich, D.R., Eds., John Wiley & Sons, New York, 1986, chap. 2.
22. Meakin, P., Computer simulation of growth and aggregation processes. in *On Growth and Form: Fractal and Non-Fractal Patterns in Physics*, Stanley, H.E. and Ostrowsky, N., Eds., Martinus Nijhoff, Boston, MA, 1986, p. 111.

23. Witten, T.M. Jr. and Sander, L.M., Diffusion-limited aggregation, a kinetic critical phenomenon, *Phys. Rev. Lett.*, 47, 1400, 1981.

24. Keefer, K.D., Growth and structure of fractally rough silica colloids, *Mater. Res. Soc. Symp. Proc.*, 73, 295, 1986.

25. Meakin, P., Formation of fractal clusters and networks by irreversible diffusion-limited aggregation, *Phys. Rev. Lett.*, 51, 1119, 1983.

26. Sakka, S. and Kamiya, Y., The sol-gel transition in the hydrolysis of metal alkoxides in relation to the formation of glass fibers and films, *J. Non-Cryst. Solids*, 48, 31, 1982

27. Sakka, S., Formation of glass and amorphous oxide fibers from solution, *Mater. Res. Soc. Symp. Proc.*, 32, 91, 1984.

28. Sakka, S., Kamiya, K., Makita, K., and Yamamoto, Y., Formation of sheets and coating films from alkoxide solutions, *J. Non-Cryst. Solids*, 63, 223, 1984.

29. Huggins, M.L., The viscosity of dilute solutions of long-chain molecules. IV. Dependence on concentration, *J. Am. Chem. Soc.*, 64, 2716, 1942.

30. Tsuchida, H., *Science of Polymers*, Baikukan Publishing Company, Tokyo, 1975.

31. Zallen, R., *The Physics of Amorphous Solids*, John Wiley & Sons, New York, 1983.

32. Stauffer, D., Coniglio, A., and Adam, M., Gelation and critical phenomena, *Adv. Polym. Sci.*, 44, 103, 1982.

33. Partlow, D.P. and Yoldas, B.E., Colloidal vs. polymer gels and monolithic transformation in glass-forming systems, *J. Non-Cryst. Solids*, 46, 153, 1981.

34. (a) Scherer, G.W., Drying gels, VII. Revision and review, *J. Non-Cryst. Solids*, 109, 171, 1989
    (b) Scherer, G.W., Theory of drying, *J. Am. Ceram. Soc.*, 73, 3, 1990.

35. Dwivedi, R.K., Drying behavior of alumina gels, *J. Mater. Sci. Lett.*, 5, 373, 1986.

36. Scherer, G.W., Drying of ceramics made by sol-gel processing, in *Drying '92*, Mujumdar, A.S., Ed., Elsevier, New York, 1992, p. 92.

37. Zarzycki, J., Prassas, M., and Phalippou, J., Synthesis of glasses from gels: the problem of monolithic gels, *J. Mater. Sci.*, 17, 3371, 1982.

38. Hench, L.L., Use of drying control chemical additives (DCCAs) in controlling sol-gel processing, in *Science of Ceramic Chemical Processing*, Hench, L.L. and Ulrich, D.R., Eds., John Wiley & Sons, New York, 1986, p. 52.

39. Prassas M, Phalippou, J, and Zarzycki, J., Synthesis of monolithic silica gels by hypercritical solvent extraction, *J. Mater. Sci.*, 19, 1656, 1984

40. Brinker, C.J. and Scherer, G.W., Sol → gel → glass: I, gelation and gel structure, *J. Non-Cryst. Solids*, 70, 301, 1985.

41. Brinker, C.J., Scherer, G.W., and Roth, E.P., Sol → gel → glass: II, physical and structural evolution during constant heating rate experiments, *J. Non-Cryst. Solids*, 72, 345, 1985.

42. Scherer, G.W., Brinker, C.J., and Roth, E.P., Sol → gel → glass: III, viscous sintering, *J. Non-Cryst. Solids*, 72, 369, 1985.

43. Brinker, C.J., Roth, E.P., Tallant, D.R., and Scherer, G.W., Structure of gels during densification, in *Science of Ceramic Chemical Processing*, Hench, L.L. and Ulrich, D.R., Eds., John Wiley & Sons, New York, 1986, chap. 3.

44. James, P.F., The gel to glass transition: chemical and microstructural evolution, *J. Non-Cryst. Solids*, 100, 93, 1988.

45. (a) Yoldas, B.E., Hydrolysis of aluminum alkoxides and bayerite conversion, *J. Appl. Chem. Biotechnol.*, 23, 803, 1973.
    (b) Yoldas, B.E., Alumina gels that form porous transparent Al_2O_3, *J. Mater. Sci.*, 10, 1856, 1975.

46. Komarneni, S., Suwa, Y., and Roy, R., Application of compositionally diphasic xerogels for enhanced densification: the system Al_2O_3–SiO_2, *J. Am. Ceram. Soc.* 69, C-155, 1986.

47. Ishmail, M.G.M.U. et al., Synthesis of mullite powder and its characteristics, *Int. J. High Technol. Ceram.*, 3, 123, 1986.

48. Riman, R.E., Ph.D. thesis, Massachusetts Institute of Technology, Cambridge, MA, 1987.

49. Jones, K., Davies, T.J., Emblem, H.G., and Parkes, P., Spinel formation from magnesium aluminum double alkoxides, *Mater. Res. Soc. Symp. Proc.*, 73, 111, 1986.

50. Dislich, H., New routes to multicomponent oxide glasses, *Angew. Chem., Int. Ed.*, 10, 363, 1971.

51. Bratton, R.J., Coprecipitates yielding $MgAl_2O_4$ spinel powders, *Am. Ceram. Soc. Bull.*, 48, 759, 1969.

52. Thomas, I.M., Method for Producing Glass Ceramics, U.S. Patent 3,791,808, February 12, 1974.

53. Yoldas, B.E., Preparation of glasses and ceramics from metal-organic compounds, *J. Mater. Sci.*, 12, 1203, 1977.

54. Brinker, C.J. and Mukherjee, S.P., Conversion of monolithic gels to glasses in a multicomponent silicate glass system, *J. Mater. Sci.*, 16, 1980, 1981.

55. Roy, R., Gel route to homogeneous glass preparation, *J. Am. Ceram. Soc.*, 52, 344, 1969.

56. Levene, L. and Thomas, I.M., Process of Converting Metalorganic Compounds and High Purity Products Obtained Therefrom, U.S. Patent 3,640,093, February 8, 1972.

57. Phalipou, J., Prassas, M., and Zarzycki, J., Crystallization of gels and glasses made from hot-pressed gels, *J. Non-Cryst. Solids*, 48, 17, 1982.

58. Yamane, M., Inoue, S., and Keiichi, N., Preparation of gels to obtain glasses of high homogeneity by low temperature synthesis, *J. Non-Cryst. Solids*, 48, 153, 1982.

59. Livage, J., Henry, M., and Sanchez, C., Sol-gel chemistry of transition metal oxides, *Prog. Solid State Chem.*, 18, 259, 1988.

60. Sanchez, C., Livage, J., Henry, M., and Babonneau, F., Chemical modification of alkoxide precursors, *J. Non-Cryst. Solids*, 100, 65, 1988.

61. Klein, L.C., Ed., *Sol-Gel Technology for Thin Films, Fibers, Preforms, Electronics and Specialty Shapes*, Noyes Publications, Park Ridge, NJ, 1988.

62. Scriven, L.E., Physics and applications of dip coating and spin coating, *Mater. Res. Soc. Symp. Proc.*, 121, 717, 1988.

63. Spiers, R.P., Subbaraman, C.V., and Wilkinson, W.L., Free coating of a Newtonian liquid onto a vertical surface, *Chem. Eng. Sci.*, 29 389, 1974.

64. Landau, L.D. and Levich, B.G., *Acta Physicochim. U.R.S.S.* 17, 42, 1942.

65. Strawbridge, I. and James, P.F., The factors affecting the thickness of sol-gel derived silica coatings prepared by dipping, *J. Non-Cryst Solids*, 86, 381, 1986.

66. Dislich, H. and Hussmann, E., Amorphous and crystalline dip coatings obtained from organometallic solutions: procedures, chemical processes and products, *Thin Solid Films*, 98, 129, 1981.

67. Bornside, D.E., Macosko, C.W., and Scriven, L.E., On the modeling of spin coating, *J. Imaging Technol.*, 13, 122, 1987.

68. Meyerhofer, D., Characteristics of resist films produced by spinning, *J. Appl. Phys.*, 49, 3993, 1978.

69. Sakka, S., Sol-gel synthesis of glasses: present and future, *Am. Ceram. Soc. Bull.*, 64, 1463, 1985.

70. Sakka, S., Fibers from the sol-gel process, in *Sol-Gel Technology for Thin Films, Fibers, Preforms, Electronics, and Specialty Shapes*, Klein, L.C., Ed., Noyes Publications, Park Ridge, NJ, 1988, p. 140.

71. Ishizaki, K., Sheppard, L., Okada, S., Hamasaki, T., and Huybrechts, B., Eds., *Porous Materials*, *Ceramic Trans.* Vol. 31, The American Ceramic Society, Westerville, OH, 1993.

6 Mixing and Packing of Powders

6.1 INTRODUCTION

In ceramic processing, it is often necessary to carry out a mixing step prior to consolidation and sintering of powders. The microstructure and engineering properties of the sintered body depend critically on how well the powders are mixed. The aim of the mixing operation is to achieve as homogeneous a mixture of the components as possible. Some consequences of inhomogeneous or inadequate mixing of powders on the microstructure of advanced ceramics were mentioned in Chapter 1 (see Figure 1.23 and Figure 1.24). Mixing operations in ceramic processing cover a wide range of systems that include dry or semidry powders, plastic mixtures of powders and organic binders, highly viscous pastes, and less viscous suspensions.

Practically, the best mixing that can be achieved is a random mixture of the components. An important property to know about a mixture is how well it is mixed, often referred to as the quality of mixing. Because of the random nature of mixing, the use of statistics is essential for understanding particulate mixing and for interpreting mixing data. Many mixing indices have been proposed in the literature, most of which are based on statistical analysis and expressed in different forms of the standard deviation. Whatever method is used for assessing the quality of mixing, the sampling procedure is very important. The sample size must be fixed in accordance with the intended use of the mixture, referred to as the *scale of scrutiny*, which is the smallest amount of material within which the quality of mixing is important.

In many systems, the particles to be mixed have different properties, such as differences in particle size, density, and shape, and they tend to segregate. Mixing and segregation occur simultaneously and are therefore competing mechanisms, particularly for free-flowing particles. Cohesive powders, such as fine powders, show little or no segregation. Mixers are often classified according to the dominant mixing mechanism: convective, diffusive, and shear. Many different types of mixers are available commercially. It is important to select a mixer with the type of mixing mechanism that is best suited to the properties of the particles to be mixed.

The microstructure and properties of the sintered body also depend critically on the packing of the particles in the green body. Severe variations in the packing density of the green body will, in general, produce microstructural heterogeneities in the sintered body, which limit the engineering properties of the fabricated article. The homogeneous packing of particles in the green body is the desired goal of the consolidation step. As the packing density controls the amount of shrinkage during firing, the achievement of a high packing density is also desirable. Geometrical particle packing concepts provide a useful basis for understanding how the structure of the consolidated powder comes about. An important practical consideration is the extent to which the parameters of the consolidation process can be manipulated to control the packing homogeneity and packing density of the green body.

6.2 MIXING OF PARTICULATE SOLIDS

Although mixing of particulate solids is a key step in many processing operations, particularly in the case of ceramics, the theory is less developed than many other unit operations. However, substantial practical experience has been accumulated over many years, which allow many mixing problems to be solved. Several texts and review articles, particularly in the powder technology and chemical engineering fields, cover the principles and practice of solids mixing, as well as mixing equipment [1–8]

6.2.1 TYPES OF MIXTURES

Several terms are used in the literature to describe the particle arrangements possible for a mixture of two or more components. Here, we shall simplify them to just three: ordered, random, and partially ordered mixtures. An *ordered mixture* is one in which the particles are arranged in a regular repeating pattern, such as a lattice structure with a unit cell. This is the most perfect mixture, but it cannot be achieved in practice. Generally, the aim is to produce a *random mixture*, defined as a mixture in which the probability of finding a particle of any component is the same at all positions in the mixture and is equal to the proportion of that component in the mixture as a whole. For particles that do not have a tendency to segregate, this is the best quality of mixture that can be achieved. A *partially ordered mixture* is one in which at least one component is ordered in its distribution, such as a coating of fine particles on the surfaces of larger particles. Figure 6.1 gives schematic examples of the three types of mixtures.

6.2.2 MIXING AND SEGREGATION MECHANISMS

The mechanisms of mixing are usually classified into three types: diffusive, shear, and convective mixing [9]. *Diffusive mixing* occurs when neighboring particles exchange places, such as when particles roll down a sloping surface. In *shear mixing*, shear stresses give rise to slip zones and mixing occurs by interchange of particles between layers within the zones. *Convective mixing* occurs by transport of groups of particles from one region of the powder mass to another. These mechanisms cannot be completely separated and all three occur to some extent in a given process. Different mixers, however, show different predominating mechanisms.

When particles to be mixed have the same important physical properties, random mixing will be achieved provided the mixing process is carried out long enough. However, in many common systems the components of a mixture have different physical properties, and tend to segregate. Particles with the same property tend to collect in some part of the mixture and random mixing is not a natural state for such a system. For dry powders, segregation occurs only if the particles are free flowing. Cohesive powders show little or no segregation. Even if the particles are originally mixed, they will segregate on handling (e.g., moving, pouring, or processing). The mixing operation should therefore be carried out as near as possible to the processing step in which the mixture is used, thereby avoiding unnecessary handling.

Mixing and segregation are competing mechanisms and occur simultaneously in a mixer. The extent of segregation depends on the difference in properties of the powders to be mixed and on the type of mixer [10]. Although differences in size, density, shape, and roughness can give rise to segregation, the difference in size is by far the most important in the mixing of dry powders. Density difference is comparatively unimportant for dry powders, but in suspensions settling occurs faster for particles of the same size as the density increases. Segregation can occur in slurries used in slip casting and tape casting, when coarser or denser particles settle preferentially if they are left unstirred for long periods.

A mixture of particles of different sizes may segregate by four different mechanisms [10,11]:

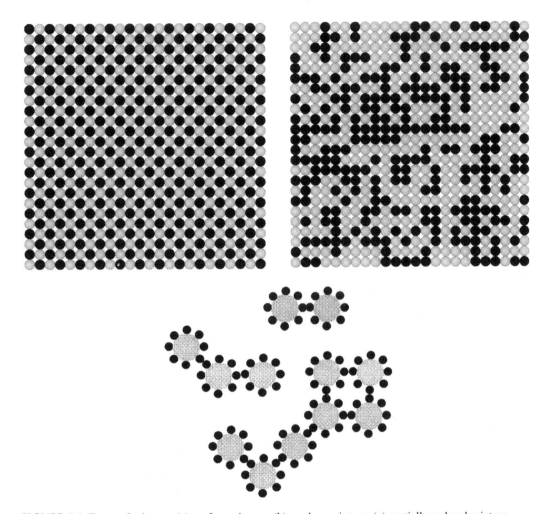

FIGURE 6.1 Types of mixture: (a) perfect mixture; (b) random mixture; (c) partially ordered mixture.

1. *Trajectory segregation:* If a small particle of radius a and density d_p, whose drag is governed by Stokes law (Equation 3.20), is projected with a velocity U into a fluid with viscosity η and density d_f, the deceleration β, defined as the retarding force divided by the mass of the particle is given by:

$$\beta = \frac{9\eta U}{2a^2 d_p} \tag{6.1}$$

The limiting distance l that a particle can travel horizontally is

$$l = \frac{Ua^2 d_p}{9\eta} \tag{6.2}$$

A particle of radius $2a$ will therefore travel four times the distance traveled by a particle of radius a. This mechanism can cause segregation when particles are caused to move through the air (Figure 6.2a) or when particles fall from the end of a conveyor belt.

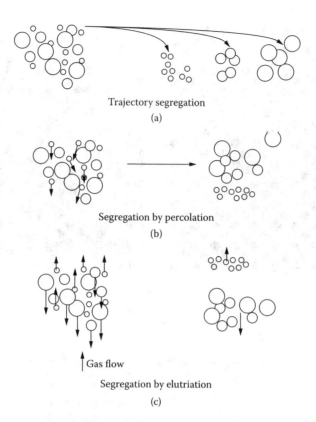

Trajectory segregation

(a)

Segregation by percolation

(b)

Gas flow

Segregation by elutriation

(c)

FIGURE 6.2 Mechanisms of segregation.

2. *Percolation of fine particles:* If a mass of particles is disturbed in such a way that individual particles move, a rearrangement in the packing of the particles occurs. Smaller particles tend to move downwards, leading to segregation (Figure 6.2b). Even a small difference in particle size can produce measurable segregation. Percolation can occur during stirring, shaking, vibration, or when pouring the particles into a heap.

3. *Rise of coarse particles on vibration:* In addition to the percolation effect, another segregation mechanism can occur when a mixture of particles of different sizes is vibrated: the larger particles move upwards, regardless of whether they are denser than the smaller particles.

4. *Elutriation segregation:* When a powder containing an appreciable fraction of particles finer than ~50 μm is charged into a storage vessel or hopper, air is displaced upwards. The air velocity may exceed the terminal velocity of the fine particles, leading to a cloud of fine particles that will eventually settle out and form a layer on top of the coarse particles (Figure 6.2c).

As segregation is caused primarily by a difference in particle size in a free-flowing powder, the difficulty of mixing two components can be reduced by making the particle size of the components as similar as possible and by reducing the particle size of both components. Experiments indicate that reducing the particle size below 50 to 75 μm will reduce segregation significantly, and below 5 to 10 μm, no appreciable segregation will occur [2]. In such fine powders, interparticle forces due to van der Waals attraction (see Chapter 4), moisture, and electrostatic charging are large compared to the gravitational and inertial forces acting on the particles. The particles stick together, preventing segregation. These powders are often referred to as *cohesive powders*. There is, however, the possibility that agglomerates will form and that they will segregate.

Segregation can also be prevented by adding small amounts of liquid to a mixture. If one of the components is fine (less than ~5 µm) and the other is coarse, mixing can cause the fines to be coated onto the surfaces of the coarse particles, giving a partially ordered mixture (Figure 6.1c) in which segregation will not occur. This type of mixture is prepared in the manufacture of vitreous-bonded grinding wheels, where the fine component that will form the glassy bond is added to the coarse abrasive grit in the mixer. For highly segregating mixtures, the possibility of using *continuous mixing* should be considered because this will give better mixing quality than batch mixing while avoiding the problems that arise from storage and handling of segregating materials [10].

6.2.3 Mixture Composition and Quality

Both the powder characteristics and the mixing method influence the mixture quality. Powder characteristics that influence mixture quality include the particle size distributions of the components in the mixture, the volume fraction of each component, the surface roughness, surface chemistry, and shape of the particles, the density of each component, and the state of agglomeration. The end use of a powder mixture determines the quality of mixture required. The end use imposes a *scale of scrutiny,* or a characteristic volume, on the mixture, which is defined as the smallest amount of material within which the quality of mixing is important. The scale of scrutiny fixes the scale or the sample size at which the mixture should be examined for the required quality of mixing. In the manufacture of pharmaceuticals, for example, the main requirement is that each tablet (or capsule) should contain the right amount of the active ingredients. The actual distribution of the ingredients within the tablet is unimportant. Therefore, the scale of scrutiny is the quantity of material making up one tablet. In the production of ceramics, on the other hand, the distribution of the particles within the body has a critical effect on the microstructure and properties, so the scale of scrutiny is often quite small. Homogeneity may be required over distances on the order of a few microns or less. Consider, for example, the mixing of fine ZrO_2 and Al_2O_3 particles for the production of ZrO_2-toughened Al_2O_3. The scale of scrutiny is on the order of the Al_2O_3 grain size because we require the ZrO_2 particles to be homogeneously distributed at the grain boundaries of the fabricated material. Failure to achieve mixing on this scale results in the formation of microstructural heterogeneities (Figure 1.24) that degrade the engineering properties of the material.

6.2.3.1 Statistical Methods

The composition and quality of a binary mixture can be assessed by removing a number of samples from different positions of the mixture and determining their composition, i.e., the proportion of one component in each mixture. The number of samples required to accurately assess mixture quality ranges from ~30 for a well-mixed system to more than 200 for a poorly mixed system [6]. Analysis of the mixture quality requires the application of statistical methods [10,11]. The statistics relevant to binary mixtures are summarized below.

Mixture composition: The true composition of a mixture is often not known, but an estimate can be found from the composition of the samples. For N samples, with the proportion of one component in the samples being $y_1, y_2, y_3, ..., y_N$, the estimate of the mixture composition is taken as the mean:

$$\bar{y} = \frac{1}{N} \sum_{i=1}^{N} y_i \tag{6.3}$$

As this estimate is based on limited information, it will not, in general, be equal to the true composition, μ. If the samples were taken from a random mixture, the proportions of one component would follow a normal distribution, and the true composition would be given by:

$$\mu = \bar{y} \pm \frac{ts}{\sqrt{N}} \tag{6.4}$$

where t is the value from Student's t-test for statistical significance (obtained from statistical tables), and s is the standard deviation of the sample composition. The value of t depends on the confidence level required. For example, at 95% confidence level, $t = 2.0$ for $N = 60$, so there is a 95% probability that the true mixture composition lies in the range $y \pm 0.258s$, or that 1 in 20 estimates of mixture variance would lie outside this range.

Standard deviation and variance: The true standard deviation σ and the true variance σ^2 of the composition of the mixture are quantitative measures of the quality of the mixture. A low standard deviation indicates a narrow spread in composition of the samples and, therefore, good mixing. Usually, σ^2 is not known, but an estimate s^2 is defined by:

$$s^2 = \frac{1}{N} \sum_{i=1}^{N} (y_i - \mu)^2 \tag{6.5}$$

if μ is known. If μ is not known, then s^2 is given by:

$$s^2 = \frac{1}{(N-1)} \sum_{i=1}^{N} (y_i - \bar{y})^2 \tag{6.6}$$

Theoretical limits of standard deviation: The true standard deviation of a random binary mixture is

$$\sigma_R = \left[\frac{f(1-f)}{n} \right]^{1/2} \tag{6.7}$$

where f and $(1-f)$ are the fractions of the two components in the mixture and n is the number of particles. Equation 6.7 indicates that if a random mixture is approached, the standard deviation of the composition of samples taken from the mixture decreases as n increases. Therefore, for a given mass of sample, the standard deviation decreases and the mixture quality increases with decreasing particle size. The true standard deviation for a completely segregated system is given by:

$$\sigma_o = [f(1-f)]^{1/2} \tag{6.8}$$

The values σ_o and σ_R give the upper and lower limits of the standard deviation for a mixture, and actual values lie between these two values.

Estimates of the standard deviation of sample composition, determined from an analysis of sets of samples taken from the mixture, will have a distribution of values. When more than 50 sets of samples are taken, the distribution of variance values can be assumed to be normal and the Student's t-test may be used. The best estimate of the variance is given by:

$$\sigma^2 = s^2 \pm [t \times E(s^2)] \tag{6.9}$$

where $E(s^2)$ is the standard error in the variance, estimated from:

$$E(s^2) = s^2(2/N)^{1/2} \tag{6.10}$$

The standard error decreases as $1/N^{1/2}$, so the precision increases as $N^{1/2}$.

When less than 50 sets of samples are taken, the variance distribution curve may not be normal and is likely to be a χ^2 (chi-squared) distribution. In this case, the limits of precision are not symmetrical, and the range of mixture variance is defined by upper and lower limits:

$$\text{Upper limit:} \quad \sigma_U^2 = \frac{s^2(N-1)}{\chi_{\alpha-1}^2} \tag{6.11a}$$

$$\text{Lower limit:} \quad \sigma_L^2 = \frac{s^2(N-1)}{\chi_{\alpha}^2} \tag{6.11b}$$

where α is the significance level. For example, for a 95% confidence range, $\alpha = 0.5(1 - 95/100) = 0.025$. The upper and lower χ^2 values, $\chi_{\alpha-1}^2$ and χ_{α}^2, for a given confidence level are found from χ^2 distribution values in statistical tables.

Mixing indices: A mixing index is often used to provide a measure of the degree of mixing. More than 40 mixing indices have been documented in the literature [6], but most are limited in their ability to discriminate between mixtures. Because of the random nature of mixing, most of the indices are based on statistical analysis and are expressed in different forms of the variance or the standard deviation. A commonly used index is the Lacey mixing index [9]:

$$M = \frac{\sigma_o^2 - \sigma^2}{\sigma_o^2 - \sigma_R^2} \tag{6.12}$$

The Lacey mixing index gives a measure of "mixing achieved" to "mixing possible." An index of 0.0 represents complete segregation and an index of 1.0 represents a completely random mixture. Practical values of this index, however, are found to lie in the range of 0.75 to 1.0, so the Lacey mixing index does not provide sufficient discrimination between mixtures. A useful mixing index [12] is

$$M = \frac{\log \sigma_o - \log \sigma}{\log \sigma_o - \log \sigma_R} \tag{6.13}$$

This index gives better discrimination between mixtures.

6.2.3.2 Measurement Techniques

The samples taken from a mixture are characterized to measure some chemical or physical property of the mixture (e.g., composition) using a direct or indirect approach. In the *direct approach*, characterization is performed after the mixing operation or after a later processing step that does not significantly alter the mixing homogeneity. The direct approach requires the selection of the sample size dictated by the scale of scrutiny. In the *indirect approach*, assessment is performed after a later processing step by determining an obvious effect on the microstructure or properties of an intermediate or final product. In the processing of ceramics, it is often difficult and time consuming to assess mixture quality using a direct, quantitative examination because the scale of scrutiny is often quite small. Therefore, the indirect approach is often used. The assessment of mixture quality can be either quantitative or qualitative in both approaches, but the qualitative methods find greater use in the indirect approach. Often, the qualitative methods form part of a quality control system for the fabrication process.

In either approach, the measurement techniques must be able to resolve some chemical or physical characteristic of the mixture on a scale equal to or smaller than the scale of scrutiny. Chemical techniques are often used in the direct approach. The methods include many of the techniques described in Chapter 3 for characterizing the chemical composition of ceramic powders. Physical techniques are most useful when the individual components have physically distinct characteristics such as size, morphology, or color. In this case, microscopy combined with image analysis is a useful technique.

The indirect approach is applicable to many systems because mixture quality often has a significant effect on the properties and microstructure of an intermediate or final product. Thus, the change in relevant properties can be followed as a function of mixing time or another parameter that influences mixing quality. In the preparation of single-phase powders by the calcination of mixed powders, x-ray diffraction of the calcined powder provides a useful method for quantitatively or qualitatively analyzing the effectiveness of the mixing process [13]. The rheology of suspensions used in slip casting and tape casting commonly decrease with increasing quality of mixing between the ceramic powder and organic additives (see Chapter 4). Microscopy of the cross-sections of the green body [14] or the sintered body (see Figure 1.24) often provides a simple, yet effective, proof test of mixing as well as other processing steps.

6.2.4 MIXING TECHNOLOGY

Mixers can be classified according to their dominant mechanisms of mixing, their design, or their operation [4–6]. As the mechanism of mixing has a more direct impact on the mixture quality, a classification based on the dominant mixing mechanism will be used here.

1. *Diffusive mixers:* Any mixer that relies on a tumbling motion, i.e., one in which the particles are in a rotating container, is mainly using diffusive mixing. This class includes rotating cylinders, cubes, double cones, and V-blenders (Figure 6.3). For free-flowing particles with different sizes, there will be a large amount of segregation in the mixer, which will limit the quality of mixing. Vanes or impellers spinning at high speed along the axis of rotation are sometimes added to improve the de-agglomeration of the powder. Baffles may be fitted inside the mixer to reduce segregation, but they have little effect.

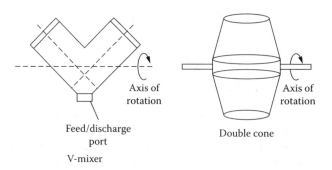

FIGURE 6.3 Examples of tumbling mixers: V mixer and double cone mixer.

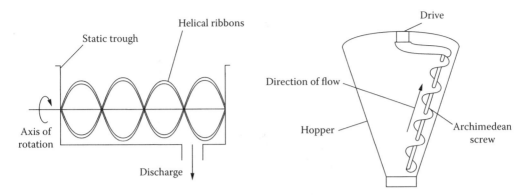

FIGURE 6.4 Examples of common convective mixers: the ribbon blender and the orbiting screw mixer (Nautamixer).

2. *Convective mixers:* Mixers in which circulation patterns are set up within a static container by rotating blades or paddles use mainly convective mixing, although this is accompanied by some diffusive and shear mixing. A convective mixer should be chosen in cases in which segregation is known to be a problem. The most common convective mixers are the *ribbon blender* and the *orbiting screw mixer* (Figure 6.4). In a ribbon mixer, helical blades or ribbons rotate on a horizontal axis inside a static cylinder or trough. An orbiting screw mixer, also referred to as a Nautamixer, consists of a screw that rotates on its own axis while simultaneously orbiting the interior of a conical vessel (or hopper). Mixing occurs as the screw lifts material from the base of the hopper to the top. *Pug mills* consist of troughs containing one or two rotating shafts equipped with heavy paddles. They are often used for mixing and de-airing plastic clay or ceramic–polymer mixtures to be used in extrusion or injection molding. The rotating paddles push material toward one end of the mixer, where it flows through a coarse extrusion nozzle or is fed directly into a screw extruder. *Impeller mixers*, consisting of a cylindrical pan in which one or more impellers rotate, are employed in a variety of configurations. They provide high shear rates that are useful for de-agglomeration but can have dead zones where only limited mixing is achieved (Figure 6.5). Dead zones are avoided in *planetary mixers*, in which the impellers are rotated in a planetary fashion. Operating the impeller blades at high speeds can facilitate shear mixing and de-agglomeration, but there is the potential for contamination because of the abrasive nature of many ceramic powders. Highly viscous slurries and pastes can often be mixed using impeller or planetary mixers.

3. *Shear Mixers:* High shear stresses are created by devices such as those used in powder comminution. The emphasis in mixing is normally on breaking down agglomerates of fine, cohesive powders rather than comminution. *Ball mills* are widely used in ceramic processing because of their cost and ease of use (see Chapter 2). They are well suited to ceramic powder mixing because the milling media provide impact and shear forces needed to break down agglomerates of fine particles. The decrease in particle size brought about by comminution also serves to increase the quality of mixing by decreasing segregation. However, contamination from the milling media can be a problem. Dry powder mixing or wet milling of slurries can be performed, but dry mixing should not be used if the powder forms compacted layers on the milling media or on the mill lining. *Roll mills* generate a large amount of shear by passing the mixture through two narrowly separated rollers, often running at different speeds to increase the shear. The material being mixed, such as a paste, must be viscous enough to adhere to the rolls. In *muller mixers*, shear forces are generated by the sliding, twisting, and grinding action of rotating wheels in a fixed or moving pan. Ploughs or blades direct the mixture under the wheels.

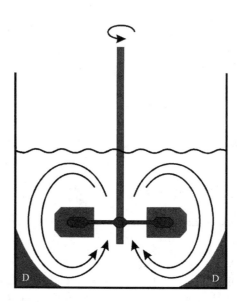

FIGURE 6.5 Schematic of an impeller mixer illustrating dead zones, D, where no active mixing occurs.

The preliminary stage of mixing plastic clay-based products is often carried out in a muller mixer. *Screw extruders* are continuous mixers commonly used for mixing plastic clay or ceramic–polymer feed material, previously mixed and de-aired in a pug mill, for extrusion or injection molding. Twin-screw extruders find more use than single-screw extruders in ceramic processing. Mixing is achieved by shear forces generated between the screws and mixer wall or between the screws themselves. Stiff pastes and plastic ceramic mixtures can be mixed in high-shear *Z-blade mixers* or *sigma blenders*. These mixers consist of two counterrotating kneading arms having a Z or sigma shape in a trough shaped to match the rotating sweep of the arms.

When selecting a mixer, the process requirements and the objectives of the mixing operation should first be established. If only a crude mixture is required, then economic considerations will generally govern the selection. On the other hand, if a homogeneous mixture is required, the characteristics of the solids to be mixed, the performance characteristics of the mixer, the operating conditions, the place of the mixer in the process, and the overall process objectives will affect the choice of mixer. If the components segregate, attempts should be made to adjust the particle size distributions of the components as close together as possible or to add a small amount of liquid. If none of these steps can be taken to reduce segregation, it is important to select a mixer that relies predominantly on convective mixing, such as a ribbon blender or an orbiting screw mixer, and to design the rest of the process to minimize the amount of segregation occurring in subsequent handling of the mixture. For mixing cohesive powders, a mixer that provides sufficient shear forces to break up agglomerates should be used. If size reduction of the particles cannot be tolerated, such as when mixing the components to make porous filters with a specified pore size or the mixing of spray-dried granules, ball mills and mixers with high-speed impellers should be excluded. In general, however, owing to the lack of quantitative information in the literature, a final choice will be made only after a test program or previous experience with a similar mixing operation.

6.3 PACKING OF PARTICLES

The packing of particles is treated in textbooks by German [15] and Cumberland and Crawford [16]. Particle packing is commonly divided into two types: (1) regular (or ordered) packing and

(2) random packing. The commonly used ceramic-forming methods produce random-packing arrangements, but regular packing is typically the first structure encountered in models, partly because of their similarity to crystalline atomic structures. Several parameters can be used to characterize the packing arrangement but two of the most widely used are (1) the *packing density* (also referred to as the packing fraction or the fractional solids content), defined as

$$Packing\ density = \frac{volume\ of\ solids}{total\ volume\ of\ the\ arrangement\ (solids + voids)} \tag{6.14}$$

and (2) the *coordination number*, which is the number of particles in contact with any given particle. The packing density is an easily measured parameter that provides much insight about the behavior of a powder.

6.3.1 Regular Packing of Monosized Spheres

The reader would be familiar with the packing of atoms in crystalline solids that produce regular, repeating, three-dimensional patterns such as the simple cubic, body-centered cubic, face-centered cubic, and hexagonal close-packed structures. The packing density and coordination number of these crystal structures for a pure metal are listed in Table 6.1.

In order to build up a three-dimensional packing pattern of particles, we can begin, conceptually, by (1) packing spheres in two dimensions to form layers and then (2) stacking the layers on top of one another. Two types of layers are shown in Figure 6.6, where the angle of intersection between the rows has limiting values of 90° (referred to as a square layer) and 60° (simple rhombic or triangular layer). Although other types of layers that have angles of intersection between these two values are possible, only the square layer and the simple rhombic layer will be considered here. There are three geometrically simple ways of stacking each type of layer on top of one another, giving rise to six packing arrangements altogether. However, examination of the arrangements will show that, neglecting the difference in orientation in space, two of the ways of stacking the square

TABLE 6.1
Packing Density and Coordination Number of Some Common Crystal Structures of a Pure Metal

Crystal Structure	Packing Density	Coordination Number
Simple cubic	0.524	6
Body-centered cubic	0.680	8
Face-centered cubic	0.740	12
Hexagonal close packed	0.740	12

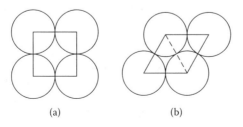

(a)   (b)

FIGURE 6.6 Two types of layers for the regular packing of monosized spheres: (a) square and (b) rhombic or triangular.

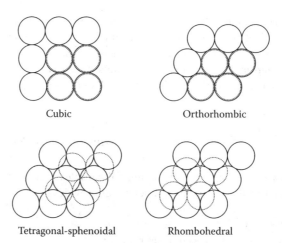

Cubic   Orthorhombic

Tetragonal-sphenoidal   Rhombohedral

FIGURE 6.7 The four packing arrangements produced by stacking square and triangular layers of monosized spheres.

TABLE 6.2
Packing Density and Coordination Number for
Regular Packing of Monosize Spheres

Packing Arrangement	Packing Density	Coordination Number
Cubic	0.524	6
Orthorhombic	0.605	8
Tetragonal-sphenoidal	0.698	10
Rhombohedral	0.740	12

layers are identical to two of the ways of stacking the simple rhombic layers. Therefore, there are only four different regular packing arrangements as shown in Figure 6.7. The packing densities and coordination numbers of these arrangements are summarized in Table 6.2.

The rhombohedral packing arrangement, which has the highest packing density, is the most stable packing arrangement. Even for monosized spherical powders, such dense packing arrangements have been achieved over only very small regions (called *domains*) of the green body when rather special consolidation procedures have been used. An example is the slow sedimentation of monosized SiO_2 particles from a stable suspension as shown earlier in Figure 4.34. The domains are separated from one another by boundaries of disorder much like the granular microstructure of polycrystallne materials. Figure 6.8 shows one problem that can arise when the green body is sintered: crack-like voids open at the domain boundaries [17]. The commonly used ceramic-forming methods produce more random-packing arrangements in the green body.

6.3.2 RANDOM PACKING OF PARTICLES

Two different states of random packing have been distinguished. If the particles are poured into a container that is then vibrated to settle the particles, the resulting packing arrangement reaches a state of highest packing density (or minimum porosity) referred to as *dense random packing*. On the other hand, if the particles are simply poured into the container so that they are not allowed to rearrange and settle into as favorable a position as possible, the resulting packing arrangement is referred to as *loose random packing*. An infinite number of packing arrangements may exist between

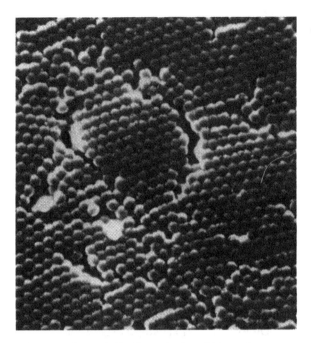

FIGURE 6.8 Partial densification of a periodically packed, multilayered arrangement of polymer spheres. Note the opening displacements at the domain boundaries. (From Lange, F.F., Powder processing science and technology for increased reliability, *J. Am. Ceram. Soc.*, 72, 3, 1989. With permission.)

these two limits. The packing densities of the particulate system after pouring and after being vibrated are commonly referred to as the *poured density* and the *tap density*, respectively.

6.3.2.1 Monosized Particles

Dense random packing of monosized spheres has been studied experimentally by shaking hard spheres in a container. The upper limit of the packing density consistently ranges from 0.635 to 0.640 [18]. Computer simulations give a value of 0.637 [19]. The maximum packing density for random packing of monosized spheres is predicted to be independent of the sphere size, and this prediction has been verified experimentally. For loose random packing of monosize spheres, theoretical simulations as well as experiments give values in the range of 0.57 to 0.61 for the packing density.

In the case of dense random packing of monosize spheres, calculations show that fluctuations in the packing density become weak beyond a distance of three sphere diameters from the center of any given sphere. For density fluctuations existing over such a small-scale, uniform densification may be achieved during sintering of the green body. Therefore, the production of regular, crystal-like particle packing, achievable at present only over very small domains, may, after all, be unnecessary from the point of view of fabrication.

Powders used in the industrial production of ceramics very rarely have spherical particles. The surfaces of the particles are also rarely smooth. Particles with rough surface textures or shapes suffer from enhanced agglomeration because of higher interparticle friction, and the packing density decreases as the particle shape departs from that of a sphere. Figure 6.9 shows the packing density for various irregular particle shapes. Spherical particles are normally desirable when a high packing density is required. However, the use of nonspherical particles does not always lead to a reduction in the packing density if the particles have a regular geometry. The highest packing density and most isotropic structures are obtained with spheres and with particles having simple, equiaxial

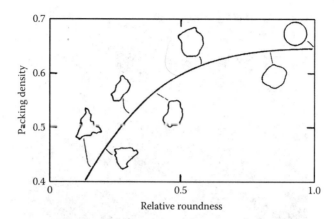

FIGURE 6.9 Packing density vs. relative roundness for randomly packed monosized particles. (From German, R.M., *Particle Packing Characteristics*, Metal Powder Industries Federation, Princeton, NJ, 1989. With permission.)

TABLE 6.3
Dense Random Packing Density for Various Particle Shapes

Particle Shape	Aspect Ratio	Packing Density
Sphere	1	0.64
Cube	1	0.75
Rectangle	2:5:10	0.51
Plate	1:4:4	0.67
Plate	1:8:8	0.59
Cylinder	5	0.52
Cylinder	15	0.28
Cylinder	60	0.09
Disk	0.5	0.63
Tetrahedron	1	0.5

Source: From German, R.M., *Particle Packing Characteristics*, Metal Powder Industries Federation, Princeton, NJ, 1989. With permission.

shapes (e.g., cubes). Anisotropic particles can be packed to high packing density if they are ordered; however, in random packing, the packing density can be quite low. Table 6.3 provides a comparison of the dense random-packing density for various particle shapes.

6.3.2.2 Bimodal Mixtures of Spheres

The packing density of an arrangement of spheres in dense random packing can be increased by filling the interstitial holes with spheres that are smaller than those of the original structure (Figure 6.10a). For this type of random packing of a binary mixture of spheres, the packing density is a function of (1) the ratio of the sphere diameters and (2) the fraction of the large (or small) spheres in the mixture. By filling the interstitial holes with a large number of very fine spheres, we can maximize the packing density of the binary mixture. Starting with an aggregate of large (coarse) spheres in dense random packing, as we add fine spheres the packing density of the mixture increases along the line CR as shown in Figure 6.11. A stage will be reached when the interstitial holes

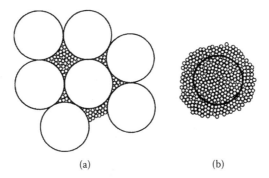

FIGURE 6.10 Increase in packing density achieved by (a) filling the interstices between large spheres with small spheres, and (b) replacing small spheres and their interstitial porosity with large spheres.

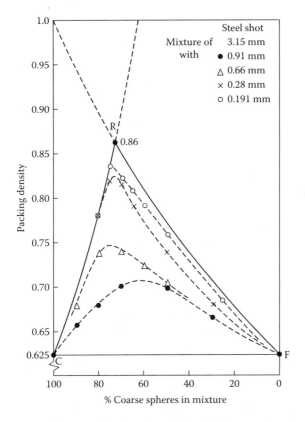

FIGURE 6.11 Binary packing of spheres showing the packing density as a function of the composition of the mixture. The curve CRF represents the theoretical predictions for dense random packing when the ratio of the large-sphere diameter to the small-sphere diameter approaches infinity. The data of McGeary for the mechanical packing of steel shot are also shown. (From McGeary, R.K., Mechanical packing of spherical particles, *J. Am. Ceram. Soc.*, 44, 513, 1961.)

between the large spheres are filled with fine spheres in dense random packing. Further additions of fine spheres will only serve to expand the arrangement of large spheres, leading to a reduction in the packing density. Assuming a packing density of 0.637 for dense random packing, the volume fraction of interstitial holes in the original aggregate of large spheres is (1 – 0.637) = 0.363. At the maximum packing density of the binary mixture, the interstitial holes are filled with a large number of fine spheres in dense random packing. The maximum packing density is therefore (0.637 + 0.363

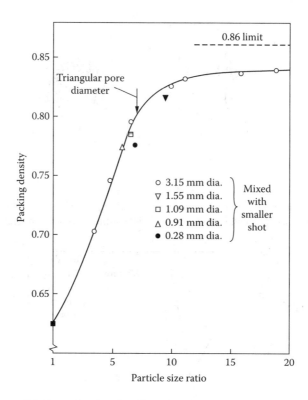

FIGURE 6.12 The data of McGeary showing the effect of particle size ratio on the maximum packing density for bimodal powder mixtures. (From McGeary, R.K., Mechanical packing of spherical particles, *J. Am. Ceram. Soc.*, 44, 513, 1961.)

× 0.637) = 0.868. The fractional volumes occupied by the large spheres and fine spheres are 0.637 and (0.868 − 0.637), respectively. The fraction (by weight or volume) of large spheres in the binary mixture is therefore 0.637/0.868 = 0.733.

In an alternative approach, we can increase the packing density of an aggregate of fine spheres in dense random packing by replacing some of them and their interstitial holes by large spheres (Figure 6.10b). In this case, the packing density of the mixture will increase along the line FR as shown in Figure 6.11. The intersection at R of the two curves CR and FR represents the state of optimum packing. Figure 6.11 also shows the experimental data of McGeary [20] for binary mixtures of spherical steel particles with different size ratios. (McGeary assumed a packing density of 0.625 for dense random packing of monosized spheres.) As the ratio of the diameters of the large sphere to the small sphere increases, the data move closer to the theoretical curve. This is illustrated more clearly in Figure 6.12, which shows McGeary's data for dense random packing of binary mixtures of spheres. The packing density increases as the size ratio increases to ~15 but is relatively unchanged at higher size ratios. A change of behavior is apparent for a size ratio of ~7, which corresponds to the size of a small particle just filling the triangular pore between the large particles.

The packing of binary mixtures of spheres is also commonly represented in terms of the apparent volume (i.e., total volume of the solid phase and porosity) occupied by unit volume of solid [21]. The apparent volume is defined as:

$$V_a = \frac{1}{1-P} \qquad (6.15)$$

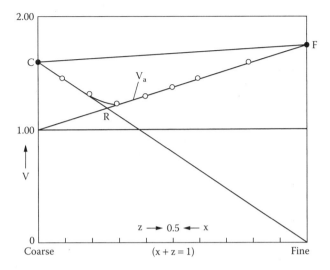

FIGURE 6.13 Binary packing of spheres plotted in terms of the apparent volume occupied by unit volume of the solid phase. In this representation, straight lines are obtained for the theoretical case CRF when the ratio of the large-sphere diameter to the small-sphere diameter approaches infinity. The data of Westman and Hugill for a diameter ratio of 50 are also shown. (From Westman, A.E.R. and Hugill, H.R., The packing of particles, *J. Am. Ceram. Soc.*, 13, 767, 1930.)

where P is the fractional volume of the voids (i.e., the porosity). As shown in Figure 6.13, the line CRF represents the theoretical curve for the packing of a binary mixture in which the size of the large spheres is very much greater than that of the small spheres.

Instead of filling the interstitial holes between the large spheres with a large number of very small spheres, another approach is to insert into each hole a single sphere with the largest possible diameter that would fit into the hole. For an aggregate of monosized spheres in dense random packing, computer simulations reveal that, with this approach, the maximum packing density of the binary mixture is 0.763 [19]. This value is smaller than the maximum packing density (0.868) obtained earlier by filling the interstices with a large number of fine spheres, but it may represent a more realistic upper limit to the packing of real powder mixtures.

6.3.2.3 Binary Mixtures of Nonspherical Particles

The mixing of different sizes of nonspherical particles also leads to an increase in the packing density, but this packing density is generally lower than for spherical particles. The greater the surface roughness, shape irregularity, and the aspect ratio of the particles, the lower the packing density. As in the case of spherical particles, the packing density of a mixture of nonspherical particles increases with increasing size ratio of the two powders and is dependent on the composition (i.e., the fraction of large to small particles). The composition that gives the maximum packing density is sensitive to the particle shape.

The packing of model mixtures consisting of cylindrical rods and spherical particles has been studied experimentally by Milewski [22]. The results provide useful insight into the key factors controlling the packing of short single-crystal fibers or whiskers (e.g., SiC and Si_3N_4) and powders (e.g., Al_2O_3) for the production of ceramic matrix composites. Packing uniformity and a high enough density remain our basic requirements for the green composite. We require a uniform distribution of the whiskers and the elimination of large voids in the composite.

As shown in Figure 6.14, experiments with cylindrical rods indicate that short fibers pack to very low densities, with the packing density decreasing with increasing aspect ratio. Whiskers with high aspect ratios (>50–100) also tend to tangle and form bundles or loose clumps, leading to poor

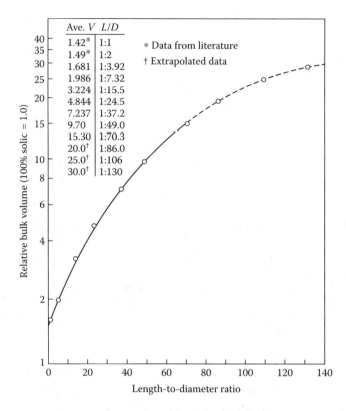

FIGURE 6.14 Packing curve for fibers with varied length-to-diameter (L/D) ratios. (From Milewski, J.V., Efficient use of whiskers in the reinforcement of ceramics, *Adv. Ceram. Mater.*, 1, 1, 1986. With permission.)

distribution of the whiskers and large voids spanned by whiskers in the green composite. Whiskers with high aspect ratios, therefore, should not be used in the production of ceramic composites. The results of Milewski for the packing of cylindrical rods and spherical particles are shown in Table 6.4. The parameter R represents the ratio of the diameter of the particle to the diameter of the rod. Efficient packing of the mixture is promoted by low volume fraction of whiskers that have low aspect ratios and are thin compared to the particles of the matrix powder.

For whisker-reinforced ceramic composites, theoretical models and experimental investigations indicate that very little enhancement in the mechanical properties is achieved for whisker aspect ratios above ~15 to 20. Whiskers with aspect ratios less than ~20 to 30 flow fairly easily and show behavior close to that of a powder. A suitable whisker aspect ratio for the production of ceramic composites is therefore ~15 to 20. Most commercial whiskers, however, contain a large fraction with aspect ratios greater than this value. Normally, ball milling is used to reduce the aspect ratio. For closer control of the aspect ratio, the ball-milled whiskers can be fractionated to produce fractions with the desired range. Good mixing of the whiskers and the matrix powder is achieved by ball milling for several hours or by colloidal dispersion techniques.

6.3.2.4 Ternary and Multiple Mixtures

The packing density of binary mixtures of spheres can be increased further by going to ternary mixtures, quaternary mixtures, and so on. For example, if each interstitial hole in the binary mixture (packing density = 0.868) is filled with a large number of very fine spheres in dense random packing, the maximum packing density becomes 0.952. Using the same approach, the maximum packing density of quaternary mixtures is 0.983. Following this packing scheme, McGeary [20] experimen-

TABLE 6.4
Experimental Packing Densities at 25%, 50%, and 75% Fiber Loading for Fiber–Sphere Packing

Fiber L/D	Percentage of Fibers	R value									
		0	0.11	0.45	0.94	1.95	3.71	6.96	14.30	17.40	∞
3.91	25	68.5	68.5	65.4	61.7	61.0	64.5	70.0	74.6	76.4	82.0
	50	76.4	74.6	67.2	61.7	60.2	64.1	67.5	72.5	74.5	75.7
	75	78.2	69.5	64.5	61.0	59.5	62.5	64.4	66.7	67.2	67.1
7.31	25	68.5	68.5	64.5	61.0	58.5	59.9	64.5	73.5	74.6	80.6
	50	76.4	71.4	67.5	58.8	55.5	56.6	58.8	65.4	67.1	67.1
	75	66.3	61.7	60.0	55.0	52.8	53.5	54.6	57.2	58.2	57.4
15.52	25	68.5	66.7	63.7	59.9	54.6	50.3	50.5	54.1	57.5	65.0
	50	61.7	55.6	51.8	50.7	45.5	42.0	42.4	44.3	44.3	48.1
	75	41.0	40.4	37.9	38.2	37.3	35.7	35.5	36.0	36.8	38.2
24.50	25	68.5	66.5[a]	61.5[a]	55.5[a]	47.5	45.5	40.2	42.7	44.7	50.5
	50	40.0	39.0[a]	38.0[a]	36.0[a]	34.0[a]	32.7	30.3	31.8	31.8	33.5
	75	26.4	26.3[a]	26.2[a]	25.8[a]	25.5[a]	25.2	24.3	25.0	25.6	26.2
37.10	25	50.0	48.0[a]	45.0[a]	42.0[a]	39.4	37.7	33.8	33.1	39.2	41.3
	50	25.7	—	—	—	—	—	22.6	22.6	22.6	25.6
	75	—	—	—	—	—	—	—	—	—	—

[a] Estimated values (extrapolated data).

Source: From Milewski, J.V., Efficient use of whiskers in the reinforcement of ceramics, *Adv. Ceram. Mater.*, 1, 1, 1986. With permission.

tally achieved packing densities of 0.90 for a ternary mixture and 0.95 for quaternary mixture of steel spheres that were compacted by vibration.

In practice, little is gained beyond the use of ternary mixtures because the finer particles do not locate into their ideal positions to maximize the packing density. Additional practical problems may arise as the number of size classes in the mixture increases. A particle size ratio of at least 7 is required for optimum packing, and for a ternary mixture of fine, medium and large particles, assuming that the fine particles are 1 μm in size, the medium and large particles will be 7 and 49 μm, respectively. The ability to produce some advanced ceramic powders with such widely different sizes is limited.

Although less severe in ternary or quaternary mixtures than in powders with a wide particle size distribution, the problem of the packing homogeneity of the green body still needs to be considered seriously. Two requirements must be satisfied for this approach to be successful. First, uniform mixing of the powder fractions by mechanical or colloidal methods must be achieved. Second, the mixture must be consolidated to produce homogeneous packing. The objective is a green body in which the small pores are fairly uniformly spaced and the large voids are eliminated.

6.3.2.5 Continuous Particle Size Distributions

Most powders used for the fabrication of ceramics have a continuous distribution of particle sizes between some minimum and maximum size. For mixtures with discrete sizes, we found that as long as the particles were very different in size, the packing density increased as the number of components in the mixture increased. Extending this concept to continuous distributions, a wide particle size distribution gives a higher packing density than a narrow particle size distribution.

The development of particle size distributions with optimum packing density has received considerable attention, particularly in the traditional ceramics sector. In porcelains, for example,

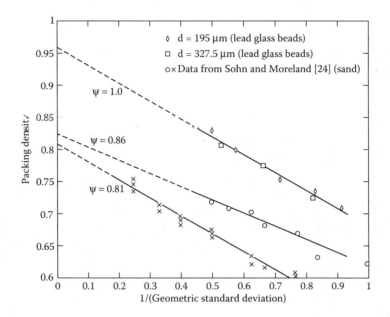

FIGURE 6.15 Packing density vs. the reciprocal of the standard deviation for particles with a log-normal size distribution. The parameter ψ is a measure of the sphericity of the particles and is equal to the reciprocal of the shape factor. (From Wakeman, R.J., Packing densities of particles with log-normal size distributions, *Powder Technol.*, 11, 297, 1975. With permission.)

clay, quartz, and feldspar, each having a continuous size distribution, are mixed to produce a continuous distribution that packs more densely. Figure 6.15 shows the results for the packing of particles with a log-normal distribution of sizes [23,24]. The packing density increases as the standard deviation of the distribution S increases (i.e., as the spread in the distribution of sizes increases) and reaches fairly high values for a wide distribution of particle sizes. The data in Figure 6.15 can be approximated by an equation of the form:

$$Packing\ density = \alpha - \frac{\beta}{S} \tag{6.16}$$

where α and β are constants for a given powder. For the spherical lead glass beads (sphericity ψ = 1), $\alpha = 0.96$, which is equal to the value predicted by Bierwagen and Saunders [25]. There is also a particle shape effect, illustrated by the data for sand, showing that the packing density decreases for the more irregular particles.

Furnas [26] considered the use of discrete particle fractions (produced by sieving) that differ in size by a constant factor of 1.414 and predicted optimum packing density when each fraction has 1.10 times the amount of powder as lower-sized sieved fraction. Although based on the use of discrete particle size fractions, the work of Furnas provides a first estimate of the shape of the particle size distribution curve that gives the optimum packing. Andreasen and Andersen [27] developed an approach to particle packing based on the use continuous particle size distributions in which optimum packing occurs when the particle size distribution can be described by a power law equation, commonly known as the *Andreasen equation*:

$$F_M(x) = \left(\frac{x}{x_L}\right)^n \tag{6.17}$$

where $F_M(x)$ is the cumulative mass fraction of particles finer than a size x, x_L is the largest particle size in the distribution, and n is an empirical constant used to fit the experimental particle size distribution. From experimental studies of many particle size distributions, Andreasen concluded that the highest packing density is achieved when n is between 1/3 and 1/2.

The Andreasen equation assumes that all particle sizes below x_L exist, including infinitely small particles. Funk and Dinger [28] modified the equation to account for the more realistic case of a distribution with a finite minimum particle size x_S:

$$F_M(x) = \frac{x^n - x_S^n}{x_L^n - x_S^n} \qquad (6.18)$$

where the exponent n is the same as that in the Andreasen equation.

Two continuous particle size distributions can be mixed to improve the packing density. Generally, the mean particle size of the two distributions should be very different, and the particle size distribution of the smaller powder should be wider than that of the larger powder. For a wide particle size distribution in which the packing density is already high, little benefit is achieved by mixing with another distribution.

In view of the high packing densities that can be achieved, it seems that we may be better off as far as the green body is concerned, with powders having a wide distribution of particle sizes. However, the packing density by itself is a misleading parameter for predicting the densification behavior and microstructural evolution during sintering. A more important consideration for advanced ceramics is the packing homogeneity or, equivalently, the spatial scale over which density fluctuations occur in the green body. Computer simulations [29] show that as the width of the particle size distribution increases, the scale over which density fluctuations occur also increases (Figure 6.16). As discussed later in the book, it is this increasing scale of density fluctuations that causes many of the problems in the sintering stage.

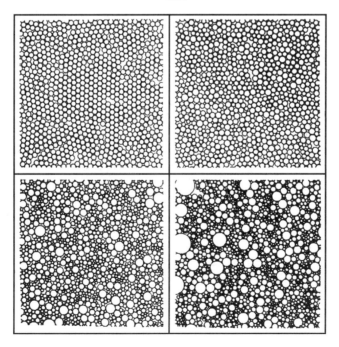

FIGURE 6.16 Four typical packings of multisized circles. (From Burk, R. and Apte, P., A packing scheme for real size distributions, *Am. Ceram. Soc. Bull.*, 66, 1390, 1987. With permission.)

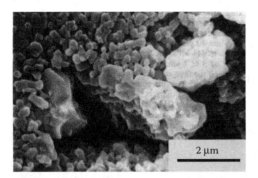 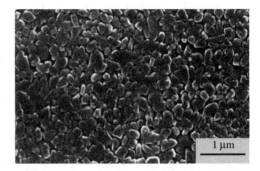

FIGURE 6.17 Scanning electron micrographs of (a) an unclassified alumina powder and (b) the powder after classification by sedimentation in water. (From Roosen, A. and Bowen, H.K., Influence of various consolidation techniques on the green microstructure and sintering behavior of alumina powders, *J. Am. Ceram. Soc.*, 71, 970, 1988. With permission.)

Commercial powders that have a wide distribution of particle sizes can be classified to provide fractions with the desired size range. One approach is to use the colloidal techniques described in Chapter 4 to disperse the powder and to remove hard agglomerates and large particles by sedimentation. The supernatant can then be decanted and fractionated into various size fractions. Although in many cases this approach may not be economical for industrial production, Figure 6.17 shows that significant improvement in the packing homogeneity can be achieved [30]. The unclassified powder (Figure 6.17a) contains large agglomerates that produce regions with very nonuniform packing, but the classified powder (Figure 6.17b) shows fairly homogeneous packing with a high packing density.

6.4 CONCLUDING REMARKS

Mixing and packing of particles are important because chemical and microstructural heterogeneities in the green body limit the ability to control the microstructure during sintering. Homogeneous mixing of the components is the desired goal of the mixing step. The quality of mixing depends on the difference in properties of the powders to be mixed and on the mixer. It can be assessed directly by taking samples and analyzing the data of a mixture property, such as composition, using statistical methods or indirectly by determining an obvious effect on the microstructure or properties of an intermediate or final product. Mixtures in ceramic processing cover a wide range, from powders to pastes or plastic bodies to free-flowing slurries. Selecting the most suitable mixer is a difficult task. Often the final choice is made with a test program or previous experience with a similar mixing operation. Homogeneous packing and high packing density are the common processing goals for ceramic green bodies. The packing density of particles with a narrow size distribution can be improved by using mixtures of particles in which a specific portion of finer particles pack in the interstices of the coarser particles, or by using a continuous particle size distribution such as the Andreasen distribution. Whether a narrow distribution or a wide distribution of particles sizes is used, the objective is a green body in which the small pores are fairly uniformly spaced and the large voids are eliminated.

PROBLEMS

6.1

Derive Equation 6.2.

6.2

When a stream of free-flowing particles with a wide distribution of sizes is charged into the center of a tank from a conveyer belt, the coarser particles tend to segregate at a greater radius, nearer the wall of the tank. Explain.

6.3

A random mixture consists of two components A and B in proportions of 30% and 70% by weight, respectively. The particles are spherical, and A and B have particle densities of 0.5 and 0.7 g/cm^3, respectively. The cumulative undersize mass distributions of the two components are given in the following table:

Size x (μm)	2057	1676	1405	1204	1003	853	699	599	500	422
$F_A(x)$	1.00	0.80	0.50	0.32	0.19	0.12	0.07	0.04	0.02	0
$F_B(x)$	—	—	1.00	0.88	0.68	0.44	0.21	0.08	0	—

If samples of 5 g are withdrawn from the mixture, what is the expected value for the standard deviation of the composition of the samples? (From Reference 10.)

6.4

Sixteen samples are removed from a binary mixture and the percentage proportions of one component by mass are:

$$41, 37, 41, 39, 45, 37, 39, 40, 41, 43, 40, 38, 39, 37, 43, 40$$

Determine the upper and lower 95% and 90% confidence limits for the standard deviation of the mixture. (From Reference 10.)

6.5

The concentrations (in wt%) of a component in 10 samples taken at random from a mixture after 10 min of mixing are: 20.2, 21.1, 21.0, 16.4, 20.3, 17.1, 22.0, 20.4, 23.4, and 20.1. After 30 min of mixing, the concentrations are 19.5, 21.0, 23.0, 18.6, 20.1, 18.2, 23.0, 19.1, 21.2, and 19.3. Calculate the standard deviation in each case. Did the additional 20 min of mixing improve the quality of mixing? (Use a 95% confidence limit.)

6.6

Find the radius r of a spherical particle that would fit precisely in the interstices of an orthorhombic packing arrangement of spheres of radius a. If all the interstices in an orthorhombic packing of spherical Al_2O_3 particles are occupied by unstabilized ZrO_2 spheres that fit precisely in the interstices, determine the composition (by weight) and the packing density of the bimodal arrangement.

6.7

Show that the representation of the random packing of binary mixtures of spheres in terms of the packing density (Figure 6.11) is equivalent to the representation in terms of the apparent volume (Figure 6.13). Explain why the CR line in Figure 6.13 extrapolates to zero, whereas the FR line extrapolates to 1.0.

6.8

For ternary packing of spheres, the maximum packing density of 0.952 is obtained by assuming that the fine spheres enter the interstices between the coarse and medium spheres. Can a similar maximum packing density be obtained by replacing some of the medium spheres with fine spheres? Explain.

6.9

Compare the Andreasen and Dinger–Funk distributions when $x_L = 100$ μm, $x_S = 0.5$ μm, and $n = 0.4$.

REFERENCES

1. Uhl, V.W. and Gray, J.B., Eds., *Mixing: Theory and Practice*, Vol. III, Academic Press, New York, 1986.
2. Harnby, N., Edwards, M.F., and Nienow, A.W., Eds., *Mixing in the Process Industries*, 2nd ed., Butterworth-Heinemann, London, 1992.
3. Kaye, B.H., *Powder Mixing*, Chapman and Hall, London, 1996.
4. Perry, R.H., Green, D.W., and Maloney, J.O., *Perry's Chemical Engineers' Handbook*, 6th ed., McGraw-Hill, New York, 1984.
5. Fan, L.T., Chen, Y.M., and Lai, F.S., Recent developments in solid mixing, *Powder Technol.*, 61, 255, 1990.
6. Poux, M., Fayolle, P., Bertrand, J., Bridoux, D., and Bousquet, J., Powder mixing: some practical rules applied to agitated systems, *Powder Technol.*, 68, 213, 1991.
7. Messer, P.F., Batching and mixing, in *Engineered Materials Handbook*, Vol. 4: Ceramics and Glasses, ASM International, Materials Park, OH, 1991, p. 95.
8. Hogg, R., Grinding and mixing of non-metallic powders, *Am. Ceram. Soc. Bull.*, 60, 206, 1981.
9. Lacey, P.M.C., Developments in the theory of particulate mixing, *J. Appl. Chem.*, 4, 257, 1954.
10. Williams, J.C., Mixing and segregation of powders, in *Principles of Powder Technology*, Rhodes, M.J., Ed., John Wiley & Sons, New York, 1990, chap. 4.
11. Rhodes, M.J., *Introduction to Particle Technology*, John Wiley & Sons, New York, 1998, chap. 9.
12. Ashton, M.D. and Valentin, F.H., The mixing of powders and particles in industrial mixers, *Trans. Inst. Chem. Eng.*, 44, T166, 1966.
13. Rossi, R.C., Fulrath, R.M., and Fuerstenau, D.W., Quantitative analysis of the mixing of fine powders, *Am. Ceram. Soc. Bulletin*, 49, 289, 1970.
14. Hunt, K.N., Evans, J.R.G., and Woodthorpe, J., The influence of mixing route on the properties of injection moulding blends, *Br. Ceram. Soc. Trans. J.*, 87, 17, 1988.
15. German, R.M., *Particle Packing Characteristics*, Metal Powder Industries Federation, Princeton, NJ, 1989.
16. Cumberland, D.J. and Crawford, R.J., *The Packing of Particles*, Elsevier, New York, 1987.
17. Lange, F.F., Powder processing science and technology for increased reliability, *J. Am. Ceram. Soc.*, 72, 3, 1989.
18. Scott, G.D., Packing of spheres, *Nature*, 188, 908, 1960.
19. Frost, H.J. and Raj, R., Limiting densities for dense random packing of spheres, *J. Am. Ceram. Soc.*, 65, C19, 1982.
20. McGeary, R.K., Mechanical packing of spherical particles, *J. Am. Ceram. Soc.*, 44, 513, 1961.
21. Westman, A.E.R. and Hugill, H.R., The packing of particles, *J. Am. Ceram. Soc.*, 13, 767, 1930.
22. Milewski, J.V., Efficient use of whiskers in the reinforcement of ceramics, *Adv. Ceram. Mater.*, 1, 1, 1986.
23. Wakeman, R.J., Packing densities of particles with log-normal size distributions, *Powder Technol.*, 11, 297, 1975.
24. Sohn, H.Y. and Moreland, C., The effect of particle size distribution on packing density, *Can. J. Chem. Eng.*, 46, 162, 1968.
25. Bierwagen, G.P. and Saunders, T.E., Studies on the effect of particle size distributions on the packing efficiency of particles, *Powder Technol.*, 10, 111, 1974.

26. Furnas, C.C., Grading aggregates I — mathematical relations for beds for broken solids for maximum density, *Ind. Eng. Chem.* 23, 1052, 1931.

27. Andreasen, A.H.M. and Andersen, J., Ueber die Beziehung Zwischen Kornabstufung und Zwischenraum in Produkten aus Losen Körnern (mit einigen experimenten), *Kolloid Z.*, 50, 217, 1930.

28. Funk, J.E. and Dinger, D.R., *Predictive Process Control of Crowded Particulate Suspensions*, Kluwer Academic Publishers, Boston, MA, 1994.

29. Burk, R. and Apte, P., A packing scheme for real size distributions, *Am. Ceram. Soc. Bull.*, 66, 1390, 1987.

30. Roosen, A. and Bowen, H.K., Influence of various consolidation techniques on the green microstructure and sintering behavior of alumina powders, *J. Am. Ceram. Soc.*, 71, 970, 1988.

7 Forming of Ceramics

7.1 INTRODUCTION

The common methods for forming ceramic powders into green bodies are described in this chapter. Table 7.1 gives a summary of the common forming methods. The specific method to be used will depend in each case on the shape and size of the green body, as well as on the fabrication cost. Because the desired goal of the forming step is the production of a green body with homogeneous particle packing and high packing density, an important practical consideration is the extent to which the parameters of the forming method can be manipulated to achieve the desired goal.

The colloidal techniques described in Chapter 4 provide considerable benefits for the control of the packing homogeneity of the green body. The production of green bodies with homogeneous microstructure from a fully stabilized colloidal suspension of spherical, fine, monodisperse particles has not been incorporated into industrial applications in which mass production is desired and fabrication cost is a key consideration. Colloidal techniques, however, play an important role in the low-cost forming methods of slip casting and tape casting, as well as in the less commonly used methods of eletrophoretic deposition and gelcasting.

Mechanical compaction of dry or semidry powders in a die is one of the most widely used forming operations in the ceramic industry. In general, the applied pressure is not transmitted uniformly because of friction between particles and the die walls, as well as between the particles themselves. The stress variations lead to density variations in the green body, limiting the packing homogeneity that can be achieved. Although the density variations can be reduced significantly by isostatic pressing, mechanical compaction provides far less control in the manipulation of the green body microstructure than the casting methods.

Plastic-forming methods, in which a mixture of the ceramic powder and additives is deformed plastically through a nozzle or in a die, provide a convenient route for the mass production of ceramic green bodies. Extrusion is used extensively in the traditional ceramics industry and to a lesser extent in the advanced ceramics sector. Injection molding received considerable attention 20 to 30 years ago, but it has not yet made any significant inroads in the forming of ceramics for industrial applications.

Additives, commonly organic in nature, play an important role in the production of the green body. They serve as aids to the forming process, playing key roles in controlling the characteristics of the feed material, achieving the desired shape, and controlling the packing homogeneity of the green body. In methods such as tape casting and injection molding, which employ a significant concentration of additives, the selection of suitable additives is often vital to the overall success of the forming process. Prior to sintering, moist green bodies must be dried and additives must be removed as completely as possible without significant disruption of the particle packing. Efficient removal of the additives must be considered at the time of their selection. Drying of moist granular solids and the removal of additives from the green body are considered in Chapter 8.

Since the 1980s, considerable attention has been devoted to a group of forming methods, referred to as *solid freeform fabrication* (SFF) or *rapid prototyping* (RP), which allow the production of objects with complex shapes directly from a computer-aided design (CAD) file without the use of traditional tools such as dies or molds. Many of the processing concerns relevant to forming by the more traditional methods, such as casting, extrusion, and injection molding, also apply to SFF.

TABLE 7.1
Feed Materials and Green Body Shapes for the Common Ceramic Forming Methods

Forming Method	Feed Material	Shape of Green Body
	Dry or Semidry Pressing	
Die compaction	Powder or free-flowing granules	Small, simple shapes
Isostatic pressing	Powder or fragile granules	Larger, more intricate shapes
	Casting of a Slurry	
Slip casting	Free-flowing slurry with low binder content	Thin, intricate shapes
Tape casting	Free-flowing slurry with high binder content	Thin, sheets
	Deformation of a Plastic Mass	
Extrusion	Moist mixture of powder and binder solution	Elongated shapes with uniform cross section
Injection molding	Granulated mixture of powder and solid binder	Small, intricate shapes

7.2 ADDITIVES IN CERAMIC FORMING

The additives used in ceramic processing are either organic or inorganic in composition. Organic additives, which can be synthetic or natural in origin, find greater use in the forming of advanced ceramics because they can be removed almost completely, commonly by pyrolysis, prior to the sintering step. Therefore, the presence of residues that can degrade the microstructure of the final product is largely eliminated. Organic additives can also be synthesized with a wide variety of compositions, providing a wide range of chemicals for specialized applications. Inorganic additives cannot generally be removed after the forming step and are used in applications, particularly in the traditional ceramics industry, in which the property requirements are not very demanding.

The additives serve a variety of specialized functions, which may be divided into four main categories: (1) solvents, (2) dispersants, (3) binders, and (4) plasticizers. Some forming methods may require, in addition, the use of other additives, such as lubricants and wetting agents. Chemical principles and practical guidelines for the selection of additives can be formulated. However, because of the wide range of available chemicals and the incomplete knowledge of the chemical structure and process mechanisms, there is often no simple way of selecting additives for a given system. Most successful additives have been found by a trial-and-error approach.

Organic additives used in ceramic processing form the subject of a book [1] in which chemical principles and applications are described. Review articles have focused on various areas of the subject, such as the use of organic binders in ceramic forming [2] and the role of additives in tape casting [3a and 3b].

7.2.1 SOLVENTS

Liquids may be classified simply into two groups: (1) aqueous and (2) organic, but a more useful classification is in terms of the polarity of the molecules: (1) nonpolar, (2) polar, and (3) hydrogen bonding. Intermolecular forces in a nonpolar liquid are small, and the attraction to a surface is both small and nonspecific, so no specific orientation of the molecules is preferred. Polar molecules have strong electric fields between any bond involving atoms that are far apart in the periodic table. Polar molecules tend to orient themselves with respect to each other, forming association groups that are the lowest energy configurations consistent with their geometry and thermal motion. Polar groups interact more strongly with a polar surface than nonpolar groups of the molecule. Adsorption of the polar groups onto the particle surface leads to a coated surface with a nonpolar exterior, which may then adsorb another layer, with the nonpolar groups adsorbed and the polar parts on

TABLE 7.2
Physical Properties of Some Liquids (at 20°C where Applicable)

Liquid	Density (g/cm³)	Dielectric Constant	Surface Tension (10^{-3} N/m)	Viscosity (10^{-3} Pa sec)	Latent Heat of Vaporization (kJ/g)	Boiling Point (°C)	Flash Point (°C)
Nonpolar or Weakly Polar							
Hexane	0.659	1.9	18	0.3	0.35	68.7	−23
Toluene	0.867	2.4	29	0.6	0.35	111	3
Xylene (*ortho*)	0.881	2	28	0.7	0.33	140	32
Benzene	0.879	2.3	24	0.65	—	80.1	−11
Trichloroethylene	1.456	3	25	0.4	0.24	87	None
Carbon tetrachloride	1.59	2.2	26	1.0	—	76.5	None
Polar							
Acetone	0.781	21	25	0.3	0.55	56.0	−17
2-Butanone (MEK)	0.805	18	25	0.4	0.44	80	−1
Cyclohexanone	0.947	18	35	0.8	0.43	155	46
Diethyl ether	0.714	4.3	17	0.24	—	34.5	−40
Ethyl acetate	0.900	6	24	0.45	0.36	77	−3
Hydrogen Bonding							
Water	0.998	80	73	1.0	2.26	100.0	None
Methanol	0.789	33	23	0.6	1.10	64.6	18
Ethanol	0.789	24	23	1.2	0.86	78.4	20
Isopropanol	0.785	18	22	2.4	0.58	82.3	21
Ethylene glycol	1.113	37	48	20.0	0.80	198	111

the exterior surface. This ordering process extends outwards until thermal motion overwhelms the decreasingly effective orientation forces. Hydrogen bonding, well known to occur in water, forms when a hydrogen atom in a polar bond comes near an atom with an unshared pair of electrons.

Liquids serve two major functions: (1) provide fluidity for the powder during mixing and forming and (2) serve as solvents for dissolving the additives to be incorporated into the powder, providing a means for uniformly dispersing the additives throughout the powder. The selection of a solvent involves basically a choice between water and an organic liquid. Organic solvents generally have higher vapor pressure, lower latent heat of vaporization, lower boiling point, and lower surface tension than water, due largely to the strong hydrogen bonding of the water molecules (Table 7.2). The actual choice of a solvent for a given application often involves the consideration of a combination of several properties, such as (1) the ability to dissolve other additives, (2) evaporation rate, (3) ability to wet the powder, (4) viscosity, (5) reactivity toward the powder, (6) safety, and (7) cost.

Generally, solubility of the solid in the liquid is enhanced if the chemicals have similar functional groups, e.g., poly(vinyl alcohol) and water, or similar molecular polarity, e.g., poly(vinyl butyral) and ethanol. Evaporation rate is an important factor in industrial tape casting, where the tape is often cast, dried, peeled off from the carrier film, and rolled up for storage in a continuous operation. Fast-drying solvents such as toluene and 2-butanone (methyl ethyl ketone, or MEK) are commonly used for tape casting, particularly for thick tapes. The evaporation rate of a liquid is determined by its latent heat of vaporization, but the boiling point is sometimes used as a rough guide. A mixture of solvents (e.g., trichloroethylene and ethanol) is sometimes used in tape casting to control the solubility and evaporation rate.

Wetting of a solid by a liquid is defined in terms of the contact angle θ (see Figure 5.19), given by:

$$\cos\theta = \frac{\gamma_{SV} - \gamma_{SL}}{\gamma_{LV}} \tag{7.1}$$

where γ_{SV}, γ_{SL}, and γ_{LV} are the interfacial tensions of the solid–vapor, solid–liquid, and liquid–vapor interfaces, respectively. Good wetting (low θ) is desirable in practice and, according to Equation 7.1, is promoted by a low value of γ_{LV} (if γ_{SL} is not changed considerably). This is often achieved by using an organic solvent (low γ_{LV}) or by adding a wetting agent to water in order to reduce its surface tension. Poor wetting can lead to undesirable effects such as foaming of the liquid during milling and an increase in the suspension viscosity. The high surface tension of water makes it more difficult for air bubbles to escape to the surface; so water-based slurries have a greater tendency towards foaming during milling. Trapped bubbles generate undesirable flaws in the green body. The use of a wetting agent can alleviate the problem.

Water has a higher viscosity than several common organic solvents. The tendency to form hydrogen bonds with –OH groups on the surfaces of oxide powders can often steepen the effect of particle concentration on the suspension viscosity. The result is often a reduction in the solids content of the suspension for the maximum usable viscosity. This difficulty with aqueous slurries has largely been alleviated, due largely to recent advances in the understanding and use of dispersants for aqueous media [4a and 4b], but it appears that high solids content is easier to achieve in a more reproducible manner with organic liquids.

The surfaces of many powders (e.g., $BaTiO_3$, AlN, and Si_3N_4) can be chemically attacked by water, leading to a change in composition and properties. For these powders, the use of an organic solvent is recommended because approaches to reducing the chemical attack in water, such as the use of a thin protective coating, are currently expensive or ineffective. An approach shown to prevent chemical attack of AlN in aqueous media involves coating the powder with a silicate layer and heating the system to form a Si-Al-O-N surface layer [5], a process that will significantly increase the cost of the powder.

Water has a distinct advantage over organic solvents when safety, cost, and waste disposal are considered. Despite the disadvantages of water outlined in the preceding text, problems with the disposal of toxic organic solvents are leading to a shift towards greater use of aqueous solvents. Toxicity and flammability are key areas of safety. A commonly used indicator of flammability is the flash point, which gives the temperature at which there is sufficient vapor, generated by evaporation, so that an already existing flame can cause a fire to start. Organic solvents such as toluene and MEK, commonly used in tape casting, have very low flash points (Table 7.2), and so precautions must be taken to avoid explosions. Many organic solvents used in forming of ceramics are toxic. Human exposure to these chemicals and waste disposal are important concerns. Trichloroethylene and toluene, two solvents that have been widely used in tape casting of ceramics, are suspected of being carcinogens.

7.2.2 DISPERSANTS

Dispersants, sometimes referred to as deflocculants, are agents employed to stabilize suspensions of solid particles in liquid systems against flocculation. Although normally used in very small concentrations (e.g., a fraction of a percent by weight), the dispersant plays a large role in maximizing the particle concentration for some usable viscosity of the slurry. The principles controlling the stabilization of colloidal suspensions were discussed in Chapter 4. Here we provide more practical information of the main types of dispersants used in forming of ceramics and how they operate to stabilize suspensions.

Dispersants have a wide range of chemical compositions [6], and for many of them, the compositions are considered to be proprietary information by the manufacturers. We will divide dispersants into three main classes, based on their chemical structure, as follows:

1. Inorganic acid salts
2. Surfactants
3. Low- to medium-molecular-weight polymers

7.2.2.1 Inorganic Acid Salts

Inorganic salts and derivatives of weak mineral acids are effective dispersants in aqueous solvents. The major families are:

1. Inorganic phosphates (e.g., sodium hexametaphosphate, $Na_6P_6O_{18}$, tetrasodium pyrophosphate, $Na_4P_2O_7$, and sodium tripolyphosphate, $Na_5P_3O_{10}$)
2. Silicates (e.g., sodium silicate, $(Na_2O)(Na_2SiO_3)_x$, where $x \approx 4$)
3. Borates (e.g., sodium tetraborate (borax),$Na_2B_4O_7$)

Sodium carbonate, Na_2CO_3, is also used as a dispersant, but it is not as powerful as the inorganic phosphates and sodium silicate. The chemical structure of a representative member of the phosphate, silicate, and borate families is given in Figure 7.1. The formulas shown are approximations of the commercial products and may vary between manufacturers.

The complex anion species adsorb strongly onto the surfaces of oxide particles. Adsorption will increase with higher molecular weight of the anion species (higher van der Waals attraction) and with the specific charge (charge per unit anion group). Adsorption coupled with the formation of a diffuse layer of the counterions (ions of opposite charge) leads to electrostatic stabilization due to repulsion between the double layers (Chapter 4). The valence and radius of the counterions can modify the repulsion between the particles and so can influence the stability of the suspension. Counterions with higher valence are more effective for causing flocculation (Schulze–Hardy rule), whereas for ions of the same valence, the smaller ions are more effective. For monovalent cations, the effectiveness of flocculation is in the order $Li^+ > Na^+ > K^+ > NH_4^+$, whereas for divalent cations, $Mg^{2+} > Ca^{2+} > Sr^{2+} > Ba^{2+}$. This sequence is known as the *Hofmeister series*. For common anions, the effectiveness of flocculation is in the order $SO_4^{2-} > Cl^- > NO_3^-$.

Sodium silicate forms one of the most effective dispersants for clays. For advanced ceramics that must meet very specific property requirements, the use of sodium silicate or other inorganic acid salts often leaves residual ions (e.g., sodium or phosphate), which even in very small concentrations can lead to the formation of liquid phases during sintering, making microstructural control more difficult.

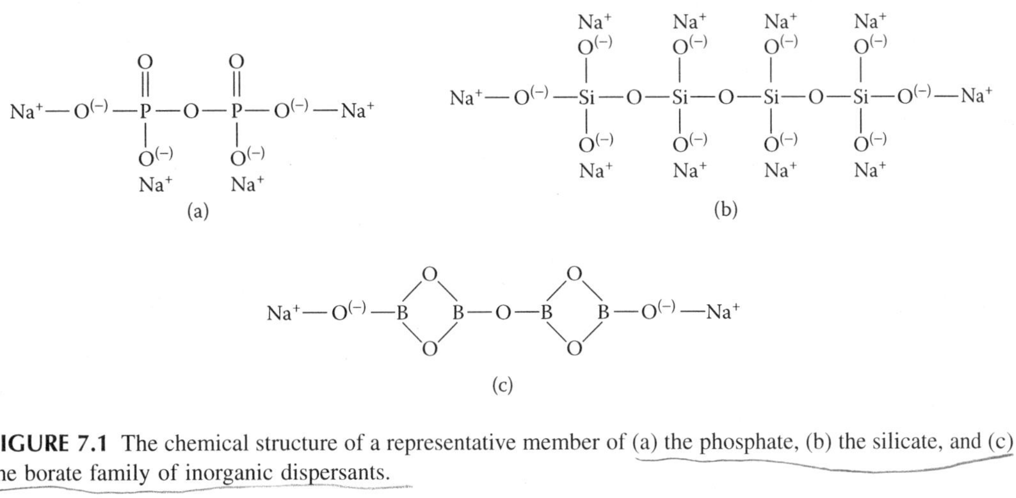

FIGURE 7.1 The chemical structure of a representative member of (a) the phosphate, (b) the silicate, and (c) the borate family of inorganic dispersants.

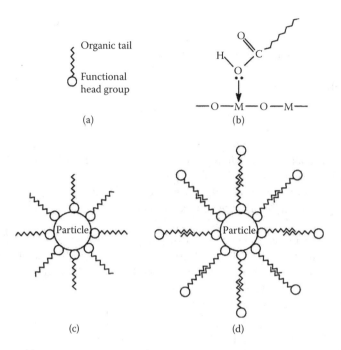

FIGURE 7.2 (a) Schematic of a surfactant molecule; (b) illustration of surfactant adsorption onto an oxide particle surface by coordinate bonding; (c) stabilization by steric repulsion between organic tails of the surfactant molecules; (d) stabilization by micelles.

7.2.2.2 Surfactants

Surfactants (shortened from the term surface active agents) assist the distribution of one phase in another. They are not limited to solid–liquid suspensions and may include liquid–liquid systems, for example. When used to stabilize suspensions of solid particles in liquids, they are called *dispersants*. Surfactants are composed of molecules that have a strong tendency to concentrate in the interfacial region rather than in the bulk of either phase. This arises because the surfactant molecular structure consists of one portion that is relatively soluble in (compatible with) the liquid and a second portion that is relatively insoluble in the liquid. The molecules preferentially orient themselves with the liquid-soluble portion sticking into the liquid and the liquid-insoluble portion adsorbed at the solid surface.

Surfactants are commonly classified in terms of the charge on the portion of the molecule that concentrates at the interface: nonionic, anionic, and cationic. A nonionic surfactant is one that has no ionizing groups. Examples are fatty alcohols ROH, fatty acids R(CO)OH, and fatty esters R(CO)OR′, where R and R′ are short-chain organic polymers. The structure of oleic acid, $C_{17}H_{34}(CO)OH$, can be said to consist of an organic tail $C_{17}H_{34}$ and a functional head group (CO)OH (Figure 7.2a). The head group is polar but does not ionize to produce a charged species. Nonionic surfactants are commonly effective in organic solvents. Adsorption onto the particle surfaces occurs either by van der Waals attraction or, more effectively, by stronger coordinate bonding. As illustrated in Figure 7.2b, using the Lewis acid–base concept, an atom in the surfactant functional group that has an unshared pair of electrons (e.g., N or O) may act as a Lewis base and form a coordinate bond with an atom (e.g., Al) on the particle surface that has an incomplete shell of electrons (Lewis acid). Stabilization most likely occurs by steric repulsion between the organic tails or micelles that are stretched out in the organic solvent (Figure 7.2c and Figure 7.2d). Examples of some commonly used nonionic surfactants are given in Figure 7.3. Menhaden fish oil is widely used as a dispersant for Al_2O_3, $BaTiO_3$, and several other oxides in organic solvents. It consists of a mixture of several

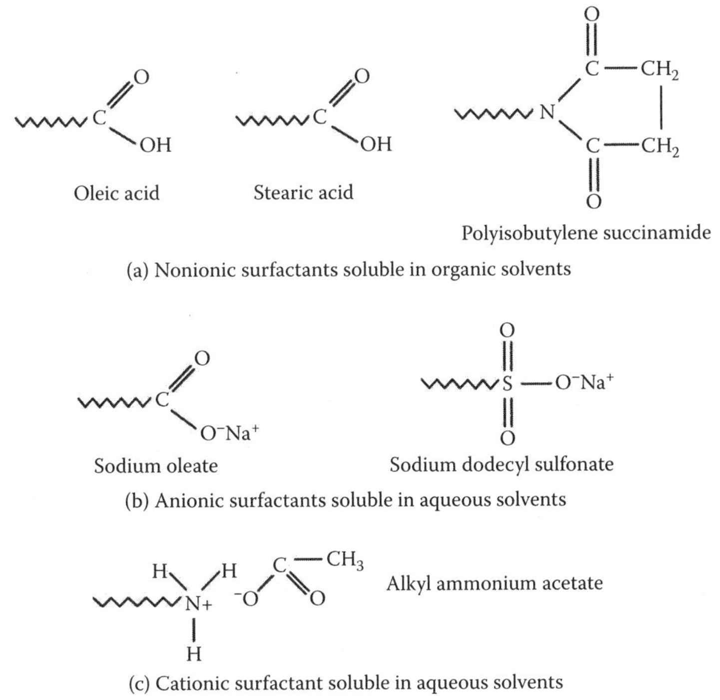

FIGURE 7.3 Functional head groups of some common surfactants.

short-chain fatty acids, R(CO)OH, with the alkyl chain (R) containing some C=C double bonds and the functional end group being a carboxylic acid (CO)OH.

Sodium oleate, $C_{17}H_{34}(CO)O^-Na^+$, is an anionic surfactant because it dissociates to a water-soluble Na^+ ion and a negatively charged oleate ion that concentrates at the interface (Figure 7.4a). Salts of the fatty acids, such as sodium oleate, sodium stearate, $C_{17}H_{35}(CO)O^-Na^+$, and the sulfonates, such as sodium alkyl sulfonate, $R(SO_2)O^-Na^+$, where R is an alkyl chain of 10 to 20 carbon atoms, are common anionic surfactants. Anion surfactants are effective in aqueous solvents. Usually, negatively charged oxygen species are formed on dissociation of anionic surfactants. Adsorption of the surfactant commonly occurs by electrostatic attraction with positively charged particle surfaces. Stabilization of the suspension occurs essentially by electrostatic repulsion between the negative charges because of the adsorbed surfactant molecules (Figure 7.4b) or micelles (Figure 7.4c).

Dodecyl ammonium acetate, $C_{12}H_{25}(N^+H_3)CH_3(CO)O^-$, is a cationic surfactant because it dissociates to a water-soluble acetate ion, $CH_3(CO)O^-$, and a positively charged (cationic) dodecyl ammonium ion, $C_{12}H_{25}(N^+H_3)$, that concentrates at the interface (Figure 7.4a). Cationic surfactants commonly consist of positively charged nitrogen species on dissociation. Except for the reversal in the sign of the charges on the surfactant and the particle surface, the mechanisms of adsorption and stabilization are similar to the anionic surfactant case.

Surfactants whose charge is positive at low pH, zero at moderate pH, and negative at high pH are called *amphoteric surfactants*. If a neutral molecule has both a positively charged group (such as a quaternary ammonium ion) and a negatively charged group (such as a carboxylate ion), it is called *zwitterionic*. Examples of zwitterionic (amphoteric) surfactants are the amino acids, $R(N^+H_2)CH_2CH_2(CO)O^-$. These may be amphoteric as well, becoming cationic at low pH and anionic at high pH. The two separated charges give the molecule a large dipole moment, and so it is strongly attracted to polarizable surfaces. These surfactants can also adsorb on surfaces of any charge, the positive group adsorbing on a negatively charged surface and the negative group on a positively charged surface.

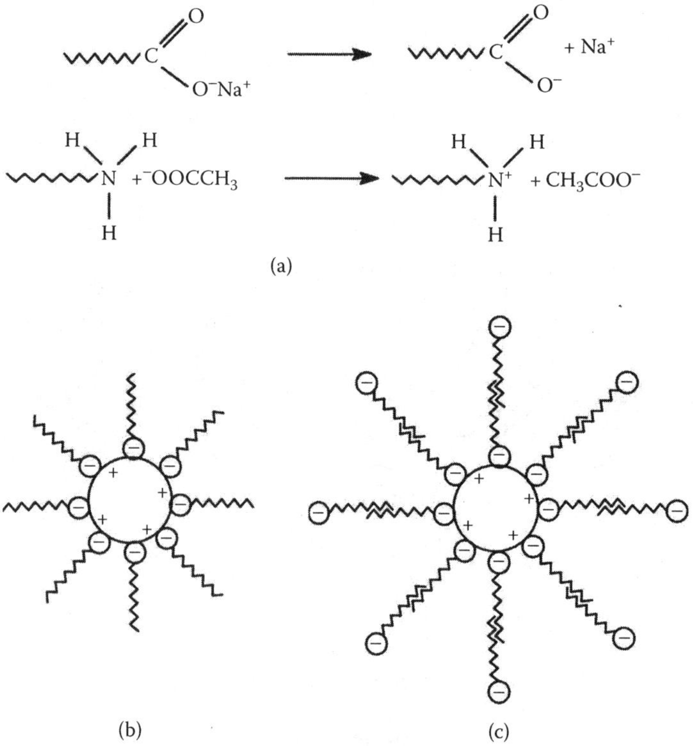

FIGURE 7.4 (a) Dissociation of anionic and cationic surfactants to form negatively charged and positively charged head groups, respectively; (b) stabilization by negatively charged surfactant; (c) stabilization by negatively charged micelles.

7.2.2.3 Low- to Medium-Molecular-Weight Polymers

These dispersants, with a molecular weight in the range of several hundred to a few thousand, are also classified into nonionic, anionic, and cationic types. Common nonionic polymeric dispersants are poly(ethylene oxide), PEO, or poly(ethylene glycol), PEG, poly(vinyl pyrrolidone), PVP, poly(vinyl alcohol), PVA, polystyrene, PS, and block copolymers of PEO–PS. With a higher molecular weight, many of these polymers are effective as binders, and their chemical compositions are given later. When the chain segment contains –OH groups or polar species, the dispersants are effective in water; otherwise, they are effective in organic solvents. In organic or aqueous solvents, adsorption of the polymers can occur by weak van der Waals bonding or more effectively by coordinate bonding (see Figure 7.2b). In aqueous solvents, hydrogen bonding can also produce very effective adsorption. Because they are uncharged, nonionic polymeric dispersants provide stabilization by steric repulsion.

Polyelectrolytes are ionic polymeric dispersants composed of ionizable repeating units. They are effective in aqueous solvents. On dissociation, the ionized groups in the chain segment can produce negatively charged species (anionic polymers) or positively charged species (cationic polymers). Some common anionic polymers are given in Figure 7.5. The sodium or ammonium salts of the polyacrylic acids have been used successfully with aqueous slurries of several oxide powders, and their use is increasing. For advanced ceramics, the use of the sodium salt of these acids should be avoided because residual Na ions, even in very small concentrations, commonly lead to the formation of an undesirable liquid phase during sintering, hindering microstructural control. An example of a cationic polymer is poly(ethylene imine), which becomes positively charged in acidic conditions but remains an undissociated weak base in basic conditions. The

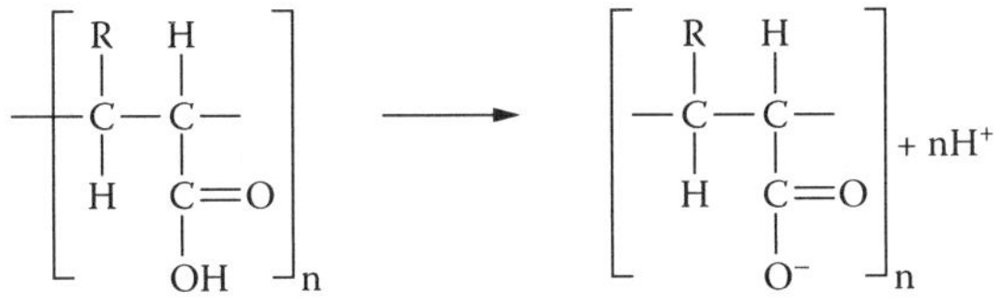

(a) Poly(acrylic acid): R=H; poly(methacrylic acid): R=CH$_3$

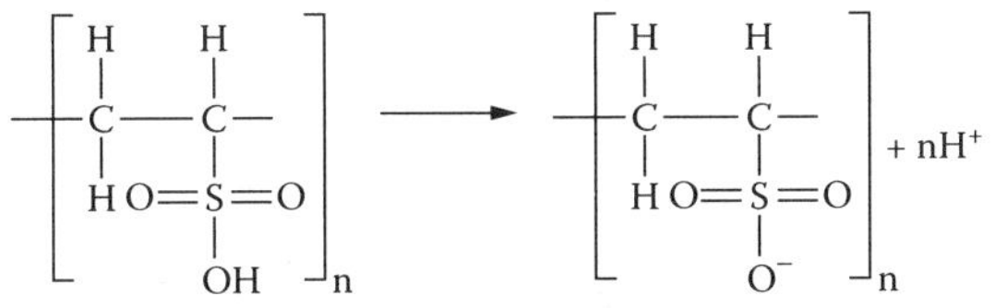

(b) Poly(vinyl sulfonic acid)

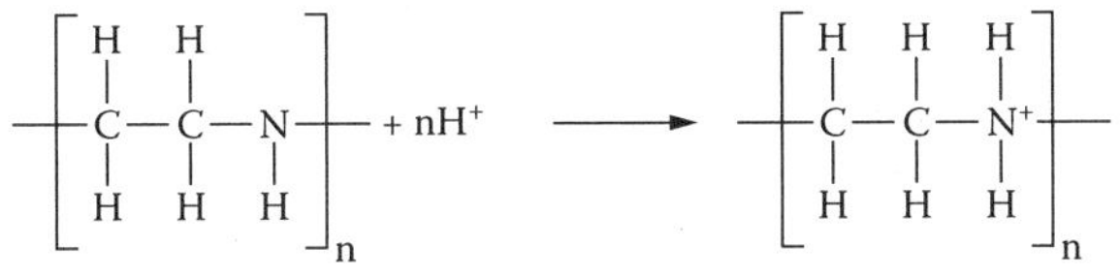

(c) Poly(ethylene imine)

FIGURE 7.5 Examples of common short-chain anionic and cationic polymers used as dispersants.

mechanisms of adsorption (electrostatic attraction) and stabilization (electrosteric) were described in Chapter 4.

7.2.3 BINDERS

Binders are typically long-chain polymers that serve the primary function of providing strength to the green body by forming bridges between the particles. In some forming methods (e.g., injection molding), they also provide plasticity to the feed material to aid the forming process. A large number of organic substances can be utilized as binders, some of which are soluble in water whereas others are soluble in organic liquids. The monomer formulas of some common synthetic binders are shown in Figure 7.6. They include the vinyls, acrylics, and the ethylene oxides (glycols). The vinyls have a linear chain backbone in which the side group is attached to every other C atom. The acrylics have the same backbone structure but may have one or two side groups attached to the C atom.

The cellulose derivatives are a class of naturally occurring binders. The polymer molecule is made up of a ring-type monomer unit having a modified α-glucose structure (Figure 7.7). The modifications to the polymer occur by changes in the side groups, R. The degree of substitution (DS) is the number of sites on which modifications are made in the monomer. Substitutions occur first at the C-5 site, followed by the C-2 site, and finally at the C-3 site. The formulas of the R groups in some common cellulose derivatives are shown in Figure 7.8.

The selection of a binder for a given forming process involves the consideration of several factors: (1) binder burnout characteristics, (2) molecular weight, (3) glass transition temperature, (4) compatibility with the dispersant, (5) effect on the viscosity of the solvent, (6) solubility in the solvent, and (7) cost. The binder as well as the other additives used to aid the forming of the green body must normally be removed as completely as possible (commonly by pyrolysis) prior to sintering. Because the concentration of the binder is commonly much greater than that of the other additives, the binder burnout characteristics is a primary consideration. The binder burnout characteristics, as discussed in Chapter 8, depend primarily on the binder chemistry and the atmosphere (oxidizing or nonoxidizing).

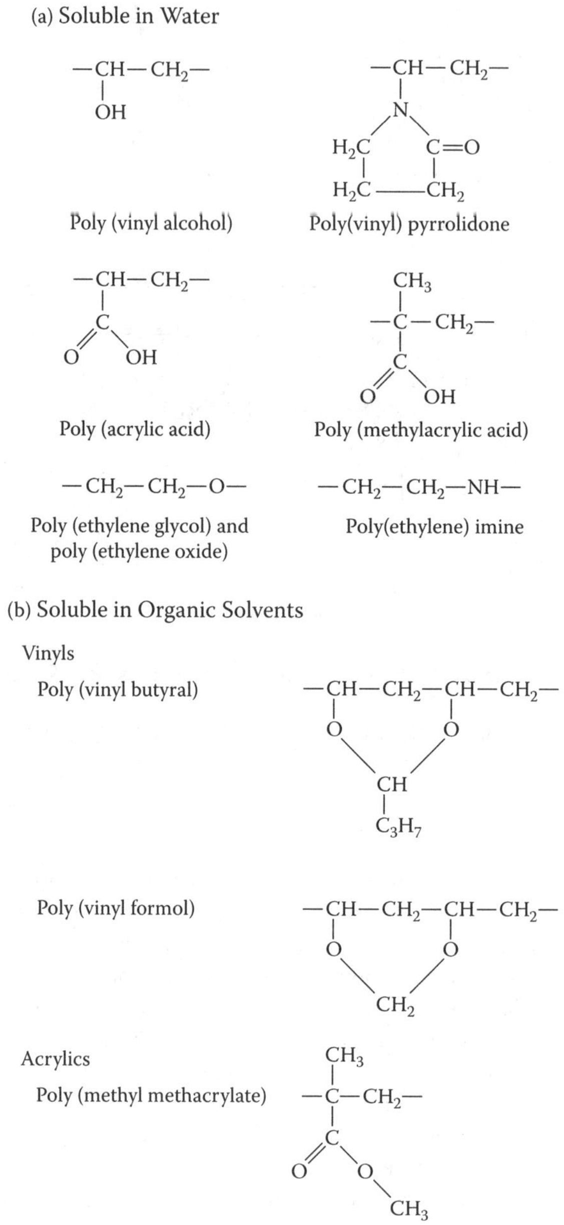

(a) Soluble in Water

$$-CH-CH_2-$$
$$\;\;\;|$$
$$\;\;OH$$

Poly (vinyl alcohol)

$$-CH-CH_2-$$
$$\;\;\;|$$
$$\;\;N$$
$$H_2C \qquad C=O$$
$$H_2C ——CH_2$$

Poly(vinyl) pyrrolidone

$$-CH-CH_2-$$
$$\;\;\;|$$
$$\;\;C$$
$$O^{\nearrow} \;\; ^{\diagdown}OH$$

Poly (acrylic acid)

$$CH_3$$
$$\;\;|$$
$$-C-CH_2-$$
$$\;\;|$$
$$\;\;C$$
$$O^{\nearrow} \;\; ^{\diagdown}OH$$

Poly (methylacrylic acid)

$$-CH_2-CH_2-O-$$

Poly (ethylene glycol) and
poly (ethylene oxide)

$$-CH_2-CH_2-NH-$$

Poly(ethylene) imine

(b) Soluble in Organic Solvents

Vinyls

Poly (vinyl butyral)

$$-CH-CH_2-CH-CH_2-$$
$$\;\;\;|\qquad\qquad\quad|$$
$$\;\;O\qquad\qquad\quad O$$
$$\qquad\diagdown\qquad\diagup$$
$$\qquad\quad CH$$
$$\qquad\quad|$$
$$\qquad\quad C_3H_7$$

Poly (vinyl formol)

$$-CH-CH_2-CH-CH_2-$$
$$\;\;\;|\qquad\qquad\quad|$$
$$\;\;O\qquad\qquad\quad O$$
$$\qquad\diagdown\qquad\diagup$$
$$\qquad\quad CH_2$$

Acrylics

Poly (methyl methacrylate)

$$CH_3$$
$$\;\;|$$
$$-C-CH_2-$$
$$\;\;|$$
$$\;\;C$$
$$O^{\nearrow} \;\; ^{\diagdown}O$$
$$\qquad\quad\diagdown CH_3$$

FIGURE 7.6 Monomer formulas of some synthetic binders: (a) soluble in water; (b) soluble in organic solvents.

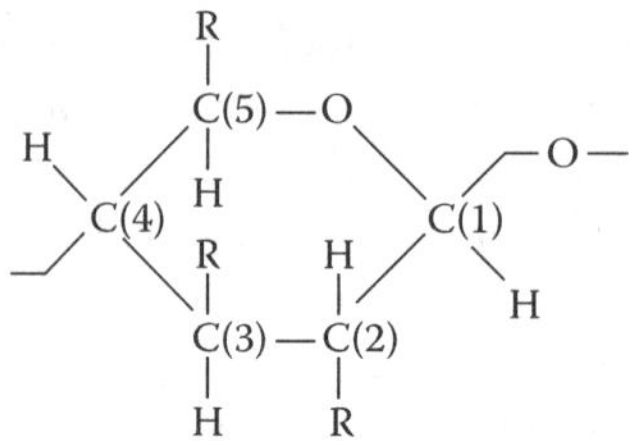

FIGURE 7.7 The modified α-glucose structure.

Binder	R group	DS
Soluble in water		
Methylcellulose	$-CH_2-O-CH_3$	2
Hydroxypropylmethylcellulose	$-CH_2-O-CH_2-CH-CH_3$ $\qquad\qquad\qquad\quad\;\;\; \|$ $\qquad\qquad\qquad\quad\; OH$	2
Hydroxyethylcellulose	$-CH_2-O-C_2H_4-O-C_2H_4-OH$ $-CH_2-O-C_2H_4-OH$	0.9–1.0
Sodium carboxymethylcellulose	$-CH_2-O-CH_2-C\overset{O}{\underset{ONa}{<}}$	
Starches and dextrins	$-CH_2-OH$	
Sodium alginate	$-C\overset{O}{\underset{ONa}{<}}$	
Ammonium alginate	$-C\overset{O}{\underset{ONH_4}{<}}$	
Soluble in organic solvents		
Ethyl cellulose	$-CH_2-O-CH_2-CH_3$	

FIGURE 7.8 Formulas of the side groups in some cellulose derivatives.

In general, high molecular weight enhances the binder strength. The glass transition temperature of the polymer T_g (the temperature that marks the transition from a rubbery state to a glassy state) must not be much higher than room temperature in order to allow the binder to deform during forming of the green body. A reduction in T_g essentially involves reducing the resistance to motion of the polymer chains. This can be achieved by the use of polymers with less rigid side groups, less polar side groups, or with a lower molecular weight. However, as discussed later, the most common approach is to reduce the intermolecular bonding between the chains through the use of a plasticizer. If a dispersant is used in the forming process, then the binder should be compatible with it. In general, the binder should not displace the dispersant from the particle surface. For oxides, this commonly means that the binder should be less polar than the dispersant.

The effect of the binder on the rheology of the solvent is a key consideration. Organic binders increase the viscosity and change the flow characteristics of the liquid. Some can even lead to the formation of a gel. In the casting methods (e.g., tape casting), increasing the binder concentration should not produce a rapid increase in the viscosity of the system because this will limit the amount of powder that can be incorporated into the suspension for some useable viscosity. On the other hand, a rapid increase in the viscosity is often desirable in extrusion to provide good green strength with a small concentration of binder.

Binders are often arbitrarily classified into low, medium, and high-viscosity grades based on how effectively they increase the viscosity of the solution. The scheme shown in Figure 7.9 has been proposed by Onoda [2], and the classification of several water-soluble binders according to this scheme is shown in Table 7.3. The binder grade depends to a large extent on the structure of the polymer chain. Polymer molecules in solution take up the conformation of a coil. Smaller coils exert less viscous drag on the molecules of the liquid and lead to a smaller increase in the viscosity with concentration. Linear chains with good flexibility in which the bonds can easily rotate, e.g.,

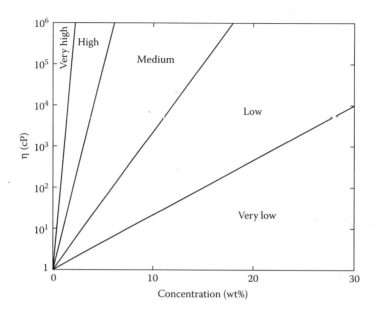

FIGURE 7.9 Criterion for viscosity grades based on viscosity–concentration relationship. (From Onoda, G.Y. Jr., The rheology of organic binder solutions, in *Ceramic Processing before Firing*, Onoda, G.Y. Jr. and Hench, L.L., Eds., John Wiley & Sons, New York, 1978, p. 241. With permission.)

vinyl, acrylic, and poly(ethylene oxide) binders, are expected to form smaller coils, and should have a lower viscosity grade than chains based on ring molecules or with large rigid side groups (e.g., some cellulose binders). Although the molecular weights of the binders shown in Table 7.3 are not known accurately, the viscosity grades appear to be consistent with this line of reasoning.

In most of the forming methods (injection molding being a well-known exception), the binder is commonly added as a solution, so its solubility in the liquid is an important factor. The backbone of the molecule consists of covalently bonded atoms such as carbon, oxygen, and nitrogen. Attached to the backbone are side groups located at frequent intervals along the length of the molecule. The chemical nature of the side groups determines, in part, what liquids will dissolve the binder. Solubility of the binder in the liquid is enhanced if they have similar functional groups or similar molecular polarity.

Sodium silicate is an inorganic binder that finds considerable use in the forming of some traditional ceramics whose properties are not deleteriously affected by Na and Si residues. The compositions used as a binder have a $Na_2O:SiO_2$ ratio in the range of ~2 to 4. Hydrolysis leads to the formation of fine SiO_2 particles that gel and form a strong bonding phase between the ceramic particles.

7.2.4 PLASTICIZERS

Plasticizers are generally organic substances with a lower molecular weights than binders. The primary function of the plasticizer is to soften the binder in the dry state (i.e., reduce the T_g of the binder), increasing the flexibility of the green body (e.g., tapes formed by tape casting). For forming processes in which the binder is introduced as a solution, the plasticizer must be soluble in the same liquid that is used to dissolve the binder. In the dry state, the binder and plasticizer are homogeneously mixed as a single substance. The plasticizer molecules get between the polymer chains of the binder, disrupting the chain alignment and reducing the van der Waals bonding between adjacent chains. This leads to softening of the binder but also reduces the strength. Some commonly used plasticizers are listed in Table 7.4.

TABLE 7.3
Viscosity Grades for Some Water-Soluble Binders

	Viscosity Grade					Electrochemical Type			
	Very Low	Low	Medium	High	Very High	Nonionic	Anionic	Cationic	Biodegradable
Gum arabic	•						X		X
Lignosulfonates	•						X		X
Lignin liquor	•						X		X
Molasses	•					X			X
Dextrins	•——	•				X			X
Polyvinylpyrrolidone	•——	•				X			
Poly(vinyl alcohol)	•————	—	•			X			
Poly(ethylene oxide)		•——	•			X			
Starch		•——	•			X			X
Acrylics		•—•					X		
Polyethylenimine (PEI)	•——	•						X	
Methycellulose		•———	——	•		X			X
Sodium carboxymethylcellulose		•——	——	•			X		X
Hydroxypropylmethylcellulose		•———	——	•		X			X
Hydroxyethylcellulose		•———	——	——	•	X			X
Sodium alginate			•——	•			X		X
Ammonium alginate			•——	•			X		X
Polyacrylamide			•——	•		X			
Scleroglucan			•			X			X
Irish moss			•				X		X
Xanthan gum			•						X
Cationic galactomanan				•				X	X
Gum tragacanth				•		X			X
Locust bean gum				•		X			X
Gum karaya				•		X			X
Guar gum		•—•				X		X	X

Source: From Onoda, G.Y. Jr., The rheology of organic binder solutions, in *Ceramic Processing before Firing*, Onoda, G.Y. Jr. and Hench, L.L., Eds., John Wiley & Sons, New York, 1978, p. 235. With permission.

TABLE 7.4
Common Plasticizers Used in Ceramic Processing

Plasticizer	Melting Point (°C)	Boiling Point (°C)	Molecular Weight
Water	0	100	18
Ethylene glycol	13	197	62
Diethylene glycol	8	245	106
Triethylene glycol	7	288	150
Tetraethylene glycol	5	327	194
Poly(ethylene glycol)	10	>330	300
Glycerol	18	290	92
Dibutyl phthalate	35	340	278
Dimethyl phthalate	1	284	194

7.2.5 OTHER ADDITIVES

The number of additives used in a given forming process should be kept to a minimum because the potential for undesirable interactions between the components increases with their number. However, small amounts of other additives are sometimes used to serve special functions. A wetting agent can be added to reduce the surface tension of the liquid (particularly water), improving the wetting of the particles by the liquid. A lubricant is commonly used in die compaction, extrusion, and injection molding to reduce the friction between the particles themselves or between the particles and the die walls. Under the application of an external pressure, the particles rearrange more easily, leading to a higher and more uniform packing density. Common lubricants are steric acid, stearates, and various waxy substances. A homogenizer such as cyclohexanone is sometimes used in tape casting to increase the mutual solubility of the components, thereby improving the homogeneity of the mixture.

7.3 FORMING OF CERAMICS

The common ceramic-forming methods, summarized in Table 7.1, will now be described, emphasizing the key process variables and how they can be manipulated to optimize the microstructure of the green body.

7.3.1 DRY AND SEMIDRY PRESSING

Uniaxial pressing in a die and isostatic pressing are commonly used for the compaction of dry powders, which typically contain <2 wt% water, and semidry powders, which contain ~5 to 20 wt% water [7,8]. Die compaction is one of the most widely used operations in the ceramics industry, allowing the formation of simple shapes rapidly and with accurate dimensions. However, the agglomeration of dry powders combined with the nonuniform transmission of the applied pressure during compaction leads to significant variations in the packing density of the green body. To minimize the density variations, die pressing should be used for the production of simple shapes (e.g., disks) with a height-to-diameter ratio less than ~0.5. Isostatic pressing produces more homogeneous packing density. It can be used to form green bodies with complex shapes and with larger height-to-diameter ratios. However, the green body has irregularities in both shape and surface quality, and often requires considerable machining.

7.3.1.1 Die Compaction

In die compaction, a powder or a granular material undergoes simultaneous uniaxial compaction and shaping in a rigid die. The overall process consists of three steps: filling of the die, powder compaction, and ejection of the compacted powder. There are three main modes of compaction, defined in terms of the relative motion of the die and the punches. In the single-action mode, the top punch moves, but the bottom punch and the die are fixed; whereas in the double-action mode, both punches move, but the die is fixed. In the floating-die mode the top punch and the die move, but the bottom punch is fixed. The double-action mode is capable of providing better packing homogeneity and is commonly used in industry.

7.3.1.1.1 Feed Material: Powders or Granules

Powders, often mixed with a small amount of binder (<5 vol%), are commonly used as the feed material in laboratory experiments. In industrial practice, the flow behavior of the feed material is an important factor when efficient die filling, fast pressing rates, and reproducible green body properties are required. Fine powders do not flow very well and are difficult to compact homogeneously; so it is often necessary to granulate them, commonly by spray drying of a slurry (Chapter 2). Additives, discussed earlier in this chapter, are important for the formulation of the slurry. For

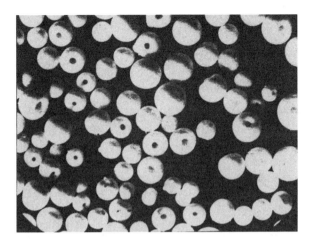

FIGURE 7.10 Example of a commercial spray-dried Al_2O_3 powder (magnification = 50×). (Courtesy of Niro Inc., Columbia, MD.)

example, an Al_2O_3 slurry will require a dispersant (e.g., ammonium polyacrylate), a binder [e.g., poly(vinyl alcohol)], a plasticizer [e.g., poly(ethylene glycol)], and a lubricant (e.g., ammonium stearate). Figure 7.10 shows a commercial spray-dried Al_2O_3 powder.

The granule characteristics are dependent on several factors, such as the particle size distribution of the initial powder, the degree of flocculation of the slurry, the type of additives, and the spray-drying conditions. The key granule characteristics are the following:

1. *Size, size distribution, and shape*: Granules prepared by spray drying in commercial equipment have nearly spherical shape and sizes in the range of ~50 to 400 μm, with the average size in the range of ~100 to 200 μm.
2. *Particle packing*: The particle packing density in the granule is controlled by the particle size distribution of the powder, as well as the particle concentration and colloidal stability of the slurry [9]. Particle packing densities of ~45 to 55% are common for granules.
3. *Particle packing homogeneity*: The particle packing homogeneity in the granule depends on the colloidal stability of the suspension and on the drying step during spray drying. Granules are commonly prepared from a partially flocculated slurry, so the particle packing is not very homogeneous. Furthermore, if the binder segregates to the granule surface during spray drying, the outer region of the granule will have a low packing density [10].
4. *Hardness*: The hardness of the granule is controlled by the particle packing density and by the nature of the binder. A high particle packing density or a hard binder (e.g., with a high glass transition temperature) leads to the formation of hard granules. On the other hand, soft granules have a lower packing density or contain a soft binder.
5. *Surface friction*: Smooth granule surfaces reduce the friction between the granules themselves and between the granules and die walls.

7.3.1.1.2 Die Filling

Flow of the granules during die filling is facilitated by a wide distribution of sizes, a spherical shape, and a smooth surface [11]. The uniformity of the die filling must also be considered because it will affect the packing homogeneity of the green body. In addition to the die filling method and the geometry of the die, the ratio of the granule size to the die diameter die can also influence filling uniformity. Narrower dies lead to a lower overall packing density in the compact because the packing density near the die walls is lower. Simulations indicate that the effect of the die walls

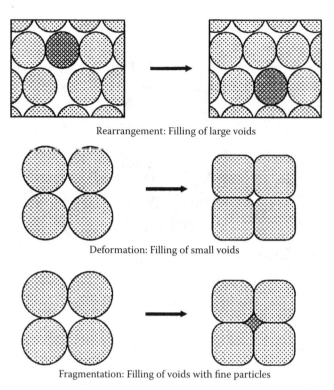

Rearrangement: Filling of large voids

Deformation: Filling of small voids

Fragmentation: Filling of voids with fine particles

FIGURE 7.11 Schematic diagram showing the stages of granule compaction.

becomes insignificant when the die diameter is greater than ~250 times the granule diameter [7]. For particle packing densities of 45 to 55% in the granules, assuming loose random packing of the granules (~60%), the actual particle packing density during die filling is only ~25 to 35%.

7.3.1.1.3   The Compaction Step

After die filling, the structure of a system of granules contains large voids on the order of the granule size and voids that are smaller than the granule size (Figure 7.11). The compaction of this system can be divided into two stages. The first stage involves reduction of the large voids by rearrangement of the granules, whereas in the second stage the small voids are reduced by deformation of the granules.

The compaction of a system of particles depends on the amount and type of agglomerates in the powder. For a system of primary particles with monomodal pores, the compaction consists essentially of one stage involving sliding and rearrangement (with some fracture at higher pressures) to reduce the voids. If the powder contains low-density weak agglomerates, then two stages of compaction may be observed: (1) rearrangement and sliding in the first stage to reduce the larger voids and (2) fracture of the agglomerates and further rearrangement and sliding in the second stage to reduce the smaller voids. For powers or granules, the particles also undergo elastic compression, which, as we shall see later, influences the ejection of the compact from the die and the creation of crack-like flaws in the green body.

Although providing only limited information, the density of the compact as a function of the applied pressure is commonly used to characterize the compaction behavior. The data can be easily measured, and they find use in process optimization and quality control. When the density is plotted vs. the logarithm of the pressure, the data for granules often show two straight lines separated by a break point (Figure 7.12). Compaction is interpreted to occur by rearrangement in the low-pressure linear region and by deformation in the high-pressure linear region. The break point scales as the

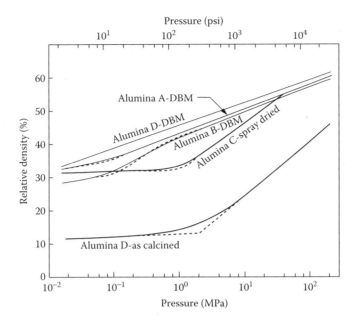

FIGURE 7.12 Compaction behavior of fine Al_2O_3 powders and spray-dried granules. DBM = dry ball milling. (From Niesz, D.E., McCoy, G.L., and Wills, R.R., Handling and green forming of fine powders, in *Materials Science Research, Vol. 11: Processing of Crystalline Ceramics*, Palmour, H., III, Davis, R.F., and Hare, T.M., Eds., Plenum, New York, 1978, p. 41. With permission.)

hardness (or the strength) of the granules. Powders consisting of primary particles show a single line, whereas an agglomerated powder may show two lines with the break point determined by the strength of the agglomerates.

The compaction process is a complex many-body problem and theoretical analysis to produce a predictive model is difficult [13]. In view of this difficulty, several empirical equations have been developed to account for the experimental data. None have been found to be generally applicable. One equation that has the advantage of simplicity and is as good as any of the others is

$$p = \alpha + \beta \ln\left(\frac{1}{1-\rho}\right) \tag{7.2}$$

where p is the applied pressure, ρ is the relative density, and α and β are constants that depend on the initial density and the nature of the material. The empirical expressions have often been criticized as being merely curve-fitting, but they have served to focus attention on the complexity of the variables and the mechanisms of the process.

The behavior of a powder changes dramatically when it is converted to granules; so the key factors influencing the compaction of particles and granules are considered separately.

7.3.1.1.4 Factors Influencing the Compaction of Particles

A severe problem in die compaction is that the applied pressure is not transmitted uniformly to the powder mainly because of friction between the powder and the die wall. The pressure gradients produce density gradients in the powder compact. The application of a uniaxial pressure p_z to the powder leads to the generation of a radial stress p_r and a shear (or tangential) stress τ at the die wall. The radial and shear stresses vary with distance along the die, so the resultant stress in the compact is nonuniform. Strijbos and coworkers [14–17] made extensive studies of the stresses occurring during die compaction using equipment designed to measure the radial pressure coefficient (equal to the ratio p_r/p_z) and the powder–wall friction coefficient (equal to the ratio τ/p_r). They investigated the effects of parameters such as the mean particle size of the powder, d_p, the roughness of the die wall, R_w, the hardness of the powder, H_p, the hardness of the wall, H_w, and the use of lubricants.

For powders of ferric oxide (Vickers hardness ≈ 600) and walls of tungsten carbide, hardened tool steel, and nonhardened tool steel (Vickers hardness ≈ 1300, 600, and 200, respectively), some data for the friction coefficient, f_{dyn}, as a function of d_p/R_w for the three values of H_p/H_w are shown in Figure 7.13. For particles smaller than the roughness, f_{dyn} is as high as 0.6. A layer of the fine particles sticks to the die wall; so there is no direct contact between the stationary powder compact and the moving wall in the friction coefficient apparatus. In this case, R_w and H_w have no effect on f_{dyn}, and the high value of f_{dyn} is a reflection of the friction between the powder particles. Failure occurs within the powder compact and not at the die wall. For particles larger than the wall roughness, f_{dyn} is between 0.2 and 0.4 and is dependent on both the powder parameters and the die wall parameters. The low values of f_{dyn} are a reflection of the friction between the particles and the relatively smooth wall. Failure in this case occurs at the powder–wall interface. The influence of the particle size on the friction coefficient is summarized schematically in Figure 7.14. The higher friction coefficient, reduced uniformity of die filling, and greater tendency for agglomeration are largely responsible for the severe density gradients in die compaction of fine powders.

The friction coefficient between the powder and a rough wall is also dependent on the direction of the grooves in the wall. The friction is lower if the grooves run in the direction of relative motion

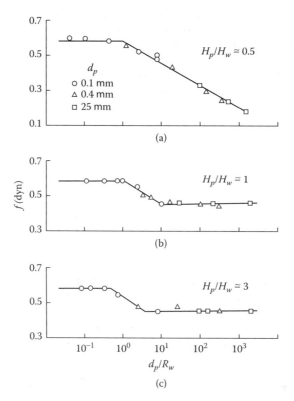

FIGURE 7.13 Dynamic powder–wall friction coefficient vs. the ratio of powder particle size to wall roughness for three values of the ratio of particle hardness to wall hardness (H_p/H_w). The powder is ferric oxide and the walls are (a) tungsten carbide, (b) hardened tool steel; and (c) nonhardened tool steel. (From Van Groenou, A.B., Pressing of ceramic powders: a review of recent work, *Powder Metallurgy Int.*, 10, 206, 1978. With permission.)

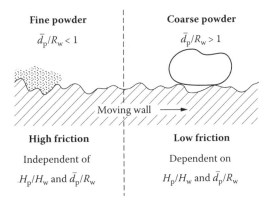

FIGURE 7.14 Sliding friction between a stationary powder compact and a moving wall. (From Strijbos, S., Powder-wall friction: the effects of orientation of wall grooves and wall lubricants, *Powder Technol.*, 18, 209, 1977.)

between the powder and the die wall. The effect of die–wall lubricants (e.g., stearic acid) can be fairly complex. For fine particles ($d_p/R_w < 1$), the coefficient of friction decreases gradually as the thickness of the lubricant increases, and the magnitude of the decrease can be fairly significant when the thickness of the lubricating layer becomes larger than the particle size. For coarse particles ($d_p/R_w > 1$), the presence of a lubricant causes only a small reduction in the die–wall friction and almost no dependence on the thickness of the lubricating layer is observed.

Density variations in powder compacts have been characterized by several techniques, including microscopy, microhardness, x-ray tomography, x-ray radiography, ultrasonic, and nuclear magnetic resonance [18]. Figure 7.15 shows the density variations in a section of a cylindrical manganese–zinc ferrite powder compact (diameter = 14 mm) produced by die compaction in the single-action mode of pressing. The large density difference in the upper and lower corners due to die–wall friction is very noticeable.

The particle size distribution also has an effect on the compaction behavior. For a powder with a wide distribution of sizes, a plot of the compact density vs. the logarithm of the applied pressure

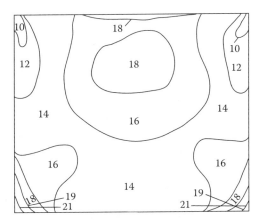

FIGURE 7.15 Density variation in a cross section of manganese ferrite powder compact produced by single-action die pressing from the top. High numbers correspond to low density, whereas low numbers correspond to high density. The numbers are the optical density of the x-ray transmission photographs of the compact. (From Welzen, J.T.A.M., Die compaction, in *Concise Encyclopedia of Advanced Ceramic Materials*, Brook, R.J., Ed., Pergamon, Oxford, 1991, p. 112. With permission.)

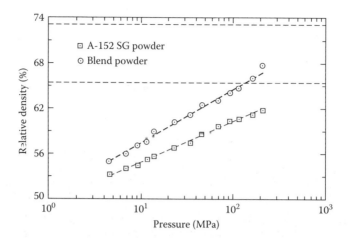

FIGURE 7.16 Density vs. pressure for powders having a different maximum particle packing density. (From Zheng, J. and Reed, J.S., Particle and granule parameters affecting compaction efficiency in dry pressing, *J. Am. Ceram. Soc.*, 71, C–456, 1988. With permission.)

has a steeper slope than that for a narrow size distribution (Figure 7.16), indicating that the deformation process occurs more easily. As outlined earlier (Figure 7.12), the type and amount of agglomerates in the powder influence its compaction behavior. The effect of the particle shape can sometimes be difficult to predict. The spherical (or equiaxial) shape is the commonly desired geometry, but flat particles with smooth surfaces can provide a higher compact density if they become aligned.

7.3.1.1.5 Factors Influencing the Compaction of Granules

The key granule characteristics that influence compaction are the hardness, the size, and the size distribution. The compaction process, as outlined earlier, can be divided into two stages: rearrangement of the granules at low pressure and deformation at higher pressure. Nominally, *hard* granules rearrange easily but, if too hard, are difficult to deform, producing a green body with large intergranular pores. These large pores are difficult to remove during sintering; so they limit the final density and produce microstructural flaws in the sintered article (Figure 7.17). Nominally, *soft* granules deform readily under pressure but, if too soft, will not rearrange sufficiently at low pressure to eliminate large packing flaws present after die filling, and so a compact with large density gradients is formed. The density gradients become magnified during sintering, often resulting in a limited final density and cracking. The requirement is, therefore, for granules with a nominal medium hardness, which can undergo rearrangement as well as deformation during compaction.

The granule hardness depends on the particle packing in the granule and on the properties of the binder. Because the particle packing in the granule is dependent on the properties of the slurry used in spray drying, control of the granule hardness actually requires an understanding of the relationship between the suspension characteristics and the granule characteristics. A stable colloidal suspension with a high particle concentration leads to nominally hard granules with a dense particle packing [9]. A partially flocculated slurry, leading to a lower particle packing density in the granules, is therefore desirable for spray drying.

The effect of granule density on compaction behavior is shown in Figure 7.18. Because the granules themselves pack to the same density during die filling, compacts formed from the denser granules have a higher green density at any pressure. However, the compact density is lower than the granule density, indicating that the large intergranular pores remain in the compact. In contrast, although the density of the compact produced from low-density granules is lower, it is higher than the granule density at any pressure above ~12 MPa, indicating that the large intergranular pores in the compacts have largely been eliminated.

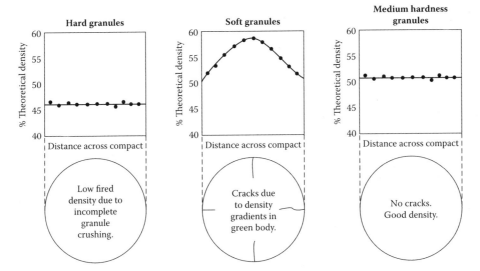

FIGURE 7.17 Qualitative results from experiments with nominally hard, soft, and medium hardness Al$_2$O$_3$ granules compacted uniaxially after irregular die filling. The upper set of graphs show density across the diameter of the green compact, whereas the lower set of illustrations are schematics showing the appearance of the sintered pellets (top view). Granule hardness was modified by granulating the Al$_2$O$_3$ powder with different organic binders. (From Glass, S.J. and Ewsuk, K.G., Ceramic powder compaction, *MRS Bull.*, 22(12), 24, 1997. With permission.)

The hardness of organic binders is determined by the glass transition temperature, T$_g$. If the compaction temperature (commonly, room temperature) is well below T$_g$, the binder is hard and brittle; so the granule is difficult to deform. On the other hand, if the temperature is well above T$_g$, the binder is soft and rubbery. If the binder is too soft, rearrangement at lower pressure is inefficient, resulting in density gradients. Methods for altering the T$_g$ of organic binders were outlined earlier in this chapter. Water is a good plasticizer for poly(vinyl alcohol), a commonly used binder in spray drying, and so the hardness of the granules can change with the humidity of the atmosphere [21,22].

Provided that it is much smaller than the die diameter, granule size does not have a significant effect on the compact density (Figure 7.18). The granule size distribution influences the packing

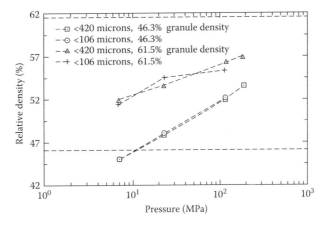

FIGURE 7.18 Density vs. pressure for a powder prepared at two granule densities. (From Zheng, J. and Reed, J.S., Particle and granule parameters affecting compaction efficiency in dry pressing, *J. Am. Ceram. Soc.*, 71, C-456, 1988. With permission.)

and the compaction behavior. A wider size distribution produces a higher packing density after die filling, but a narrow size distribution is found to produce a higher compact density after pressing [8]. Binder segregation to the granule surface and macroscopic flaws in the granule (e.g., holes) produce packing heterogeneities in the compact that normally remain as microstructural flaws in the sintered body.

7.3.1.1.6 Ejection of the Powder Compact

During compaction, the particles also undergo elastic compression. When the pressure is released, the stored elastic energy leads to an expansion of the compact. This expansion is referred to as *springback*, *strain recovery*, or *strain relaxation*. Springback is almost instantaneous on release of the pressure. The amount of springback depends on several factors, such as the powder, the organic additives, the applied pressure, the rate of pressing, and the gas permeability of the powder compact. Generally, it is larger for higher amounts of organic additives and for higher applied pressure. Whereas a small amount of strain recovery is desirable to cause the compact to separate from the punch, an excessive amount can lead to flaws. Ejection of the powder compact from the die is resisted by friction between the compact and the die wall. Lubricants added to reduce die–wall friction during the compaction process also serve to reduce the pressure required for ejection.

7.3.1.1.7 Compaction Defects

On completion of die pressing, we require the green body be free of macroscopic flaws and density gradients be as low as possible. Density gradients lead to the development of crack-like voids in the sintered body and can also lead to cracking and warping during sintering. They also enhance the formation of flaws in the compact on ejection from the die. Several factors can be adjusted to reduce the extent of density gradients in the powder compact. Uniform die filling reduces the amount of internal movement of the powder during compaction. Lubricants reduce the friction between the particles as well as die–wall friction. Stress gradients (and hence density gradients) due to die–wall friction are reduced as the ratio of the length to diameter (L/D) of the compact is decreased. For the single-action mode of die compaction, L/D should be less than ~0.5, whereas for the double-action mode, it should be less than ~1.

The common flaws in compacts formed by die pressing are illustrated in Figure 7.19. They are caused mainly by springback and by friction at the die walls. The use of a binder to increase the compact strength, reduction of the applied pressure to reduce the extent of the springback, slow release of the pressure to reduce the rate of springback, and the use of a lubricant to reduce die–wall friction can significantly reduce the tendency for flaw formation.

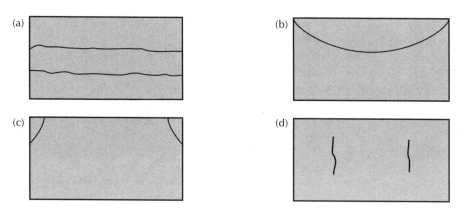

FIGURE 7.19 Illustrations of typical defects in die compaction of dry or semidry powders: (a) delamination, (b) end capping, (c) ring capping, and (d) vertical cracks.

7.3.1.2 Isostatic Compaction

Isostatic pressing involves the application of a uniform hydrostatic pressure to the powder contained in a flexible rubber container. There are two modes of isostatic pressing: wet-bag pressing and dry-bag pressing (Figure 7.20). In wet-bag pressing, a flexible rubber mold is filled with the powder, submerged in a pressure vessel filled with oil, and pressed. After pressing, the mold is removed from the pressure vessel and the green body is retrieved. Wet-bag pressing is used for the formation of complex shapes and for large sizes. In dry-bag pressing, the mold is fixed in the pressure vessel and need not be removed. The pressure is applied to the powder situated between a fairly thick rubber mold and a rigid core. After release of the pressure, the powder compact is removed from the mold. Dry-bag pressing is easier to automate than wet-bag pressing. It has been used for the formation of spark plug insulators by compressing a porcelain powder mixture around a metal core, as well as for plates and hollow tubes. Compared to die compaction, the formation of flaws in isostatically pressed compacts is much less severe, but delamination and fracture (caused by springback) can still occur if the applied pressure is released too rapidly.

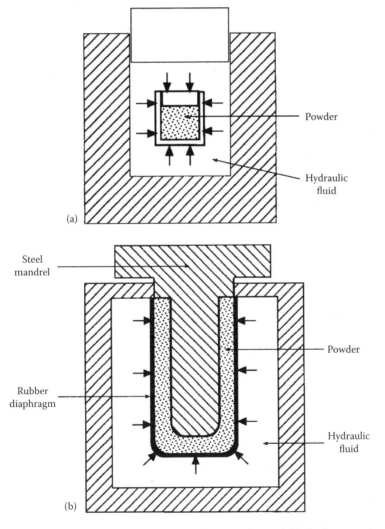

FIGURE 7.20 Two modes of isostatic pressing: (a) wet-bag pressing and (b) dry-bag pressing.

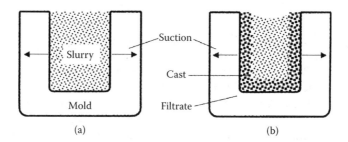

FIGURE 7.21 Schematic diagram of the slip-casting system: (a) initial system; (b) after the formation of a thin cast.

7.3.2 CASTING METHODS

The common casting methods are slip casting, pressure casting, and tape casting. They are based on colloidal systems in which removal of the liquid is used to consolidate particles suspended in a slurry. In slip casting and pressure casting, consolidation of the particles is accomplished as the liquid flows through a porous medium under a pressure gradient. In tape casting, evaporation of the liquid leads to consolidation. *Gelcasting* is a recently introduced method in which the particles in a slurry are immobilized by polymerization and cross-linking of a monomer solution to form a gel, after which the liquid is evaporated. *Electrophoretic deposition* is a process in which ceramic layers are deposited on an electrode by applying a direct current (dc) electric field to a colloidal suspension.

Homogeneous particle packing and as high a green density as possible remain the common requirements for the green body (the cast). The slurry used to produce these characteristics must have not only a high particle concentration but also the right rheological properties for adequate flow in order to provide a high enough casting rate for economical production. The attainment of the slurry characteristics requires an understanding of (1) the colloidal interactions between particles in a suspension and (2) the factors that control the rheological behavior of suspensions (Chapter 4). An understanding of particle packing concepts (Chapter 6) and the use of organic additives in processing, described earlier in this chapter, are also important.

Although the casting methods have the capability for producing homogeneous particle packing in the green body, they are generally limited to the production of relatively thin articles. Slip casting offers a route for the production of complex shapes and is widely used in the traditional clay-based industry. It has been steadily introduced over the past 50 years or so to the production of advanced ceramics. Tape casting is widely used for the production of thin sheets, substrates, and multilayer components for the electronic packaging industry.

7.3.2.1 Slip Casting

In slip casting [23,24], a slurry is poured into a permeable mold commonly made from gypsum ($CaSO_4 \cdot 2H_2O$). The microporous nature of the mold provides a capillary suction pressure, on the order of ~0.1 to 0.2 MPa, which draws the liquid (the filtrate) from the slurry into the mold. A consolidated layer of solids, referred to as a *cast* (or *cake*), forms on the walls of the mold (Figure 7.21). After a sufficient thickness of the cast has formed, the surplus slip is poured out and the mold and cast are allowed to dry. Normally, the cast shrinks away from the mold during drying and can be easily removed. Once fully dried, the cast is heated to burn out any organic additives and then sintered to produce the final article. Examples of slip-casting compositions are given in Table 7.5.

7.3.2.1.1 Slip-Casting Mechanics

Some of the early analyses incorrectly treated the mechanics of slip casting in terms of a *diffusion* process. The slip-casting process involves the flow of liquid through a porous medium, which is described by Darcy's law. In one dimension, Darcy's law can be written as:

TABLE 7.5
Examples of Slip-Casting Compositions

Whiteware		Alumina	
Material	Concentration (vol%)	Material	Concentration (vol%)
Clay, silica, feldspar	45–50	Alumina	40–50
Water	50	Water	50–60
Sodium silicate, polyacrylate, or lignosulfate (dispersant)	<0.5	Ammonium polyacrylate (dispersant)	0.5–2
Calcium carbonate (flocculant, if required)	<0.1	Ammonium alginate or methyl cellulose (binder)	0–0.5

$$J = \frac{K(dp/dx)}{\eta_L} \tag{7.3}$$

where J is the flux of liquid, K is the permeability of the porous medium, dp/dx is the pressure gradient in the liquid, and η_L is the viscosity of the liquid. In slip casting, the pressure gradient that causes flow arises from the capillary suction pressure of the mold. As the consolidation of the particles proceeds, the filtrate passes through two types of porous media: (1) the consolidated layer and (2) the mold (Figure 7.21).

Several authors have adopted the model of Adcock and McDowall [25], which neglects the resistance of the mold to liquid flow, treating the process in terms of liquid flow through the porous consolidated layer. In this case, an application of Darcy's law leads to a parabolic relation for the increase in the thickness of the cast L_c with time t:

$$L_c^2 = \frac{2K_c pt}{\eta_L(V_c / V_s - 1)} \tag{7.4}$$

where K_c is the permeability of the cast, p is the pressure difference across the cast (assumed to be constant and equal to the suction pressure of the mold), V_c is the volume fraction of solids in the cast (assumed to be incompressible), and V_s is the volume fraction of solids in the slurry. The rate of consolidation decreases with time, and this limits slip casting to a certain thickness of the cast above which further increases in the thickness are very time consuming.

Models in which the resistance of both media to flow is taken into account have been developed more recently [26,27]. The capillary suction pressure p is given by:

$$p = \Delta p_c + \Delta p_m \tag{7.5}$$

where Δp_c and Δp_m are the hydraulic pressure drops in the cast and in the mold, respectively (Figure 7.22). If η_L, V_c, V_s, K_c, the porosity P_m, and the permeability K_m of the mold do not change with time, the pressure drops Δp_c and Δp_m are linear. The flux of the liquid must be the same in the cast and in the mold, and so for this case:

$$J = \frac{K_c}{\eta_L L_c} \Delta p_c = \frac{K_m}{\eta_L L_m} \Delta p_m \tag{7.6}$$

where L_c is the thickness of the cast and L_m is the thickness of the mold saturated with liquid. Integration of Equation 7.6 subject to the appropriate boundary conditions gives a parabolic equation [27]:

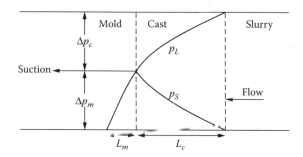

FIGURE 7.22 Hydraulic pressure distribution across the cast and the mold in slip casting.

$$L_c^2 = \frac{2Hpt}{\eta_L} \tag{7.7}$$

where the function H depends on the properties of both the cast and the mold, given by:

$$H = \frac{1}{(V_c/V_s - 1)\left[\dfrac{1}{K_c} + \dfrac{(V_c/V_s - 1)}{P_m K_m}\right]} \tag{7.8}$$

When the resistance to liquid flow in the mold can be neglected, Equation 7.4 is recovered from Equation 7.7 and Equation 7.8. This situation applies when $(V_c/V_s - 1)/P_m K_m$ in Equation 7.8 is very small compared to $1/K_c$, i.e., when

$$\frac{K_c(V_c/V_s - 1)}{P_m K_m} \ll 1 \tag{7.9}$$

According to Equation 7.9, the hydraulic resistance of the mold can be neglected when the mold porosity P_m, the volume fraction of solids in the slurry V_s, and the ratio of the mold permeability to the cast permeability K_m/K_c are high.

7.3.2.1.2 Effect of Permeability of the Cast

For a given system, Equation 7.4 predicts that the rate of consolidation increases with an increase in the permeability of the cast K_c. Various models have been put forward to account for the permeability of porous media. One of the most popular, based on its simplicity and its ability to outline the key parameters, is the Carman–Kozeny equation:

$$K = \frac{P^3}{\alpha(1-P)^2 S^2 \rho_S^2} \tag{7.10}$$

where P is the porosity, S is the specific surface area (surface area per unit mass of the solid phase), ρ_S is the density of the solid phase, and α is a constant, equal to 5 for many systems, that defines the shape and tortuosity of the channels. For a cast consisting of spherical monodisperse particles with a diameter D, the permeability can be expressed as

$$K_c = \frac{D^2(1-V_c)^3}{180V_c^2} \qquad (7.11)$$

It is clear that in order to increase K_c, we should increase D or reduce V_c. For advanced ceramics, an increase in D is often not desirable because this leads to a rapid decrease in the sintering rate. On the other hand, varying V_c, by controlling the colloidal stability of the slurry, can provide an effective method for controlling the consolidation rate (Chapter 4).

7.3.2.1.3 Effect of Mold Parameters

Equation 7.4 and Equation 7.7 indicate that the consolidation rate increases with the capillary suction pressure p of the mold. If there were no other effects, an increase in p would always lead to a shorter time for a given thickness of the cast. The capillary suction pressure varies inversely as the pore radius of the mold, and it might be thought that a decrease in the pore radius would lead to an increase in the casting rate. However, the permeability of the mold K_m also decreases with a decrease in the pore radius of the mold; so there should be an optimum pore size to give the maximum rate of casting [27].

7.3.2.1.4 Effect of Slurry Parameters

The colloidal stability of the slurry has the strongest influence on the microstructure of the cast. A flocculated slurry leads to a cast with a high porosity. Moreover, in this case, the effective pressure in the cast (p_s in Figure 7.22) decreases rapidly and nonlinearly from the mold–cast interface. The compressible nature of the highly porous cast coupled with the variation in p_s leads to a rapid reduction in the density of the cast as it builds up and to an almost constant high-porosity region in the bulk of the cast [25]. Heterogeneities in the green body microstructure, we will recall, hinder microstructural control during sintering. A well-dispersed slip containing no agglomerates and stabilized by electrostatic or steric repulsion leads to the formation of a cast with high packing density and homogeneous microstructure. In practice, the dense cast formed from a well-dispersed slip has a low permeability and so the rate of casting is low. For industrial operations in which such low casting rates are uneconomical, the slip is partially deflocculated.

Equation 7.4 indicates that the consolidation rate increases with the solids concentration of the slurry V_s. For monodisperse particles, V_s is not expected to have a significant effect on the resulting green density of the cast. In the case of a wide distribution of particle sizes, however, the effect of V_s on the green density is somewhat more complex. When V_s is low, a reduction in the green density may be expected if the sedimentation rate of the particles is significant compared to the casting rate, leading to segregation of particles of different size. For more concentrated suspensions, the sedimentation rate is diminished significantly and high green densities can be achieved if the fine particles fill the interstices between the large ones.

Other parameters such as the size, size distribution, and shape of the particles can also influence the compressibility and, hence, the packing density of the cast. Figure 7.23 summarizes the effects of particle size, particle shape, and the degree of flocculation of the slurry on the packing density of the cast. For coarse particles (greater than ~10 to 20 μm), colloidal effects are insignificant and the degree of particle dispersion has no effect on the packing density. Large spheres of the same size produce casts with a packing density of ~0.60 to 0.65, close to that for dense random packing, whereas irregular particles produce casts with a lower packing density. As the particle size decreases below ~10 μm, colloidal effects control the packing density. At one extreme, well-dispersed slips produce casts with high packing density, whereas at the other extreme, flocculated slips yield a low packing density. As outlined earlier, particles with a distribution in sizes may give casts with a packing density that is higher than that for monodisperse particles if segregation does not take place.

The rate of consolidation also increases with a decrease in the viscosity of the filtrate η_L (Equation 7.4). A reduction in η_L is achieved by increasing the temperature of the slurry or, in a

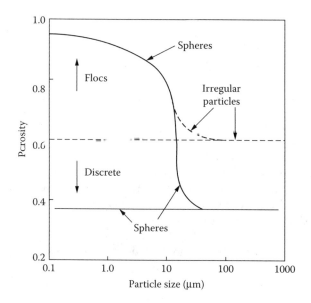

FIGURE 7.23 Schematic diagram showing the effects of particle size, shape, and degree of flocculation on the porosity of the cast produced in slip casting. (From Tiller, F.M. and Tsai, C.-D., Theory of filtration of ceramics: I, slip casting, *J. Am. Ceram. Soc.*, 69, 882, 1986. With permission.)

less practical way, by using a liquid with a lower viscosity. For aqueous slips, an increase in the temperature improves the stability of the slip and leads to a cast with a higher packing density. This decreases the consolidation rate because the permeability of the cast K_c decreases. However, the reduction in η_L with increasing temperature has a far greater effect and the overall result is an increase in the casting rate.

7.3.2.1.5 Microstructural Flaws in Slip-Cast Green Bodies

Several types of microstructural flaws can be present in green articles formed by slip casting. They arise during the casting operation and are generally related to the properties of the slurry. Large voids due to air bubbles in the slurry are a common occurrence, and they can be avoided by improving the wetting characteristics of the particles by the liquid, proper de-airing of the slurry, and the avoidance of turbulent flow of the slurry during the casting operation. Elongated (aniso-metric) particles can be preferentially aligned along certain directions, usually parallel to the mold surface. Segregation, in which the larger particles settle faster than the smaller, can be alleviated by improving the colloidal stability of the slurry.

7.3.2.2 Pressure Casting

Equation 7.4 indicates that for a given slurry, as the filtration pressure p increases, the time taken to produce a given thickness of cast L decreases. The casting can, therefore, be speeded up by the application of an external pressure to the slurry. This is the principle of *pressure casting*, also referred to as *pressure filtration* or *filter pressing*. The gypsum molds used in slip casting are weak and cannot withstand pressures greater than ~0.5 MPa; so plastic or metal molds must commonly be used in pressure casting.

A schematic of the main features of a laboratory-scale pressure-casting device is shown in Figure 7.24. Particles in the slurry form a consolidated layer (the cast) on the filter as the liquid is forced through the system. The cast provides a much greater resistance to flow of the liquid when compared to the filter, so the kinetics of pressure casting can be described by Equation 7.4.

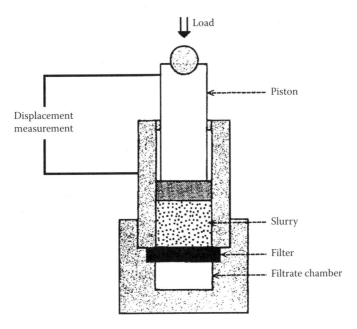

FIGURE 7.24 A laboratory-scale pressure-casting apparatus.

The kinetics and mechanics of pressure casting have been studied by Fennelly and Reed [28] and more recently by Lange and Miller [29]. As observed for slip casting, the colloidal stability of the slurry has the strongest influence on the microstructure of the cast. Figure 7.25 shows that the highest packing density is obtained with well-dispersed slurries, and for these slurries the packing density obtained is independent of the applied pressure above ~0.5 MPa. Dynamic models for particle packing that incorporate rearrangement processes have not been developed. However, the high packing densities obtained with dispersed slurries at such low applied pressures indicate

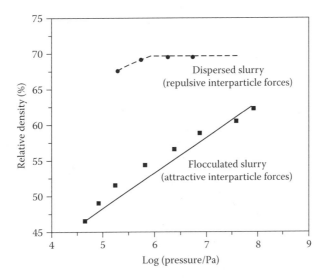

FIGURE 7.25 Relative density of different bodies produced from the same Al_2O_3 powder by filtration at different applied pressures. Bodies were consolidated from either dispersed (pH = 2) or flocculated (pH = 8) aqueous slurries containing 20 vol% solids. (From Lange, F.F., Powder processing science and technology for increased reliability, *J. Am. Ceram. Soc.*, 72, 3, 1989. With permission.)

that the repulsive forces between the particles must facilitate rearrangement. For flocculated slurries, the packing density is dependent on the applied pressure, and the approximate linear relationship between the relative density and the logarithm of the applied pressure shows the same trend observed earlier for the die pressing of dry powders.

Problems have been observed with pressure-cast bodies of advanced ceramics produced from either a dispersed or a flocculated slurry. As the last portion of the slurry is consolidated during pressure casting, the pressure gradient across the cast becomes zero and the total applied pressure is transferred onto the cast. On removal of the applied pressure, the cast expands, undergoing strain recovery due to the stored elastic energy. The nature of the strain recovery is, however, different from that observed in the die pressing of dry powders. The strain recovery for compacts produced by pressure casting is time dependent. This phenomenon of time-dependent strain recovery arises because the fluid (liquid or gas) must flow into the compact to allow the particle network to expand and relieve the stored strain. The magnitude of the recovered strain increases with increasing consolidation pressure in a nonlinear manner and can be described by a Hertzian elastic stress–strain relation of the form:

$$p = \beta \varepsilon^{3/2} \tag{7.12}$$

where p is the stress, ε is the strain, and β is constant for a given particulate system. For Al_2O_3, the recovered strain can be fairly large (2 to 3%) for moderately low pressures in the range of 50 to 100 MPa. A consequence of the fairly large strain recovery is the tendency for cracking to occur in the compact. This problem can be alleviated to some extent by reducing the consolidation pressure and by using a small amount of binder (less than ~2 wt%). None of these problems have been reported for pressure-cast bodies of clay-based materials, presumably because of the significant plasticity of clays.

Compared to slip casting, pressure casting offers the advantages of greater productivity through shorter consolidation times and the requirement of a smaller amount of floor space for the setting up of the molds. On the other hand, pressure-casting molds are more expensive. Slip casting will remain an important forming method for ceramics, but the trend is toward increasing use of pressure casting, particularly for fine powders of advanced ceramics.

7.3.2.3 Tape Casting

The principles and practice of tape casting are covered in a recent book [31] and in several review articles [32–35]. In tape casting, sometimes referred to as the *doctor-blade process*, the slurry is spread over a surface covered with a removable sheet of paper or plastic using a carefully controlled blade (the doctor blade). For the production of long tapes, the blade is stationary and the surface moves (Figure 7.26), whereas for the production of short tapes in the laboratory, the blade is

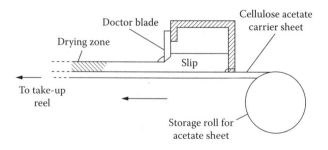

FIGURE 7.26 Schematic diagram of the tape-casting process.

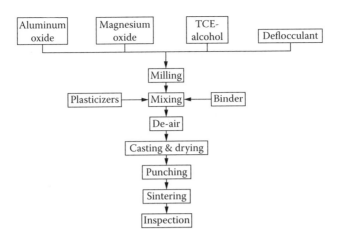

FIGURE 7.27 Schematic flow diagram for the tape casting of an alumina substrate material.

commonly pulled over a stationary surface. Drying occurs by evaporation of the solvent to produce a tape, consisting of particles bonded by the polymeric additives, which adheres to the carrier surface. The flexible green tape can be stored on take-up reels or stripped from the carrier surface and cut into the desired lengths for subsequent processing. Sheets with thicknesses from ~10 μm to ~1 mm can be prepared by tape casting. The steps in the production of a MgO-doped Al_2O_3 component by tape casting are summarized in Figure 7.27.

7.3.2.3.1 Slurry Preparation

The preparation of the slurry is a critical step in the tape casting process. The factors that govern the selection of solvent, dispersant, binder, plasticizer and other additives were described earlier in this chapter. Most tape casting operations currently use organic solvents but the trend is towards aqueous-based systems. In addition to the ability to dissolve the processing additives, selection of a solvent is dependent on the thickness of the tape to be cast, the drying rate of the tape, and the surface on which the tape is cast. In laboratory batch casting, thin tapes can be cast from highly volatile solvent systems whereas thicker tapes (> 0.25 mm) can be cast from slower drying solvents. For continuous casting in industry, the drying rate is an important consideration, since the tape must often be dried and stored on take-up reels prior to subsequent processing.

The dispersant may be the most important processing additive in that it serves to lower the viscosity of the slurry, allowing the use of a high particle concentration for a given useable viscosity. The selection of the binder–plasticizer combination is also important because a high binder concentration is used in tape-casting slurries. The binder and plasticizer must provide the required strength and flexibility of the green tape and must also be easily burnt out prior to sintering of the tape. Many organic systems can easily satisfy these requirements if the binder burnout process is carried out in an oxidizing atmosphere. However, several applications require that the additives be removed in a nonoxidizing atmosphere. A binder–plasticizer system that leaves as little residue as possible after the burnout process must be selected. Typical slurry formulations for tape casting are given in Table 7.6.

7.3.2.3.2 The Tape-Casting Process

As illustrated in Figure 7.26, a key component of the tape-casting equipment is the doctor-blade assembly. It consists of an adjustable doctor blade mounted in a frame with a reservoir to hold the slurry before it is metered out under the blade to form the thin layer of slurry on the carrier surface. The flow behavior of the slurry during the casting of the tape has been analyzed theoretically to determine the influence of the casting parameters on the thickness of the tape [36]. Assuming a

TABLE 7.6
Examples of Tape-Casting Compositions (Concentrations in wt%)

Powder	Solvent	Binder	Plasticizer	Dispersant	Other Additives
		Nonaqueous Formulation for Use in Oxidizing Atmospheres			
Al_2O_3 (59.5)	Ethanol (8.9)	PVB (2.4)	Octylphthalate (2.2)	Fish oil (1.0)	
MgO (0.1)	TCE (23.2)		PEG (2.6)		
		Nonaqueous Formulation for Use in Nonoxidizing Atmospheres			
$BaTiO_3$ (69.9)	MEK (7.0)	30 wt% solution	PEG (2.8)	Fish oil (0.7)	Cyclohexanone
	Ethanol (7.0)	of acrylic in	Butyl benzyl		(homogenizer)
		MEK (9.3)	phthalate (2.8)		(0.5)
		Aqueous Formulation			
Al_2O_3 (69.0)	Deionized water (14.4)	Acrylic emulsion (cross-linkable) (6.9)	Acrylic emulsion (low T_g) (9.0)	Ammonium polyacrylate (0.6)	Poly (oxyalkyl-enediamine) (0.1)

Source: From Mistler, R.E. and Twiname, E.R., *Tape Casting — Theory and Practice*, The American Ceramic Society, Westerville, OH, 2000; Mistler, R.E., Shanefield, D.J., and Runk, R.B., Tape casting of ceramics, in *Ceramic Processing Before Firing*, Onoda, G.Y. Jr. and Hench, L.L., Eds., John Wiley & Sons, New York, 1978, p. 411; Mistler, R.E., The principles of tape casting and tape casting applications, in *Ceramic Processing*, Terpstra, R.A., Pex, P.P.A.C., and DeVries, A.H., Eds., Chapman and Hall, London, 1995, p. 147.

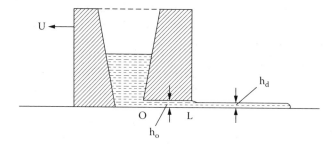

FIGURE 7.28 Section of a tape-casting unit.

Newtonian viscous slurry and laminar flow in a simple casting unit (Figure 7.28), the thickness of the dry tape, h_d, is given by:

$$h_d = \frac{\alpha\beta}{2} \frac{\rho_w}{\rho_d} h_o \left(1 + \frac{h_o^2 \Delta p}{6\eta UL} \right)$$
(7.13)

where α (<1) and β (<1) are correction factors, ρ_w and ρ_d are the densities of the slurry and the dry tape, respectively, h_o and L are the height and thickness of the doctor blade, respectively, Δp is the pressure difference (determined by the height of the slurry in the reservoir), η is the viscosity of the slurry, and U is the velocity of the doctor blade relative to the casting surface.

According to Equation 7.13, the thickness of the dried tape would simply be proportional to the height h_o of the doctor blade if the second term in the brackets is much smaller than unity. For values of h_o smaller than ~200 μm, this can be achieved if the parameters η, U, L, and Δp are kept within certain ranges. Large values of h_o can lead to significant deviations from this simple relationship. The effect is more pronounced for small values of η, U, and L. A knife-edged doctor blade (very small value of L) appears to be unsuitable for tape casting. While a variety of blade designs and shapes are used in the tape-casting sector, a flat-bottomed doctor blade is effective practically and has some theoretical support based on Equation 7.13. A double doctor blade is often found to be useful when a uniform thickness must be maintained over a long length of tape.

In practice, the casting speed is largely determined by the type of casting process: continuous or batch. For a continuous process, the speed is determined by the length of the casting machine, the thickness of the tape, and the volatility of the solvent. Typical casting speeds can vary from 15 cm/min for a continuous process to 50 cm/min for a batch process. Several carrier surfaces have been used in tape casting, ranging from glass plates in laboratory machines employing a moving doctor blade, to polyester and coated polyester carriers in continuous industrial machines. The type of carrier surface to be used is determined by the interaction of the solvent–binder system with the carrier surface and by how well the dried tape can be peeled off.

Drying occurs by evaporation of the solvent from the surface and the tape adheres to the carrier surface, and so the shrinkage during drying occurs in the thickness of the tape. Typically, the dried tape has a thickness equal to approximately half the doctor-blade height and consists of approximately 50 vol% ceramic particles, 30 vol% organic additives, and 20 vol% porosity. Despite the relatively low density of the tape after binder burnout, the sintered tape can reach nearly full density. However, the occurrence of microstructural flaws and warping of the tape are common problems resulting from inadequate or improper processing.

7.3.2.3.3 Microstructural Flaws in Tape-Cast Sheets

The main types of microstructural flaws in green tapes are similar to those outlined earlier for slip-cast bodies. Large voids are caused by air bubbles in the slurry or by rapid evaporation of the liquid during drying. They can be avoided by proper de-airing of the slurry or controlling the drying rate. Segregation due to differential settling of particles with different sizes is not a severe problem unless the tape is thick, the rate of drying is slow, or the colloidal stability of the slurry is poor. Because of the high concentration of binder in the slurry, binder segregation to the surface of the tape during drying can lead to a gradient in the packing density. The effect of particle or binder segregation is to produce warping (or even cracking) of the tape during sintering. If the particles in the slurry are elongated (anisometric), flow under the doctor blade produces preferential alignment of the particles along the direction of flow. This alignment of elongated particles can be used to produce ceramics with textured or aligned grain microstructure. However, for electronic substrate applications, particle alignment in tape-cast sheets is undesirable because the resulting anisotropic shrinkage during sintering makes dimensional control difficult.

7.3.2.4 Gelcasting

Gelcasting is a recently introduced process based on techniques taken from the traditional ceramics industry and the polymer industry [37–40]. As summarized in Figure 7.29, a slurry of ceramic particles dispersed in a monomer solution is poured into a mold, and the monomer is polymerized to form a gel-like bonding phase, immobilizing the particles. The system is removed from the mold while still wet, dried by evaporation of the liquid, heated to burn out the organic additives, and finally sintered.

Gelcasting commonly employs aqueous solvents (although organic solvents can also be used), dispersants, and processing methods similar to those used in traditional slip casting to produce the normally required slurry properties for casting: stability against flocculation, high particle concentration (~50 vol% for gelcasting), and low viscosity. The key element of the process is the addition of an organic monomer to the solution that is polymerized *in situ* to form a strong, cross-linked gel. In addition to preventing segregation or settling of the particles, the gel gives strength to the body to withstand capillary stresses during drying, so thick as well as thin parts can be formed.

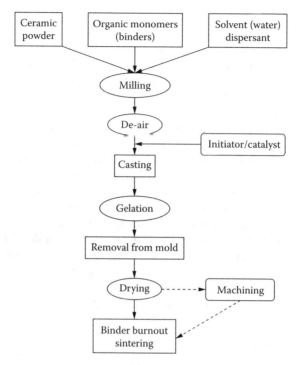

FIGURE 7.29 Schematic flow diagram of the gelcasting process.

There is only 2 to 4 wt% of polymeric material in the dried body, so binder burnout is not normally a limiting step in the fabrication process.

7.3.2.4.1 Monomers and Polymerization

The monomer solution consists of the solvent (commonly water), a chain-forming monomer, a chain-branching (cross-linking) monomer, and a free radical initiator. Commonly used chain formers are methacrylamide (MAM), hydroxymethacrylamide (HMAM), *N*-vinyl pyrrolidone (NVP), and methoxy poly(ethylene glycol) monomethacrylate (MPEGMA). Sometimes two monomers may be used in combination (e.g., MAM and NVP, or MPEGMA and HMAM). The cross-linkers commonly used are methylene bisacrylamide (MBAM) and poly (ethylene glycol) dimethacrylate (PEGDMA). The free radical initiator used most often is ammonium persulfate (APS) with tetramethylethylene diamine (TEMED) used as a catalyst.

The choice of the monomer system depends on several factors, such as the reactivity of the system (including the reaction temperature), the strength, stiffness and toughness of the gel, and the strength and machinability of the green body. One system used for several ceramics is MAM-MBAM, in which the total monomer concentration in the solution is 10 to 20 wt% and the MAM to MBAM ratio is 2 to 6. Another system is MAM-PEGDMA, in which the monomer concentration is also 10 to 20 wt% but the MAM to PEGDMA ratio is 1 to 3. Examples of gelcasting compositions are given in Table 7.7.

The formation of the gel occurs in two stages: initiation and polymerization. In the initiation stage, the viscosity does not change and almost no heat is generated. The addition of the initiator at room temperature allows a reasonable time (30 to 120 min) for de-airing of the slurry and mold filling. The polymerization step is often conducted at an elevated temperature (40 to 80°C), and the resulting faster reaction rate produces gelation in a short time. Because the reaction is exothermic, its progress can be followed by monitoring the temperature of the system.

TABLE 7.7
Examples of Compositions for Gelcasting

Ceramic powder[a]	Dispersant	Monomer solution[c]	Initiator
Al_2O_3	Ammonium polyacrylate[b]	MAM-MBAM or MAM-PEGDMA	APS/TEMED
Si_3N_4	Poly(acrylic acid)[b]	MAM-MBAM or MAM-PEGDMA	APS/TEMED
SiC	Tetramethyl ammonium hydroxide (pH > 11)	MAM-MBAM or MAM-PEGDMA	APS/TEMED

[a] Approximately 50 vol%.

[b] 0.5–2 vol%.

[c] Approximately 50 vol% (monomer concentration in the solution = 10–20 wt%); MAM-MBAM ratio = 2 to 6; MAM-PEGDMA ratio = 1 to 3.

Source: From Omatete, O.O., Janney, M.A., and Strehlow, R.A., Gelcasting — a new ceramic forming process, *Am. Ceram. Soc. Bull.*, 70, 1641, 1991; Young, A.C. et al., Gelcasting of alumina, *J. Am. Ceram. Soc.*, 74, 612, 1991.

7.3.2.4.2 Mold Materials

The commonly used mold materials for gelcasting are aluminum, glass, polyvinylchloride, polystyrene, and polyethylene. Aluminum and especially anodized aluminum are used widely for permanent production molds, whereas glass and polymeric materials are useful for laboratory experiments. The gelcasting system can react with the contact surfaces, so the mold surfaces are often coated with mold release agents, such as the commercial mold releases employed in the polymer processing industry.

7.3.2.5 Electrophoretic Deposition

The method of electrophoretic deposition (EPD) is shown schematically in Figure 7.30. A dc electric field causes the charged particles in a colloidal suspension to move toward, and deposit on, the oppositely charged electrode. EPD involves a combination of electrophoresis (Chapter 4) and particle deposition on the electrode. Successful EPD to form a deposit with homogeneous particle packing and high packing density requires a stable suspension. Agglomerated particles in an unstable suspension form a low-density deposit with an inhomogeneous microstructure. Therefore, an understanding of colloidal interactions in a suspension (Chapter 4) is important for EPD.

The rate of deposition is controllable by manipulating the applied electric field between the electrodes and the suspension concentration. The suspension can be stabilized electrostatically, sterically, or electrosterically, but the particles must carry a net charge in order to respond to the electric field. If an organic dispersant is used to stabilize the suspension, its concentration would be very small, and so binder burnout after the forming process is not a limiting step. Because the EPD process involves charged species, water (dielectric constant ≈80) would be an ideal liquid for the suspension, but because of electrolysis, organic solvents such as methanol, ethanol, propanol, and acetone are often used. The deposit can be formed on the outside, the inside, or both the outside and the inside of the electrode. The electrode used for deposition can be shaped but simple geometries, such as a flat plate or a cylinder, are often used. Often, the electrically conducting electrode does not form part of the required article and must be removed from the green body. Polishing the electrode or using a shape of conducting paper or plastic can facilitate electrode removal. Although deposition by EPD can be fast, the technique is best used for forming coatings and thin objects.

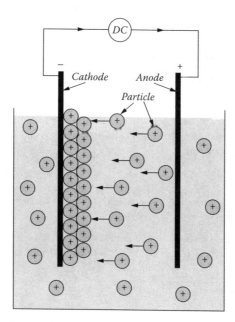

FIGURE 7.30 Schematic drawing of an electrophoretic deposition cell showing the process.

7.3.2.5.1 Kinetics and Mechanisms of Deposition

Although electrophoresis is well understood, the deposition mechanism of the particles on the electrode has been the subject of much controversy, and several explanations have been put forward to account for the phenomenon [41–45]. Figure 7.31 illustrates a recent explanation of the deposition process [46]. As a positively charged oxide particle with its surrounding double layer of counterions (the lyosphere) moves toward the cathode in an EPD cell, the electric field coupled with the motion of the charged particle through the liquid causes a distortion of the double-layer envelope: the envelope becomes thinner ahead of and thicker behind the particle. Compared to a particle far away from the electrode, the ζ potential is greater for the leading hemisphere of the particle and smaller for the trailing hemisphere. Cations in the liquid also move to the cathode along with the positively charged particles. The counterions in the trailing lyosphere tails tend to react with the high concentration of cations surrounding them, leading to a thinning of the double layer around the trailing surface of the particle. The next incoming particle, which has a thin double layer around the leading surface, can approach close enough so that van der Waals attractive forces dominate. The result is coagulation and deposition on the electrode.

The kinetics of EPD are important for controlling the thickness of the deposited layer. There are two modes of operating the system. In constant-voltage EPD, the voltage between the electrodes is maintained constant. Because deposition requires a steeper electric field than electrophoresis, the electric field decreases as the deposition thickness (and consequently the electrical resistance) increases. The particle motion and, hence, the rate of deposition decrease. Under constant-current EPD, on the other hand, the electric field is maintained constant by increasing the total potential difference between the electrodes, so the limited deposition in constant-voltage EPD is avoided.

Assuming that the suspension is homogeneous and the change in concentration is due to EPD only, the mass of particles m deposited on the electrode is equal to that removed from the suspension. Therefore:

$$\frac{dm}{dt} = AvC \qquad (7.14)$$

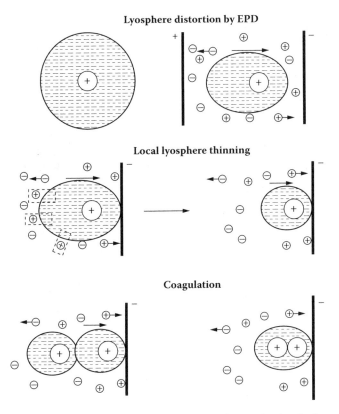

FIGURE 7.31 Schematic illustration of the electrophoretic deposition mechanism by lyosphere distortion and thinning. (From Sarkar, P. and Nicholson, P.S., Electrophoretic deposition (EPD): mechanisms, kinetics, and applications to ceramics, *J. Am. Ceram. Soc.* 79, 1987, 1996. With permission.)

where A is the area of the electrode, v is the velocity of the particle, C is the concentration of particles in the suspension, and t is the deposition time. For a concentrated suspension, the velocity of the particles is given by the Helmholtz–Smoluchowski equation (Chapter 4):

$$v = \frac{\varepsilon \varepsilon_o \varsigma E}{\eta} \qquad (7.15)$$

where ε is the dielectric constant of the liquid, ε_o is the permittivity of free space, ς is the zeta potential of the particles, E is the applied electric intensity, and η is the viscosity of the liquid. If m_o is the initial mass of the particles in the suspension, then

$$m = m_o - VC \qquad (7.16)$$

where V is the volume of the suspension. Combining Equation 7.14 and Equation 7.15 subject to the boundary condition of Equation 7.16 gives:

$$m = m_o(1 - e^{-\alpha t}) \qquad (7.17)$$

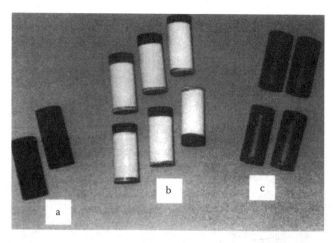

FIGURE 7.32 Photograph showing (a) porous LaMnO$_3$ tubes (1 in diameter × 2 in), (b) Y$_2$O$_3$-stabilized ZrO$_2$ coatings (white) deposited on LaMnO$_3$ tubes, and (c) ZrO$_2$ coatings on LaMnO$_3$ tubes after sintering at 1300°C, giving dense, translucent coatings. (From Basu, R.N., Randall, C.A., and Mayo, M.J., Fabrication of dense electrolyte films for tubular solid oxide fuel cells by electrophoretic deposition, *J. Am. Ceram. Soc.*, 84, 33, 2001. With permission.)

$$\frac{dm}{dt} = m_o \alpha e^{-\alpha t} \tag{7.18}$$

where $\alpha = Av/V$. According to Equation 7.18, the rate of deposition decreases exponentially with time and is controlled by the parameter α. If α is larger, the initial rate of EPD is higher but the rate decreases more rapidly.

7.3.2.5.2 Applications of EPD

EPD is a well-established technique that has been used to deposit inorganic coatings on metal wires for valve filaments and cathodes [47], monolithic ceramics such as Al$_2$O$_3$ and MgO [48,49], and beta-alumina electrolytes for sodium–sulfur batteries [50,51]. More recently, it has been applied to the deposition of high Tc superconductors [52–54], Y$_2$O$_3$-stabilized ZrO$_2$ (YSZ) electrolytes for solid oxide fuel cells [55], and laminated composites of Y$_2$O$_3$-stabilized ZrO$_2$ and Al$_2$O$_3$ [56]. Figure 7.32 shows an example of YSZ coatings deposited by EPD on porous LaMnO$_3$ substrates with a cylindrical cross section and sintered to almost full density.

7.3.3 PLASTIC-FORMING METHODS

Plastic deformation of a moldable powder–additive mixture is employed in several ceramic-forming methods. Extrusion is used extensively in the traditional ceramics sector to form moist clay–water mixtures into green articles with regular cross sections, such as solid and hollow cylinders, tiles, and bricks. The method is also used to form advanced ceramics for several applications, such as catalyst supports, capacitor tubes, and electrical insulators. Injection molding of a ceramic–polymer mixture is a useful method for the mass production of small ceramic articles with complex shapes, but it has not seen significant industrial application largely because of two factors: (1) high tooling costs relative to other common forming methods and (2) binder removal prior to sintering remains a limiting step for green bodies with a thickness greater than ~1 cm. Extrusion and injection molding of ceramics have benefited considerably from the principles and technology developed in the plastics industry. Extruders and molding machines used in the plastics industry are employed, but some modification of the machines is required for ceramic systems (e.g., hardening of the contact surfaces).

Two basic requirements must be satisfied for successful plastic forming: (1) the mixture must flow plastically (above a certain yield stress) for the desired shape to be formed and (2) the shaped article must be strong enough to resist deformation under the force of gravity or under stresses associated with handling. The selection of additives and the formulation of the mixture to be used in forming are critical steps in meeting these requirements.

7.3.3.1 Extrusion

In extrusion, a powder mixture in the form of a stiff paste is compacted and shaped by forcing it through a nozzle in a piston extruder or a screw-fed extruder [57]. The piston extruder is simple in design, consisting of a barrel, a piston, and a die. In contrast, the screw extruder is complex (Figure 7.33), and considerable attention goes into the design of the extruder barrel and screw [58]. The screw has to mix the powder and other additives into a homogeneous mass and generate enough pressure to transport the mixture against the resistance of the die. Shaping of the extruded body is achieved with the head of the extruder screw and the die. The extruder screw head changes the rotational flow of the mixture produced by the screw into an axial flow for extrusion and to produce uniform flow in the die. In the release of the body from the extruder, the die must generate the required cross section, allow uniform flow across the entire cross section, and ensure a smooth surface.

The main approaches used for imparting the required plastic properties to the feed material are: (1) manipulating the characteristics of the powder–water system, commonly used for clays, and (2) adding a binder solution to the powder, commonly used for advanced ceramics. Clay particles develop desirable plastic characteristics when mixed with a controlled amount of water (15 to 30 wt% depending on the type of clay). The plasticity arises from two main factors: (1) particle-to-particle bonding due to the charged particle surfaces and intervening charges (the "house of cards" structure described in Chapter 4) and (2) surface tension effects due to the presence of

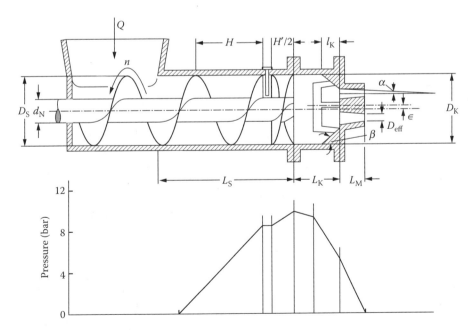

FIGURE 7.33 Pressure variation along a screw extrusion press. (From Pels Leusden, Extrusion, in *Concise Encyclopedia of Advanced Ceramic Materials*, Brook, R.J., Ed., Pergamon, Oxford, 1991, p. 131. With permission.)

TABLE 7.8
Examples of Compositions Used in Extrusion

| Whiteware | | Alumina | |
Material	Concentration (vol%)	Material	Concentration (vol%)
Kaolin	16	Alumina	45–50
Ball clay	16	Water	40–45
Quartz	16	Ammoniul polyacrylate (dispersant)	1–2
Feldspar	16	Methylcellulose (binder)	5
Water	36	Glycerin (plasticizer)	1
$CaCl_2$ (flocculant)	<1	Ammonium stearate (lubricant)	1

water between the particles. Surface charge and surface tension do not play a significant role in coarse ceramic powders, but some degree of plasticity may be developed by adding fine particles (such as clay or boehmite) provided they are chemically compatible with the coarse powder.

The powders of advanced ceramics, when mixed with water, do not possess the desirable plastic characteristics found in the clay–water system. For this reason, they are often mixed with a viscous solution containing a few weight percent of an organic binder to achieve the desired plastic characteristics. The solvent is commonly water but nonaqueous solvents (e.g., alcohols and mineral spirits) can also be used. As the extruded body must have sufficient green strength, the binder is generally selected from the medium to high-viscosity grades, e.g., methylcellulose, hydroxyethyl cellulose, poly(acrylimides), or poly(vinyl alcohol). Methylcellulose undergoes thermal gelation, a property that offers considerable benefits for extrusion of ceramics [59]. At higher temperatures, methylcellulose forms a gel, reflected by a sharp rise in viscosity, but returns to its original consistency at lower temperatures. Examples of compositions used in extrusion are given in Table 7.8. Clay-based systems are often slightly flocculated using a small concentration of a flocculant, such as $MgCl_2$, $AlCl_3$, or $MgSO_4$. A lubricant, such as a stearate, silicone, or petroleum oil, is commonly used to reduce die–wall friction.

7.3.3.1.1 Extrusion Mechanics

The flow pattern of the feed material through the extruder influences the quality of the shaped article. The rheology of concentrated ceramic suspensions can be divided into four classes: ideal plastic, Bingham, shear thinning, and shear thickening (Chapter 4). Bingham-type behavior is widely observed and forms a useful approximation for theoretical analysis [60,61]. A characteristic feature of the velocity profile is the occurrence of differential flow with a central plug (Figure 7.34). The velocity is constant with radius in the central plug, but decreases with radius between the central plug and the inner wall of the tube. In more extreme situations, slippage flow or complete plug flow can occur, in which the velocity of the material through the extruder is independent of the radius of the tube.

7.3.3.1.2 Extrusion Flaws

A variety of flaws can occur in the extruded body [62]. The common macroscopic flaws are laminations, tearing, and segregation (Figure 7.35). Laminations are cracks that generally form a pattern or orientation, particularly in a screw-fed extruder, because of incomplete reknitting of the feed material around the auger. Tearing consists of surface cracks that form as the material leaves the die and is caused by poor die design or by low plasticity of the mixture. Segregation involves separation of the liquid and the solid phases of the mixture during extrusion and is often caused by poor mixing. Microscopic flaws such as pores (caused by trapped air) and inclusions (due to contamination) can also occur.

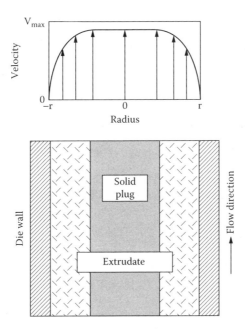

FIGURE 7.34 Differential flow with a central plug commonly observed in extrusion.

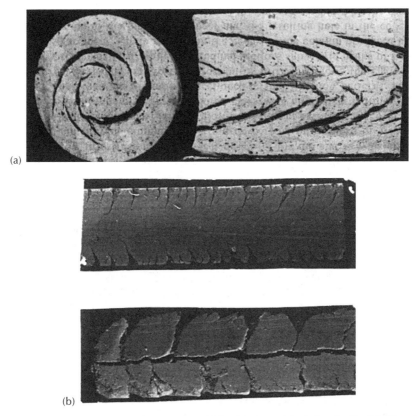

FIGURE 7.35 Typical defects observed in extrusion: (a) lamination cracks in longitudinal (left) and transverse (right) section; (b) edge tearing. (From Robinson, G.C., Extrusion defects, in *Ceramic Processing Before Firing*, Onoda, G.Y. Jr. and Hench, L.L., Eds., John Wiley & Sons, New York, 1978, p. 391. With permission.)

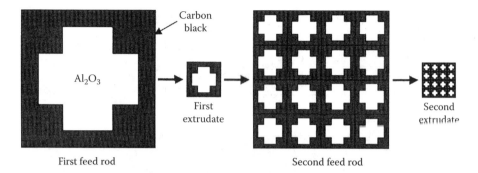

FIGURE 7.36 Illustration of the formation of fine-scale features by coextrusion.

7.3.3.2 Coextrusion

A method based on repeated coextrusion of a powder-filled thermoplastic polymer has been developed recently to form ceramics with a textured microstructure [63] or with fine-scale features [64]. A schematic of the method is shown in Figure 7.36. A mixture of ceramic particles (e.g., Al_2O_3), thermoplastic polymer (e.g., ethylene vinyl acetate), and processing aid, e.g., low-molecular-weight poly(ethylene glycol) as a plasticizer, containing ~50 vol% particles, is formed into a rodlike feed material (the feed rod) with the required arrangement of the ceramic phase by extrusion in a piston extruder or by lamination. Extrusion of the feed rod, commonly at temperatures in the range of 100 to 150°C, through a die in a piston extruder produces an extrudate with a smaller cross section (e.g., a diameter that is approximately five times smaller than the diameter of the feed rod) and with a corresponding reduction in the scale of the structural features. The material produced after the first extrusion is cut into suitable lengths and reassembled to form a new feed rod. After repeated extrusion, an extrudate with the required small-scale structural features is formed. Successful coextrusion is dependent on controlling the rheological properties of the feed material and the extrusion parameters to produce an extrudate with the required uniformity of structural features. After the final extrusion, the extrudate is compacted into the required shape, heated to decompose the binder and other additives, and sintered to produce the final article.

7.3.3.3 Injection Molding

The principles and practice of injection molding of metals and ceramics are covered in two books [65,66] and in several review articles [67–69]. The production of a ceramic article by injection molding involves the following steps: selection of the powder and the binder, mixing the powder with the binder, production of a homogeneous feed material in the form of granules, injection molding of the green body, removal of the binder (debinding), and finally sintering. The debinding step, involving features common to other forming methods (e.g., tape casting), is considered in the next chapter. Here, we shall consider key factors in the preparation of the feed material and the injection molding of the green body.

7.3.3.3.1 Powder Characteristics

When the forming step is considered in combination with other processing steps such as debinding and sintering, the desirable powder characteristics for injection molding are generally no different from those described in Chapter 2 (see Table 2.1). A smaller particle size is beneficial for shape retention during debinding and for ease of sintering, but the viscosity of the powder–binder mixture is higher and binder removal from the molded article is slower. A wider distribution of particle sizes gives a higher packing density, higher green strength, and lower shrinkage during sintering. On the other hand, the rate of binder removal is slower, the system is prone to segregation, and microstructural evolution during sintering can often be inhomogeneous. A more spherical (or

TABLE 7.9
Thermoplastic Binder Systems for Ceramic Injection Molding

Major Binder	Minor Binder	Plasticizer	Other Additives
Polypropylene	Microcrystalline wax	Dimethyl phthalate	Stearic acid
Polyethylene	Paraffin wax	Diethyl phthalate	Oleic acid
Polystyrene	Carnauba wax	Dibutyl phthalate	Fish oil
Poly(vinyl acetate)		Dioctyl phthalate	Organo silane
Poly(methyl methacrylate)			Organo titanate

equiaxial) shape leads to higher packing density, lower viscosity of the powder–binder mixture, improved flow during forming, a lower green strength, and often to slumping during binder removal. Generally, an equiaxial powder, free from hard agglomeration, with an average particle size smaller than 10 μm, a narrow or wide particle size distribution, and a packing density greater than ~60% is often suitable for most ceramic injection-molding operations.

7.3.3.3.2 Binder System

Careful selection of the binder system is vital to the success of injection molding. The binder must provide the desired rheological properties to the feed material to allow formation of the powder into the desired shape and then must be removed completely from the shaped article, prior to sintering, without disrupting the particle packing or react chemically with the powder. A good binder must, therefore, have the right rheological, chemical, and debinding characteristics. It must also possess other qualities suitable for manufacturing, such as environmental safety and low cost.

It is often difficult to find a single binder that will satisfy all the desired characteristics; so in practice, the binder system is generally a blend of at least three components: a major binder, a minor binder, and a processing aid. The major binder controls the rheology of the feed material during injection molding to form a body free from defects. It also controls the strength of the green body and the debinding behavior. The minor binder is used to modify the flow properties of the feed material for good filling of the mold. It may also provide benefits in the debinding stage by extending the range of conditions over which the binder is removed. For example, the removal of the minor binder (by dissolution or by pyrolysis) before the major binder will create a network of porosity through which the decomposition products of the major binder can be removed more easily. The processing aid may include small amounts of a plasticizer to reduce the glass transition temperature of thermoplastic binders, a wetting agent to improve the wetting between the particle surfaces and the polymer melt, and a lubricant to reduce interparticle and die–wall friction.

Several binder systems are used in ceramic injection molding, and they can be classified into five types, based on the composition of the major binder phase: (1) thermoplastic compounds, (2) thermosetting compounds, (3) water-based systems, (4) gelation systems, and (5) inorganics. Of these, the thermoplastic compounds are the most widely used and understood. They include most of the common commercial polymers, such as polystyrene, polyethylene, polypropylene, poly(vinyl-acetate), and poly(methyl methacrylate). Examples of materials used for thermoplastic binder systems are given in Table 7.9.

7.3.3.3.3 Powder–Binder Mixture

The ratio of powder to binder is a key parameter for successful injection molding. Too little binder in the mixture leads to a high viscosity and to the formation of trapped air pockets, both of which make molding difficult. On the other hand, too much binder leads to microstructural heterogeneities in the molded article and to slumping during binder burnout.

If f_m is the maximum packing density of the particles, defined as the volume fraction at which the particles touch so that flow is not possible, then the actual particle volume fraction f used in

TABLE 7.10
Examples of Ceramic Injection-Molding Feed Materials

Component	Composition (wt%)	
Powder	1-μm Al$_2$O$_3$ (85)	20-μm Si (82)
Major binder	Paraffin wax (14)	Polypropylene (12)
Minor binder	—	Microcrystalline wax (4)
Other additives	Oleic acid (1)	Stearic acid (2)

injection molding is 5 to 10 vol% lower than f_m. The remainder of the mixture consists of the binder and other additives. This means that for a well-dispersed powder, the particles are separated from their neighbors by a thin layer of polymer (~50 nm thick) during molding, which is required to achieve flow of the mixture. The volume fraction of particles f incorporated into the binder, therefore, depends on the particle size distribution, the particle shape, and, for particles finer than ~1 μm, the particle size. For equiaxial particles greater than ~1 μm, f is the range of 0.60 to 0.75.

The volume fraction of powder that can be reasonably incorporated into the mixture is best determined from viscosity measurements using a capillary rheometer over the range of conditions expected in the forming operation. Data for the relative viscosity (the viscosity of the mixture divided by the viscosity of the unfilled polymer) vs. particle concentration can be well fitted by the following equation [70]:

$$\eta_r = \left(\frac{1 - 0.25 f / f_m}{1 - f / f_m} \right)^2 \tag{7.19}$$

Inhomogeneous mixing and incomplete breakdown of the powder agglomerates influence the viscosity and cause serious problems in the molding step. The use of high-shear-stress mixers can serve to remove residual agglomeration during mixing. After the mixing step, the cooled mixture is passed though a cutting mill, where it is converted into granules (a few millimeters in diameter) that form the feed material for molding. Examples of the feed-material composition are given in Table 7.10.

7.3.3.3.4 Molding

When compared to injection molding of plastics, control of the molding process in ceramics is more difficult because of the higher density, viscosity, thermal conductivity, and elastic modulus associated with the particle-filled polymer. The higher modulus and higher cooling rate (due to the higher thermal conductivity) coupled with the lower fracture toughness of the particle-filled polymer means that the molded article is prone to cracking due to the development of residual stress during the solidification process. Attempts have been made to model the residual stress development in ceramic injection molding, but a more practical approach, involving a systematic variation of the process variables, is often used to optimize the molding conditions.

Figure 7.37 illustrates the principle of operation of an injection molding machine and some of the parameters that influence its functioning. The feed material in the form of granules is fed into the machine and transported by a screw or plunger to the injection chamber, where it is heated to produce a viscous mass and then injected, under pressure, into the mold cavity. When the cavity is filled, the mold is cooled and the green body is ejected. It is obvious from Figure 7.37b that a given machine variable can influence more than one of the fundamental variables. This, combined with the large number of variables, makes optimization of the process a considerable task. In general, injection molding is best applied selectively to the forming of small articles with complex shapes.

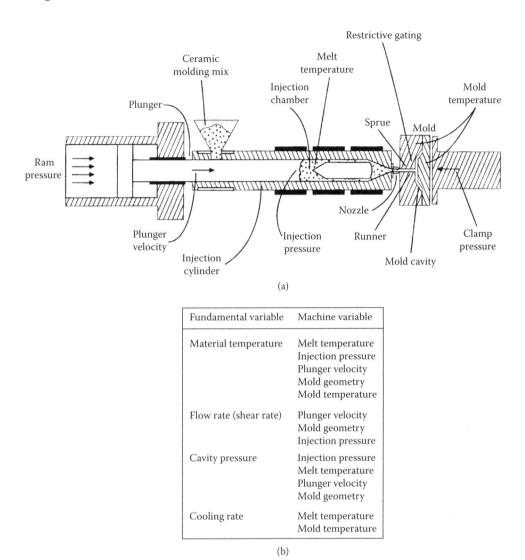

FIGURE 7.37 (a) Schematic of a plunger injection-molding machine identifying the principal machine variables; (b) relation of fundamental injection-molding variables to plunger machine variables. (From Mangels, J.A. and Trela, W., Ceramic components by injection molding, in *Advances in Ceramics, Vol. 9: Forming of Ceramics*, Mangels, J.A., Ed., The American Ceramic Society, Columbus, OH, 1984, p. 220. With permission.)

7.4 SOLID FREEFORM FABRICATION

Solid freeform fabrication (SFF) is a term used to describe processing technologies that allow parts to be formed with the required geometrical complexity directly from a computer-aided design (CAD) file, without the use of traditional forming tools such as dies or molds. There is no general agreement on the term, and several other terms, of which rapid prototyping is the most common, are also used. The basic approach of SFF is illustrated in Figure 7.38. A prototype, which represents the physical component to be built, is modeled on a CAD system. The model is next converted into a format that is analyzed by a computer, which slices the model into cross sections. The cross sections are systematically recreated in the computer-aided manufacturing step, employing suitable equipment and materials, to produce an actual three-dimensional prototype.

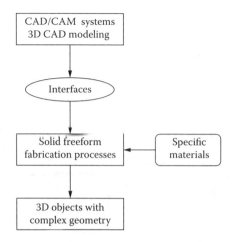

FIGURE 7.38 Basic principle of solid freeform fabrication.

SFF technology is not intended to replace traditional ceramic manufacturing technologies. A key benefit of the technology is the rapid production of prototypes of the object to be copied or developed, for limited production runs, evaluation of design, performance of limited testing, and improvement of the manufacturing process. A characteristic feature of SFF techniques is the layerwise deposition of material only where it is required: so the methods also provide significant flexibility for the production of small, complex-shaped components with unique structural features that are not achievable with other ceramic-forming methods.

A wide variety of techniques, employing liquids, powders and laminates, is available for SFF. The main features of the techniques and details of the available commercial equipment are described in recent texts [71–73]. The more mature techniques, such as stereolithography, laminated object manufacturing, fused deposition modeling, and selective laser sintering, have been developed and commercialized for the production of polymer or plastic components. Since the 1980s, these techniques have undergone rapid development and application to the forming of ceramics [74]. SFF of ceramics has also benefited from the rapid development of techniques based on the use of suspensions [75].

In general, the ceramic SFF techniques provide methods for forming complex-shaped green bodies from common feed materials, such as particle-filled polymers, powders, and suspensions. Many of the processing issues that have been discussed earlier, such as polymeric additives, colloidal interactions, rheological behavior, mixing of the constituents, and particle packing in the green body, are still applicable. The suspensions used in SFF often contain a considerable amount of liquid, and so drying of the as-deposited suspension may present problems in controlling the macroscopic shape of the component and the uniformity of the component walls. The particle-filled polymers used as feed materials in many of the SFF techniques can contain 40 to 70 vol% organic material, which results in debinding problems that limit the thickness of the component. Drying and debinding of ceramic green bodies are discussed in Chapter 8.

The key features of ceramic SFF techniques are discussed in three sections based on the type of feed material: particle-filled polymers, powders, and suspensions.

7.4.1 PARTICLE-FILLED POLYMER METHODS

7.4.1.1 Fused Deposition Modeling

Fused deposition modeling (FDM) is a technique that builds plastic objects by extrusion of a polymer–wax filament through a nozzle (Figure 7.39). The technique is essentially a hot extrusion process. A modification of the technique, referred to as *fused deposition of ceramics* (FDC), has

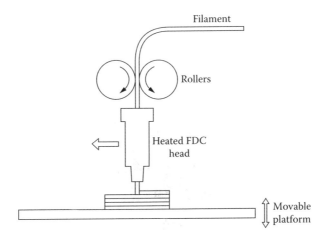

FIGURE 7.39 Schematic of the fused deposition process.

been developed to create ceramic components from particle-filled polymer filaments [76,77]. The ceramic–polymer mixture (50 to 60 vol% particles) is first extruded to form filaments with a diameter of ~2 mm, after which the spooled filaments are fed into a computer-controlled, heated extrusion head (100 to 150°C). Extrusion of the plastic mixture through a nozzle (diameter 0.25 to 0.64 mm), according to a computer-controlled pattern, is used to form the object layer by layer. The green body is then subjected to debinding and sintering steps to produce a dense object.

Ceramic–polymer mixtures used in FDC have to be optimized to produce desirable flow properties for the hot extrusion process [78] in a manner similar to that described for injection molding. The filaments must also have enough flexibility to allow winding and unwinding of continuous lengths on a spool, as well as sufficient stiffness to act as a piston for extrusion of the molten material through the fine nozzles. There must be good adhesion between each deposited layer. Removal of the large amount of polymeric binder material as cleanly as possible must also be considered. Inadequate processing of the ceramic–polymer feed material and limitations of the deposition process can lead to internal and surface flaws that degrade the strength of the final sintered object [79].

7.4.1.2 Laminated Object Manufacturing

In laminated object manufacturing (LOM), layers of a thin sheet are sequentially bonded to each other and cut by a laser according to a computer-generated model (Figure 7.40). A modified form of the process, referred to as *computer-aided manufacturing of laminated engineering materials* (CAM-LEM), has been developed to produce complex shapes directly from tape-cast ceramics [80]. In CAM-LEM, individual slices are cut from the tape-cast sheet using a laser and stacked to assemble the computer-aided design. Following lamination of the slices and binder removal, the body is sintered to produce a ceramic component. Modifications of the LOM process have also been developed to produce multilayered ZrO_2–Al_2O_3 composites [81] and fiber-reinforced ceramic matrix composites [82].

7.4.2 POWDER METHODS

7.4.2.1 Selective Laser Sintering

In selective laser sintering (SLS), components are built layer by layer by scanning a laser beam over a thin layer of powdered material [73]. For polymers and some metals, interaction of the laser beam with the powder raises the temperature to the point of melting, resulting in particle bonding

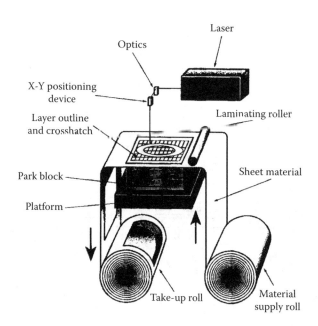

FIGURE 7.40 Illustration of laminated object manufacturing.

and fusion of the particles to themselves as well as to the previous layer to form a solid object. Crystalline ceramics cannot be formed directly by SLS because matter transport by solid-state diffusion is insignificant during the short time of laser scanning. An alternative way is to use a polymeric binder mixed with the ceramic powder, which provides the bonding phase for forming by SLS. Following binder removal, the body is sintered to produce a dense object.

7.4.2.2 Three-Dimensional Printing

In three-dimensional printing (3DP), complex-shaped parts are formed by sequentially depositing a thin layer of ceramic powder followed by ink-jet printing of a binder solution to fix the powder in place and to selectively define the geometry of the part [83]. A thin layer of powder can be formed by roll compaction, but more homogeneous particle packing and higher packing density are obtained by deposition from a well-dispersed suspension (e.g., through a nozzle 100 to 200 μm in diameter) followed by drying. After application of the binder solution to fix the powder, the layer is heated to remove excess liquid (water). Once a single layer is complete, the sequential slurry and binder deposition processes are repeated until the part is completed. The binder is then cured to develop adequate strength, and the unwanted powder is redispersed in a liquid to recover the part. Finally, the shaped part is heated to decompose the binder and sintered to produce a ceramic object.

Three key aspects of the 3DP process can be identified. First, the colloidal properties of the suspension and the drying of the deposited layer control the structure of the deposited powder layer [84]. Second, the interaction of the binder solution with the powder layer must be optimized to control the shape uniformity of the printed part [85]. Third, the redispersion of the unwanted powder to retrieve the printed part is controlled by the chemistry and colloidal properties of the suspension [86].

7.4.3 Suspension Methods

7.4.3.1 Stereolithography

A schematic of the stereolithography technique is shown in Figure 7.41. In the formation of ceramics, the liquid monomer solution used for polymer fabrication is replaced by a highly

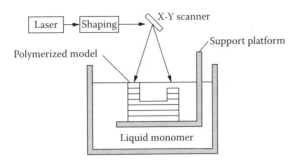

FIGURE 7.41 Diagram of the stereolithography apparatus.

concentrated suspension of ceramic particles in a monomer solution. An ultraviolet (UV) laser beam is scanned on the surface of the suspension, and the monomer is cured to form a polymer layer that binds the particles together. Some UV-curable solutions used in stereolithography of ceramics are similar to the gelcasting solutions described earlier, but with the thermal initiators replaced by photoinitiators. When the first layer is completed, the support platform is lowered by a depth equal to the layer thickness, and the suspension flows over the polymerized layer. The laser is scanned over the new surface to form the second layer, and the process is repeated until the component is completed. Ceramic components are obtained after binder removal and sintering [87].

If a dense ceramic object is the goal of the fabrication process, then the green body and the starting suspension must have many of the desirable characteristics discussed earlier for the casting methods, e.g., the suspension must have a high concentration of particles (≥50 vol%) to produce a high packing density in the green body and a low viscosity for ease of flow during the stereolithography process. Thus, low-viscosity monomer solutions and effective dispersants to control the colloidal interactions between the particles (and hence the viscosity of the suspension) are important requirements for successful processing. The layer depth of curing during stereolithography is also an important parameter. With currently available equipment, an individual layer thickness of 100 to 200 μm is required to achieve successful lamination and adhesion of the layers. The presence of ceramic particles in the monomer solution enhances photon scattering and leads to a reduction in the mean transport length of the photons through the solution. Thus, maximizing the curing depth in the monomer solution (e.g., through the use of appropriate monomers and photoinitiators) is important for achieving the required layer thickness [88].

7.4.3.2 Direct Ceramic Ink-Jet Printing

In *direct ceramic ink-jet printing* (DCIJP), the ceramic powder is contained in an ink, which is cast through a printer nozzle [74]. The ink is essentially a well-dispersed suspension, containing 10 to 15 vol% particles, ~5 vol% organic additives (dispersant + binder), and 80 to 85 vol% liquid (Table 7.11). Typically, the nozzle diameter is ~50 μm, and the suspension is deposited in the form of drops with a diameter of ~100 μm. Figure 7.42 shows an example of the complex structures formed by DCIJP [89]. Successful forming by DCIJP is critically dependent on the properties of the ink [90–92]. The inks must have a sufficient concentration of particles for rapid deposition, as well as the appropriate viscosity and surface tension for consistent droplet formation. Good colloidal stability of the suspension, achieved through the effective use of polymeric additives, is essential for avoiding sedimentation in the printing device and clogging of the nozzles. The inks should also have a high drying rate for enhancing the printing speed and, on drying, the structure should have sufficient strength for subsequent handling. As discussed earlier for tape casting, these properties depend on the particle characteristics, the solvent and organic additives, and adequate processing procedures.

TABLE 7.11

Example of an Ink Composition Used in Direct Ceramic Ink-Jet Printing (DCIJP)

Material	Concentration (vol%)
ZrO$_2$ powder[a]	14
Solsperse 13940[b]	12
Isopropanol	14
Octane	57
Wax	3

[a] Average particle size = 0.45 μm.

[b] Dispersant solution containing 40 wt% dispersant and 60 wt% solvent.

Source: From Zhao, X, Evans, J.R.G., and Edirisinghe, M.J., Direct inkjet-printing of vertical walls, *J. Am. Ceram Soc.*, 85, 2113, 2002.

7.4.3.3 Robocasting

Robocasting is a method based on computer-controlled, layer-by-layer deposition of highly concentrated ceramic suspensions by extrusion through a narrow orifice [93]. Figure 7.43 shows schematic illustrations of a robocasting apparatus. The orifice openings can range from a few tenths to a few millimeters. A typical suspension contains 50 to 65 vol% particles, 35 to 50 vol% solvent (commonly, water), and 1 to 5 vol% organic additives. In the process, individual layers are sequentially deposited after the previous layer has had sufficient time to dry. In some features, robocasting is similar to slip casting or gel casting but without the use of molds.

The structural integrity of the component formed by robocasting is dependent on the rheological properties of the suspension and on the drying of the deposited layers. The slurry must be pseudoplastic enough to flow through a narrow orifice and yet must transform to a solid-like mass after deposition. If drying is too slow, slumping may occur because of the accumulated mass of several layers, providing a stress greater than the yield stress of the pseudoplastic layer. On the other hand, too fast a drying may lead to cracking, warping, and delamination. To overcome limitations on the shape uniformity of the deposited material and to reduce macroscopic defects, robocasting of gelcasting suspensions has been investigated [94]. Gelation of the organic system to form a polymeric network results in better shape retention and imparts sufficient strength to the component for subsequent processing.

7.4.3.4 Freeze Casting

Freezing of suspensions, commonly aqueous, in a mold at low temperature (less than –40°C) has been used on a limited basis to form ceramics with complex geometry [95]. The green body is obtained after demolding and sublimation of the frozen solvent (freeze drying). Benefits of this technique include the elimination of capillary drying stresses and ease of debinding due to the small amount of organic additives used in the process. The use of freeze casting as an SFF technique has received some recent attention. The suspension, deposited by a computer-controlled nozzle, is frozen layer-by-layer to form the desired shape, and the green body is obtained after freeze drying. Disadvantages of the process include the effort needed to achieve the cold temperature for the freezing step and the slow freeze drying below 0°C. The use of nonaqueous sublimable vehicles that allow freeze casting to be performed near room temperature has been reported recently [96a and 96b].

FIGURE 7.42 (a) Example of a complex structure (a maze) built by direct ceramic ink-jet printing; (b) top surface of the vertical walls of three-dots width after sintering. (From Zhao, X, Evans, J.R.G., and Edirisinghe, M.J., Direct inkjet-printing of vertical walls, *J. Am. Ceram Soc.*, 85, 2113, 2002. With permission.)

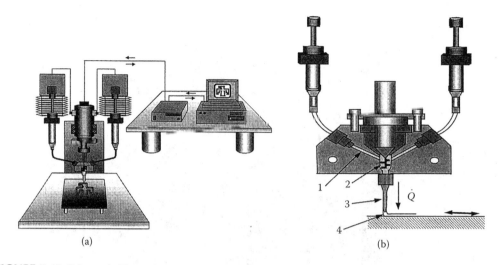

FIGURE 7.43 Schematic illustration of (a) robocasting apparatus, and (b) two-nozzle delivery system showing four shear zones in the mixing chamber: (1) pumping from the syringe, (2) mixing via paddle mixer, (3) extrusion from the nozzle tip, and (4) deposition onto a moving table. (From Morissette, S.L. et al., Solid freeform fabrication of aqueous alumina-poly(vinyl alcohol) gelcasting suspensions, *J. Am. Ceram. Soc.*, 83, 2409, 2000. With permission.)

7.5 CONCLUDING REMARKS

In this chapter we considered the common methods for forming ceramic powders into a shaped green body. Forming is a critical step prior to the sintering process. Heterogeneities present in the green body cannot be easily removed during sintering, and so control of the green body microstructure provides considerable benefits for microstructural control of the final article. The selection of a forming method depends on the size and shape of the article, as well as on the cost. Careful manipulation of the powder characteristics and the consolidation parameters provides a useful approach for optimizing the homogeneity of the green microstructure. Organic additives play an important role in the forming process, and selection of the right additives is often vital to the successful production of a green body. The key features of solid freeform fabrication methods were described. These methods have been the subject of considerable research and development since the 1980s, and their application is expected to grow significantly in the future.

PROBLEMS

7.1

For each of the following surfactants: sodium stearate, oleic acid, and alkyl ammonium acetate,

    a. Draw the structure of the functional end group, indicating whether the surfactant is nonionic, cationic, or anionic.

    b. State whether the surfactant is soluble in aqueous or organic solvents.

    c. Sketch and briefly explain the mechanism of adsorption of the surfactant onto the surfaces of Al_2O_3 particles.

    d. State and briefly explain the dominant mechanism by which the surfactant can stabilize a suspension of Al_2O_3 particles.

7.2

Draw the structure of the repeating unit in the dispersants poly(acrylic acid) and poly(ethylene imine).

   a. State whether each dispersant is soluble in aqueous solvents or organic solvents.
   b. State whether each polymer is nonioinic, cationic, or anionic.
   c. Which of the two dispersants would be more effective for stabilizing a suspension of SiO_2 particles? Explain why.

7.3

Sodium silicate is commonly used as a dispersant for clays and for some oxide ceramics. It is also used as a binder in some ceramic-forming methods. Compare and explain the mechanism by which sodium silicate functions as a dispersant and as a binder.

7.4

Draw the structure of the repeating unit in ethyl cellulose and hydroxyethyl cellulose. Ethyl cellulose is not soluble in aqueous solvents but hydroxyethyl cellulose is. Suggest an explanation for the difference in solubility of these two binders.

7.5

Explain what is meant by the term viscosity grade of a binder. Draw the structure of the repeating unit in poly(vinyl pyrrolidone) and ammonium alginate. In Table 7.3, poly(vinyl pyrrolidone) is listed as a low-viscosity binder, whereas ammonium alginate is listed as a high-viscosity binder. Suggest an explanation for the difference in viscosity grade of these two binders.

7.6

Menhaden fish oil and poly(vinyl butyral) are often used as a dispersant and binder combination for Al_2O_3 particles.

   a. Draw the structure of the functional end group in Menhaden fish oil and the repeating unit in poly(vinyl butyral).
   b. Is this dispersant and binder combination soluble in aqueous or organic solvents?
   c. Suggest an explanation why this dispersant and binder combination is effective for Al_2O_3.

7.7

Explain what is meant by the term *glass transition temperature* of an organic binder. Poly(vinyl alcohol) and poly(ethylene glycol) are often used as a binder and plasticizer combination for oxide particles.

   a. Draw the structure of the repeating unit in poly(vinyl alcohol) and poly(ethylene glycol).
   b. Is this combination of binder and plasticizer soluble in aqueous or organic solvents?
   c. Suggest an explanation why poly(ethylene glycol) is an effective plasticizer for poly(vinyl alcohol).

7.8

Discuss how the viscosity of an Al_2O_3 suspension would change depending on the order in which poly(acrylic acid) and poly(vinyl alcohol) are added to the suspension.

7.9

The particle packing density in spray-dried granules is 50% and the granules themselves achieve a packing density of 55% after die filling. Determine the particle packing density of the material after die filling. During compaction, the granules rearrange to a packing density of 62.5% prior to any significant deformation. Estimate the particle packing density of the material after the rearrangement stage. After compaction, the particle packing density of the green body is 65%. Assuming the original granules are spherical and of the same size, estimate the degree of flattening of the granules during compaction, giving the answer as a ratio of the diameters along the axial and radial directions of the die.

7.10

What would be the effect on slip-casting behavior when using:

  a. A cold mold and a warm slip?
  b. A warm mold and a cold slip?
  c. A warm mold and a warm slip?

7.11

An aqueous Al_2O_3 slip has a density of 2.5 g/cm³. Estimate the concentration of Al_2O_3 in the slip on a weight basis and on a volume basis. The slip is pressure cast under an applied pressure of 1.5 MPa. Assuming that the density of the cast is equal to that for dense random packing of monosize spheres and the particle size of the powder is 1 μm, use Equation 7.4 and Equation 7.11 to estimate the time required for the formation of a 1-cm-thick cast.

7.12

A slip for tape casting contains, on a weight basis, 100 parts Al_2O_3, 10 parts of nonvolatile organics, and 35 parts toluene. After drying, the adherent tape is 60% of the original thickness of the cast. Estimate the porosity of the tape after drying and after binder burnout. After sintering, there is a shrinkage of 7.5% in the thickness of the tape and the relative density of the tape is 95.0%. Estimate the linear shrinkage in the plane of the tape, assuming homogeneous shrinkage in the plane.

7.13

Consider an injection-molded article consisting of ceramic powder and polymeric binder:

  a. Develop a relationship between the ceramic solids content of the article and its linear shrinkage when sintered to a specific end-point density, assuming that the binder is removed completely and that the shrinkage is homogeneous.
  b. Repeat the calculation assuming that, on pyrolyis, the binder produces an 85% yield of ceramic powder with the same composition as the starting powder.
  c. Assuming a sintered density equal to the theoretical density of the ceramic, plot the linear shrinkage vs. ceramic solids content for the cases considered in (a) and (b). (The density of the binder and the theoretical density of the ceramic are 1.2 and 4.0 g/cm³, respectively.)

REFERENCES

1. Shanefield, D.J., *Organic Additives and Ceramic Processing*, 2nd ed., Kluwer Academic Publishers, Boston, MA, 2000.
2. Onoda, G.Y. Jr., The rheology of organic binder solutions, in *Ceramic Processing Before Firing*, Onoda, G.Y. Jr. and Hench, L.L., Eds., John Wiley & Sons, New York, 1978, p. 235.
3. (a) Moreno, R., The role of slip additives in tape-casting technology, Part I: solvents and dispersants, *Am. Ceram. Soc. Bull.*, 71, 1521, 1992.
   (b) Moreno, R., The role of slip additives in tape-casting technology, Part II: binders and plasticizers, *Am. Ceram. Soc. Bull.*, 71, 1647, 1992.
4. (a) Cesarano, J., III, Aksay, I.A., and Bleier, A., Stability of aqueous α-Al_2O_3 suspensions with poly(methacrylic acid) polyelectrolyte, *J. Am. Ceram. Soc.*, 71, 250, 1988.
   (b) Cesarano, J., III and Aksay, I.A., Processing of highly concentrated aqueous α-alumina suspensions stabilized with polyelectrolytes, *J. Am. Ceram. Soc.*, 71, 1062, 1988.
5. Howard, K.E., Method of Making Moisture Resistant Aluminum Nitride Powder and Powder Produced Thereby, U.S. Patent 5,234,712, 1993.
6. Bernhardt, C., Preparation of suspensions for particle size analysis: methodical recommendations, liquids, and dispersing agents, *Adv. Colloid Interface Sci.*, 29, 79, 1988.
7. Bortzmeyer, D., Die pressing and isostatic pressing, in *Materials Science and Technology, Vol. 17A: Processing of Ceramics*, Pt. I, Brook, R.J., Ed., VCH, New York, 1996, p. 127.
8. Glass, S.J. and Ewsuk, K.G., Ceramic powder compaction, *MRS Bull.*, 22(12), 24, 1997.
9. Takahashi, H., Shinohara, N., Okumiya, M., Uematsu, K., Junichiro, T., Iwamoto, Y., and Kamiya, H., Influence of slurry flocculation on the character and compaction of spray-dried silicon nitride granules, *J. Am. Ceram. Soc.*, 78, 903, 1995.
10. Uematsu, K., Immersion microscopy for detailed characterization of defects in ceramic powder and green bodies, *Powder Technol.*, 88, 291, 1996.
11. Funk, J.E. and Dinger, D.R., *Predictive Process Control of Crowded Particulate Suspensions*, Kluwer Academic Publishers, Boston, MA, 1994.
12. Niesz, D.E., McCoy, G.L., and Wills, R.R., Handling and green forming of fine powders, in *Materials Science Research, Vol. 11: Processing of Crystalline Ceramics*, Palmour, H., III, Davis, R.F., and Hare, T.M., Eds., Plenum, New York, 1978, p. 41.
13. Aydin, I, Briscoe, B.J., and Ozkan, N., Modeling of powder compaction: a review, *MRS Bull.*, 22(12), 45, 1997.
14. Strijbos, S., van Groenou, A.B., and Vermeer, P.A., Recent progress in understanding die compaction of powders, *J. Am. Ceram. Soc.*, 62, 57, 1979.
15. Van Groenou, A.B., Pressing of ceramic powders: a review of recent work, *Powder Metallurgy Int.*, 10, 206, 1978.
16. Strijbos, S., Rankin, P.J., Klein Wassink, R.J., Bannink, J., and Oudemans, G.J., Stresses occurring during one-sided die compaction of powders, *Powder Technol.*, 18, 187, 1977.
17. Strijbos, S., Powder-wall friction: the effects of orientation of wall grooves and wall lubricants, *Powder Technol.*, 18, 209, 1977.
18. Lannutti, J.J., Characterization and control of compact microstructure, *MRS Bulletin*, 22(12), 38, 1997.
19. Welzen, J.T.A.M., Die compaction, in *Concise Encyclopedia of Advanced Ceramic Materials*, Brook, R.J., Ed., Pergamon, Oxford, 1991, p. 112.
20. Zheng, J. and Reed, J.S., Particle and granule parameters affecting compaction efficiency in dry pressing, *J. Am. Ceram. Soc.*, 71, C–456, 1988.
21. Nies, C.W. and Messing, G.L., Effect of glass transition temperature of poly(ethylene glycol)-plasticized poly(vinyl alcohol) on granule compaction, *J. Am. Ceram. Soc.*, 67, 301, 1984.
22. Di Milia, R.A. and Reed, J.S., Stress transmission during the compaction of a spray-dried alumina powder in a steel die, *J. Am. Ceram. Soc.*, 66, 667, 1983.
23. Fries, R. and Rand, B., Slip casting and filter pressing, in *Materials Science and Technology, Vol. 17A: Processing of Ceramics*, Pt. I, Brook, R.J., Ed., VCH, New York, 1996, p. 153.
24. Kingery, W.D., Ed., *Ceramic Fabrication Processes*, MIT Press, Cambridge, MA, 1958, p. 5.
25. Adcock, D.S. and McDowall, I.C., The mechanism of filter pressing and slip casting, *J. Am. Ceram. Soc.*, 40, 355, 1957.

26. Aksay, I.A. and Schilling, C.H., Mechanics of colloidal filtration, in *Advances in Ceramics, Vol. 9: Forming of Ceramics*, Mangels, J.A., Ed., The American Ceramic Society, Columbus, OH, 1984, p. 85.

27. Tiller, F.M. and Tsai, C.-D., Theory of filtration of ceramics: I, slip casting, *J. Am. Ceram. Soc.*, 69, 882, 1986.

28. Fennelly, T.J. and Reed, J.S., Mechanics of pressure casting, *J. Am. Ceram. Soc.*, 55, 264, 1972.

29. Lange, F.F. and Miller, K.T., Pressure filtration: consolidation kinetics and mechanics, *Am. Ceram. Soc. Bull.*, 66, 1498, 1987.

30. Lange, F.F., Powder processing science and technology for increased reliability, *J. Am. Ceram. Soc.*, 72, 3, 1989.

31. Mistler, R.E. and Twiname, E.R., *Tape Casting — Theory and Practice*, The American Ceramic Society, Westerville, OH, 2000.

32. Williams, J.C., Doctor-blade process, in *Treatise on Materials Science and Technology*, Vol. 9, Wang, F.F.Y., Ed., Academic Press, New York, 1976, p. 331.

33. Mistler, R.E., Shanefield, D.J., and Runk, R.B., Tape casting of ceramics, in *Ceramic Processing Before Firing*, Onoda, G.Y. Jr. and Hench, L.L., Eds., John Wiley & Sons, New York, 1978, p. 411.

34. Mistler, R.E., The principles of tape casting and tape casting applications, in *Ceramic Processing*, Terpstra, R.A., Pex, P.P.A.C., and DeVries, A.H., Eds., Chapman and Hall, London, 1995, p. 147.

35. Hellebrand, H., Tape casting, in *Materials Science and Technology, Vol. 17A: Processing of Ceramics*, Pt. I, Brook, R.J., Ed., VCH, New York, 1996, p. 189.

36. Chou, Y.T., Ko, Y.T., and Yan, M.F., Fluid flow model for ceramic tape casting, *J. Am. Ceram. Soc.*, 70, C–280, 1987.

37. Janney, M.A., Method for molding ceramic powders, U.S. Patent No. 4,894,194, January 16, 1990.

38. (a) Janney, M.A. and Omatete, O.O., Method for Molding Ceramic Powders Using a Water-Based Gelcasting, U.S. Patent No. 5,028,362, July 2, 1991.
    (b) Janney, M.A. and Omatete, O.O., Method for Molding Ceramic Powders Using a Water-Based Gelcasting Process, U.S. Patent No. 5,145,908, September 8, 1992.

39. Omatete, O.O., Janney, M.A., and Strehlow, R.A., Gelcasting — a new ceramic forming process, *Am. Ceram. Soc. Bull.*, 70, 1641, 1991.

40. Young, A.C., Omatete, O.O., Janney, M.A., and Menchhofer, P.A., Gelcasting of alumina, *J. Am. Ceram. Soc.*, 74, 612, 1991.

41. Hamaker, H.C. and Verwey, E.J.W., The forces between the particles in electrodeposition and other phenomena, *Trans. Faraday Soc.*, 35, 180, 1940.

42. Koelmans, H. and Overbeek, J.Th.G., Stability and electrophoretic deposition of suspensions in non-aqueous media, *Disc. Faraday Soc.*, 18, 52, 1954.

43. Grillon, F., Fayeulle, D., and Jeandin, M., Quantitative image analysis of electrophoretic coatings, *J. Mater. Sci. Lett.*, 11, 272, 1992.

44. Shimbo, M., Tanzawa, K., Miyakawa, M., and Emoto, T., Electrophoretic deposition of glass powder for passivation of high voltage transistors, *J. Electrochem. Soc.*, 132, 393, 1985.

45. Mizuguchi, J., Sumi, K., and Muchi, T., A highly stable nonaqueous suspension for the electrophoretic deposition of powdered substances, *J. Electrochem. Soc.*, 130, 1819, 1983.

46. Sarkar, P. and Nicholson, P.S., Electrophoretic deposition(EPD): mechanisms, kinetics, and applications to ceramics, *J. Am. Ceram. Soc.* 79, 1987, 1996.

47. Birks, J.B., Electrophoretic deposition of insulating materials, in *Progress in Dielectrics*, Vol. 1, Birks, J.B. and Schulman, J.H., Eds., Heywood and Company, London, 1959, p. 273.

48. Andrews, J.M., Collins, A.H., Cornish, D.C., and Dracass, J., The forming of ceramic bodies by electrophoretic deposition, *Proc. Br. Ceram. Soc.*, 12, 211, 1969.

49. Krishna Rao, D.U., and Subbarao, E.C., Electrophoretic deposition of magnesia, *Am. Ceram. Soc. Bull.*, 58, 467, 1979.

50. Kennedy, J.H. and Foissy, A., Fabrication of beta-alumina tubes by electrophoretic deposition from suspension in dichloromethane, *J. Electrochem. Soc.*, 122, 482, 1975.

51. Powers, R.W., Ceramic aspects of forming beta-alumina by electrophoretic deposition, *Am. Ceram. Soc. Bull.*, 65, 1270, 1986.

52. Chu, C.T. and Dunn, B., Fabrication of $YBa_2Cu_3O_{7-x}$ superconducting coatings by electrophoretic deposition, *Appl. Phys. Lett.*, 55, 492, 1989.

53. Maiti, H.S., Datta, S., and Basu, R.N., High T_c superconductor coating on metal substrates by electrophoretic technique, *J. Am. Ceram. Soc.*, 72, 1733, 1989.

54. Sarkar, P. and Nicholson, P.S., Fabrication of textured Bi-Sr-Ca-Cu-O thick film by electrophoretic deposition, *J. Appl. Phys.*, 69, 1775, 1991.

55. Basu, R.N., Randall, C.A., and Mayo, M.J., Fabrication of dense electrolyte films for tubular solid oxide fuel cells by electrophoretic deposition, *J. Am. Ceram. Soc.*, 84, 33, 2001.

56. Sarkar, P., Huang, X., and Nicholson, P.S., Structural ceramic laminates by electrophoretic deposition, *J. Am. Ceram. Soc.*, 75, 2907, 1992.

57. Janney, M.A., Plastic forming of ceramics: extrusion and injection molding, in *Ceramic Processing*, Terpstra, R.A., Pex, P.P.A.C., and DeVries, A.H., Eds., Chapman and Hall, London, 1995, p. 174.

58. Pels Leusden, C.O., Extrusion, in *Concise Encyclopedia of Advanced Ceramic Materials*, Brook, R.J., Ed., Pergamon, Oxford, 1991, p. 131.

59. Schuetz, J.E., Methylcellulose polymers as binders for extrusion of ceramics, *Am. Ceram. Soc. Bull.*, 65, 1556, 1986.

60. Buckingham, E., On plastic flow through capillary tubes, *Proc. Am. Soc. Test. Mater.*, 21, 1154, 1921.

61. Reiner, M., Ueber die Strömung Einer Elastischen Flüssigkeit durch Eine Kapillare, *Kolloid Z.*, 39, 80, 1926.

62. Robinson, G.C., Extrusion defects, in *Ceramic Processing before Firing*, Onoda, G.Y. Jr. and Hench, L.L., Eds., John Wiley & Sons, New York, 1978, p. 391.

63. Brady, G.A, Hilmas, G.E., and Halloran, J.W., Forming textured ceramics by multiple co-extrusion, *Ceram. Trans.*, 51, 301, 1995.

64. Van Hoy, C., Barda, A., Griffith, M., and Halloran, J.W., Microfabrication of ceramics by co-extrusion, *J. Am. Ceram. Soc.*, 81,152, 1998.

65. German, R.M. and Bose, A., *Injection Molding of Metals and Ceramics*, Metal Powder Industries Federation, Princeton, NJ, 1997.

66. Mutsuddy, B.C. and Ford, R.G., *Ceramic Injection Molding*, Chapman & Hall, New York, 1995.

67. Evans, J.R.G., Injection Molding, in *Materials Science and Technology, Vol. 17A: Processing of Ceramics*, Pt. I, Brook, R.J., Ed., VCH, New York, 1996, p. 267.

68. German, R.M., Hens, K.F., and Lin, S.-T.P., Key issues in powder injection molding, *Am. Ceram. Soc. Bull.*, 70, 1294, 1991.

69. Mangels, J.A. and Trela, W., Ceramic components by injection molding, in *Advances in Ceramics, Vol. 9: Forming of Ceramics*, Mangels, J.A., Ed., The American Ceramic Society, Columbus, OH, 1984, p. 220.

70. Chong, J.S., Christianson, E.B., and Baer, A.D., Rheology of concentrated suspensions, *J. Appl. Polym. Sci.*, 15, 2007, 1971.

71. Kochan, D., Ed., *Solid Freeform Manufacturing*, Elsevier, New York, 1993.

72. Kai, C.C. and Fai, L.K., *Rapid Prototyping*, John Wiley & Sons, New York, 1997.

73. Beaman, J.J., *Solid Freeform Fabrication: A New Direction in Manufacturing*, Kluwer Academic Publishers, Boston, MA, 1997.

74. Halloran, J.W., Freeform fabrication of ceramics, *Br. Ceram. Trans.*, 98, 299, 1999.

75. Tay, B.Y., Evans, J.R.G., and Edirisinghe, M.J., Solid freeform fabrication of ceramics, *Internat. Mater. Rev.*, 48, 341, 2003.

76. Agarwala, M.K., Bandyopadhyay, A., van Weeren, R., Safari, A., Danforth, S.C., Langrana, N.A., Jamalabad, V.R., and Whalen, P.J., FDC, rapid fabrication of structural components, *Am. Ceram. Soc. Bull.*, 75(11), 60, 1996.

77. Bandyopadhyay, A., Panda, R.K., Janas, V.F., Agarwala, M.K., Danforth, S.C., and Safari, A., Processing of piezocomposites by fused deposition technique, *J. Am. Ceram. Soc.*, 80, 1366, 1997.

78. McNulty, T.F., Mohammadi, F., Bandyopadhyay, A., Shanefield, D.J., Danforth, S.C., and Safari, A., Development of a binder formulation for fused deposition of ceramics, *Rapid Prototyping J.*, 4(4), 144, 1998.

79. Agarwala, M., Jamalabad, V.R., Langrana, N.A., Safari, A., Whalen, P.J., and Danforth, S.C., Structural quality of parts processed by fused deposition, *Rapid Prototyping J.*, 2(4), 4, 1996.

80. Cawley, J.D., Heuer, A.H., Newman, W.S., and Mathewson, B.B., Computer-aided manufacturing of laminated engineering materials, *Am. Ceram. Soc. Bull.*, 75(5), 75, 1996.

81. Griffin, E.A., Mumm, D.R., and Marshall, D.B., Rapid prototyping of functional ceramic composites, *Am. Ceram. Soc. Bull.*, 75(7), 65, 1996.

82. Klosterman, D.A., Chartoff, R.P., Osborne, N.R., Graves, G.A., Lightman, A., Han, G., Bezeredi, A., Rodrigues, S., Pak, S., Kalmanovich, G., Dodin, L., and Tu, S., Direct fabrication of ceramics, CMCs by rapid prototyping, *Am. Ceram. Soc. Bull.*, 77(10), 69, 1998.

83. Sachs, E., Cima, M., Williams, P., Brancazio, D., and Cornie, J., Three-dimensional printing: rapid tooling and prototypes directly from a CAD model, *J. Eng. Ind.*, 114, 481, 1992.

84. Grau, J.E., Uhland, S.A., Moon, J., Cima, M.J., and Sachs, E.M., Controlled cracking of multilayer ceramic bodies, *J. Am. Ceram. Soc.*, 82, 2080, 1999.

85. Moon, J., Grau, J.E., Knezevic, V., Cima, M.J., and Sachs, E.M., Ink-jet printing of binders for ceramic components, *J. Am. Ceram. Soc.*, 85, 755, 2002.

86. Moon, J., Grau, J.E., and Cima, M.J., Slurry chemistry control to produce easily redispersible ceramic powder compacts, *J. Am. Ceram. Soc.*, 83, 2401, 2000.

87. Griffith, M.L. and Halloran, J.W., Freeform fabrication of ceramics via stereolithography, *J. Am. Ceram. Soc.*, 79, 2601, 1996.

88. Lee, J.H., Prud'homme, R.K., and Aksay, I.A., Cure depth in photopolymerization: experiments and theory, *J. Mater. Res.*, 16, 3536, 2001.

89. Zhao, X, Evans, J.R.G., and Edirisinghe, M.J., Direct inkjet-printing of vertical walls, *J. Am. Ceram Soc.*, 85, 2113, 2002.

90. Teng, W.D., Edirisinghe, M.J., and Evans, J.R.G., Optimization of dispersion and viscosity of a ceramic jet printing ink, *J. Am. Ceram. Soc.*, 80, 486, 1997.

91. Teng, W.D. and Edirisinghe, M.J., Development of ceramic inks for direct continuous jet printing, *J. Am. Ceram. Soc.*, 81, 1033, 1998.

92. Song, J.H., Edirisinghe, M.J., and Evans, J.R.G., Formulation and multilayer jet printing of ceramic inks, *J. Am. Ceram. Soc.*, 82, 3374, 1999.

93. Cesarano J., III, Segalman, R., and Calvert, P., Robocasting provides moldless fabrication from slurry deposition, *Ceram. Industry*, 148(4), 94, 1998.

94. Morissette, S.L., Lewis, J.A., Cesarano, J., III, Dimos, D.B., and Baer, T., Solid freeform fabrication of aqueous alumina-poly(vinyl alcohol) gelcasting suspensions, *J. Am. Ceram. Soc.*, 83, 2409, 2000.

95. Sofie, S.W. and Dogan, F., Freeze casting of aqueous alumina slurries with glycerol, *J. Am. Ceram. Soc.*, 84, 1459, 2001.

96. (a) Araki, K. and Halloran, J.W., New freeze-casting technique for ceramics with sublimable vehicles, *J. Am. Ceram. Soc.*, 87, 1859, 2004.
    (b) Araki, K. and Halloran, J.W., Room-temperature freeze casting for ceramics with non-aqueous sublimable vehicles in the naphthalene-camphor eutectic system, *J. Am. Ceram. Soc.*, 87, 2014, 2004.

8 Drying, Debinding, and Microstructural Characterization of Green Bodies

8.1 INTRODUCTION

Prior to sintering, green bodies formed by the methods described in the previous chapter often need to be subjected to additional processing operations, of which drying and removal of organic processing additives are the most important. Liquids used to aid the forming process must be removed from the article prior to sintering. If the green body contains too much liquid, it will develop flaws during heating, and in extreme cases it will be disrupted by the evolution of vapor. *Drying* generally refers to the removal of liquid from a solid by evaporation. Commonly, it involves transfer of heat from the surrounding environment, such as flowing warm air, to the moist article, and the simultaneous transfer of vapor in the reverse direction. The process is accompanied by the movement of moisture, either liquid or vapor, within the pores of the body and often by shrinkage of the article.

The process of removing binders and other organic additives from the green body is referred to as *debinding*. Most often, debinding is achieved by heating the green body at temperatures lower than the sintering temperature to decompose the binder, a process referred to as *binder burnout* or *thermal debinding*. Ideally, we would like to remove the organic additives as completely as possible without disrupting the particle packing or producing any new microstructural flaws in the green body. Residual contaminants, such as carbon residue, and flaws, such as cracks and large voids, generally have an adverse effect on the microstructural evolution during sintering and, hence, on the properties of the fabricated body. Debinding can be a critical step in ceramic processing, especially for green bodies containing a significant concentration of binder, such as those formed by tape casting, injection molding, and a few solid freeform fabrication techniques.

The development of stresses in the body during drying and binder burnout can lead to problems with dimensional control, cracking, and the growth of microstructural flaws. The main sources of drying stresses are pressure gradients in the liquid as it flows from the interior to replace liquid evaporated from the surface. Drying and binder burnout can lead to the evolution of large volumes of gaseous products, and so gas pressure gradients are another source of stress. There is also thermal stress due to temperature gradients during heating or cooling of the body in drying or binder burnout. In tape-cast and injection-molded articles, for which the binder concentration is normally significant, binder burnout can often be a limiting step in the fabrication process, particularly for thicker samples. The objective is to accomplish drying and binder removal at a "safe" rate to avoid the problems aforementioned but not at too slow a rate that would make the process uneconomical.

Because the green microstructure has a significant influence on the microstructure development during sintering, some knowledge of the green body microstructure is useful. Measurement of the green density, together with characterization of the pore size distribution by mercury porosimetry, and observation of a fractured surface in the scanning electron microscope (SEM) are simple to perform and provide useful information on the packing density and packing homogeneity, factors that have a strong effect on microstructural evolution during sintering. The presence of processing flaws can also be easily determined from the SEM observation.

8.2 DRYING OF GRANULAR CERAMICS

In granular ceramics, the solid phase consists of a cohesive mass of rigid solid particles or grains, typically with micrometer sizes. The general principles of drying discussed in Chapter 5 for gels are also applicable to granular solids, but there are some important differences. Granular ceramics do not exhibit the large drying shrinkages observed for polymeric gels. Linear shrinkages of less than 10% are often found, rather than values of 50% or more for gels. The pores in granular ceramics are also much larger, and so the permeability is higher, liquid flow is easier, and capillary stresses are smaller. It is recommended that the reader become familiar with the principles of drying discussed in Chapter 5 and reviewed in Reference 1.

8.2.1 DRYING OF GRANULAR LAYERS

Investigations of the drying of liquid drops containing a dilute concentration of colloidal particles (particle concentration = 10^{-4}) show considerable migration of the particles [2–5], leading to the deposition of a ring (Figure 8.1). The characteristic ring deposit is caused by capillary flow. Pinning of the contact line between the drying drop and the substrate ensures that liquid evaporating from the edge is replenished from the interior. In the initial stage of drying, particles at the outer edge of the drop undergo consolidation to form a ring. As drying proceeds, liquid flows to the outer region to maintain a saturated state (i.e., the voids between the particles are completely

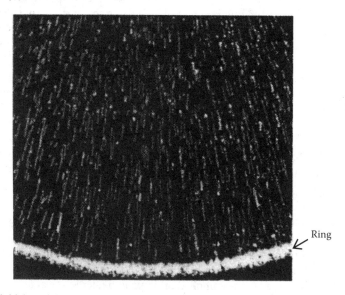

FIGURE 8.1 Colloidal particle migration during drying of a liquid drop containing a dilute concentration of colloidal particles. The arrow denotes the ring deposit that forms at the edge of the drop, and the background streaks denote the migration of individual particles to the edge as a result of capillary-induced fluid flow. (From Deegan, R.D. et al., Capillary flow as the cause of ring stains from dried liquid drops, *Nature*, 389, 827, 1997. With permission.)

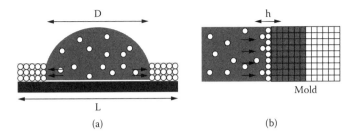

FIGURE 8.2 Schematic illustration of (a) drying of a granular film and (b) the slip-casting process. Arrows indicate the direction of fluid flow. (From Guo, J.J. and Lewis, J.A., Aggregation effects on the compressive flow properties and drying behavior of colloidal silica suspensions, *J. Am. Ceram. Soc.*, 82, 2345, 1999. With permission.)

filled with liquid), a process driven by the capillary suction pressure of the ring (Figure 8.2). This outward flow carries the particles to the edge, depositing them between the saturated ring and the supersaturated suspension. This mechanism predicts a power law growth of the ring mass, m, with time, t, according to $m \approx t^{1.3}$, a law independent of the particular substrate, carrier fluid, or deposited solids. This law has been verified by microscopic observations of ring growth of dilute colloidal suspensions.

The structure of the ring deposit depends on several factors such as the colloid stability of the suspension and the concentration of particles. Stable suspensions with low particle concentration produce monolayer ring deposits with densely packed structures. More concentrated suspensions or less stable suspensions produce multilayer, low-density deposits with solid deposits inside the ring. Ring deposits provide a potential means to write or deposit a fine pattern onto a surface.

Films deposited on a rigid substrate experience biaxial tensile stresses during drying because they cannot shrink in the plane of the film. The in-plane tensile stress develops during drying as a result of the capillary tension in the pore liquid as it stretches to cover the dried surface exposed by the evaporation of the liquid. The capillary tension in the liquid would cause the film to shrink but the rigid substrate prevents the shrinkage in the plane of the film, so a tensile in-plane stress develops in the coating. The drying of coatings produced from electrostatically stabilized, binder-free suspensions (0.5-μm Al_2O_3 particles) are observed to crack spontaneously regardless of the drying rate [6,7] when the film thickness is above a certain critical value (~50 μm). The existence of a critical cracking thickness can be explained by a linear elastic fracture model [8], which states that a constrained film will crack when subjected to a stress only if the strain energy released in the process exceeds the energy required to form the crack. The critical cracking thickness depends on several factors, of which the ceramic particle size, the liquid surface tension, and the colloid stability of the suspension have the most pronounced effect. Because cracking is dependent on the capillary-induced tensile stress in the plane of the coating, the critical cracking thickness increases with increasing particle size and with decreasing surface tension of the liquid. It also increases for a flocculated suspension but at the expense of a lower particle packing density in the dried film, and with the use of a binder.

In situ measurements of the drying stresses in coatings produced from electrostatically stabilized, binder-free suspensions of Al_2O_3 particles indicate that the cracking behavior is consistent with a capillary-induced biaxial tensile stress acting on the entire coating rather than a differential stress generated by a moisture gradient over the thickness of the coating [7]. A stress does not develop in the coating until the supersaturated region has disappeared, but it rises to a maximum soon after this point and decreases monotonically thereafter (Figure 8.3). The maximum value of the stress σ_{max} is ~2 MPa for coatings of 0.4-μm particles. It is found to be proportional to the surface tension of the liquid γ_{LV} and inversely proportional to the particle size. For homogeneously packed solids, the pore size is commonly proportional to the particle size, so

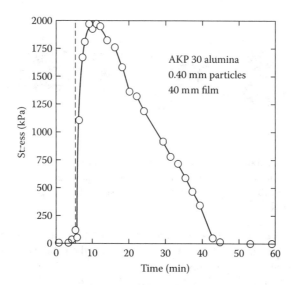

FIGURE 8.3 Stress history of a film, 40-µm thick, prepared from 0.4-µm Al_2O_3 particles cast on a 9×9 mm^2 silicon substrate. The drying rate was 7.2×10^{-6} kg m^{-2} s^{-1}. (From Chiu, R.T, Garino, T.J., and Cima, M.J., Drying of granular ceramic films: II, drying stress and saturation uniformity, *J. Am. Ceram. Soc.*, 76, 2769, 1993. With permission.)

σ_{max} can be related to the capillary tension p in the liquid of the saturated coating, given by the Young and Laplace equation:

$$p = -2\gamma_{LV} / r \qquad (8.1)$$

where r is the radius of curvature of the liquid meniscus.

Investigations of the drying behavior of SiO_2 coatings (particle size ≈ 0.5 µm) indicate that the maximum stress in the coating and the critical cracking thickness are strongly dependent on the colloid stability of the suspension [9]. The critical cracking thickness increases from 40 to 50 µm for highly stable suspensions to 650 to 660 µm for strongly flocculated suspensions. For stabilized suspensions, a dimple or depression is also observed to form at the center of the coating after drying. The formation mechanism of the dimple is similar to that for ring formation described earlier. The buildup of the ring mass for the coating exhibits a square root dependence on time, as observed for slip-cast layers whose thickness follows a similar time dependence (see Equation 7.4). As Figure 8.2 illustrates, the processes leading to dimpling or ring formation in colloidal coatings are analogous to those in slip casting. Dimpling of the coating can be suppressed by hindering the motion of the particles in the suspension, such as by increasing the particle concentration or by decreasing the colloidal stability.

8.2.2 DRYING OF GRANULAR SOLIDS

Drying of solids is one of the oldest and most common operations found not only in ceramics but also in the food, chemical, agricultural, pulp and paper, mineral, polymer, and textile industries. The complexity of the process coupled with difficulties and deficiencies in its mathematical description means that it is not well understood. Drying is therefore an amalgam of science and technology, as well as an art, developed from experimental observations and operating experience. Drying of solids in ceramic and other industries is treated in several texts, covering the theory, principles, and practice [10–12]. In this section, our focus will be on the drying of moist green bodies formed by the processing methods described in Chapter 7, such as cast or extruded articles. A qualitative description of the drying of moist ceramics is given by Ford [13].

8.2.2.1 Driving Forces for Shrinkage and Moisture Movement

Capillary pressure, osmotic pressure, and disjoining pressure were discussed in Chapter 5 as possible driving forces for the shrinkage of gels. These driving forces can also operate in the drying of granular solids. It has been argued that electrostatic repulsion between clay particles leads to tension in the liquid, which draws the liquid from the interior of the drying body [14], thereby providing an additional driving force for moisture movement. Even for clays, in which these phenomena are most evident, it has been argued that that osmotic pressure is less important than capillary pressure during drying, because moisture gradients persist for long periods when evaporation is prevented [15]. In addition, it has been shown that the final drying shrinkage of kaolinite clay is directly related to the surface tension of the pore liquid [16]. The swelling pressure of clays in water is estimated at <10 MPa [17], which is comparable to the capillary pressure in pores with radii >14 nm, if γ_{LV} for water in Equation 8.1 is assumed to be 0.07 J/m². These observations provide support for capillary pressure being the dominant driving force in the drying of clays and fine-grained ceramics.

The concept of moisture stress (or moisture potential) has been introduced as a global parameter for studying moisture movement and other physical properties during drying of clays [18]. *Moisture stress* is defined as the work done by the water per unit mass of water when a small quantity of water is transported from the clay–water system to a free water surface at the same temperature and height as the clay. Thermodynamically, it is the partial Gibbs free energy of a liquid in a porous medium, and is given by:

$$\psi = \left(\frac{RT}{\rho_L V_m} \right) \ln \left(\frac{p_V}{p_o} \right) \tag{8.2}$$

where R is the gas constant, T is the absolute temperature, ρ_L and V_m are the density and molar volume of the liquid, p_V is the vapor pressure of the liquid in the system, and p_o is the vapor pressure over a flat surface of the pure liquid. The moisture stress subsumes all the driving forces, because the vapor pressure can be reduced by factors such as capillary pressure, osmotic pressure, disjoining pressure, hydration forces, and adsorption forces. It can be obtained by measuring the vapor pressure of the liquid in the system [15]. A difficulty with the application of the moisture stress concept is that capillary pressure produces flow, whereas concentration gradients (that give rise to osmotic pressure) cause diffusion; so it is necessary to apply portions of the total moisture stress to different processes.

8.2.2.2 Stages of Drying

The stages of drying were discussed decades ago by Sherwood [19–21]. The drying curve, often plotted to show the weight of the sample (or its moisture content) as a function of time (Figure 8.4a), may be converted to a curve relating the rate of drying (weight loss per unit time per unit area) to the moisture content of the sample (Figure 8.4b). The moisture content is often expressed as a percentage of the dry weight of the solid (dry basis), defined by the equation:

$$Moisture\ content\,(\%) = \frac{(Wet\,weight - Dry\,weight) \times 100}{Dry\,weight} \tag{8.3}$$

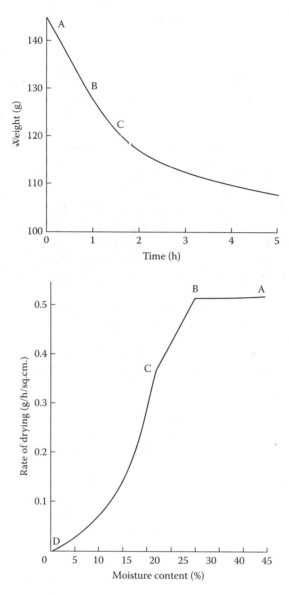

FIGURE 8.4 Representation of drying in terms of (a) weight of moist article vs. time and (b) rate of drying vs. moisture content.

Articles formed by casting or extrusion typically have a moisture content of 20 to 40%. Under conditions of constant air temperature, humidity, and velocity, the drying of moist granular solids can be divided into two or three distinct periods [15]:

1. The constant-rate period, A–B, during which the rate of drying is independent of the moisture content
2. The first falling-rate period, B–C, during which the rate of drying is approximately a linear function of moisture content
3. The second falling-rate period, C–D, with a curvilinear relationship between drying rate and moisture content

The constant-rate period and the first falling-rate period end at moisture contents B and C, respectively, termed the first critical and second critical moisture content, respectively. In many cases, depending on the material and the drying conditions, the first and second falling-rate periods cannot be distinguished, and the drying curve then consists of a constant-rate period and a falling-rate period. In this case, the moisture content at the end of the constant-rate period is called the critical moisture content.

8.2.2.2.1 Constant-Rate Period (CRP)

When the mechanism of heat transfer from the surrounding air to the drying surface is convection, the rate of evaporation during the CRP is independent of the material being dried [15]. The evaporation rate is close to that for an open dish of liquid. It is evident that during the CRP water can flow from the interior to the evaporation surface at a sufficient rate to keep the surface wet. However, the evaporation rate can still remain constant even when dry patches form on the surface of the body [22]. There is a stagnant or slowly moving boundary layer of vapor over the drying surface, and if the breadth of the dry patches is small compared to the thickness of the boundary layer, diffusion parallel to the surface homogenizes the boundary layer at the equilibrium concentration of vapor. The rate of evaporation, $\dot{V}_E$, is controlled by the transport of vapor across the boundary layer and is constant. It is proportional to the difference between the vapor pressure in equilibrium with the drying surface of the body, p_w, and the ambient vapor pressure, p_a:

$$\dot{V}_E = H(p_w - p_a) \tag{8.4}$$

where H is a mass-transfer coefficient, which depends on the temperature, air velocity, and geometry of the system (Figure 8.5). The vapor pressure of the liquid is related to the capillary tension in the liquid p by:

$$p_w = p_o \exp\left(-\frac{pV_m}{RT}\right) \tag{8.5}$$

where p_o is the vapor pressure over a flat surface of the pure liquid, V_m is the molar volume of the liquid, R is the gas constant, and T is the absolute temperature. Using Equation 8.1, Equation 8.4, and Equation 8.5, we see that evaporation will continue as long as

$$p_a < p_o \exp\left(\frac{2V_m\gamma_{LV}}{RTr}\right) \tag{8.6}$$

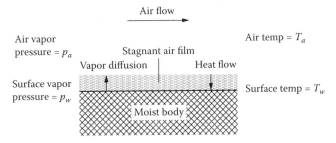

FIGURE 8.5 Equilibrium conditions at a wet surface during drying by convection.

If a body of liquid is subjected to a stream of air of constant velocity, temperature, and humidity, a constant rate of evaporation is rapidly attained, indicating that a state of equilibrium has been reached between the flow of heat to the surface and the rate of loss of latent heat due to the evaporation of water. When evaporation starts, the temperature at the surface falls because of the evaporation of the liquid. This fall in temperature causes heat to flow to the surface, increasing the rate of evaporation, and this feedback process equilibrates when the heat flow to the surface equals the rate of loss of latent heat associated with the evaporated water. The drying surface reaches a constant temperature T_w, called the *wet-bulb temperature* (Figure 8.5). The exterior surface of a drying body is at the wet-bulb temperature during the CRP. If heat transferred to the drying surface from the air is the only source of energy available for the evaporation, then the evaporation rate is also given by:

$$\dot{V}_E = L(T_a - T_w) \tag{8.7}$$

where L is a constant and T_a is the temperature of the drying air. Any heat the surface receives by radiation or by conduction through the mass will increase the temperature and therefore the vapor pressure of the water, thereby resulting in a faster rate of drying.

During the CRP, evaporation of the liquid is accompanied by shrinkage of the body (Figure 8.6). The tension in the liquid, given by Equation 8.1, is supported by the solid phase, which therefore goes into compression. The compressive stresses cause the body to contract and the liquid meniscus remains at the surface. Any water evaporated from the surface is replaced by water migrating from the interior of the body, so the decrease in volume is equal to the volume of water evaporated. As drying proceeds during the CRP, the particles achieve a denser packing and the body becomes stiffer. The liquid meniscus at the surface deepens and the tension in the liquid increases. Eventually, the particles, surrounded by a thin layer of bound water, touch and shrinkage stops. This marks the end of the CRP. At this point, the radius of the liquid meniscus is equal to the radius of the pores in the solid, and the liquid exerts the maximum possible stress. The moisture content at end of the CRP is called the *first critical moisture content* or, in clay technology, the *leatherhard* moisture content.

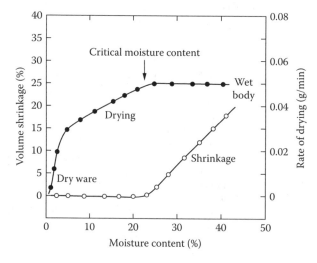

FIGURE 8.6 Change in volume accompanying the drying of a ceramic body. (From Kingery, W.D., *Introduction to Ceramics*, Wiley-Interscience, New York, 1960, p. 50. With permission.)

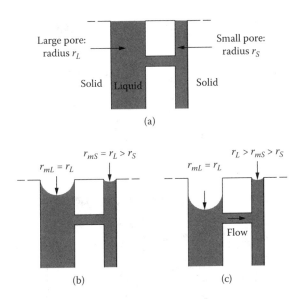

FIGURE 8.7 Sketch illustrating evaporation and liquid flow during drying of a solid with different pore sizes (a). The large pore empties first, while the meniscus is maintained at the surface of the small pore (b), (c) and then the small pore empties. (r_m is the radius of the meniscus.)

8.2.2.2.2 First Falling-Rate Period (FRP1)

When shrinkage stops, further evaporation drives the liquid meniscus into the body and air enters the pores. In the FRP1, the rate of evaporation decreases and the temperature of the surface rises above the wet-bulb temperature. Most of the evaporation still occurs at the exterior surface of the body, so the surface remains below the ambient temperature, and the rate of evaporation is sensitive to changes in air velocity, temperature, and humidity [15]. The liquid in the pores near the surface remains in the funicular condition (see Figure 5.22a), so contiguous pathways exist along which liquid flow to the surface can occur. There is also some liquid evaporation within the unsaturated pores, and the vapor is transported to the surface by diffusion.

In a granular solid with a distribution of pore sizes, the drying front (the liquid–vapor interface) can move into the body in a highly irregular manner. To see how this arises, let us consider a simple model in which interconnected pores of two different radii are present (Figure 8.7a). Even though the pores have different radii (r_L and r_S), initially liquid evaporates from them at the same rate so that the radii of the menisci (r_m) are equal. The capillary tension in the liquid is given by Equation 8.1 with $r = r_m$. If the radii of the menisci were different, the capillary tension would also be different and liquid would flow from one pore to the other until the menisci became equal again. As evaporation from the surface proceeds, the radius of the menisci decreases. However, almost no shrinkage of the body occurs because the particles are practically touching. A point is reached where the radius of the meniscus is equal to the radius of the large pore, i.e., $r_m = r_L$ (Figure 8.7b). Further evaporation forces the liquid to retreat into the large pore. However, the radius of the meniscus, r_m, will continue to decrease in the small pore and the capillary tension will suck liquid from the large pore (Figure 8.7c). In this way, the large pore empties first and the small pore remains full of liquid. After the large pore has been emptied, the small pore starts to empty. In practice, we will have a distribution of pore sizes and pore shapes, but the same principles will apply. Regions containing thousands of pores can empty while the surrounding pores remain full.

In practice, considerable redistribution of the liquid would be expected to occur during drying, and the drying front is not expected to move uniformly into the body. For a moist green body consisting of monosized 0.5-μm SiO_2 particles, Shaw [23,24] showed that that the drying front is fractally rough (Figure 8.8a) on the scale of the pores but smooth on a much larger scale (Figure

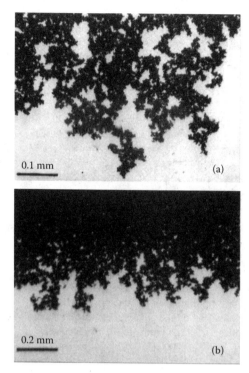

FIGURE 8.8 The drying front in a green body composed of monosized 0.5-μm SiO$_2$ particles. Saturated pores are white and empty pores are black. The drying front is (a) fractally rough on the scale of the particles but (b) smooth on a larger scale. (From Shaw, T.M., Movement of a drying front in a porous material, *Mater. Res. Soc. Symp. Proc.*, 73, 215, 1986. With permission.)

8.8b). Because irregularity of the drying front is on the scale of the pores, it is very small compared to the dimensions of the body, and is quite smooth on a macroscopic scale.

8.2.2.2.3 Second Falling-Rate Period (FRP2)

As the drying front recedes further into the body, if the body is thick enough, flow of liquid becomes so slow that liquid near the surface of the body becomes isolated in pockets. The liquid in this region is said to enter the pendular condition (see Figure 5.22b). Flow of liquid to the surface stops and liquid is removed from the body only by diffusion of its vapor. This marks the start of the FRP2, in which evaporation occurs inside the body. The surface of the body is nearly dry with the temperature approaching that of the ambient and the rate of evaporation becomes less sensitive to external conditions (air velocity, temperature, and humidity). The drying front is drained by flow of funicular liquid, which evaporates at the boundary of the funicular or pendular regions. In the pendular regions, vapor is in equilibrium with isolated pockets of liquid and adsorbed films, and the main transport process is expected to be diffusion.

As the drying front recedes further from the surface, the body may expand slightly as the compressive forces on the solid phase is gradually relieved [13,15]. There is also a buildup of differential strain because the solid network is being compressed more in the saturated region than near the surface. For a plate drying from one side, this may cause warping because faster contraction of the wet side makes the plate convex toward the drying side. As the saturated region becomes thinner, its contraction is more effectively prevented by the larger unsaturated region, raising the tension of the solid network of the saturated region.

Analysis of the drying in the falling-rate periods involves coupled equations for flow of heat and liquid, and diffusion of vapor. There are several reviews of the many theories that have been proposed to describe the FRP1 [25–28]. One of the most complete descriptions is provided by

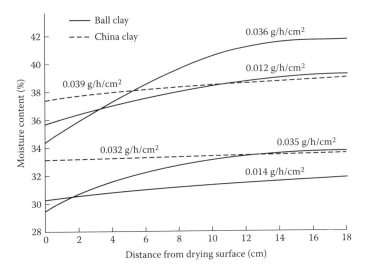

FIGURE 8.9 Moisture gradients during the constant-rate period. (From Ford, R.W., *Ceramics Drying*, Pergamon Press, New York, 1986, p. 53. With permission.)

Whitaker [25,26]. The use of Whitaker's model requires knowledge of several parameters such as the permeability, thermal conductivity, and diffusivity of vapor. A successful test of the model has been performed for the drying of porous sandstone [29,30].

8.2.2.3 Moisture Distribution and Movement

The moisture distribution in the body during drying is important, particularly if cracking and warping are to be avoided. Many studies have been performed to determine the moisture distribution when simple shapes are dried, particularly for clay-based materials [13,15]. Bars of clay, initially with uniform moisture content and dried under constant air conditions from one side only, have received the most attention, and several attempts have been made to devise mathematical theories to account for the experimental data.

In the CRP, water evaporating from the surface is replaced by water migrating from the interior. Moisture gradients will develop, and, for a given thickness of the body, the steepness of the gradient will depend on the relative rates of evaporation and liquid flow. Because the liquid flow rate depends on the permeability of the body, fine-grained materials are expected to develop steeper moisture gradients. Figure 8.9 shows how the moisture content changes along plastic clay bars when drying is restricted to one end. The curves tend to be parabolic, with the steepest gradients at the drying surface.

In the FRP1, migration of liquid to the surface coupled with a decrease in the evaporation rate may cause the moisture distribution curve to flatten out to some extent [13]. However, in the FRP2, the surface of the body is nearly dry and the main transport process is diffusion of vapor from the interior of the porous body to the surface. Figure 8.10 shows the moisture distribution curves for a china clay during drying in a constant environment. The curves for an average moisture content of 23% and below have approximately the same shape and are approximately parallel. This indicates that a stage in the FRP2 is reached when, at any particular instant, water is evaporating at the same rate from all planes parallel to the surface. The rate of evaporation is the same at any distance from the surface and decreases through the thickness of the body at the same rate. These conditions apply until the moisture content of the surface reaches zero (or is in equilibrium with the surrounding air) when the dry zone starts to extend inward from the surface.

Movement of moisture has also been observed to occur if a temperature gradient is set up within a moist clay irrespective of whether water is being removed by evaporation or not [13,15].

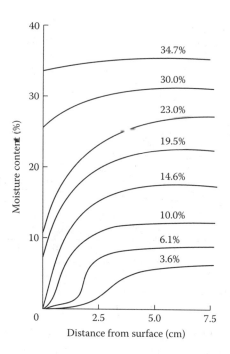

FIGURE 8.10 Moisture gradients during the falling-rate period (china clay). (From Ford, R.W., *Ceramics Drying*, Pergamon Press, New York, 1986, p. 59. With permission.)

Water moves from warmer to cooler positions, a phenomenon referred to as *thermo-osmosis*. The mechanism of moisture movement depends on whether the pores are completely saturated or only partially saturated with water, that is, whether the moisture content is above or below the first critical moisture content (or the leatherhard moisture content). With complete saturation, in which there are no drained or partially drained pores available, it is unlikely that water can migrate as a vapor, so movement must be by liquid flow. Changes in the moisture content of a clay above the leatherhard moisture content will lead to corresponding changes in volume. If the surface temperature is raised, moisture is caused to migrate toward the cooler center, producing differential shrinkage, and the article may crack. The effects of thermo-osmosis in moist clays having moisture contents below the leatherhard value are perhaps of less practical importance because the differential shrinkage caused by nonuniform moisture content is less.

During drying, moisture moves from the interior toward the surface irrespective of whether there is a temperature gradient in the clay or not. If temperature gradients are present, this movement will tend to be reinforced or opposed, depending on whether the surface is cooler or warmer than the interior.

8.2.2.4 Drying Stresses

The general principles of stress development during drying discussed in Chapter 5 for gels are also applicable to granular solids. We do not wish to repeat the discussion here, but we shall outline the salient features of the discussion for readers who are not interested in the theoretical aspects.

If evaporation of liquid exposed the solid phase, the solid–liquid interface (specific energy = γ_{SL}) would be replaced by a solid–vapor interface (specific energy = γ_{SV}), thereby raising the energy of the system because $\gamma_{SV} > \gamma_{SL}$. Liquid would flow from the interior to prevent exposure of the solid, and as it stretches toward the exterior, the liquid goes into tension. The consequences are that liquid tends to flow from the interior along the pressure gradient, and the tension in the liquid is balanced by a compressive stress on the solid network, which causes shrinkage. The lower the permeability of the body, the more difficult it is to draw liquid from the interior, and so the pressure gradient is greater. As the pressure gradient increases, so does the variation in the strain rate, with the surface contacting faster than the interior. It is this differential strain (the spatial variation of the strain for an elastic body or strain rate for a viscous body) that produces the stresses during the drying of a body. The development of drying stresses because of pressure gradients is analogous to the development of thermal stresses in a body subjected to temperature gradients [1,31].

For a flat plate, the stress in the solid phase in the plane of the plate (*xy* plane) is given by [1]:

$$\sigma_x = \sigma_y = \langle p \rangle - p \tag{8.8}$$

where *p* is the tension (negative pressure) in the liquid and $\langle p \rangle$ is the average pressure in the liquid. If the tension in the liquid is uniform, $p = \langle p \rangle$, there is no stress on the solid phase. However, when *p* varies through the thickness, the network tends to contact more where *p* is high, and this differential strain can cause warping or cracking. If the evaporation rate is high, *p* can achieve its maximum value, given by:

$$\sigma_x = \frac{2\gamma_{LV} \cos \theta}{a} \tag{8.9}$$

where γ_{LV} is the surface tension (or specific surface energy) of the liquid–vapor interface, θ is the contact angle, and *a* is the radius of the pore. For clay–water mixtures with a particle size in the range 0.5 to 1 μm, assuming that $\theta = 0$, $\gamma_{LV} = 0.07$ J/m², and the pore size is equal to half the particle size, we get $\sigma_x \approx 0.5$ MPa. Unfired granular ceramics containing little binder are very weak, and this stress is sufficiently high to cause cracking. For a plate with a half thickness of *L* drying from both sides, if the evaporation rate $\dot{V}_E$ is slow, then the stress at the drying surface is given by [1]:

$$\sigma_x \approx \frac{L\eta_L \dot{V}_E}{2K} \tag{8.10}$$

where η_L is the viscosity of the liquid and K is the permeability of the porous body.

8.2.2.5 Warping and Cracking

Warping, cracking, and anisotropic shrinkage are due to differential strain (or shrinkage) of the body, which may be caused by several factors such as:

1. Pressure gradients in the liquid during drying (discussed earlier)
2. Inhomogeneities in the body resulting from processing operations prior to drying, such as uneven distribution of the moisture in the body, as well as orientation or segregation of the particles
3. Mechanical restraints imposed on the drying body, such as friction between the body and the surface on which it is placed

It is commonly observed that cracking during drying is more likely if the pore (or particle) size of the body is fine, the body is thick, or the drying rate is fast, and that cracks generally appear at the critical point, when shrinkage stops and the liquid–vapor interface moves into the body. These observations can be well explained by Scherer's theory of drying discussed in Chapter 5. The tendency to fracture is expected to increase with σ_x, which according to Equation 8.9 and Equation 8.10, varies as the thickness of the body and the evaporation rate, and inversely as the

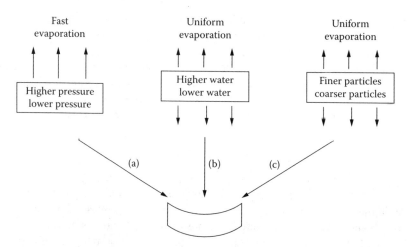

FIGURE 8.11 Schematic diagram illustrating warping during drying due to (a) pressure gradient in the liquid, (b) a moisture gradient in the body, and (c) segregation of particles.

particle size and the permeability. At the critical point, the radius of the meniscus reaches it lowest value, equal to the pore radius, so σ_x reaches its highest value.

Even if the drying were performed in conditions under which pressure gradients in the liquid were unimportant, inhomogeneities due to inadequate processing can still lead to warping, cracking, or anisotropic shrinkage (Figure 8.11). According to Figure 8.6, in the CRP, regions of the body having different moisture contents will experience different shrinkages, giving rise to stresses. Particles that are not equiaxial (such as clay particles) can undergo preferred orientation during forming operations (such as casting and extrusion), leading to anisotropic shrinkage. Differential settling of particles having a wide distribution of sizes leads to segregation during forming operations such as casting, resulting in differential shrinkage, warping, and cracking. In the drying of large articles on a floor, the shrinkage of the surface in contact with the floor is restrained by frictional forces operating over the contact area, so the shrinkage at this position may be reduced. In extreme cases, cracking in this surface may occur. These drying problems can be alleviated by improvement in the processing method, allowance in the initial relative dimensions of the green article, protection of certain areas of large objects to reduce evaporation from these positions, and judicious support of the article.

8.2.2.5.1 Avoidance of Warping and Cracking

During the CRP, the boundary condition at the surface of the body is given by Equation 5.47, relating the evaporation rate $\dot{V}_E$ to the pressure gradient in the liquid ∇p:

$$\dot{V}_E = \frac{K}{\eta_L} \nabla p \Big|_{surface} \qquad (8.11)$$

where K is the permeability of the body and η_L is the viscosity of the liquid. According to Equation 8.11, fast evaporation rate leads to high ∇p. To avoid cracking or warping, the body must be dried slowly. However, the "safe" drying rates may be so slow that uneconomically long drying times are needed.

To increase the safe drying rates, a few procedures can be used. For a given $\dot{V}_E$, ∇p decreases with higher K and lower η_L. The permeability K increases roughly as the square of the particle (or pore) size (Equation 7.11). One approach is to use a larger particle size or to mix a coarse filler with the particles. This approach is often impractical because it can lead to a reduction in densification rate during sintering. Another approach, based on a decrease in η_L, is used practically in high-humidity drying, in which the process is carried out at somewhat elevated temperatures (~70°C) and high ambient humidity in the drying atmosphere. Increasing the temperature leads to a decrease in η_L (by a factor of ~2 at 70°C) and to an increase in the drying rate, but the increase in the drying rate is counteracted by increasing the ambient humidity (Equation 8.4). In this way a reasonable drying rate is achieved while keeping ∇p small. The sequence of operation in a high-humidity dryer involves increasing the ambient humidity followed by increasing the temperature and, after drying, decreasing the humidity followed by decreasing the temperature.

8.2.2.6 Drying Technology

Several texts provide detailed descriptions of dryers and drying technology for solid materials in the ceramic, agricultural, chemical, food, pharmaceutical, pulp and paper, mineral, polymer, and textile industries [10,12,32]. We shall pay particular attention to the drying of moist ceramic bodies formed by the methods described in Chapter 7. Drying equipment can be classified in several ways, but the two most useful classifications are based on (1) the method in which heat is transferred to the moist solid and (2) the manner in which the moist material is handled within the dryer [32]. The first method of classification reveals differences in dryer design and operation, whereas the second method is most useful in the selection of dryers for preliminary consideration in a given drying problem. Further subclassification, such as continuous or batchwise operation, is possible but generally unnecessary.

Convection (or *direct*) *dryers* are the most common types of dryers used for ceramics. Drying is accomplished by passing a current of warm gas, commonly air, over the surface of the moist body. The warm air supplies the heat required for evaporation by direct contact between the flowing air and the moist body (i.e., heat transfer by convection), and it sweeps away the evaporated moisture from the surface and transports it out of the drying system. Increasing the velocity v of the air stream increases the rate of evaporation and several empirical relationships have been developed. The evaporation rate is found to increase as v^n, where $n \approx 0.7$–0.8. Examples of convection dryers are the belt conveyer dryer and the tray dryer [32].

Conduction dryers, also referred to as indirect or contact dryers, are more appropriate for thin articles. Drying is accomplished by placing the solid on a warm floor or, less commonly, against a warm wall. Much of the heat required for evaporation of the liquid is conducted through the body from the surface resting on the warm floor, but further heat is conveyed to the drying surfaces of the body by the currents of warm air rising from the floor around its edges. The vaporized liquid is removed independently of the heating medium. The drying of moist clay bodies on pallets in a current of warm air provides a good example of the effect that the conduction of heat to the drying surface can have on the drying rate [13]. For pallets made of wood, a poor conductor, little heat from the warm air stream passes through the pallet into the clay. On the other hand, for steel pallets, the drying surface receives additional heat by conduction through the steel and clay. When wooden pallets were replaced by steel pallets, the drying rate of 19-mm slabs of clay was found to increase by 40%.

In *radiant heat dryers*, heat is conveyed to the drying surface by radiation of energy from a hot source close to the moist article, without significant heating of the intervening air. For normally available temperatures, the hot sources used in the drying of solids emit radiation predominantly in the infrared range (a wavelength range of ~0.8 μm to 1 mm), so radiant heat dryers are often referred to as *infrared dryers*. Sources of infrared radiation used for drying of ceramics are (1) sheathed-element heaters, with an operating temperature of ~750°C, consisting of a resistance wire running down the center of a cylindrical metal sheath from which it is insulated by a layer of MgO and (2) gas-heated refractory reflectors, often called dull emitters, operating at a temperature of ~550°C. Whereas radiation can produce more rapid heating of the drying surface than heat transfer by convection, the radiated heat travels in a straight path, and so the surface must be exposed directly to the infrared radiation if it is to be heated. Infrared drying is therefore used for glazes and for simple shapes. *Dielectric heat dryers* operate on the principle of heat generation within the moist solid by placing the body in high-frequency electric field. *Microwave heating* has been used for the drying as well as the sintering of a variety of ceramics, and interest in this technique is growing. Coupling of the water molecules in a moist ceramic with the microwave field can lead to uniform heating, resulting in uniform drying. Equipment and operating costs are high; so microwave heating is used mainly for drying high-unit-value products.

8.3 BINDER REMOVAL (DEBINDING)

The term *binder removal* or *debinding* refers to the removal of binders as well as other organic additives, such as plasticizers, dispersants, and lubricants, from the green body. The mass transport processes responsible for binder removal involve liquid or gas flow; so the number of ways of removing the binder from the green body is limited. In general, debinding can be accomplished by three methods: (a) extraction by capillary flow into a porous surrounding material, (b) solvent extraction, and (c) thermal decomposition. By far the most commonly used method is thermal decomposition, referred to as *binder burnout* or *thermal debinding*.

Debinding of green bodies formed by the methods described in Chapter 7 presents varying degrees of difficulty. Die-pressed articles often contain less than 5 to 10 vol% binder, and the presence of a large volume of connected porosity makes binder burnout straightforward. On the other hand, the void space in tape-cast sheets after drying is nearly filled with binders, and the void space in injection-molded articles are completely filled; so debinding can often be a difficult and time-consuming step [33,34]. Our discussion of binder removal is therefore most closely identified with green bodies formed by methods such as tape casting, injection molding, and some solid freeform fabrication routes.

8.3.1 EXTRACTION BY CAPILLARY FLOW

In extraction by capillary flow, also referred to as *wicking*, the green body is heated in a packed powder bed or on a porous substrate that absorbs the molten binder. The smaller pores in the powder bed or substrate draw out the liquid binder from the larger pores of the body because of the difference in capillary pressure. For an injection-molded article in which porosity is absent, the net time for binder removal by wicking, t, from a flat plate with a thickness L is given by [35]:

$$t \approx \frac{5L^2 \eta V_s^2 D_w}{\gamma_{LV}(1-V_s)^3 D(D-D_w)} \tag{8.12}$$

where η and γ_{LV} are the viscosity and specific surface energy, respectively, of the molten binder, V_s is the particle packing density of the body, D is the particle diameter of the body, and D_w is the particle diameter of the powder bed. Rapid binder removal by wicking is promoted by small D_w and small L. The viscosity of a molten polymer increases rapidly with its molecular weight M (approximately as $M^{3.4}$); so wicking is generally useful for low molecular weight waxes but not for high-molecular-weight polymers. The amount of binder removed scales as the square root of time.

All of the binder cannot be removed by wicking. When the molten binder enters the pendular state, in which contiguous pathways for liquid flow to the surface no longer exist, binder removal stops. Typically, up to ~70% of the binder in an injection-molded article can be removed by wicking, leaving continuous, open porosity to facilitate subsequent gas transport when the remaining binder is burned out.

8.3.2 SOLVENT EXTRACTION

Solvent extraction involves immersing the component in a liquid that dissolves at least one binder phase, leaving an open pore structure for subsequent binder burnout. Full debinding is possible but not practical because the resulting powder mass will have almost no strength. The time t for debinding is given by [35]:

$$t \approx \frac{L^2}{\alpha} \ln\left(\frac{V_B}{1-V_s}\right) \exp\left(\frac{Q}{RT}\right) \tag{8.13}$$

where L is the thickness of the body, α is a factor that depends on the solubility of the binder in the solvent, V_B is the fraction of the binder to be removed, V_s is the particle packing density of the body, Q is the activation energy (per mole) for the dissolution of the binder in the solvent, R is the gas constant, and T is the absolute temperature.

In injection molding, binder systems for solvent extraction consist of at least two components, one soluble in the solvent and the other insoluble, to hold the particles in place after extraction of the soluble component. The soluble component must be at least ~30 vol% of the binder system in order to have sufficient interconnectivity for extraction. Interaction between the solvent and the binder can involve both swelling and dissolution, and so it may be necessary to choose a good solvent and perform the extraction above the theta temperature to reduce swelling (see Chapter 4).

8.3.3 Thermal Debinding

In thermal debinding, the binder is removed as a vapor by heating at ambient pressure in an oxidizing or nonoxidizing atmosphere, or under a partial vacuum. The process is influenced by both chemical and physical factors. Chemically, the composition of the binder and the atmosphere determine the decomposition temperature and the decomposition products. Physically, the removal of the binder is controlled by heat transfer into the body and mass transport of the decomposition products out of the body. Binder systems used in ceramic forming often consist of a mixture of at least two components that differ in volatility and chemical decomposition behavior. The ceramic powder can also alter the decomposition of the pure polymer. In view of the complexity of real systems, we first consider the basic features of thermal debinding for a green body containing a simplified binder system, such as a single thermoplastic binder, and later outline key factors related to more practical systems.

8.3.3.1 Stages and Mechanisms of Thermal Debinding

In general, for a thermoplastic binder, thermal debinding can be roughly divided into three stages (Figure 8.12). Stage 1 involves the initial heating of the binder to a point where it softens (~150 to 200°C). Chemical decomposition and binder removal are negligible in this stage, but the occurrence of several other processes such as shrinkage, deformation, and bubble formation can seriously affect the ability to control the shape and structural uniformity of the body. Shrinkage occurs by a rearrangement process as the particles try to achieve a denser packing under the action of the surface tension of the polymer melt. The magnitude of the shrinkage increases with decreasing particle packing density in the green body. Deformation is enhanced by a lower particle packing density, higher binder content, and lower melt viscosity. Bubble formation results from the decomposition of the binder as well as from residual solvent, dissolved air, or air bubbles trapped within the green body during forming, and it provides a possible source of failure or flaw formation during thermal debinding [36].

In stage 2, typically covering a temperature range of 200 to 400°C, most of the molten binder is removed by evaporation. If the molten binder has sufficient mobility, appreciable capillary flow of the binder can accompany the evaporation process. The nature of the decomposition reactions depends on the chemical composition of the binder and on the atmosphere. Evaporation of low-molecular-weight binders can occur readily in inert atmospheres such as nitrogen or argon. In the

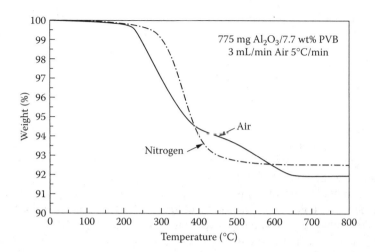

FIGURE 8.12 Thermogravimetric results for Al_2O_3/poly(vinyl butyral), PVB, mixtures (7.7 wt% PVB) heated in air and in nitrogen at 5°C/min. (From Scheiffele, G.W. and Sacks, M.D., Pyrolysis of poly(vinyl butyral) binders: II, effects of processing variables, *Ceram. Trans.*, 1, 559, 1988. With permission.)

case of multicomponent binders, evaporation of the low-molecular-weight constituents such as plasticizers and waxes can leave a concentration gradient within the high-molecular-weight polymer [37,38].

High-molecular-weight polymers can neither flow nor evaporate until they have been degraded to lower-molecular-weight segments. The degradation products diffuse through the molten polymer and evaporate from the surface. In inert atmospheres (such as argon), two main thermal degradation mechanisms can occur. Polymers such as polyethylene undergo thermal degradation by chain scission at random points in the main chain to form smaller segments (Figure 8.13a), leading to a reduction in the polymer viscosity and an increase in the volatility. Other polymers such as poly(methyl methacrylate) undergo depolymerization reactions to produce a high percentage of volatile monomers (Figure 8.13b). In oxidizing atmospheres (such as air), degradation occurs predominantly by oxidation, but there is also some thermal degradation. Activation energies for oxidative degradation of polymers are generally lower than activation energies for thermal degradation; so oxidative degradation often occurs at lower temperatures or at a faster rate. Oxidative degradation starts at the surface of the body and moves inward, and is controlled by oxygen transport to the surface of the binder front. The products contain a high percentage of volatile low-molecular-weight compounds such as water, carbon dioxide, and carbon monoxide.

In stage 3, the small amount of binder still remaining in the body is removed by evaporation and decomposition at temperatures above ~400°C. Binder removal is facilitated by the highly porous nature of the body, but a small amount of carbon residue generally remains within the sample. For a given powder–binder system, the amount of carbon residue is generally higher for inert atmospheres than for oxidizing atmospheres. Temperatures greater than 600°C are often required for reducing the concentration of carbon residue.

8.3.3.2 Models for Thermal Debinding

As described above, high-molecular-weight thermoplastic binders undergo thermal debinding by evaporation of low-molecular-weight species produced by thermal or oxidative degradation. Low-molecular-weight binders simply undergo evaporation without the need for any significant degradation. There are, therefore, three possible models for thermal debinding: (1) evaporation, (2) thermal degradation followed by evaporation, and (3) oxidative degradation followed by evaporation. The evaporation model differs from the thermal degradation model only as far as the concen-

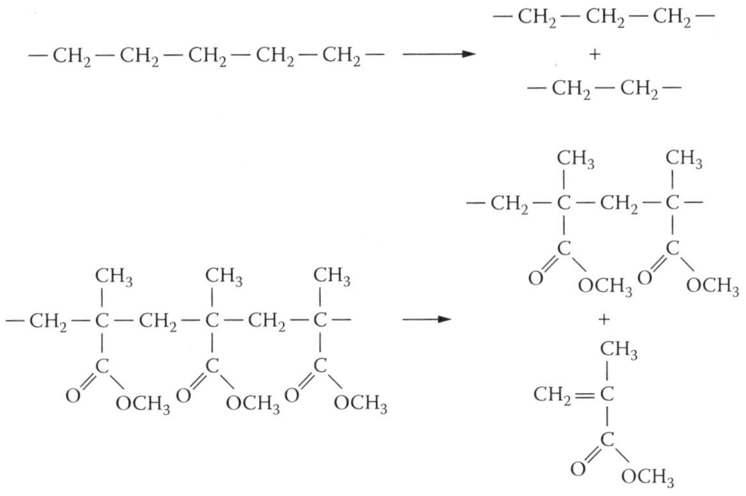

FIGURE 8.13 Mechanisms of thermal degradation of polymers illustrating (a) chain scission at random points in the chain, e.g., poly(ethylene), and (b) depolymerization (unzipping) to produce monomers, e.g., poly(methyl methacrylate).

tration of the volatile species is concerned. In the evaporation model, the volatile species are present in the binder at some initial concentration and the concentration decreases as debinding proceeds, whereas in the thermal degradation model, the volatile species have a concentration that is initially zero but increases with degradation of the binder and then decreases as the volatile species evaporate. The thermal degradation model is, therefore, expected to describe the main features of the evaporation of low-molecular-weight binders.

Assuming that the temperature is uniform, thermal degradation produces volatile low-molecular-weight species throughout the molten binder. Removal of the binder occurs by diffusion of these low-molecular-weight species through the molten polymer and evaporation at the surface. The rate of evaporation and the rate of transport of the degradation products through the body determine the concentration profile of the products. Volatile products present, for example, in the center of the body must not be allowed to reach temperatures above their boiling point because this will lead to the formation of bubbles and, hence, microstructural flaws.

Models have been developed to analyze the combined degradation reaction kinetics and diffusional mass transport during thermal debinding [39,40], and used to predict favorable properties of the polymer and degradation product for controlling the formation of internal flaws such as bubbles [41]. These models have limited predictive capability for the selection of binders and can only be used as a rough guide. A high enthalpy of vaporization is desirable to increase the boiling point of the monomer. A low activation energy for diffusion is favorable to ensure that flaws do not occur at low temperatures. As the parameters that control the diffusivity also control the dependence of the viscosity on temperature, a slight dependence of the polymer viscosity on temperature is suggested. The degradation of the polymer should take place at high temperature at which stage high diffusivity of the monomer prevails. The maximum permissible heating rate for debinding is predicted to increase if a gas overpressure is applied.

In oxidative degradation, the reaction occurs at the molten binder–gas interface that recedes into the body as degradation proceeds. The weight loss can be described by shrinking core kinetics [42]. The reaction rate is controlled by diffusion of oxygen to the binder surface and transport of the degradation products outward. A simple quantitative model for oxidative degradation has been developed, which assumed a single binder and a planar binder–gas interface that receded uniformly into the body as debinding progressed [35]. The gaseous reaction products can be removed at ambient pressure via permeation or under partial vacuum via diffusion through the porous outer

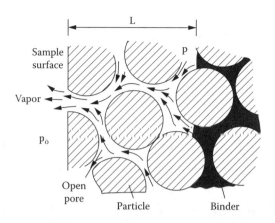

FIGURE 8.14 Schematic diagram of the model for thermal debinding by oxidative degradation, where the binder–vapor interface is at a distance L from the surface of the compact. (From German, R.M., Theory of thermal debinding, *Int. J. Powder Metall.*, 23, 237, 1987. With permission.)

layer (Figure 8.14). Whether diffusion or permeation is rate controlling or not depends essentially on the mean free path of the gas molecules. In diffusion control, the collisions are mainly with the pore structure, whereas in permeation control, the collisions are mainly with other gas molecules. The model predicts that the debinding time t for the diffusion-controlled process is given by [35]:

$$t \approx \frac{L^2}{2D(1-V_s)^2(p-p_o)} \frac{(M_w kT)^{1/2}}{V_M} \tag{8.14}$$

where L is the thickness of the body, D is the particle size, V_s is the particle packing density, p is the pressure in the pores, p_o is the ambient pressure, M_w is the molecular weight and V_M is the molecular volume of the vapor, k is the Boltzmann constant, and T is the absolute temperature. For the case of permeation control, the debinding time is given by [35]:

$$t \approx \frac{20L^2\eta}{D^2 F} \frac{V_s^2}{(1-V_s)^3} \frac{p}{(p^2-p_o^2)} \tag{8.15}$$

where η is the viscosity of the vapor, F is the volume change associated with the burnout of the binder, and the other terms are as defined for Equation 8.14.

For both diffusion and permeation control, the debinding time varies as the square of the component thickness. Small particle size and high particle packing density reduce the rate of binder removal; on the other hand, they are beneficial for the subsequent sintering step. A conflict, therefore, exists between rapid removal of the binder and the achievement of high density during sintering. Process optimization or the use of a sintering aid that enhances the densification process can provide a solution to the conflicting requirements. Equation 8.14 and Equation 8.15 also indicate that a low ambient pressure or a vacuum serves to reduce the time for binder removal. A vacuum, however, does not lead to oxidative degradation. Furthermore, temperature control and transport of heat are poor in a vacuum. The use of an oxidizing gas at reduced ambient pressure may provide adequate degradation as well as good thermal transport.

Binder removal by thermal degradation has some features that are similar to those encountered in the drying of moist granular solids. For a system having a distribution of pore sizes, considerable distribution of the molten binder occurs (see Figure 8.7), driven by capillary pressure, and the binder–gas interface does not move uniformly into the body. Pore channels first develop deep into the body as the binder from the larger pores is drawn into the smaller pores. These trends in porosity development and binder redistribution have been observed for tape-cast ceramic sheets in which the void space was nearly filled with a two-component thermoplastic binder system [43,44].

Computer simulations have been performed to investigate the effects of capillary-driven molten binder redistribution and diffusion of volatile species in the binder during thermal debinding [45]. The removal of plasticizer species from granular solids in which the void space is filled with two-component binder systems consisting of volatile and nonvolatile components has been studied. Binder removal times were found to decrease by an order of magnitude or more when capillary-driven binder redistribution was allowed to occur. The simulations can provide insight, at the microscopic level, into the transition from a completely filled void space to one in which a percolating network of porosity develops.

8.3.3.3 Thermal Debinding in Practice

Practical binder systems, particularly in the case of injection molding, consist of a least two components that differ in their volatility and decomposition behavior. A key to efficient debinding is to select components that do not have a significant overlap between the temperature ranges for removal. In this case, removal of the first binder at lower temperature creates a network of porosity through which the decomposition products of the second component can be removed more easily. Plasticizers, oils, and waxes have lower melting temperatures than most high-molecular-weight thermoplastic polymers, and they form a useful minor component in thermoplastic binder mixtures.

The decomposition of the polymeric binder in a ceramic green body is more complex than that of the pure binder. Detailed studies of powder and atmosphere effects on the binder burnout and decomposition chemistry of ceramic green bodies containing a single binder have been performed by Masia et al. [46] and Sacks et al. [47a–47c]. Figure 8.15 shows that oxide powders can catalyze the decomposition or pyrolysis reaction of poly(vinyl butyral), PVB, leading to a decrease in the decomposition temperature. Cerium oxide (CeO_2), for example, reduces the temperature for the greatest weight loss by ~200°C. The catalytic effect of the oxide powders is present in both oxidizing and nonoxidizing atmospheres, but the effect in nonoxidizing atmospheres is reduced, indicating that oxygen plays a role in the catalysis.

Binders that normally burn out completely in the pure state often leave a small amount of carbon residue that cannot be easily removed from the particle surfaces [46,48]. The amount of

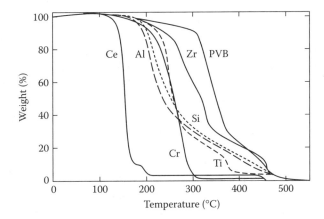

FIGURE 8.15 Thermogravimetric plots of the decomposition of poly(vinyl butyral), PVB, from films containing 2 wt% PVB and 98 wt% of various oxides during heating in air at 5°C/min. (From Masia, S. et al., Effect of oxides on binder burnout during ceramics processing, *J. Mater. Sci.*, 24, 1907, 1989. With permission.)

carbon residue depends on several factors, such as the binder composition, the powder composition, the gaseous atmosphere, and the structure and chemistry of the powder surface. The amount of carbon residue is often higher in inert atmospheres than in oxidizing atmospheres. In air, whereas pure PVB left no measurable carbon residue, heating oxide powders with 2 wt% PVB left 0.2 to 0.4 wt% of the original mass of PVB after heating at 5°C/min to 600°C. Subsequent removal of the carbon residue above 600°C depended on the composition of the oxide powder, with CeO_2, TiO_2, and Al_2O_3 being very effective.

It is essential to achieve some balance between the time for debinding and the prevention of flaws during debinding. Normally, this is achieved by controlling the heating cycle. A very low heating rate makes the thermal-debinding process time consuming, whereas a high heating rate leads to bubble formation, rapid melting of the binder, and distortion of the body. Initially, the heating rate should be low (less than ~1°C/min) and isothermal stages should be appropriately incorporated until the pores are partially opened, after which the heating rate can be increased. Thermogravimetric analysis (TGA) of a sample of the green body can provide valuable information in the development of a heating cycle (see Figure 8.12). For some injection-molded bodies, the use of solvent extraction to partially open the pores, followed by thermal debinding, can serve to reduce the difficulties of the debinding step.

8.4 GREEN MICROSTRUCTURES AND THEIR CHARACTERIZATIONS

Methods used on a more or less routine basis for characterizing the microstructure of ceramic green bodies are the following:

1. Density determination by measuring the mass and dimensions or by the Archimedes technique
2. Determination of the pore size distribution and the volume of open porosity by mercury porosimetry or gas adsorption
3. Microstructural observation using scanning electron microscopy (SEM)

The density gives information about the particle packing density, whereas the pore size distribution is used as a rough guide to the packing homogeneity. SEM is often used to check for extreme packing variations and, for a mixture of two or more solid phases, for mixing variations, as well as for processing flaws. Detecting the presence of large pores and crack-like voids in the green body is particularly important. If present, these flaws cannot be easily removed during the subsequent sintering stage, and so they often remain in the final article, leading to deterioration of the properties.

The measured density is most often the *bulk density*, D, equal to the mass divided by the external volume, but the density is best expressed as the *relative density*, ρ equal to D/D_t, where D_t is the theoretical density of the solid (the density of the fully dense or pore-free solid). The relative density gives a density value that is independent of the material, making it more useful for comparing the densification properties of different materials. The bulk density is obtained directly by measuring the mass and the external volume of the body. For a body with a regular shape, the external volume is determined from the dimensions. If the body has an irregular shape, an immersion technique is appropriate. Prior to immersion, the pores must be filled with water or with a water-insoluble liquid, such as paraffin oil, or must be sealed with a thin impervious coating of paraffin wax. The density of a porous material can also be determined by the Archimedes technique. The sample is first weighed in air (mass = m_1), weighed again after fully impregnated with a liquid (m_2), and finally weighed while immersed in the liquid (m_3). The density can be found from the equation:

$$D = \frac{m_1 D_l}{m_2 - m_3} \qquad (8.16)$$

where D_l is the density of the liquid. For water, the density (in g/cm³) is given by:

$$D_l = 1.0017 - 0.0002315T \qquad (8.17)$$

where T is the temperature in °C. The theoretical density of the solid is most easily obtained from a handbook [49], but if not found for the required material, it can be determined by measuring the density of the fully dense material or calculated from the unit cell parameters.

The porosity of the body, equal to $(1 - \rho)$, gives the fraction of the total volume of the body that consists of voids. The open porosity and the pore size distribution are often determined using mercury porosimetry or, for very fine pores, by gas adsorption. The techniques are identical to those described in Chapter 3 for determining the pore characteristics of particles. The importance of pore size distribution of the green body on sintering has received considerable attention [50,51]. When the procedure is carried out carefully, differences in the pore size determined by mercury porosimetry can be related to differences in the particle packing in the green bodies. Figure 8.16 shows the pore size distributions determined by mercury porosimetry for green bodies formed by slip casting stable suspensions of alumina powders with a broad and a narrow distribution of particle sizes. The body formed from the broad size distribution powder had a higher density ($\rho = 0.73$) and a smaller median pore channel radius (~30 nm) compared to the narrow size distribution body ($\rho = 0.65$ and median pore radius ≈50 nm). The results indicate that the fine particles are filling the interstitial voids between the large particles in the broad-size-distribution body.

SEM is most easily performed by observing a fractured surface of the green body. Often, the green body is too weak for manipulation, but this difficulty can be overcome by lightly presintering the body to develop adequate strength.

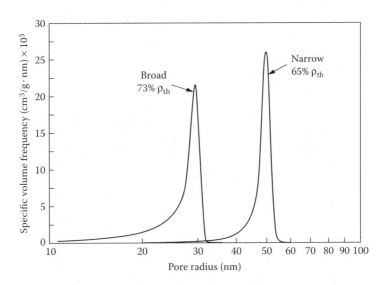

FIGURE 8.16 Plots of specific pore volume frequency vs. pore channel radius obtained by mercury porosimetry for slip-cast samples prepared with broad and narrow particle size distributions. (From Yeh, T-S. and Sacks, M.D., Effect of particle size distribution on the sintering of alumina, *J. Am. Ceram. Soc.*, 71, C-484, 1988. With permission.)

Additional information can be obtained by using other techniques [52]. Impregnation of the green body with epoxy resin, followed by sectioning, polishing, and observation in the SEM provides quasi-three-dimensional information of the green microstructure [53], but the method is time consuming and only the features present on the polished surface are observed. A liquid immersion technique has been developed to characterize the internal structure of granules and green bodies [54]. In this technique, the specimens are made transparent by immersion in a liquid with a matching refractive index, and observed in an optical microscope under transmitted light. The technique can be used to observe low concentrations of microstructural defects (one to tens of microns in size). Limitations of the method include the use of thin sections, less than a few tenths of a millimeter, and the difficulty of finding immersion liquids with high enough refractive indices to match those of many common ceramics. These limitations can be alleviated to some extent by the use of infrared microscopy to observe the specimens [55]. Figure 8.17 shows IR photomicrographs of (a) Si_3N_4 granules and (b) a green compact formed by pressing at only ~20 MPa, followed by heating for 2 h at 1300°C to burn out the binder and to develop adequate strength for handling.

FIGURE 8.17 Liquid immersion infrared photomicrographs of (a) Si_3N_4 granules and (b) Si_3N_4 green body formed by pressing the granules under a pressure of 20 MPa. (From Uematsu, K. et al., Infrared microscopy for examination of structure in spray-dried granules and compacts, *J. Am. Ceram. Soc.*, 84, 254, 2001. With permission.)

The micrographs show the presence of dimples in the center of the granules, as well as boundaries between the granules in the green body.

Small-angle neutron scattering (SANS) provides a nondestructive method for characterizing the porosity of green or sintered ceramic bodies [56,57], but SANS equipment is not available in many research laboratories. The technique has been used to characterize pores and inhomogeneities with sizes in the range of 1 nm to 10 μm. Techniques for characterizing density variations in green compacts were mentioned in Chapter 7. They include microhardness, x-ray tomography, x-ray radiography, ultrasonic, and nuclear magnetic resonance [58]. The use of these techniques is unnecessary in most processing experiments.

8.5 CONCLUDING REMARKS

In this chapter we considered drying and debinding of ceramic green bodies formed by the methods described in Chapter 7. Drying and debinding are the major heat treatment steps of the green body prior to sintering. They must be carefully controlled to limit further creation of microstructural flaws in the body and to prevent warping and cracking. We described the stages and mechanisms of drying, capillary-driven liquid flow, and stress development during drying of granular layers and green bodies. Methods to avoid warping and cracking during drying were outlined. Binder removal from green bodies were discussed, paying particular attention to thermal debinding of green bodies formed by injection molding or tape casting, in which the void space between the particles are nearly or completely filled with organic additives. Models developed to analyze thermal debinding of green bodies with simple binder systems were described and used to assess favorable properties of the polymer and degradation product for limiting the formation of internal flaws. Finally, methods for characterizing the green body microstructure were described.

PROBLEMS

8.1

Determine the relative humidity above which water will not evaporate from pores with a radius of 5 nm at 25°C. ($\gamma_{LV} = 0.07$ J/m^2 for water at 25°C.)

8.2

Consider a moist body in the early stages of drying when the pores are filled with a liquid. In which direction would liquid flow occur if (a) the temperature is higher at the surface than in the interior and (b) the body has been uniformly heated at 100% humidity and the humidity is quickly reduced?

8.3

Assuming isotropic shrinkage, derive a relationship between the volume shrinkage and the linear shrinkage of a body.

8.4

A plastic mass of clay has a moisture content of 30%. The clay is extruded and cut into 20-cm-long bricks. The drying characteristics of the clay was measured previously and found to be as follows: (a) leatherhard moisture content = 15%, (b) apparent solid volume of 100 g of dry clay = 30 cm^3, and (c) the volume of a dry test piece containing 100 g of clay = 45 cm^3. Assuming isotropic shrinkage during drying, determine the dried length of the bricks.

8.5

A 10-cm^3 Al$_2$O$_3$ green body has a particle packing density of 65% and the void space between the particles is filled completely with poly(methyl methacrylate), PMMA. Assuming that all the PMMA is converted to a volatile monomer, determine the volume of the volatile monomer. The molecular weight and density of PMMA are 10^4 and 1.2 g/cm^3, respectively.

8.6

Differential thermal analysis (DTA) data show that the decomposition of poly(methyl methacrylate), PMMA, has one exothermic peak, whereas the decomposition of poly(methacrylic acid), PMAA, has a lower temperature endothermic trough and two exothermic peaks at higher temperature. Compare and discuss the structural and chemical changes occurring during the decomposition of PMMA and PMAA.

8.7.

When layers of different ceramic powders containing 2 wt% poly(vinyl butyral), PVB, were heated at 5°C/min in air, the amount of carbon residue (in ppm) at 700°C were found to be: Al$_2$O$_3$: 260; CeO$_2$: 90; TiO$_2$: 70; ZrO$_2$: 160; SiO$_2$: 400; Cr$_2$O$_3$: 210. Plot a graph showing the carbon content vs. the isoelectric point of the oxides, and discuss the influence of the powder on the carbon content resulting from burnout of PVB.

REFERENCES

1. Scherer, G.W., Theory of drying, *J. Am. Ceram. Soc.*, 73, 3, 1990.
2. Deegan, R.D., Bakajin, O., Dupont, T.F., Huber, G., Nagel, S.R., and Witten, T.A., Capillary flow as the cause of ring stains from dried liquid drops, *Nature*, 389, 827, 1997.
3. Deegan, R.D., Bakajin, O., Dupont, T.F., Huber, G., Nagel, S.R., and Witten, T.A., Contact line deposits in an evaporating drop, *Phys. Rev. E*, 62, 756, 2000.
4. Deegan, R.D., Pattern formation in drying drops, *Phys. Rev. E*, 61, 475, 2000.
5. Denkov, N.D., Velev, O., Kralchevski, P., Ivanov, I., Yoshimura, H., and Nagayama, K., Mechanism of formation of two-dimensional crystals from latex particles on substrates, *Langmuir*, 8, 3183, 1992.
6. Chiu, R.T, Garino, T.J., and Cima, M.J., Drying of granular ceramic films: I, effect of processing variables on cracking behavior, *J. Am. Ceram. Soc.*, 76, 2257, 1993.
7. Chiu, R.T, Garino, T.J., and Cima, M.J., Drying of granular ceramic films: II, drying stress and saturation uniformity, *J. Am. Ceram. Soc.*, 76, 2769, 1993.
8. Hu, M.S., Thouless, M.D., and Evans, A.G., Decohesion of thin films from brittle substrates, *Acta Metall.*, 36, 1301, 1988.
9. Guo, J.J. and Lewis, J.A., Aggregation effects on the compressive flow properties and drying behavior of colloidal silica suspensions, *J. Am. Ceram. Soc.*, 82, 2345, 1999.
10. Keey, R.B., *Drying Principles and Practice*, Pergamon Press, New York, 1972.
11. Mujumdar, A.S., Ed., *Advances in Drying*, Vol. 1, 1980; Vol. 2, 1982; Vol. 3, 1984, Hemisphere Publishing, New York.
12. Mujumdar, A.S., Ed., *Handbook of Industrial Drying*, Marcel Dekker, New York, 1987.
13. Ford, R.W., *Ceramics Drying*, Pergamon Press, New York, 1986.
14. Macey, H.H., Clay-water relationships and the internal mechanism of drying, *Trans. Br. Ceram. Soc.*, 41, 73, 1942.
15. Moore, F., The mechanism of moisture movement in clays with particular reference to drying: a concise review, *Trans. Br. Ceram. Soc.*, 60, 517, 1961.
16. Kingery, W.D. and Francl, J., Fundamental study of clay: XIII. Drying behavior and plastic properties, *J. Am. Ceram. Soc.*, 37, 596, 1954.

17. Van Olphen, H., *An Introduction to Clay Colloid Chemistry*, 2nd ed., John Wiley & Sons, New York, 1977.

18. Packard, R.Q., Moisture stress in unfired ceramic bodies, *J. Am. Ceram. Soc.*, 50, 223, 1967.

19. Sherwood, T.K., The drying of solids — I, *Ind. Eng. Chem.*, 21, 12, 1929.

20. Sherwood, T.K., The drying of solids — II, *Ind. Eng. Chem.*, 21, 976, 1929.

21. Sherwood, T.K., The drying of solids — III, *Ind. Eng. Chem.*, 22, 132, 1930.

22. Suzuki, M. and Maeda, S., On the mechanism of drying of granular beds — mass transfer from discontinuous source, *J. Chem. Eng. Jpn.*, 1, 26, 1968.

23. Shaw, T.M., Movement of a drying front in a porous material, *Mater. Res. Soc. Symp. Proc.*, 73, 215, 1986.

24. Shaw, T.M., Drying as an immiscible displacement process with fluid counterflow, *Phys. Rev. Lett.*, 59, 161, 1987.

25. Whitaker, S., Simultaneous heat, mass, and momentum transfer in porous media: a theory of drying, *Adv. Heat Transfer*, 13, 119, 1977.

26. Whitaker, S., Heat and mass transfer in porous granular media, in *Advances in Drying*, Vol. 1, Mujumdar, A.S., Ed., Hemisphere Publishing, New York, 1980, chap. 2.

27. Fortes, M. and Okos, M.R., Drying theories: their bases and limitations as applied to foods and grains, in *Advances in Drying*, Vol. 1, Mujumdar, A.S., Ed., Hemisphere Publishing, New York, 1980, chap. 5.

28. Van Brakel, J., Mass transfer in convective drying, in *Advances in Drying*, Vol. 1, Mujumdar, A.S., Ed., Hemisphere Publishing, New York, 1980, chap. 5.

29. Wei, C.K., Davis, H.T., Davis, E.A., and Gordon, J., Heat and mass transfer in water-laden sandstone: convective heating, *AIChE J.*, 31, 1338, 1985.

30. Wei, C.K., Davis, H.T., Davis, E.A., and Gordon, J., Heat and mass transfer in water-laden sandstone: microwave heating, *AIChE J.*, 31, 842, 1985.

31. Cooper, A.R., Quantitative theory of cracking and warping during the drying of clay bodies, in *Ceramic Processing Before Firing*, Onoda, G.Y. Jr. and Hench, L.L. Eds., John Wiley & Sons, New York, 1978, chap. 21.

32. Porter, H.F., Schurr, G.A., Wells, D.F., and Semrau, K.T., Solids drying and gas-solid systems, in *Perry's Chemical Engineers' Handbook*, 6th ed., Perry, R.H. and Green, D., Eds, McGraw-Hill, New York, 1984, p. 20.

33. German, R.M. and Bose, A., *Injection Molding of Metals and Ceramics*, Metal Powder Industries Federation, Princeton, NJ, 1997.

34. Evans, J.R.G., Injection moulding, in *Materials Science and Technology, Vol. 17A: Processing of Ceramics*, Pt. I, Brook, R.J., Ed., VCH, New York, 1996, chap. 8.

35. German, R.M., Theory of thermal debinding, *Int. J. Powder Metall.*, 23, 237, 1987.

36. Dong, C. and Bowen, H.K., Hot-stage study of bubble formation during binder burnout, *J. Am. Ceram. Soc.*, 72, 1082, 1989.

37. Angermann, H.H., Yang, F.K., and Van der Biest, O., Removal of low molecular weight components during thermal debinding of powder compacts, *J. Mater. Sci.*, 27, 2534, 1992.

38. Angermann, H.H. and Van der Biest, O., Low temperature debinding kinetics of two-component model systems, *Int. J. Powder Metall.*, 29, 239, 1993.

39. Calvert, P. and Cima, M., Theoretical models for binder burnout, *J. Am. Ceram. Soc.*, 73, 575, 1990.

40. Evans, J.R.G., Edirisinghe, M.J., Wright, J.K., and Crank, J., On the removal of organic vehicle from moulded ceramic bodies, *Proc. R. Soc. Lond.*, A432, 321, 1991.

41. Matar, S.A., Evans, J.R.G., Edirisinghe, M.J., and Twizell, E.H., The influence of monomer and polymer properties on the removal of organic vehicle from ceramic and metal moldings, *J. Mater. Res.*, 10, 2060, 1995.

42. Wright, J.K. and Evans, J.R.G., Kinetics of the oxidative degradation of ceramic injection-moulding vehicle, *J. Mater. Sci.*, 26, 4897, 1991.

43. Cima, M.J., Lewis, J.A., and Devoe, A.D., Binder distribution in ceramic greenware during thermolysis, *J. Am. Ceram. Soc.*, 72, 1192, 1989

44. Lewis, J.A. and Cima, M.J., Mass transfer processes during multicomponent binder thermolysis, *Mater. Res. Soc. Symp. Proc.*, 249, 363, 1992.

45. Lewis, J.A, Galler, M.A., and Bentz, D.P., Computer simulations of binder removal from 2-D and 3-D model particulate bodies, *J. Am. Ceram. Soc.*, 79, 1377, 1996.

46. Masia, S., Calvert, P.D., Rhine, W.E., and bowen, H.K., Effect of oxides on binder burnout during ceramics processing, *J. Mater. Sci.*, 24, 1907, 1989.

47. (a) Sun, Y.-N., Sacks, M.D., and Williams, J.W., Pyrolysis behavior of acrylic polymers and acrylic polymer/ceramic mixtures, *Ceram. Trans.*, 1, 538, 1988.
    (b) Shih, W.-K., Sacks, M.D., Scheiffele, G.W., Sun, Y.-N., and Williams, J.W., Pyrolysis of poly(vinyl butyral) binders: I, degradation mechanisms, *Ceram. Trans.*, 1, 549, 1988.
    (c) Scheiffele, G.W. and Sacks, M.D., Pyrolysis of poly(vinyl butyral) binders: II, effects of processing variables, *Ceram. Trans.*, 1, 559, 1988.

48. Cima, M.J. and Lewis, J.A., Firing-atmosphere effects on char content from alumina–poly(vinyl butyral) films, *Ceram. Trans.*, 1, 567, 1988.

49. *CRC Handbook of Chemistry and Physics*, CRC Press, Cleveland, OH, 2005–2006.

50. Roosen, A. and Bowen, H.K., Influence of various consolidation techniques on the green microstructure and sintering behavior of alumina powders, *J. Am. Ceram. Soc.*, 71, 970, 1988.

51. Yeh, T-S. and Sacks, M.D., Effect of particle size distribution on the sintering of alumina, *J. Am. Ceram. Soc.*, 71, C-484, 1988.

52. Bonekamp, B.C. and Veringa, H.J., Green microstructures and their characterization, in *Materials Science and Technology, Vol. 17A: Processing of Ceramics,* Pt. I, Brook, R.J., Ed., VCH, New York, 1996, chap. 10.

53. Weeks, M.J. and Laughner, J.W., Characterization of unfired ceramic microstructures, *Adv. Ceram.*, 21, 793, 1987.

54. Uematsu, K., Immersion microscopy for detailed characterization of defects in ceramic powder and green bodies, *Powder Technol.*, 88, 291, 1996.

55. Uematsu, K., Uchida, N., Kato, Z., Tanaka, S., Hotta, T., and Naito, M., Infrared microscopy for examination of structure in spray-dried granules and compacts, *J. Am. Ceram. Soc.*, 84, 254, 2001.

56. Hardman-Rhyne, K.A., Berk, N.F., and Fuller, E.R., Microstructural characterization of ceramic materials by small angle neutron scattering techniques, *J. Res., Natl. Bur. Stand. (U.S.)*, 89, 17, 1984.

57. Frase, K.G. and Hardman-Rhyne, K.A., Porosity in spinel compacts using small-angle neutron scattering, *J. Am. Ceram. Soc.*, 71, 1, 1988.

58. Lannutti, J.J., Characterization and control of compact microstructure, *MRS Bull.*, 22(12), 38, 1997.

9 Sintering and Microstructure Development

9.1 INTRODUCTION

The heat treatment step in which the dried, debinded green body is converted to a useful solid with the required microstructure is referred to as *sintering* (or *firing*). In Chapter 1, we outlined the essentials of the sintering process for the production of ceramics from powders. The reader will recall the four categories of sintering: solid-state sintering, liquid-phase sintering, viscous sintering, and vitrification, as well as sintering under an externally applied pressure, pressure-assisted sintering, of which hot pressing and hot isostatic pressing are common examples. A list of some ceramic compositions and the way they are sintered is given in Table 9.1.

In order to produce ceramics with controlled microstructures that must meet exacting property requirements, we need to look more closely at the sintering process, to understand how the process works on a detailed scale, even down to the atomic scale, to identify the important parameters, and to find out how to manipulate them. The theory, principles, and practice of sintering are covered in great depth in Reference 1 and Reference 2 and are summarized in a recent review [3]. In this chapter, we outline the basic principles of sintering and how they are applied practically to the production of ceramics with controlled microstructure.

The driving force for sintering identifies the factors that cause the body to sinter, whereas the mechanisms of sintering identify how matter is transported. Sintering models attempt to predict the rate at which the porous body densifies, and how the densification rate depends on key process parameters such as particle size, temperature, and pressure. The models generally assume drastic simplifications of matter transport and microstructure and, at best, provide only a qualitative understanding of the sintering process in practical systems. Analysis of densification during sintering is, by itself, not very useful. We must also understand how the microstructure of the porous body evolves. Densification of porous polycrystalline systems is accompanied by grain growth and pore growth, leading to coarsening of the microstructure. The occurrence of coarsening reduces the driving force for densification, a situation often described as a competition between sintering (densification) and coarsening. Control of grain growth during sintering is a primary consideration if products with high density are to be achieved.

A comprehensive sintering model should be capable of describing the entire sintering process as well as the accompanying microstructural evolution, such as grain size, pore size, and the distribution of grain and pore sizes. In view of the complex nature of the microstructural changes that occur in practical powder compacts, it is unlikely that such a theory will be developed. A more realistic approach is to develop a fundamental understanding of the densification and coarsening phenomena separately, and then explore the consequences of their interaction. This approach provides a framework for process optimization without extensive trial and error, and for developing improved processing techniques.

The common requirements for the sintered product are high density and small grain size, because these microstructural characteristics enhance most engineering properties. The particle size and particle packing of the green body often have the most significant effect on sintering.

TABLE 9.1
Some Ceramic Compositions and the Way They Are Sintered

Composition	Sintering Process	Application
Al_2O_3	Solid-state sintering with MgO additive	Sodium vapor arc lamp tubes
	Liquid-phase sintering with a silicate glass	Furnace tubes, refractories
MgO	Liquid-phase sintering with a silicate glass	Refractories
Si_3N_4	Liquid-phase sintering with oxide additives (e.g., Al_2O_3 and Y_2O_3) under nitrogen gas pressure or under an externally applied pressure	High-temperature structural ceramics
SiC	Solid-state sintering with B and C additives; liquid-phase sintering with Al, B and C or oxide additives	High-temperature structural ceramics
ZnO	Liquid-phase sintering with Bi_2O_3 and other oxide additives	Electrical varistors
$BaTiO_3$	Liquid-phase sintering with TiO_2-rich liquid	Capacitor dielectrics, thermistors
Pb (Zr,Ti)O_3 (PZT) (Pb,La)(Zr,Ti)O_3 (PLZT)	Sintering with a lead-rich liquid phase, hot pressing	Piezoelectric actuators and electro-optical devices
ZrO_2/(3–10 mol% Y_2O_3)	Solid-state sintering	Electrical conducting oxide for fuel cells
Mn–Zn and Ni–Zn ferrites	Solid-state sintering under a controlled oxygen atmosphere	Soft ferrites for magnetic applications
Porcelain	Vitrification	Electrical insulators, tableware
SiO_2 gel	Viscous sintering	Optical devices

However, assuming that proper powder preparation and consolidation procedures are in effect, successful fabrication is still strongly dependent on the ability to control the microstructure through manipulation of the process variables in the sintering stage. The wide variety of sintering techniques developed to obtain ceramics with the required density, microstructure, and composition generally involves the manipulation of some combination of the heating schedule, atmosphere, and applied pressure.

9.2 SOLID-STATE SINTERING

When a porous polycrystalline powder compact is heated at a constant rate, the density, taken as the mass divided by the external volume, often shows a sigmoidal dependence on the temperature (Figure 9.1). At some temperature, the density starts to increase slowly, then enters a second region where the density increases markedly, and after reaching a value of 90–95% of the theoretical value, a third region is reached where the increase in density slows and finally stops. The final limiting density will depend on several factors, such as the composition and particle size of the powder, particle packing in the green body, the rate of heating, and the atmosphere in which the body is sintered. For example, the final density is often higher for a green body with more homogeneous particle packing.

Observation of the microstructure will reveal that the green compact consists of identifiable particles in contact, with porosity between them (Figure 9.2a). This porosity consists of a continuous, three-dimensional network with a complex geometry. As sintering starts, the particles become joined together at the contact points, forming necks, due to matter transport to the necks by diffusion, and the strength of the body increases. Further sintering leads to an increase in the neck diameter and smoothing of the pore surfaces, giving a three-dimensional network of solid matter that appears markedly different from the structure of the green compact. The microstructure can be described

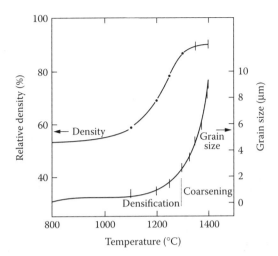

FIGURE 9.1 The density and grain size of a TiO_2 powder compact as a function of the sintering temperature. (From Yan, M.F., Microstructural control in the processing of electronic ceramics, *Mater. Sci. Eng.*, 48, 53, 1981. With permission.)

as an interconnected network of grains intertwined with a continuous pore network (Figure 9.2b). Matter transport into the pores leads to further shrinkage and a reduction in porosity. Eventually, a stage is reached when the continuous pore network starts to break up into individual, isolated pores. Reduction of porosity continues, but at a slower rate, until densification stops (Figure 9.2c). Ideally, all the pores would be removed, giving a fully dense product, but this rarely happens unless special techniques are used. During all of this, the grains continue to grow, and this grain growth can lead to the coalescence of neighboring pores, so the average pore size often increases, even though the porosity is decreasing. Grain growth becomes more rapid in the later stage of densification when the pores become isolated (Figure 9.1), and in many cases, the rapidly growing grains may engulf many of the pores, separating then from the grain boundaries. This situation is undesirable, because these pores are almost impossible to remove with further heating.

9.2.1 DRIVING FORCE FOR SINTERING

Sintering is an irreversible process, and as with all irreversible processes it is accompanied by a reduction in the free energy of the system. The factor that causes the powder compact to sinter, leading to a reduction in the free energy, is referred to as the *driving force* for sintering. For single-phase particles, the reduction of the free energy associated with the elimination of the internal surface area of the powder compact provides the driving force for sintering (Figure 9.3). To see this, let us consider, for example, one mole of powder, consisting of spherical particles with a radius a. The number of particles is

$$N = \frac{3M}{4\pi a^3 d} = \frac{3V_m}{4\pi a^2} \tag{9.1}$$

where d is the density of the particles, which are assumed to contain no internal porosity, M is the molecular weight, and V_m is the molar volume. The surface area of the system of particles is

$$S_A = 4\pi a^2 N = \frac{3V_m}{a} \tag{9.2}$$

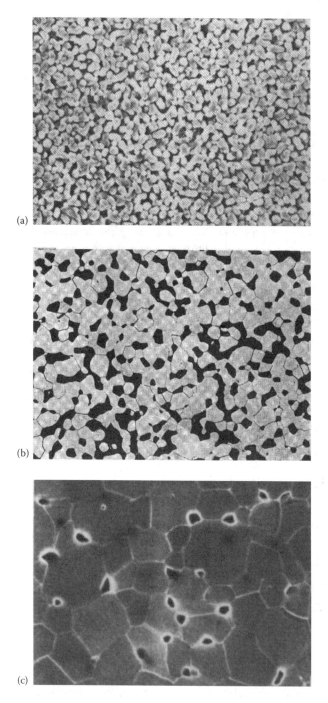

FIGURE 9.2 Examples of real microstructures (planar sections) for (a) initial stage of sintering, (b) intermediate stage, and (c) final stage.

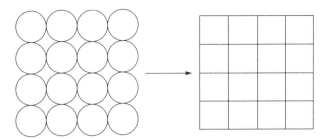

FIGURE 9.3 Schematic diagram illustrating the driving force for sintering due to the reduction in surface free energy.

If γ_{SV} is the specific surface energy (i.e., the surface energy per unit area) of the particles, then the surface free energy associated with the system of particles is

$$E_S = \frac{3\gamma_{SV}V_m}{a} \tag{9.3}$$

E_S represents the decrease in free energy of the system if a fully dense body were to be formed from the mole of particles. Equation 9.3 indicates that the driving force for sintering increases inversely with the particle radius, so finer particles have a greater incentive for sintering. The magnitude of the driving force can be estimated. For Al_2O_3 particles with a radius $a \approx 1$ μm, γ_{SV} ≈ 1 J/m^2, and $V_m \approx 25 \times 10^6$ m^3, we find that $E_S \approx 75$ J/mol. Compared to other processes, such as chemical reactions, this decrease in energy is rather small, but the distance matter has to be transported during sintering is also small (on the order of the particle size), so sintering occurs at a reasonable rate at sufficiently high temperatures.

The decrease in free energy given by Equation 9.3 assumes that the dense solid contains no grain boundaries, which would be the case for a glass. As outlined later, for polycrystalline materials, the grain boundaries play a role that must be included in determining the magnitude of the driving force. The decrease in surface free energy is accompanied by an increase in energy associated with the boundaries between the grains, so the net change in energy is determined by the magnitudes of these two changes.

9.2.2 Effects of Surface Curvature

Equation 9.3 indicates that sintering is driven by changes in the curvature (the inverse of the radius of curvature) of the surface. For a liquid drop, the reader will recall the difference in pressure between the liquid and the outside air. As the liquid drop tries to shrink to reduce its surface energy, it exerts a pressure on the liquid. Curved solid surfaces are also important in sintering, so it is useful to understand their effects.

9.2.2.1 Stress on the Atoms under a Curved Surface

The specific energy and curvature of the particle surfaces provide an effective stress on the atoms under the surface, given by the equation of Young and Laplace:

$$\sigma = \gamma_{SV}\left(\frac{1}{a_1} + \frac{1}{a_2}\right) \tag{9.4}$$

where γ_{SV} is the specific energy of the surface with principal radii of curvature a_1 and a_2. Defining the curvature of the surface as

$$K = \frac{1}{a_1} + \frac{1}{a_2} \qquad (9.5)$$

the effective stress can be written as:

$$\sigma = \gamma_{SV} K \qquad (9.6)$$

For a spherical surface, $a_1 = a_2 = a$, giving $\sigma = 2\gamma_{SV}/a$. The curvature of a "convex" surface is taken to be positive, so the effective stress on the atoms is greater than zero (compressive), whereas the curvature of a "concave" surface is taken to be negative, so the effective stress is less than zero (tensile).

9.2.2.2 Chemical Potential of the Atoms under a Curved Surface

In certain categories of sintering, such as pressure sintering, the driving force can arise from surface curvature as well as other factors, such as an applied pressure. The chemical potential, μ, provides a framework for incorporating these different driving forces. In general, sintering is driven by differences in the chemical potential. The effects of surface curvature on the difference between the chemical potentials of atoms under a flat surface and those under a curved surface can be found by analyzing the work done in transferring a small number dn of atoms from a flat, semiinfinite surface to a curved surface, such as the surface of a sphere of radius a (Figure 9.4). This work done is simply the surface energy γ_{SV} times the change in surface area, dS. The change in volume of the sphere dV is equal to the atomic volume Ω times dn. Thus:

$$dV = 4\pi a^2 da = \Omega dn \qquad (9.7)$$

The work done in transferring dn atoms is, by definition, the chemical potential change $\Delta\mu$ times dn, which is equal to the accompanying change in the surface energy of the sphere:

$$\Delta\mu dn = \gamma_{SV}(8\pi a da) \qquad (9.8)$$

Using Equation 9.7, we get:

$$\Delta\mu = \frac{2\gamma_{SV}\Omega}{a} \qquad (9.9)$$

In deriving Equation 9.9, it is assumed that the removal of a small number of atoms from the flat surface of the semiinfinite solid does not result in any change in the surface energy of that surface. This is because replacing these missing surface atoms by atoms from the bulk of the solid does not involve any shape change and hence any expenditure of energy. We can assign a constant chemical potential μ_o to the atoms under this reference surface, so the chemical potential of the atoms under the curved surface can be written as:

$$\mu_a = \mu_o + \frac{2\gamma_{SV}\Omega}{a} \qquad (9.10)$$

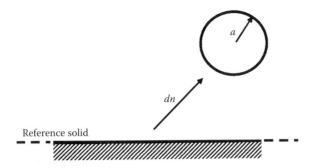

FIGURE 9.4 Transport of *dn* atoms from the flat surface of a semiinfinite reference solid to the curved surface of a solid sphere of radius *a*.

For any curved surface with principal radii of curvature a_1 and a_2, Equation 9.10 can be generalized to

$$\mu_a = \mu_o + \gamma_{SV}\Omega K \qquad (9.11)$$

where the curvature K is given by Equation 9.5.
Using Equation 9.6 and Equation 9.11, we can establish a connection between the chemical potential and the mechanical stress acting on the atoms under a curved surface:

$$\sigma = \frac{\Delta\mu}{\Omega} \qquad (9.12)$$

For a porous solid in which the pores are simplified as uniform spheres with radius *r*, and ignoring the changes occurring in the grain boundary area during sintering, Equation 9.12 would define the driving force for sintering, taken as the equivalent externally applied stress that would have the same effect on densification as the pore curvature, given by:

$$\sigma = \frac{2\gamma_{SV}}{r} \qquad (9.13)$$

Equation 9.13 would be a good expression for the driving force (referred to as the *sintering stress*, *sintering potential*, or *sintering pressure*) of a porous glass, consisting of spherical pores with the same radius *r*, uniformly distributed in the glass matrix. The equation for the driving force is more complicated for a polycrystalline system [4]. For an idealized system of spherical particles of (radius *a*) and spherical pores (radius *r*) in the late stage of densification, one expression is [5]:

$$\sigma = \frac{2\gamma_{SV}}{r} + \frac{2\gamma_{gb}}{a} \qquad (9.14)$$

where γ_{gb} is the specific energy of the grain boundary.

9.2.2.3 Vacancy Concentration under a Curved Surface

The compressive stress exerted on the atoms under a convex surface squeezes out vacancies, so the vacancy concentration below a convex surface will be below normal relative to a flat surface

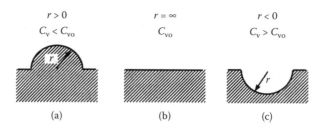

FIGURE 9.5 The equilibrium vacancy concentration under (a) a curved surface with positive curvature is lower than that for (b) a flat surface, which is lower than that under (c) a surface with negative curvature.

(Figure 9.5). On the other hand, the atoms under a concave surface are relaxed, so the vacancy concentration under a concave surface will be greater than normal, relative to a flat surface. This difference in vacancy concentration drives the diffusional flux of vacancies from a concave surface to a convex surface. The flux of atoms is equal and opposite to that of the vacancies, so atoms diffuse from the convex surface to the concave surface. Several models for predicting the sintering of idealized systems are based on an analysis of the diffusional flux of vacancies. The concentration of vacancies under a curved surface with principal radii of curvature a_1 and a_2 is given by [1]:

$$C_v = C_{vo} \exp\left[-\frac{\gamma_{SV}\Omega K}{kT}\right]$$ (9.15)

where C_{vo} is the vacancy under a flat reference surface, Ω is the volume of the vacancy (taken to be equal to the atomic volume), k is the Boltzmann constant, T is the absolute temperature, and K is the curvature given by Equation 9.5. When $\gamma_{SV}\Omega K \ll kT$, Equation 9.15 becomes:

$$C_v - C_{vo} = \Delta C_v = -\frac{C_{vo}\gamma_{SV}\Omega K}{kT}$$ (9.16)

where ΔC_v is the vacancy concentration difference between the curved surface and a flat surface.

9.2.2.4 Vapor Pressure over a Curved Surface

The pressure exerted on the atoms under a curved surface alters the chemical potential of the atoms and the vapor pressure of the atoms in equilibrium with the surface. A convex surface has a higher vapor pressure than a flat surface, whereas a concave surface has a lower vapor pressure than a flat surface. By considering the free energy change associated with removing a fixed number of atoms from the vapor phase and adding them to the curved surface, with a consequent reduction in the vacancy concentration beneath the surface, the vapor pressure over a curved surface is given by [1]:

$$p_{vap} = p_o \exp\left[\frac{\gamma_{SV}\Omega K}{kT}\right]$$ (9.17)

where p_o is taken as the vapor pressure over a flat, reference surface. Equation 9.17 is commonly known as the *Kelvin equation*. When $\gamma_{SV}\Omega K \ll kT$, Equation 9.17 becomes:

$$p_{vap} - p_o = \Delta p = \frac{p_o \gamma_{SV} \Omega K}{kT} \tag{9.18}$$

where Δp is the vapor pressure difference between a curved surface and a flat surface.

9.2.3 Grain Boundary Effects

As a first step toward assessing the effects of grain boundaries in porous polycrystalline powder compacts, let us consider the hypothetical situation of a pore connected to three grains in an infinite solid (Figure 9.6). Equilibrium will impose two important requirements on the geometry. First, the chemical potential of the atoms under the pore surface must be the same everywhere, which is equivalent to saying that the curvature of the pore surface must be the same everywhere. Therefore, the pore surface must consist of circular arcs in two-dimensional models. Second, no net force must be present at the junction of the pore surface and the grain boundary. This leads to a specific angle of intersection between the surface and the grain boundary at the junction. Representing the interfacial tensions as vectors, with magnitudes equal to the surface and grain boundary tensions, and with directions tangential to the pore surface or in the direction of the grain boundary, then force balance gives:

$$2\gamma_{SV} \cos \frac{\psi}{2} = \gamma_{gb} \tag{9.19}$$

where γ_{SV} is the surface energy, γ_{gb} is the grain boundary energy, and ψ is the dihedral angle. Surface and grain boundary energies are susceptible to changes due to impurities and crystal orientation, so the dihedral angle may not be the same everywhere. Careful measurements of thermal grooves in MgO and Al_2O_3 show a wide range of dihedral angles in each material [6]. The distribution for undoped Al_2O_3 is particularly broad but becomes considerably narrower upon doping with MgO (Figure 9.7). The average dihedral angle is ~120°, but many of the angles are less than 120°, corresponding to $\gamma_{SV} < \gamma_{gb}$.

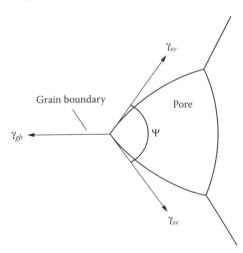

FIGURE 9.6 Equilibrium shape of a pore surrounded by three grains in two dimensions, and the balance of interfacial forces where the boundary intersects the pore. ψ is the dihedral angle, γ_{SV} is the solid–vapor interfacial tension, and γ_{gb} is the grain boundary tension.

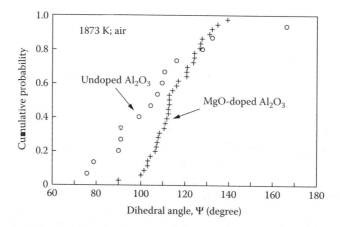

FIGURE 9.7 Cumulative distribution of dihedral angles in undoped Al_2O_3 and MgO-doped Al_2O_3. (From Handwerker, C.A. et al., Dihedral angles in magnesia and alumina: distributions from surface thermal grooves, *J. Am. Ceram. Soc.*, 73, 1371, 1990. With permission.)

For porous polycrystalline solids, as sintering takes place, there will be a decrease in free energy, ΔE_s, resulting from a reduction in the pore surface area, but an increase in energy, ΔE_{gb}, because of an increase in the grain boundary area. Whether the pore shrinks or not is dependent on the relative magnitudes of ΔE_s and ΔE_{gb}. Geometrically, it can be shown that if the sides of the pore are concave to the pore, then ΔE_s is greater in magnitude than ΔE_{gb}, and the pore will shrink. On the other hand, if the sides of the pore are convex, then the pore will grow. Let us consider, in two dimensions, a pore with a dihedral angle $\psi = 120°$ (Figure 9.8) that is surrounded by N grains. The number N is called the *pore coordination number*. The pore has straight sides if $N = 6$, "convex" sides for $N < 6$, and "concave" sides for $N > 6$. The surface of the pore will move toward its center of curvature, so the pore with $N < 6$ will shrink whereas the one with $N > 6$ will grow. The pore is metastable for $N = 6$, and this number is called the *critical pore coordination number*, denoted N_c. In fact, it can be shown that for the pore with "convex" sides ($N < 6$), the decrease in the pore surface energy

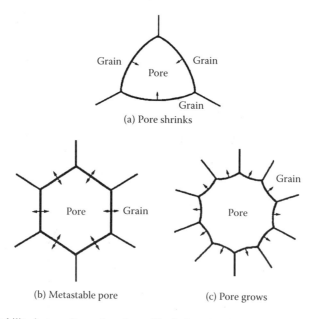

FIGURE 9.8 Pore stability in two dimensions for a dihedral angle of 120°.

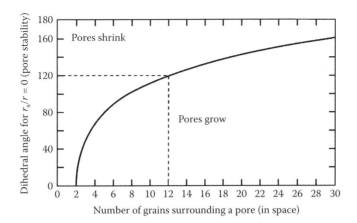

FIGURE 9.9 Conditions for pore stability in three dimensions as a function of pore coordination number. (From Kingery, W.D. and Francois, B., Sintering of crystalline oxides, I. Interactions between grain boundaries and pores, in *Sintering and Related Phenomena*, Kuczynski, G.C., Hooton, N.A., and Gibbon, G.F., Eds., Gordon and Breach, New York, 1967, p. 471.)

is greater than the increase in the grain boundary energy, whereas for the pore with "concave" sides ($N > 6$), the decrease in pore surface energy is smaller than the increase in grain boundary energy. For the metastable situation ($N = 6$), the decrease in pore surface energy exactly balances the increase in grain boundary energy. We can go on to consider other dihedral angles; for example, $N_c = 3$ for $\psi = 60°$. The general result is that N_c decreases with the dihedral angle. The two parameters are connected by a simple geometrical relationship given by $\psi = (180N_c - 360)/N_c$.

The geometrical considerations can be extended to three dimensions, in which case the pore is a polyhedron. The analysis has been carried out by Kingery and Francois [7]. The balance between the reduction in pore surface area and the increase in the grain boundary area leads to a criterion that prescribes the maximum pore coordination number that will permit a pore to shrink. As shown in Figure 9.9, for a given dihedral angle (e.g., 120°), pores with a coordination number less than a certain critical value ($N_c = 12$) will shrink but pores with $N > N_c$ will grow. A refinement of the analysis indicates that the pore with $N > N_c$ may not grow continuously but may reach some limiting size.

From this discussion, we see that a poorly compacted powder containing pores that are large compared to the grain size, would be difficult to densify, especially if the dihedral angle is low, because the pore coordination number is large. In forming powder compacts for sintering, we should make the pore size smaller than the grain size. This means preparing compacts with high green density and uniform pore size distribution by cold isostatic pressing or by colloidal methods, so that the fraction of pores with $N > N_c$ is minimized. For a system with fine grains and large pores, densification would not occur unless some grain growth were allowed to occur, making $N < N_c$. This does not mean that we would seek to promote grain growth to achieve densification. As discussed later, grain growth leads to a significant reduction in the densification rate, as well as to an increase in the pore size, so no long term benefit for densification is achieved. The use of a limited amount of grain growth to make $N < N_c$ is most relevant for large pores that may have been introduced inadvertently into the green body.

9.2.4 MECHANISMS OF SINTERING

Solid-state sintering of polycrystalline materials occurs by diffusion of atoms (or ions) along definite paths that define the mechanisms of sintering. We will recall that matter is transported from regions of higher chemical potential (the source of matter) to regions of lower chemical potential (the sink).

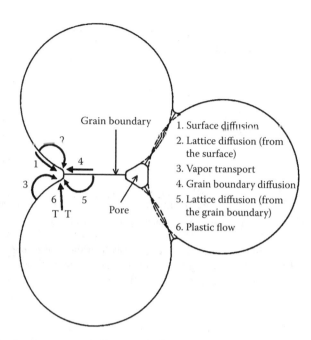

FIGURE 9.10 Six mechanisms can contribute to the sintering of a porous powder compact of crystalline particles. Only mechanisms 4 to 6 lead to densification, but all cause the necks to grow and so influence the rate of densification.

There are at least six different mechanisms of sintering in polycrystalline materials, as shown schematically in Figure 9.10 for a system of three sintering particles. They all lead to bonding and growth of necks between the particles, so the strength of the powder compact increases during sintering. Only certain of the mechanisms, however, lead to shrinkage or densification, and a distinction is commonly made between densifying and nondensifying mechanisms. Surface diffusion, lattice diffusion from the particle surfaces to the neck, and vapor transport (mechanisms 1, 2, and 3) lead to neck growth without densification and are referred to as nondensifying mechanisms. Grain boundary diffusion and lattice diffusion from the grain boundary to the pore (mechanisms 4 and 5) are the most important densifying mechanisms in polycrystalline ceramics. Diffusion from the grain boundary to the pore permits neck growth as well as densification. Plastic flow by dislocation motion (mechanism 6) also leads to neck growth and densification but is more common in the sintering of metal powders.

The nondensifying mechanisms cannot simply be ignored. They lead to coarsening of the microstructure, thereby reducing the driving force for the densifying mechanisms. Sintering is, therefore, said to involve a competition between densification and coarsening. We will recall from Chapter 1 that the production of ceramics with high density would require choosing sintering conditions under which the coarsening mechanisms are not very active, whereas a highly porous body would be favored when the coarsening mechanisms dominate (see Figure 1.18).

In addition to the alternative mechanisms, there are additional complications arising from the diffusion of the different ionic species making up the compound [1]. To preserve change neutrality of the local composition, the flux of the different ionic species will be coupled. A further complication arises because each ionic species can diffuse along different paths. It is the *slowest diffusing species along its fastest path* that controls the rate of sintering. Another complicating factor is that the rate-controlling mechanism for a given material can change with changing conditions of the process variables such as temperature, grain size, and composition [8]. As an example, let us consider the creep of Al_2O_3, a process in which slow deformation under an applied uniaxial stress occurs by lattice or grain boundary diffusion. Figure 9.11 shows a schematic plot of the log (creep

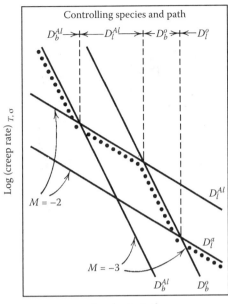

FIGURE 9.11 Diffusional creep in polycrystalline Al_2O_3 (schematic). The solid lines show the creep rate dependence on grain size for each assumed species and path. The dotted line traces the rate-limiting mechanism, given by the slower diffusing species along its fastest path. (From Coble, R.L. and Cannon, R.M., Current paradigms in powder processing, *Mater. Sci. Res.*, 11, 151, 1978. With permission.)

rate) vs. log (grain size) at constant temperature and applied stress. For Al_2O_3, the relative values of the grain boundary and lattice diffusion coefficients are known: $D_l^{Al} > D_l^O$ and $D_b^O > D_b^{Al}$. The two lines with slope = −2 show the creep rate by lattice diffusion if Al and O are the rate-limiting ions whereas the two lines with slope = −3 give the corresponding creep rate by grain boundary diffusion. The controlling species and path are given by the dotted line. We notice that each combination of species and path is controlling over a specific grain size range and that in each of these regimes, it is not the topmost line, giving the highest creep rate, that is rate-controlling. The rate-controlling species and path is given by the second line, representing the other ion along its fastest path.

For amorphous materials (glasses), viscous flow is the dominant mechanism by which neck growth and densification occur. The matter transport path is not as clearly defined as for polycrystalline ceramics. Figure 9.12 shows as an example the sintering of two glass spheres by viscous flow.

9.2.5 STAGES OF SINTERING

The microstructure of real powder compacts, consisting initially of identifiable particles in contact, changes continuously, as well as drastically during sintering. It is therefore difficult to find a single geometrical model that adequately represents the entire process, yet still provide the simplicity for theoretical analysis of sintering. Following Coble [9], it is common to divide the process into three idealized, sequential stages, with the geometry of each stage corresponding roughly to the major microstructural features of a real powder compact. A stage represents an interval of time or density over which the microstructure is considered to be reasonably well defined. The three stages of sintering are called (1) the *initial stage*, (2) the *intermediate stage*, and (3) the *final stage*. Figure 9.13 shows the geometrical models suggested by Coble for the three stages.

The *initial stage* would begin as soon as some degree of atomic mobility is achieved and, during this stage, sharply concave necks form between the individual particles, assumed to be

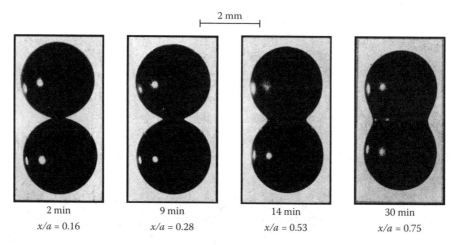

FIGURE 9.12 Two-particle model made of glass spheres (3 mm in diameter) sintered at 1000°C for different times, showing the increase in the neck radius x relative to the particle radius a. (From Exner, H.E., Principles of single phase sintering, *Rev. Powder Metall. Phys. Ceram.*, 1, 1, 1979.)

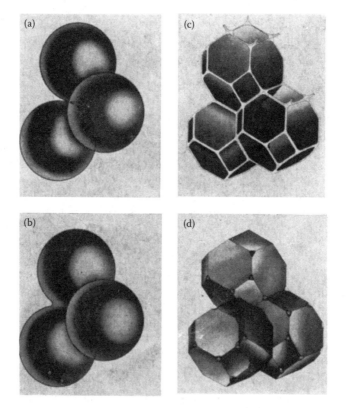

FIGURE 9.13 Idealized models for the three stages of sintering: (a) initial stage: model structure represented by spheres in tangential contact, (b) near the end of the initial stage: spheres have begun to coalesce. The neck growth indicated is for center-to-center shrinkage of 4%, (c) intermediate stage: dark grains have adopted the shape of a tetrakaidecahedron, enclosing white pore channels at the grain edges, (d) final stage: pores are tetrahedral inclusions at the corners where four tetrakaidecahedra meet. (From Coble, R.L., Sintering crystalline solids. I. Intermediate and final state diffusion models, *J. Appl. Phys.*, 32, 787, 1961. With permission.)

monosized spheres (Figure 9.13a and Figure 9.13b). The major microstructural change is neck growth between the particles, with the amount of densification being small, typically the first 3 to 5% of linear shrinkage, so this stage ends when the relative density reaches a value of ~0.65. In the *intermediate stage*, the high curvatures of the initial stage are assumed to have been eliminated, and the microstructure consists of a three-dimensional interpenetrating network of solid grains with the shape of a *tetrakaidecahedron*, and continuous channel-like pores with a circular cross section sitting on the edges of the grains (Figure 9.13c). Densification is assumed to occur by the pores simply shrinking to reduce their radius. This stage is considered valid to a relative density of ~0.90, and therefore covers most of the densification. In the final stage, the pores are assumed to pinch off and become isolated at the grain corners (Figure 9.13d). In this idealized model, the pores are assumed to shrink continuously and may disappear altogether.

9.2.6 SINTERING MODELS AND THEIR PREDICTIONS

Several theoretical models have been developed to analyze the sintering of porous powder compacts. The analytical models and numerical simulations have received the most attention.

9.2.6.1 Analytical Models

The analytical models for polycrystalline materials assume a simplified, highly idealized geometry for each stage of sintering, and for each mechanism, the diffusional flux equations for atomic transport are solved analytically to provide equations for the sintering kinetics. Models can provide only a *qualitative* understanding of the sintering process in practical systems because of they make drastic simplifications of the microstructure and matter transport processes, such as monodisperse particles, uniform packing, a single mechanism operating, and absence of grain growth. For amorphous materials, the kinetic equations are based on an energy balance concept put forward by Frenkel [10], in which the rate of energy dissipation by viscous flow is equated to the rate of energy gained by the reduction in surface area.

9.2.6.1.1 Initial Stage Sintering Equations

As an example of the analytical models, let us consider the initial stage sintering by the mechanism of grain boundary diffusion. The geometrical model commonly consists of two contacting spheres with the same radius, referred to as the two-sphere model. Figure 9.14 shows the geometrical parameters of the model after some time t of sintering, when a neck of radius X and a concave surface of radius r, has developed between the two particles of radius a. The flux of atoms from the grain boundary (the source) into the neck (the sink), taken as opposite to the flux of vacancies, is given by Fick's law [1]:

$$J_a = \frac{D_v}{\Omega} \frac{dC_v}{dx} \tag{9.20}$$

where D_v is the vacancy diffusion coefficient, Ω is the volume of an atom or vacancy, dC_v/dx is the vacancy concentration gradient (in one dimension), and C_v is the *fraction* of sites occupied by the vacancies. The volume of matter transported into the neck per unit time is

$$\frac{dV}{dt} = J_a A_{gb} \Omega \tag{9.21}$$

where A_{gb} is the cross sectional area over which diffusion occurs. Grain boundary diffusion is assumed to occur over a constant thickness δ_{gb} so that $A_{gb} = 2\pi X \delta_{gb}$. Combining Eqs. (9.20) and (9.21) and substituting for A_{gb} give

$$\frac{dV}{dt} = D_v 2\pi X \delta_{gb} \frac{dC_v}{dx} \qquad (9.22)$$

Since the neck radius increases radially in a direction orthogonal to a line joining the centers of the spheres, a one dimensional solution is adequate. Assuming that the vacancy concentration between the neck surface and the center of the neck is constant, then $dC_v/dx = \Delta C_v/X$, where ΔC_v is the difference in vacancy concentration between the neck surface and the center of the neck, given by Equation 9.16. From Figure 9.14, the two principal radii of curvature of the neck surface are $r_1 = r$ and $r_2 = -X$, and it is assumed that $X \gg r$. Substituting into Equation 9.22 gives

$$\frac{dV}{dt} = \frac{2\pi D_v C_{vo} \delta_{gb} \gamma_{SV} \Omega}{kTr} \qquad (9.23)$$

Using the relations given in Figure 9.14b for V and r, and putting the grain boundary diffusion coefficient D_{gb} equal to $D_v C_{v0}$ we obtain

$$\frac{\pi X^3}{2a} \frac{dX}{dt} = \frac{2\pi D_{gb} \delta_{gb} \gamma_{SV} \Omega}{kT} \left(\frac{4a}{X^2} \right) \qquad (9.24)$$

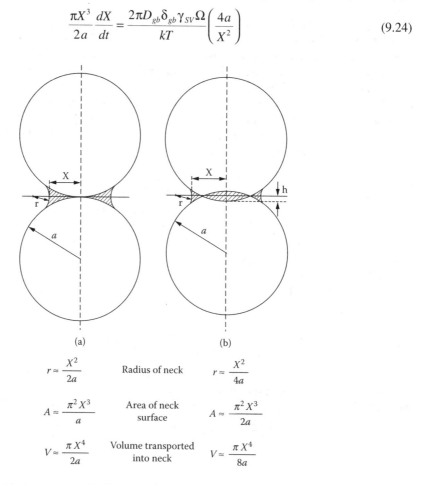

$r \approx \dfrac{X^2}{2a}$	Radius of neck	$r \approx \dfrac{X^2}{4a}$
$A \approx \dfrac{\pi^2 X^3}{a}$	Area of neck surface	$A \approx \dfrac{\pi^2 X^3}{2a}$
$V \approx \dfrac{\pi X^4}{2a}$	Volume transported into neck	$V \approx \dfrac{\pi X^4}{8a}$

FIGURE 9.14 Geometrical parameters for the two-sphere model used in the derivation of the initial stage sintering equations for crystalline particles. The geometries shown correspond to those for (a) the nondensifying mechanisms and (b) the densifying mechanisms.

Rearranging, Equation 9.24 gives

$$X^5 dX = \frac{16 D_{gb} \delta_{gb} \gamma_{SV} \Omega a^2}{kT} dt \tag{9.25}$$

After integration and application of the boundary conditions $X = 0$ at $t = 0$, Equation 9.25 becomes

$$X^6 = \frac{96 D_{gb} \delta_{gb} \gamma_{SV} \Omega a^2}{kT} t \tag{9.26}$$

We may also write Equation 9.26 in the form:

$$\frac{X}{a} = \left(\frac{96 D_{gb} \delta_{gb} \gamma_{SV} \Omega}{kTa^4} \right)^{1/6} t^{1/6} \tag{9.27}$$

Equation 9.27 predicts that the ratio of the neck radius to the sphere radius increases as $t^{1/6}$.

For this densifying mechanism, the linear shrinkage, defined as the change in length, ΔL, divided by the original length, L_0, can also be found. As a good approximation (see Figure 9.14) we can write

$$\frac{\Delta L}{L_o} = -\frac{h}{a} = -\frac{r}{a} = -\frac{X^2}{4a^2} \tag{9.28}$$

where h is half the interpenetration distance between the spheres. Using Equation 9.27 we obtain

$$\frac{\Delta L}{L_o} = -\left(\frac{3 D_{gb} \delta_{gb} \gamma_{SV} \Omega}{2kTa^4} \right)^{1/3} t^{1/3} \tag{9.29}$$

so the shrinkage is predicted to increase as $t^{1/3}$.

The equations for neck growth and, for the densifying mechanisms, shrinkage, can be expressed in the general form [1]:

$$\left(\frac{X}{a} \right)^m = \frac{H}{a^n} t \tag{9.30}$$

$$\left(\frac{\Delta L}{L_0} \right)^{m/2} = -\frac{H}{2^m a^n} t \tag{9.31}$$

where m and n are numerical exponents that depend on the mechanism of sintering and H is a function that contains the geometrical and material parameters of the powder system. Depending on the assumptions made in the models, a range of values for m, n and the numerical constant in H have been obtained. The values given in Table 9.2 represent the most plausible values for each mechanism [11,12].

TABLE 9.2
Plausible Values for the Constants Appearing in Equation 9.30 and Equation 9.31 for the Initial Stage of Sintering

Mechanism	m	n	H[b]
Surface diffusion[a]	7	4	$56D_s\delta_s\gamma_{sv}\Omega/kT$
Lattice diffusion from the surface[a]	4	3	$20D_l\gamma_{sv}\Omega/kT$
Vapor transport[a]	3	2	$3p_0\gamma_{sv}\Omega/(2\pi mkT)^{1/2}kT$
Grain boundary diffusion	6	4	$96D_{gb}\delta_{gb}\gamma_{sv}\Omega/kT$
Lattice diffusion from the grain boundary	5	3	$80\pi D_l\gamma_{sv}\Omega/kT$
Viscous flow	2	1	$3\gamma_{sv}/2\eta$

[a] Denotes nondensifying mechanism, i.e., $\Delta L/L_0 = 0$.

[b] D_s, D_l, D_{gb} are the diffusion coefficients for surface, lattice, and grain boundary diffusion. δ_s and δ_{gb} are the thicknesses for surface and grain boundary diffusion. γ_{sv} is the specific surface energy. p_0 is the vapor pressure over a flat surface. m is the mass of an atom. k is the Boltzmann constant. T is the absolute temperature. η is the viscosity.

The form of the neck growth equations indicates that a plot of log (X/a) vs. log t yields a straight line with a slope equal to $1/m$, so by fitting the theoretical predictions to experimental data, the value of m can be found. A similar procedure can be applied to the analysis of shrinkage, if it occurs during sintering. Data for validating the models are commonly obtained by measuring the neck growth in simple systems (e.g., two spheres, a sphere on a plate or two wires) or the shrinkage in a compacted mass of spherical particles. As m is dependent on the mechanism of sintering, at first sight it may seem that the measurement of m would provide information on the mechanism of sintering. However, the basic assumption in the models of a single dominant mass transport mechanism is not valid in most polycrystalline powder systems. When more than one mechanism operates simultaneously, the measured exponent may correspond to an entirely different mechanism. The other simplifying assumptions of the models must also be remembered. The extension of the two-sphere geometry to real powder compacts is valid only if the particles are spheres of the same size arranged in a uniform pattern. In practice, this system is, at best, approached only by the uniform consolidation, by colloidal methods, of monodisperse powders.

Sintering diagrams (or sintering maps) have been constructed for a few materials to show the dominant mechanisms as functions of the sintering temperature and density [13,14]. The construction of the maps relies on the sintering equations derived for the analytical models (Table 9.2) and data for the material constants, such as the diffusion coefficient, surface energy, and atomic volume. The form that a sintering diagram can take is shown in Figure 9.15 for the sintering of copper spheres with a radius of 57 μm. The axes are the neck radius X, normalized to the radius of the sphere a, and the homologous temperature T/T_M, where T_M is the melting temperature of the solid. The diagram is divided into various fields, and within each field, a single mechanism is dominant, i.e., it contributes most to neck growth. At the boundary between two fields (shown as solid lines), two mechanisms contribute equally to the sintering rate. Superimposed on the fields are contours of constant sintering time. Because of the drastic simplifications of the models and the inadequacy of the database, the applicability of the maps is limited. Nevertheless, they have proved useful for visualizing conceptual relationships between the various mechanisms and changes in the sintering behavior under different sintering conditions.

9.2.6.1.2 Intermediate- and Final-Stage Equations

The geometrical models commonly used for the intermediate and final stages of sintering of polycrystalline systems consist of a space-filling array of equal-sized *tetrakaidecahedra*, each of

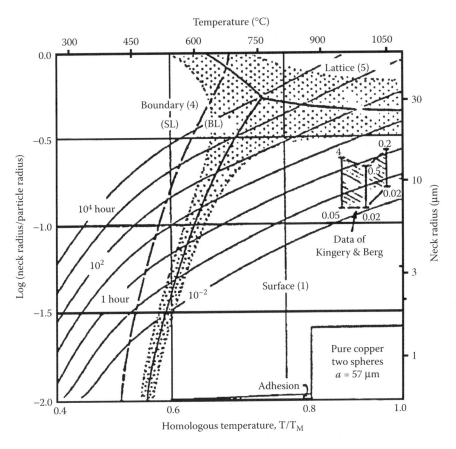

FIGURE 9.15 Neck size sintering diagram for copper spheres. (From Ashby, M.F., A first report on sintering diagrams, *Acta Metall.*, 22, 275, 1974. With permission.)

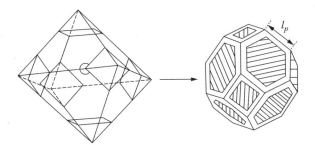

FIGURE 9.16 Sketch illustrating the formation of a tetrakaidecahedron from a truncated octahedron.

which represents one particle. For the intermediate stage, the porosity is cylindrical, with the axis of the cylinder coinciding with the edge of the tetrakaidecahedra (Figure 9.13c), whereas for the final stage, the porosity is spherical, centered at the grain corners (Figure 9.13d).

A tetrakaidecahedron is constructed from an octahedron by trisecting each edge and joining the points to remove the six edges (Figure 9.16). The resulting structure has 36 edges, 24 corners and 14 faces (8 hexagonal and 6 square). The volume of the tetrakaidecahedron is

$$V_t = 8\sqrt{2}l_p^3 \tag{9.32}$$

where l_p is the edge length of the tetrakaidecahedron. For the intermediate stage, if r is the radius of the pore, then the total volume of the porosity per unit cell is

$$V_p = \frac{1}{3}(36\pi r^2 l_p)$$

(9.33)

The porosity of the unit cell, V_p/V_t, is therefore

$$P_c = \frac{3\pi}{2\sqrt{2}}\left(\frac{r^2}{l_p^2}\right)$$

(9.34)

Since the models assume that the pore geometry is uniform, the nondensifying mechanisms cannot operate. This is because the chemical potential is the same everywhere on the pore surface. We are left with the densifying mechanisms: lattice diffusion and grain boundary diffusion. Plastic flow is not expected to operate in any significant extent for ceramic systems.

Equations developed by Coble [9] for the intermediate and final-stage sintering of polycrystalline systems can be expressed in a form relating the densification rate to the process variables [1]:

$$\frac{1}{\rho}\frac{d\rho}{dt} = A\left(\frac{D\Omega}{G^m kT}\right)\left(\frac{\alpha\gamma_{SV}}{r}\right)$$

(9.35)

where ρ is the relative density at time t, A is a constant that depends on the geometrical model, D is the diffusion coefficient, Ω is the atomic volume of the rate-controlling species, G is the grain size, m is an exponent that depends on the mechanism of diffusion, α is a constant that depends on the geometry of the pore, γ_{SV} is the surface energy, and r is the pore radius. For sintering by the grain boundary diffusion mechanism, the diffusion coefficient $D = D_{gb}\delta_{gb}$, where D_{gb} is the grain boundary diffusion coefficient and δ_{gb} is the grain boundary width, whereas for lattice diffusion, $D = D_l$, the lattice diffusion coefficient. Values for A, m, and α derived by Coble [9] are given in Table 9.3.

Models for *viscous sintering* of glasses have been developed by Scherer [15] for the intermediate stage, and by Mackenzie and Shuttleworth [16] for the final stage. The Mackenzie and Shuttleworth model consists of isolated spherical pores of radius r in a dense glass matrix. The rate of densification can be expressed as [1]:

$$\frac{1}{\rho}\frac{d\rho}{dt} = \frac{3}{4}\frac{(1-\rho)}{\rho\eta}\left(\frac{2\gamma_{SV}}{r}\right)$$

(9.36)

where η is the viscosity of the solid glass. For most of the intermediate stage, the predictions of the Scherer model are not significantly different from those of the Mackenzie and Shuttleworth model, so Equation 9.36 should also provide a reasonable approximation for the intermediate stage.

TABLE 9.3
Values for the Constants Appearing in the Densification Rate Equation (Equation 9.35)

Mechanism	Intermediate stage	Final stage
Lattice diffusion	$A = 40/3$; $m = 2$; $\alpha = 1$	$A = 40/3$; $m = 2$; $\alpha = 2$
Grain boundary diffusion	$A = 95/2$; $m = 3$; $\alpha = 1$	$A = 15/2$; $m = 3$; $\alpha = 2$

9.2.6.2 Predictions of the Analytical Models

One of the most useful features of the analytical models is what they predict for the influence of the principal processing parameters, such as grain (or particle) size, temperature, and applied pressure, on the rate of sintering. For the densifying mechanisms, Equation 9.35 and Table 9.3 indicate that particle size has a strong effect on the densification rate. For a fixed temperature and composition, and assuming that the pores size r scales as the grain size G, the equations predict that

$$\dot{\rho} = \frac{1}{\rho} \frac{d\rho}{dt} = \frac{H}{G^n} \tag{9.37}$$

where H is a constant, and $n = 3$ for lattice diffusion, and $n = 4$ for grain boundary diffusion. Because of the larger exponent, n, grain boundary diffusion is the dominant densification mechanism for smaller particle sizes, whereas lattice diffusion is dominant at larger particle sizes.

We can use the initial stage sintering equations to predict the influence of particle size on the relative rates of sintering by the different mechanisms, including the nondensifying mechanisms. According to Equation 9.30 and Table 9.2, for a fixed temperature and composition, the time t taken to reach a given microstructural change, such as the same X/a ratio, is proportional to a^n. For two geometrically similar systems with particle sizes a_1 and a_2, the times taken to reach the same microstructural change is:

$$\frac{t_2}{t_1} = \left(\frac{a_2}{a_1}\right)^n = \lambda^n \tag{9.38}$$

where λ is the ratio of the particle sizes. Equations of the same form as Equation 9.38 have been derived in a more elegant way by Herring [17] for the different sintering mechanisms, and they are often referred to as *Herring's scaling laws*.

To determine the relative rates of the different mechanisms, it is more useful to write Equation 9.38 in terms of a rate. For a given change, the rate is inversely proportional to the time, so

$$\frac{(Rate)_2}{(Rate)_1} = \lambda^{-n} \tag{9.39}$$

As an example, let us suppose that grain boundary diffusion and vapor transport are the dominant mass transport mechanisms in a given system. Then the rates of sintering by these two mechanisms vary with the scale of the system according to

$$(Rate)_{gb} \sim \lambda^{-4} \tag{9.40}$$

and

$$(Rate)_{ec} \sim \lambda^{-2} \tag{9.41}$$

The variation of the rates of sintering with λ for the two mechanisms is illustrated in Figure 9.17. The cross-over point for the two lines is arbitrary but this does not affect the validity of the results. We see that for small λ, i.e., as the particle size becomes smaller, the rate of sintering by

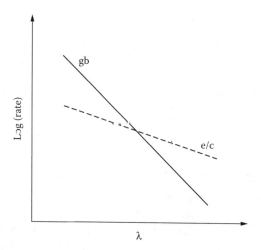

FIGURE 9.17 Schematic diagram of the relative rates of sintering by grain boundary diffusion (gb) and by vapor transport (e/c) as a function of the scale (i.e., particle size) of the system.

grain boundary diffusion is enhanced compared to that for vapor transport. Conversely, the rate of sintering by vapor transport dominates for larger λ, i.e., for larger particle sizes. Smaller particle size is therefore beneficial for densification when grain boundary diffusion and vapor transport are the dominant mechanisms. As this example illustrates, to use the scaling laws effectively, we must know the dominant mechanisms of sintering. Often, this information is difficult to determine or not available. The geometrical similarity of the microstructural changes, assumed in the scaling laws, is also difficult to maintain in real powder compacts.

Assuming no grain growth, the influence of temperature on sintering is seen through the dependence of the rate equations on the diffusion coefficient, defined by

$$D = D_o \exp\left(-\frac{Q}{RT}\right) \tag{9.42}$$

where D_o is a constant for a given system, Q is the activation energy for diffusion, R is the gas constant, and T is the absolute temperature. For many systems, activation energies for lattice diffusion, Q_l, grain boundary diffusion, Q_{gb}, and surface diffusion, Q_s, scale as

$$Q_l > Q_{gb} > Q_s \tag{9.43}$$

Because of its higher activation energy, lattice diffusion tends to dominate at higher temperatures, whereas surface diffusion becomes more dominant with decreasing temperature. As surface diffusion leads to coarsening, the attainment of high density would be favored by higher sintering temperatures (where the densifying mechanisms dominate) and reduced heating times at lower temperatures (where the coarsening mechanisms dominate).

Equation 9.36 predicts a weaker dependence of the densification rate on particle size for viscous sintering than for solid-state sintering. For a given temperature and composition, assuming that the pore size scales as the particle size, the densification rate is predicted to be proportional to the inverse of the particle size. The influence of temperature is seen from the dependence of the glass viscosity, η, on temperature, T, which, for many oxide glasses, is well represented by the Fulcher equation:

$$\eta = A \exp\left(\frac{B}{T - T_o} \right) \qquad (9.44)$$

where A, B, and T_o are constants for a given composition.

9.2.6.3 Effect of Applied Pressure

A compressive stress enhances the chemical potential of the atoms at the grain boundaries but has no effect on the chemical potential of the atoms on the pore surface [18]. This enhancement of the chemical potential difference leads to an increase in the driving force for densification. Equation (9.35) now becomes

$$\frac{1}{\rho} \frac{d\rho}{dt} = A\left(\frac{D\Omega}{G^m kT} \right)\left(\frac{\alpha \gamma_{SV}}{r} + p_a \phi \right) \qquad (9.45)$$

where p_a is the hydrostatic component of the applied stress, and ϕ is a geometrical factor, called the stress intensification factor, which accounts for the effective stress on the grain boundary being greater than the applied stress because of the presence of porosity in the body. For spherical pores randomly distributed in a solid with relative density ρ, ϕ is given by

$$\phi = 1/\rho \qquad (9.46)$$

At a given value of ρ, the rate of densification by diffusion mechanisms is predicted to increase linearly with applied pressure.

9.2.6.4 Numerical Simulations of Sintering

When compared to the analytical models, numerical simulations provide an improved description of some of the complexities of sintering, such as the use of more realistic geometrical models and the treatment of simultaneous mechanisms [19,20]. The simulations can be complex, and in many cases the results cannot be easily cast into a form showing the dependence of the sintering rate on the key parameters. On the other hand, they provide good insight into how matter is transported and how the microstructure changes during sintering. In particular, finite element modeling has been used very effectively in the analysis of a variety of sintering problems [21,22]. As an example, Figure 9.18 shows the results of finite element simulations for the viscous sintering of two rigid particles coated with an amorphous layer. The relative density is plotted as a function of reduced time (proportional to actual time) for different coating thicknesses, s. For s greater than 0.2 of the particle radius a, full density is predicted at a rate comparable to that for the Mackenzie and Shuttleworth (M–S) model, and for a packing without a rigid core [21].

9.3 GRAIN GROWTH AND COARSENING

Grain growth is the term used to describe the increase in the average grain size of a polycrystalline material. Grain growth occurs in both dense and porous polycrystalline solids at sufficiently high temperatures. For the conservation of matter, the sum of the individual grain sizes must remain constant so an increase in the average grain size is accompanied by the disappearance of some grains, usually the smaller ones. The grain boundary is a region with a complex structure, 0.5 to 1.0 nm wide, between two crystalline domains (the grains). The atoms in the grain boundary have

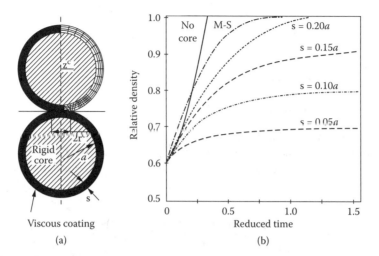

Viscous coating
(a)   (b)

FIGURE 9.18 Finite element simulation of viscous sintering of two rigid particles coated with an amorphous layer: (a) rigid particles of radius a coated with a viscous layer of thickness s and showing the finite element mesh, (b) predictions for the relative density vs. reduced time for different coating thicknesses, and for the Mackenzie and Shuttleworth (M–S) model. (From Jagota, A. and Dawson, P.R., Simulation of viscous sintering of two particles, *J. Am. Ceram. Soc.*, 73, 173, 1990. With permission.)

a higher energy than those in the bulk of the grains, so the grain boundary is characterized by a specific energy, γ_{gb}, typically on the order of 0.2 to 1.0 J/m². The reduction in the grain boundary area, and hence the energy associated with the grain boundaries, provides the *driving force* for grain growth.

Grain growth occurs as atoms (or ions) diffuse less than an interatomic distance from one side of the boundary to new positions on the other side, resulting in one grain growing at the expense of another (Figure 9.19). The atoms move from the "convex" surface on one side of the grain

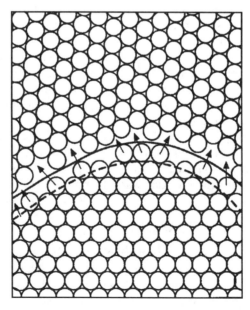

FIGURE 9.19 Classical picture of a grain boundary. The boundary migrates downward as atoms move less than an interatomic spacing from the convex side of the boundary to the concave side.

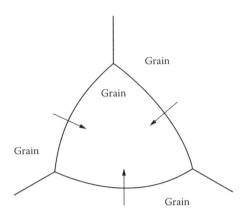

FIGURE 9.20 Movement of the grain boundary toward its center of curvature.

boundary to the "concave" surface on the other side more readily than in the reverse direction because the chemical potential of the atoms under the convex surface is higher than that for the atoms under the concave surface. The result of this net flux is that the boundary migrates toward its center of curvature (Figure 9.20).

Grain growth in ceramics is generally divided into two types: (1) normal grain growth and (2) abnormal grain growth, which is sometimes referred to as exaggerated grain growth or discontinuous grain growth. In normal grain growth (Figure 9.21a), the average grain size increases with time but the grain size distribution remains almost self-similar (invariant in time). Abnormal grain growth is the process whereby a few large grains grow rapidly at the expense of the smaller grains, giving a bimodal grain size distribution (Figure 9.21b). Anisotropic grain growth is a type of abnormal grain growth in which the abnormal grains grow in an elongated manner, often with faceted (straight) sides.

The term *coarsening* is frequently used to describe the process in porous ceramics in which the increase in the average grain size is accompanied by an increase in the average pore size. Coarsening reduces the driving force for sintering and increases the diffusion distance for matter transport, thereby reducing the rate of sintering. In porous ceramics, abnormal grain growth is accompanied by breakaway of the boundaries from the pores. Figure 9.22 shows examples of normal and abnormal grain growth in dense and porous ceramics.

The suppression of coarsening mechanisms forms a key requirement for the achievement of high density [23]. Another requirement is that the microstructure be stabilized such that the pores and the grain boundaries remain attached (Figure 9.23). Pores trapped in the abnormal grains are difficult to remove, limiting the final density. The large, abnormal grains also lead to a deterioration in the engineering properties of the material. Control of normal grain growth and, in particular, abnormal grain growth, depends primarily on the ability to reduce the intrinsic (pore-free) mobility of the grain boundaries, so an understanding of grain growth control in dense ceramics forms a key step toward microstructure control during sintering.

Coarsening of precipitates or particles in a solid, liquid or gaseous medium is commonly referred to as *Ostwald ripening*. The process is especially important in liquid-phase sintering, but many features of grain growth and coarsening in solid-state sintering are also shared by the Ostwald ripening process.

9.3.1 NORMAL GRAIN GROWTH

Normal grain growth in pure, dense, single-phase solids has been analyzed using a number of different approaches [24]. In one of the earliest models, Burke and Turnbull [25] analyzed grain growth by diffusion of atoms across an isolated boundary under the driving force of the pressure

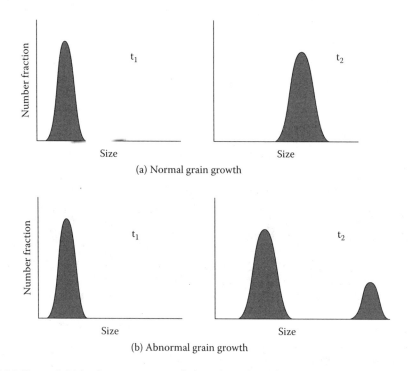

FIGURE 9.21 For an initial microstructure consisting of a unimodal distribution of grain sizes, (a) normal grain growth results in an increase in the average grain size whereas the grain size distribution remains almost self-similar; (b) abnormal grain growth is characterized by a few large grains growing rapidly at the expense of the surrounding matrix grains and may lead to a bimodal distribution of grain sizes.

difference across the curved boundary. More correctly, the chemical potential gradient across the boundary should be used. The grain boundary energy γ_{gb} is assumed to be isotropic and independent of the crystallographic direction, and the grain boundary width δ_{gb} is assumed to be constant.

    If the average rate of grain boundary migration, v_b (sometimes called the *grain boundary velocity*) is taken to be approximately equal to the instantaneous rate of grain growth, we can write

$$v_b \approx dG / dt \qquad (9.47)$$

where G is the average grain size. It is also assumed that v_b can be represented as the product of the driving force for boundary migration F_b and the boundary mobility M_b, so

$$v_b = M_b F_b \qquad (9.48)$$

where M_b includes effects arising from the mechanism of migration. The pressure difference across the boundary is given by the equation of Young and Laplace

$$\Delta p = \gamma_{gb} \left(\frac{1}{a_1} + \frac{1}{a_2} \right) \qquad (9.49)$$

where and a_1 and a_2 are the principal radii of curvature of the boundary. Assuming that the radius of the boundary is proportional to G, then

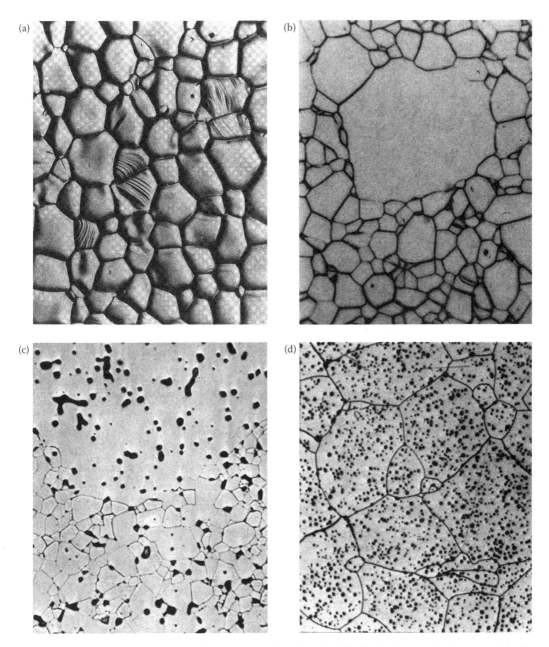

FIGURE 9.22 Microstructures illustrating (a) normal grain size distribution in an alumina ceramic, (b) initiation of abnormal grain growth in an alumina ceramic, (c) normal and abnormal grain growth in a porous nickel–zinc ferrite, (d) an alumina ceramic that has undergone considerable abnormal grain growth. [(a) and (b) from Brook, R.J., Controlled grain growth, in *Treatise on Materials Science and Technology*, Wang, F.F.Y., Ed., Academic Press, New York, 1976, p. 331. With permission; (c) and (d) courtesy of J.E. Burke.]

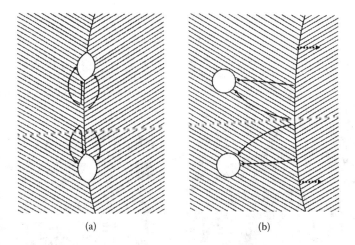

(a)   (b)

FIGURE 9.23 (a) Densification mechanisms for porosity attached to a grain boundary. The arrows indicate paths for atom diffusion. (b) Densification mechanisms for porosity separated from a grain boundary. The solid arrows indicate paths for atom diffusion and the dashed arrows indicate the direction of boundary migration. (From Brook, R.J., Controlled grain growth, in *Treatise on Materials Science and Technology*, Wang, F.F.Y., Ed., Academic Press, New York, 1976, p.331. With permission.)

$$\left(\frac{1}{a_1} + \frac{1}{a_2} \right) = \frac{\alpha}{G} \tag{9.50}$$

where α is a geometrical constant that depends on the shape of the boundary. Taking the driving force for atomic diffusion across the boundary to be equal to the gradient in the chemical potential, we have:

$$F_b = \frac{d\mu}{dx} = \frac{d}{dx}(\Omega\Delta p) = \frac{1}{\delta_{gb}}\left(\frac{\Omega\gamma_{gb}\alpha}{G} \right) \tag{9.51}$$

where Ω is the atomic volume and $dx = \delta_{gb}$ is the grain boundary width. The flux of atoms across the boundary is [1]:

$$J = \frac{D_a}{\Omega kT}\frac{d\mu}{dx} = \frac{D_a}{\Omega kT}\left(\frac{\Omega\gamma_{gb}\alpha}{\delta_{gb}G} \right) \tag{9.52}$$

where D_a is the diffusion coefficient for atomic motion across the grain boundary. The boundary velocity becomes:

$$v_b \approx \frac{dG}{dt} = \Omega J = \frac{D_a}{kT}\left(\frac{\Omega}{\delta_{gb}} \right)\left(\frac{\alpha\gamma_{gb}}{G} \right) \tag{9.53}$$

If the driving force is taken as the pressure difference across the boundary, $\alpha\gamma_{gb}/G$, as in the Burke and Turnbull analysis, v_b is given by:

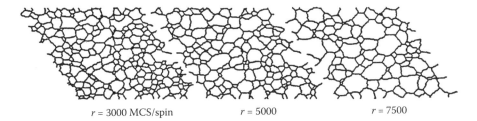

$r = 3000$ MCS/spin $\qquad r = 5000 \qquad r = 7500$

FIGURE 9.24 Computer simulation of normal grain growth, showing the evolution of the microstructure for a $Q = 64$ model on a triangular lattice. (From Anderson, M.P. et al., Computer simulation of grain growth — I. Kinetics, *Acta Metall.*, 32, 783, 1984. With permission.)

$$v_b \sim \frac{dG}{dt} = M_b \left(\frac{\alpha \gamma_{gb}}{G} \right) \tag{9.54}$$

Comparing Equation 9.53 and Equation 9.54, M_b is defined as:

$$M_b = \frac{D_a}{kT} \left(\frac{\Omega}{\delta_{gb}} \right) \tag{9.55}$$

Integrating Equation 9.54 gives:

$$G^2 = G_o^2 + Kt \tag{9.56}$$

where G_o is the initial grain size ($G = G_0$ at $t = 0$) and K is a temperature-dependent growth factor given by:

$$K = 2\alpha \gamma_{gb} M_b \tag{9.57}$$

Equation 9.56 predicts a parabolic growth law for grain growth. The growth factor K has an Arrhenius dependence on temperature T, given by $K = K_o\ exp\ (-Q/RT)$, where K_o is a constant, R is the gas constant, and Q is the activation energy (per mole) for grain growth.

The boundary mobility M_b depends on the diffusion coefficient D_a for the atomic jumps across the boundary of the pure material and is termed the intrinsic boundary mobility. For an ionic solid, in which both cations and anions must diffuse, D_a represents the diffusion coefficient of the rate-limiting (or slowest) species. In ceramics, the experimentally determined boundary mobility is rarely as high as M_b given by Equation 9.55 because, as described later, segregated solutes, inclusions, pores and second-phase films can also exert a drag force on the boundary.

More recently, the use of computer simulations has provided a valuable technique for the analysis of grain growth. Simulations employing a Monte Carlo method show a remarkable ability to provide realistic pictures of grain growth (Figure 9.24), and to provide a good fit to some experimental data [26,27].

Many grain growth models do not predict a parabolic law. In general, the grain growth equation can be written in the form

$$G^m = G_o^m + Kt \qquad\qquad (9.58)$$

where m is an exponent that has values in the range 2 to 4. A nonintegral value $m = 2.44$ is predicted by the computer simulations depicted in Figure 9.24.

9.3.2 Abnormal Grain Growth

Microstructures of polycrystalline ceramics that have been heated for some time at a sufficiently high temperature often show very large (abnormal) grains in a matrix of finer grains. A well-known microstructure in the ceramic literature involves the growth of a relatively large single-crystal Al_2O_3 in a fine-grained Al_2O_3 matrix, which appears to show the single-crystal grain growing much faster than the matrix grains (Figure 9.25). Earlier explanations considered that the grain size distribution of the starting material was the major factor leading to abnormal grain growth. They were based on the Hillert theory of grain growth [28] which predicted that any grain with a size greater than twice the average grain size would be predisposed to growing abnormally. This explanation is not supported by recent computer simulations [29] and theoretical analysis [30], which show that although the large grains grow, they do not outstrip the normal grains. The normal grains grow at a faster relative rate so that the large (abnormal) grains eventually return to the normal size distribution (Figure 9.26). The size effect is therefore not a sufficient criterion for abnormal grain growth. Inhomogeneities in chemical composition, liquid phases, and particle packing have long been suggested as possible causes of abnormal grain growth.

Computer simulations and theoretical analysis indicate that true, abnormally rapid grain growth can occur if the boundary of the abnormal grain has a higher mobility or a lower energy than the surrounding matrix grains [31–33]. Grain orientation and grain boundary structure are two factors that can lead to variations in the mobility and energy of grain boundaries. Boundaries of large abnormal grains often show faceting on low index crystallographic orientations. Frequently, the growth planes are those with low surface energy. Low-energy boundaries often have a lower mobility. Growth that is slow normal to and rapid parallel to the low-energy surfaces often lead to platelike abnormal grains, a process commonly referred to as *anisotropic grain growth*.

Abnormal grain growth in Al_2O_3 has received considerable attention, largely based on attempts to understand the role of MgO in suppressing abnormal grain growth in this system [34]. As discussed

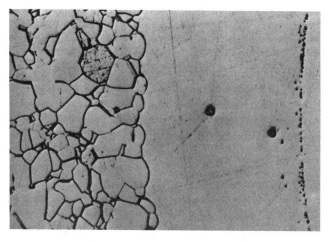

FIGURE 9.25 Example of abnormal grain growth: a large Al_2O_3 crystal growing into a matrix of uniformly sized grains. (Magnification 495×) (Courtesy of R.L. Coble.)

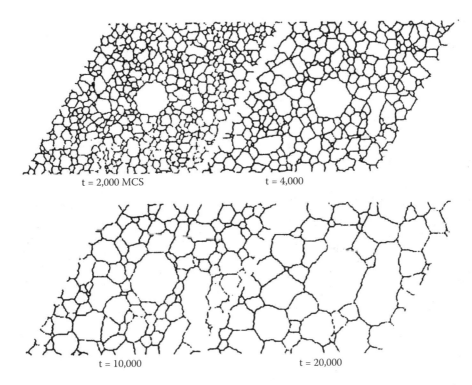

t = 2,000 MCS   t = 4,000

t = 10,000   t = 20,000

FIGURE 9.26 Computer simulation of the growth of a large grain in an otherwise normal grain growth microstructure for an isotropic system. The microstructure was produced by running the normal grain growth simulation procedure for 1000 MCS and the large gain was then introduced as a circular grain with an initial size equal to five times the average grain size of the matrix grains. (From Srolovitz, D.J., Grest, G.S., and Anderson, M.P., Computer simulation of grain growth — V. Abnormal grain growth, *Acta Metall.*, 33, 2233, 1985. With permission.)

later, the single most important role of MgO is the reduction of the boundary mobility, M_b, but it is also recognized that MgO plays an important role in reducing the anisotropies in the surface and grain boundary energies and in the boundary mobilities [6,35,36]. On the other hand, doping Al_2O_3 with TiO_2 has been observed to enhance the formation of elongated, faceted grains [37,38].

Although the suppression of abnormal grain growth forms a key goal in sintering, the ability to exploit abnormal grain growth in a controlled manner can provide significant property advantages in some ceramic systems. The growth of a controlled distribution of elongated (abnormal) grains in a matrix of finer grains (Figure 9.27a) has been used to enhance the fracture toughness of Si_3N_4, SiC, and mullite [39–41]. Alignment of the abnormally growing grains (Figure 9.27b) by the process of templated grain growth [42], hot pressing, or sinter forging has been used to achieve anisotropic dielectric properties in layered-structured ferroelectrics such as bismuth titanate [43,44]. Abnormal grain growth has been used effectively for many years to produce some single-crystal ferrites by seeding a fine-grained material with a large single crystal, and more recently to grow $BaTiO_3$ and lead-magnesium niobate single crystals from polycrystalline materials [45,46].

9.3.3 OSTWALD RIPENING

Let us consider a system consisting of a dispersion of spherical precipitates with different radii in a medium in which the particles have some solubility (Figure 9.28). Due to their higher chemical potential, the atoms under the surface of the sphere have a higher solubility in the surrounding

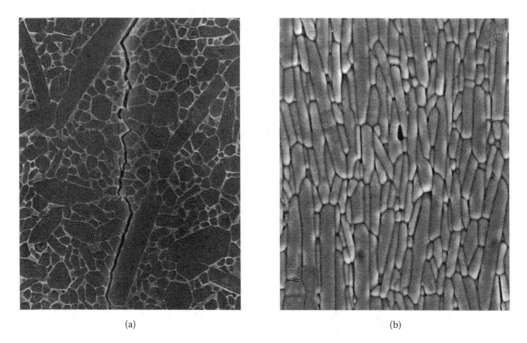

(a) (b)

FIGURE 9.27 Exploitation of controlled anisotropic abnormal grain growth: (a) self-reinforced silicon nitride, showing the interaction of a propagating crack with the microstructure (magnification 2000×) (Courtesy of M.J. Hoffmann.), (b) bismuth titanate with a microstructure of aligned anisotropic grains, giving anisotropic dielectric properties (magnification 1000×).

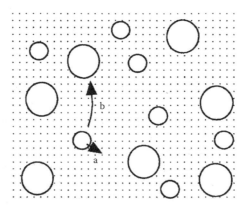

FIGURE 9.28 Coarsening of particles in a medium by matter transport from the smaller to the larger particles. Two separate mechanisms can control the rate of coarsening: (a) reaction at the interface between the particles and the medium and (b) diffusion through the medium.

medium than the atoms under a flat surface. Making use of the relation between chemical potential and concentration [1], and assuming ideal solutions, we can write:

$$\mu - \mu_o = kT \ln \frac{C}{C_o} \tag{9.59}$$

where C is the concentration of solute atoms surrounding a particle of radius a, C_0 is the concentration over a flat surface, k is the Boltzmann constant and T is the absolute temperature. Using Equation 9.10, and assuming that $\Delta C = C - C_0$ is small, Equation 9.59 becomes:

$$\frac{\Delta C}{C_o} = \frac{2\gamma\Omega}{kTa} \qquad (9.60)$$

where γ is the specific energy of the interface between the precipitates and the medium. The higher solute concentration around a precipitate with smaller radius gives rise to a net flux of matter from the smaller precipitates to the larger ones, so the dispersion coarsens by a process in which the smaller precipitates dissolve and the larger ones grow. This type of coarsening process is referred to as *Ostwald ripening*, in honor of the work of Ostwald in this area around 1900. The driving force for the process is the reduction in the interfacial area between the precipitates and the medium.

The basic theory of Ostwald ripening was developed independently by Greenwood [47], Wagner [48], and Lifshitz and Slyozov [49], and is generally referred to as the *LSW theory*. The theory applies strictly to a dilute dispersion of particles in a solid, liquid, or gaseous medium. At any given time, particles smaller than a critical size will dissolve, surrounding themselves with a zone of excess solute that will find its way to particles larger than the critical size. The rate-limiting step can be either diffusion between the particles or the deposition or dissolution of atoms at the particle surfaces (reaction controlled growth). In either case, the theory predicts that at some later time, a self-similar, steady state particle size distribution is approached asymptotically in which the growth of the critical radius, a^*, follows a simple law of the form:

$$(a^*)^m = (a_o^*)^m + Kt \qquad (9.61)$$

where a_o^* is the initial radius, K is a temperature-dependent rate constant, t is the time, and m is an exponent that depends on the rate-controlling mechanism: $m = 3$ for diffusion control and $m = 2$ for surface reaction control. Equation 9.61 has the same form as the normal grain growth equation (Equation 9.58). The steady state particle size distribution is shown in Figure 9.29.

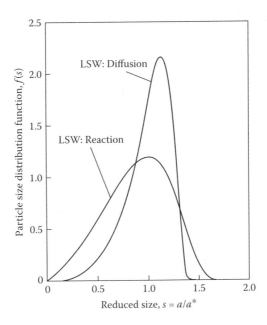

FIGURE 9.29 The particle size distribution function plotted vs. the reduced size for Ostwald ripening controlled by interface reaction or diffusion.

TABLE 9.4
Examples of Systems in Which Dopants Have Been Used Successfully

Host	Dopant	Concentration (at%)
Al_2O_3	Mg	0.025
$BaTiO_3$	Nb, Co	0.5–1.0
ZnO	Al	0.02
CeO_2	Y, Nd, Ca	3–5
Y_2O_3	Th	5–10
SiC	(B + C)	0.3 B + 0.5 C

The LSW theory has been the subject of considerable study and revision [1]. A key factor is the interaction between the particles when the dispersion of particles in the medium is no longer dilute. Ardell [50] found that for diffusion control, the exponent $m = 3$ is still valid but the self-similar size distribution approaches that for reaction control (i.e., it is broader and more symmetrical) when the volume fraction of particles is greater than ~0.2. Analysis of the approach to the steady state LSW distribution indicates that for distributions with a broad width, the standard deviation of the distribution function decreases to the characteristic time-invariant value, but for systems with a narrow width, the standard deviation increases to the time-invariant value [51]. This behavior during the transient regime indicates that the self-similar state acts as a strong attraction for the evolution of the particle size distribution.

9.3.4 CONTROL OF GRAIN GROWTH

As outlined earlier, the approach to grain growth control is commonly based on reducing the boundary mobility. The most effective method involves the use of additives (referred to as *dopants*) that are incorporated into the particles to form a solid solution. Examples of systems where dopants have been used successfully are given in Table 9.4. The most celebrated example was reported by Coble [52] who showed that the addition of 0.25 wt% of MgO to Al_2O_3 produced a polycrystalline translucent material with theoretical density (Lucalox). Figure 9.30 shows the effectiveness of MgO for controlling the grain size of Al_2O_3. Another effective approach, but one that is less widely used, involves the use of fine, inert second-phase particles at the grain boundaries. An example is the use of ZrO_2 particles for controlling the grain growth of Al_2O_3 (see Figure 1.24).

9.3.4.1 Effect of Dopants: Solute Drag

The concentration of the dopant (the solute) is often believed to be well below the solid solubility limit in the ceramic (the host), but this is not clear in some systems. It is often found that the effectiveness of the dopant in suppressing grain growth is dependent on its ability to segregate at the grain boundaries. The concentration of the solute segregated at the grain boundary is higher than that in the bulk of the grain, but a solid solution still exists. The major driving forces leading to segregation of an equilibrium concentration of solute at the grain boundaries are: (1) a reduction in the elastic strain energy of the crystal lattice due to a size difference between the solute atom and the host atom for which it substitutes, and (2) an electrostatic potential of interaction between aliovalent solutes and charged grain boundaries [53]. Segregation driven by elastic strain energy is limited to the core of the grain boundary, typically to a width of <1 nm. In contrast, for segregation driven by an electrostatic interaction, a compensating space charge layer adjacent to the charged grain boundaries produces a greater width of segregation, typically 1 nm to a few tens of nanometers (Figure 9.31).

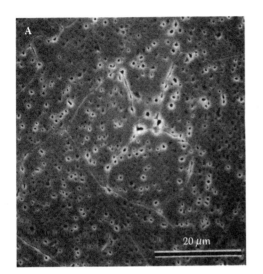
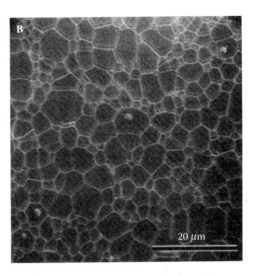

FIGURE 9.30 Microstructures of sintered Al$_2$O$_3$: (A) undoped material showing pore–grain boundary separation and abnormal grain growth, (B) MgO-doped material showing high density and equiaxial grain structure. (From Harmer, M.P., A history of the role of MgO in the sintering of α-Al$_2$O$_3$, *Ceram. Trans.*, 7, 13, 1990. Courtesy of M.P. Harmer.)

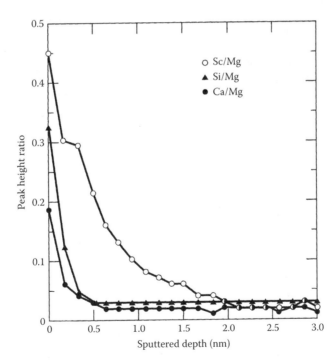

FIGURE 9.31 Concentration vs. depth of solutes at a grain boundary in MgO. Results from Auger spectroscopy during ion sputtering of a grain boundary exposed by fracture. A wider distribution is observed for the trivalent solute Sc segregated in a space charge layer, compared to Ca and Si, which adsorb at the core of the boundary. (From Chiang, Y.-M. et al., *J. Am. Ceram. Soc.*, 64, 383, 1981. With permission.)

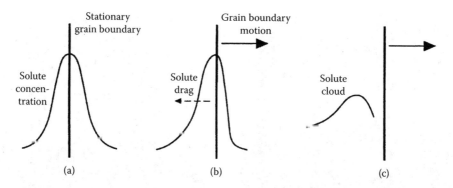

FIGURE 9.32 Sketch of the solute drag effect produced by the segregation of dopants to the grain boundaries: (a) symmetrical distribution of the dopant in the region of a stationary grain boundary, (b) for a moving boundary, the dopant distribution becomes asymmetric if the diffusion coefficient of the dopant atoms across the boundary is different from that of the host atoms. The asymmetric distribution produces a drag on the boundary, (c) breakaway of the boundary from the dopant leaving a solute cloud behind.

The inhibition of grain growth is believed to occur by a mechanism of solute drag [54]. A strong interaction is assumed to exist between the segregated solute and the grain boundary, so the solute must be carried along with the moving boundary. For a hypothetical stationary boundary, the concentration profile of the solute ions will be symmetrical (Figure 9.32a). The force of interaction due to the solute ions to the right of the boundary balances that due to the ions to the left of the boundary, so the net force of interaction is zero. If the boundary now starts to move, the dopant concentration profile becomes asymmetric, as the diffusivity of the solute ions across the boundary is expected to be different from that of the host (Figure 9.32b). This asymmetry results in a retarding force or drag on the boundary which reduces the driving force for migration. If the driving force for boundary migration is high enough, the boundary will break away from the high concentration of solute (sometimes called a solute cloud) and when this occurs (Figure 9.32c), its mobility will approach the intrinsic value, given by Equation 9.55.

Initially the mobility due to solute drag is constant, but as the velocity increases the boundary continually sheds solute, and for sufficiently high velocity, is able to break away from the solute cloud and migrate at the intrinsic velocity. As shown in Figure 9.33 for the theoretical relationship

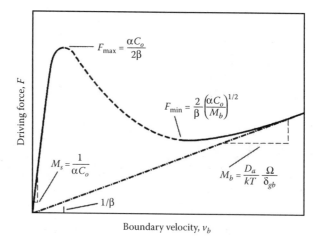

$$F_{max} = \frac{\alpha C_o}{2\beta}$$

$$F_{min} = \frac{2}{\beta}\left(\frac{\alpha C_o}{M_b}\right)^{1/2}$$

$$M_s = \frac{1}{\alpha C_o}$$

$$M_b = \frac{D_a}{kT}\frac{\Omega}{\delta_{gb}}$$

$$1/\beta$$

FIGURE 9.33 Driving force–velocity relationship for boundary migration controlled by solute drag and in the intrinsic regime. (From Yan, M.F., Cannon, R.M., and Bowen, H.K., Grain boundary migration in ceramics, in *Ceramic Microstructures '76*, Fulrath, R.M. and Pask, J.A., Eds., Westview Press, Boulder, CO, 1977, p. 276.)

between velocity and driving force, transitions are predicted to occur from the solute drag limited velocity to the intrinsic velocity over a range of driving forces [55]. Direct observations indicate that the grain boundary motion is often not uniform but starts and stops as if making transitions between the solute drag and intrinsic regimes [56].

The boundary mobility M'_b can be written in terms of the intrinsic component M_b and the solute drag component M_s:

$$M'_b = \frac{v_b}{F} = \left(\frac{1}{M_b} + \frac{1}{M_s} \right)^{-1} \tag{9.62}$$

where v_b is the boundary velocity, and F is the total drag force on the boundary, equal to the intrinsic drag F_b and the drag due to the solute F_s. For situations where the solute segregates to the grain boundary and the center of the boundary contributes most heavily to the drag effect, M_s is given by:

$$M_s = \frac{D_b \Omega}{4kT\delta_{gb}QC_o} \tag{9.63}$$

where D_b is the diffusion coefficient for the solute atoms across the boundary of width δ_{gb}, Ω is the atomic volume of the host atoms, C_o is the concentration of the solute atoms in the lattice, and Q is the partition coefficient (>1) for the dopant distribution between the boundary and the lattice.

According to Equation 9.63, dopants are predicted to be most effective for reducing the boundary mobility when the diffusion coefficient of the rate-limiting species D_b is low and the segregated solute concentration QC_o is high. This indicates that aliovalent solutes with larger ionic radii than the host would be effective for suppressing grain growth. Whereas some systems show behavior consistent with this prediction, as for example CeO_2 doped with trivalent solutes [57,58], the selection of an effective dopant is, in general, complicated because the dopant can have multiple roles.

9.3.4.2 Effect of Fine, Inert Second-Phase Particles

Let us consider a system of fine second-phase particles, inclusions or precipitates dispersed randomly in a polycrystalline solid in which they are insoluble and immobile. If a grain boundary moving under the driving force of its curvature encounters an inclusion, it will be held up by the particle until the motion elsewhere has proceeded sufficiently far for it to break away. If there is a sufficient number of particles, we might expect that the boundary will be pinned when it encounters the particles and boundary migration will therefore cease. This mechanism was suggested by Zener in a communication to Smith [59]. Zener assumed that the inclusion particles are monosize, spherical, insoluble, immobile, and randomly distributed in the polycrystalline solid. Taking a grain boundary with principal radii of curvature a_1 and a_2, the driving force (per unit area) for boundary motion is

$$F_b = \gamma_{gb} \left(\frac{1}{a_1} + \frac{1}{a_2} \right) \tag{9.64}$$

Assuming that a_1 and a_2 are proportional to the grain size, G, then

$$F_b = \alpha\gamma_{gb} / G \tag{9.65}$$

where α is a geometrical shape factor (e.g., $\alpha = 2$ for a spherical grain). When the grain boundary intersects an inclusion, its further movement is hindered (Figure 9.34a, b). A dimple is formed in

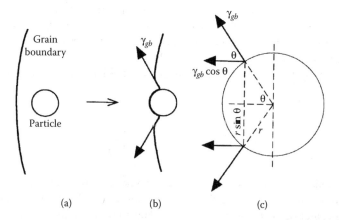

FIGURE 9.34 Interaction of a grain boundary with an immobile particle: (a) approach of the boundary toward the particle, (b) interaction between the grain boundary and the particle leading to a retarding force on the boundary, (c) detailed geometry of the particle–grain boundary interaction.

the grain boundary and, compared to the inclusion-free boundary, extra work must be performed for the equivalent motion of the boundary. This extra work manifests itself as a retarding force on the boundary. If r is the radius of the inclusion (Fig. 9.34c), then the retarding force exerted by the inclusion on the boundary is

$$F_r = (\gamma_{gb} \cos\theta)(2\pi r \sin\theta) \qquad (9.66)$$

i.e., the retarding force is the grain boundary tension, resolved in the direction opposite to that of the grain boundary motion, times the perimeter of contact. The retarding force is a maximum when $\theta = 45°$, so $\sin\theta = \cos\theta = 1/\sqrt{2}$, and

$$F_r^{max} = \pi r \gamma_{gb} \qquad (9.67)$$

If there are N_A inclusions per unit area of the grain boundary, then the maximum retarding force (i.e., per unit area of the boundary) is

$$F_d^{max} = N_A \pi r \gamma_{gb} \qquad (9.68)$$

N_A is difficult to determine but is related to N_V, the number of inclusions per unit volume, by the relation:

$$N_A = 2r N_V \qquad (9.69)$$

If the volume fraction of the inclusions in the solid is f, then

$$N_V = \frac{f}{(4/3)\pi r^3} \qquad (9.70)$$

Substituting for N_A in Equation 9.68 gives:

$$F_d^{max} = \frac{3f\gamma_{gb}}{2r} \qquad (9.71)$$

The net driving force per unit area of the boundary is

$$F_{net} = F_b - F_d^{max} = \gamma_{gb}\left(\frac{\alpha}{G} - \frac{3f}{2r}\right) \qquad (9.72)$$

When $F_{net} = 0$, boundary migration will cease and this occurs when

$$G_L = \frac{2\alpha}{3}\frac{r}{f} \qquad (9.73)$$

where G_L is the limiting grain size. Equation 9.73, referred to as the *Zener relationship*, indicates that a limiting grain size will be reached, the magnitude of which is proportional to the inclusion size and inversely proportional to the inclusion volume fraction. Changes in temperature would not affect this equilibrium relationship but would affect only the *rate* at which the system approaches the equilibrium condition. Though it may be difficult to achieve in practice, if a limiting grain size is reached, further grain growth can only occur if (1) the inclusions coarsen by Ostwald ripening, (2) the inclusions go into solid solution into the matrix or (3) abnormal grain growth occurs. Recent work indicates that the pinning particles are not distributed as randomly as assumed in the Zener model, so Equation 9.73 requires some modification to take into account the fraction of pinning particles actually at the grain boundaries [60].

9.3.5 GRAIN GROWTH AND PORE EVOLUTION IN POROUS COMPACTS

Grain growth in porous ceramics is complex but a few limiting cases can be discussed. Grain growth in very porous ceramics characteristic of the early stage of sintering is limited (Figure 9.1), but some coarsening of the microstructure can occur by processes such as surface diffusion and vapor transport. These changes can have a significant effect on microstructural evolution in the later stages. For a model consisting of two originally spherical particles separated by a grain boundary, taken as a rough approximation to the early stage microstructure, migration of the boundary is difficult. As sketched in Figure 9.35, the balance of interfacial tensions requires that the equilibrium dihedral angle of the grain boundary groove at the surface of the sphere be maintained. Migration of the boundary would actually involve a significant increase in the grain boundary area. The occurrence of grain growth is energetically unfavorable unless other processes that significantly reduce this energy barrier come into play.

One way in which grain growth can occur was suggested by Greskovich and Lay [61] on the basis of observations of the coarsening of Al_2O_3 powder compacts. The model is shown in Figure 9.36 for two particles and for a cluster of particles. For two particles that differ in size, surface diffusion is assumed to assist in the rounding of the particles and the growth of the necks between the particles. Whether or not the boundary can migrate depends on whether or not the structure permits it, i.e., there must be a decrease in the total energy of the system for an incremental movement of the boundary. As sketched for the two particles, it is assumed that surface diffusion produces the structural changes for the movement of the boundary to be favorable. The rate of migration of the boundary depends on the difference in initial size between the particles: the greater the difference in size, the greater the curvature of the boundary and the greater the driving force for the boundary to sweep through the smaller grain. Neck growth between the grains is

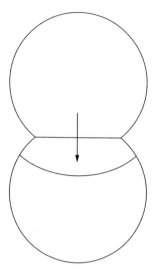

FIGURE 9.35 For an idealized initial stage microstructure, grain growth increases the total grain boundary area between two spheres. The dihedral angle constraint at the surface of the sphere creates a boundary curvature that opposes grain growth.

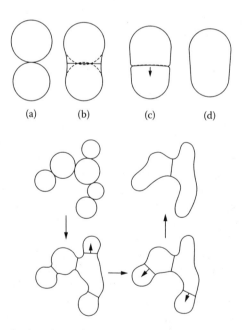

FIGURE 9.36 Qualitative mechanism for grain growth in porous powder compacts. Top: Two particles — (a) particles of slightly different size in contact, (b) neck growth by surface diffusion between the particles, (c) grain boundary migrating from the contact plane, and (d) grain growth. Bottom: A cluster of particles. Arrows on the grain boundaries indicate the direction of boundary movement. (From Greskovich, C. and Lay, K.W., Grain growth in very porous Al$_2$O$_3$ compacts, *J. Am. Ceram. Soc.*, 55, 142, 1972. With permission.)

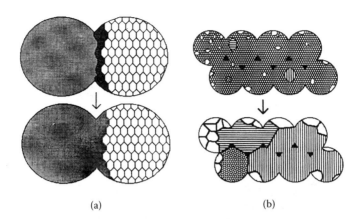

FIGURE 9.37 Growth of intra-agglomerate grains through polycrystalline necks: (a) growth of a large grain proceeds across a neck by incremental consumption of smaller grains, (b) exaggerated intra-agglomerate crystalline growth results in the development of large grains in the early stage of sintering. The resulting grain structure may contain some large grains that grow abnormally during inter-agglomerate sintering. (From Edelson, L.H. and Glaeser, A.M., Role of particle substructure in the sintering of monosized titania, *J. Am. Ceram. Soc.*, 71, 225, 1988. With permission.)

likely to be much slower than the migration of the boundary, so it controls the rate of the overall coarsening process.

Greskovich and Lay considered surface diffusion to be the dominant neck growth mechanism, but vapor transport may also be important in systems with fairly high vapor pressures. Furthermore, coarsening controlled by neck growth is not expected to be the only process operating in very porous systems. A heterogeneously packed powder compact will, after some sintering, contain porous regions as well as dense regions. For this type of structure, coarsening is likely to be controlled by two separate mechanisms: (1) neck growth in the porous regions and (2) curvature-driven boundary migration in the dense regions.

Another way in which grain growth can occur in porous compacts was described by Edelson and Glaeser [62]. As sketched in Figure 9.37a, if large grain size differences exist at the necks between the particles, elimination of grain boundaries by the advancing boundary releases enough energy to make the overall process favorable. In this way, the finer neighboring grains can be consumed in an incremental growth process. Figure 9.37b shows that the growing grain can also entrap porosity, a process that can limit the density of the final article. However this process is less likely to occur in the earlier stages of sintering because the fairly large, continuous pores that are closely spaced provide a significant drag on the boundary and limit its mobility.

The grain growth rate increases in the intermediate stage of sintering and migration of the boundaries leads to coalescence of the pores [63], so the average pore size also increases (Figure 9.38). Grain growth is most pronounced in the final stage when the pores pinch off and become isolated. Normal grain growth has been analyzed using an idealized model consisting of a nearly spherical pore on an isolated grain boundary. The kinetics of normal grain growth can be described by Equation 9.58, in which the value of the exponent m (2–4) now depends on mechanism of the pore motion and the extent of the drag produced by the pores on the boundaries [23]. Several mechanisms have the same value of m so that an unambiguous determination of the grain growth mechanism from experimental data is difficult.

An understanding of pore evolution during sintering is important for a broader understanding of the microstructural evolution, but very few investigations have been carried out due largely to the difficulty of quantitative analysis and the time-consuming nature of detailed stereological characterization. Rhines and DeHoff [64] found that the pore network changed by collapse of the network and the reforming of a new network with lower connectivity. Measurements of the open

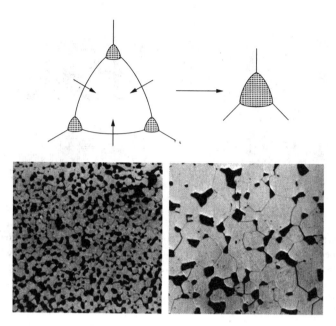

FIGURE 9.38 (a) Schematic illustration of grain growth accompanied by pore coalescence. (b) Grain growth and pore coalescence in a sample of UO$_2$ after 2 min, 91.5% dense and 5 h, 91.9% dense, at 1600°C. (Magnification 400×) (From Kingery, W.D. and Francois, B., Grain growth in porous compacts, *J. Am. Ceram. Soc.*, 48, 546, 1965. With permission.)

and closed porosity during the sintering of a heterogeneously packed UO$_2$ powder showed that the volume of closed porosity started to increase when the open porosity had decreased to 15 vol% [65]. For a homogeneously packed powder, it is expected that the formation of closed porosity would occur more suddenly and when the volume of open porosity had decreased to a value of less than 10%.

9.3.6 INTERACTIONS BETWEEN PORES AND GRAIN BOUNDARIES

The attainment of high density during sintering, as outlined earlier, depends on the ability to stabilize the microstructure such that the pores and the grain boundaries remain attached. For an idealized final-stage microstructure consisting of an isolated, spherical pore at a grain boundary, the interaction between the pore and the boundary has been used to determine the conditions for pore attachment and breakaway.

9.3.6.1 Pore Mobility

Small, isolated pores situated at the grain boundary can be dragged along by the moving grain boundary. The reason is that the boundary moving under the influence of its curvature applies a force on the pore causing the pore to change its shape [66,67]. The leading surface of the pore becomes less strongly curved than the trailing surface (Figure 9.39). The difference in curvature leads to a chemical potential difference that drives the flux of atoms from the leading surface to the trailing surface. The result is that the pore moves forward in the direction of the boundary motion. Matter transport from the leading surface to the trailing surface can occur by three separate mechanisms: vapor transport (evaporation/condensation), surface diffusion, and lattice diffusion.

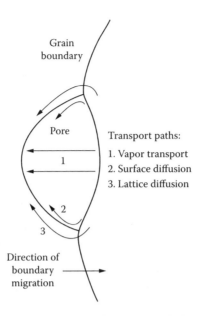

FIGURE 9.39 Possible transport paths for a pore moving with a grain boundary: 1, vapor transport (evaporation and condensation); 2, surface diffusion; 3, lattice diffusion.

The flux of matter from the leading surface to the trailing surface of the pore can be analyzed to derive an equation for the pore mobility M_p. By analogy with the case of a moving boundary, the pore velocity v_p is defined by a force-mobility relationship,

$$v_p = M_p F_p \tag{9.74}$$

where F_p is the driving force acting on the pore. Let us consider a pore of average radius r in which matter transport occurs by surface diffusion from the leading surface to the trailing surface. The net atomic flux is

$$J_s A_s = \left(\frac{D_s}{\Omega kT} F_a \right) 2\pi r \delta_s \tag{9.75}$$

where J_s is the flux of atoms, A_s is the area over which surface diffusion occurs, D_s is the surface diffusion coefficient, Ω is the atomic volume, F_a is the driving force on an atom, δ_s is the thickness for surface diffusion, k is the Boltzmann constant and T is the absolute temperature. If the pore moves forward a distance dx in a time dt, the volume of matter which must be moved per unit time is $\pi r^2 (dx/dt)$.

Equating the number of atoms moved to the net flux gives:

$$\frac{\pi r^2}{\Omega} \frac{dx}{dt} = - \frac{D_s 2\pi r \delta_s}{\Omega kT} F_a \tag{9.76}$$

where the negative sign is inserted because the flux is opposite to the direction of motion. The work done in moving the pore a distance dx is equal to that required to move $\pi r^2 dx/\Omega$ atoms a distance $2r$, so

$$F_p dx = -F_a \frac{\pi r^2 dx}{\Omega} 2r \tag{9.77}$$

Substituting for F_a from Equation 9.76 and rearranging gives

$$\Gamma_p - \frac{\pi r^4 kT}{D_s \delta_s \Omega} \frac{dx}{dt} \tag{9.78}$$

Putting the velocity of the pore v_p equal to dx/dt, the pore mobility is given by

$$M_p = \frac{D_s \delta_s \Omega}{\pi kT r^4} \tag{9.79}$$

The pore mobility for matter transport by vapor transport or by lattice diffusion can be derived by a similar procedure and the formulas are summarized in Table 9.5. For all three mechanisms,

---

TABLE 9.5
Pore and Boundary Parameters

	Mobilities[a]
M_p **Mobility of spherical pore; migration by faster of**	
Surface diffusion	$\dfrac{D_s \delta_s \Omega}{kT \pi r^4}$
Lattice diffusion	$\dfrac{D_l \Omega}{fkT \pi r^3}$
Vapor transport	$\dfrac{D_g d_g \Omega}{2kT d_s \pi r^3}$
M_b **Mobility of boundary**	
Pure system	$\dfrac{D_a \Omega}{kT \delta_{gb}}$
Solute drag	$\dfrac{\Omega}{kT \delta_{gb}} \left(\dfrac{1}{D_a} + \dfrac{4QC_o}{D_b} \right)^{-1}$
	Forces[b]
F_p **Maximum drag force of pore**	$\pi r \gamma_{gb}$
F_b **Force per unit area of pore-free boundary**	$\dfrac{\alpha \gamma_{gb}}{G}$

[a] f = correlation factor; d_g = density in the gas phase of the rate-controlling species; d_s = density in the solid phase of the rate-controlling species.
[b] α = geometrical constant depending on the grain shape (e.g., $\alpha = 2$ for spherical grains).

Source: Adapted from Brook, R.J., Controlled grain growth, in *Treatise on Materials Science and Technology*, Wang, F.F.Y., Ed., Academic Press, New York, 1976. With permission.

M_p is found to have a strong dependence on r, increasing rapidly with decreasing r. Fine pores are highly mobile and therefore better able to remain attached to the moving boundary during grain growth.

9.3.6.2 Kinetics of Pore–Boundary Interactions

The condition for the pore to separate from the boundary is $v_p < v_b$, which can also be written as:

$$F_p M_p < F M_b \tag{9.80}$$

where F is the effective driving force on the boundary. If F_d is the drag force exerted by a pore, then a balance of forces requires that F_d is equal and opposite to F_p. Considering unit area of the boundary in which there are N_A pores, Equation 9.80 can be written

$$F_p M_p < (F_b - N_A F_p) M_b \tag{9.81}$$

where F_b is the driving force on the pore-free boundary due to its curvature. Rearranging Equation 9.81, the condition for pore separation can be expressed as:

$$F_b > N_A F_p + \frac{M_p F_p}{M_b} \tag{9.82}$$

The condition for pore attachment to the boundary is $v_p = v_b$, which can also be written as:

$$F_p M_p = (F_b - N_A F_p) M_b \tag{9.83}$$

Putting $v_p = F_p M_p = v_b$ and rearranging gives:

$$v_b = F_b \frac{M_p M_b}{N_A M_b + M_p} \tag{9.84}$$

Two limiting conditions can be defined. When $N_A M_b \gg M_p$, then

$$v_b = F_b M_p / N_A \tag{9.85}$$

The driving force on the boundary is nearly balanced by the drag of the pores and the boundary motion is limited by the pore mobility. This condition is referred to as *pore control*. The other limiting condition is when $N_A M_b \ll M_p$, in which case:

$$v_b = F_b M_b \tag{9.86}$$

The presence of the pores has almost no effect on the boundary velocity, a condition referred to as *boundary control*. The condition separating the pore control and boundary control regions is defined by:

$$N_A M_b = M_p \tag{9.87}$$

which defines the equal mobility curve.

9.3.6.3 Microstructural Maps

Brook [68] developed a method to visualize the interaction between the pore and the boundary during sintering. Two situations can be visualized: (1) the pore remains attached to the migrating boundary (referred to as *attachment*) or (2) the boundary breaks away, leaving the pore behind (*separation*). The conditions under which each situation becomes important are determined as functions of the grain size and the pore size, and the results are plotted on a diagram, often referred to as *a microstructural map*.

To illustrate how the map is constructed, let us assume that pore migration occurs by surface diffusion. Using the relations for M_b and M_p from Table 9.5, and the approximation that $N_A \approx 1/X^2$, where the interpore distance $X \approx G$, Equation 9.87 gives:

$$G_{em} = \left(\frac{D_a \pi}{D_s \delta_s \delta_{gb}} \right)^{1/2} r^2 \tag{9.88}$$

where G_{em} is the grain size defined by the equal mobility condition. Using logarithmic axes for a plot of G vs. $2r$ (Figure 9.40), the equal mobility condition is represented by a straight line with a slope of 2.

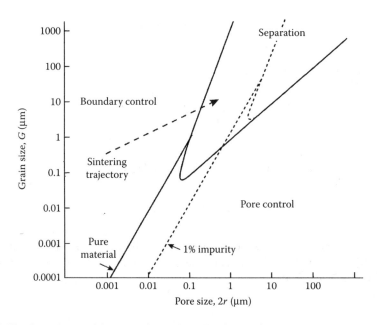

FIGURE 9.40 The dependence of the type of pore-boundary interaction on microstructural parameters when pores migrate by surface diffusion. The interpore spacing is assumed to be equal to the grain size. (From Brook, R.J., Pore-grain boundary interactions and grain growth, *J. Am. Ceram. Soc.*, 52, 56, 1969. With permission.)

To determine the conditions for separation of the boundary from the pore, we recall that the maximum force exerted by the grain boundary on a pore to drag it along is given by Equation 9.67. The maximum velocity that the pore can attain is therefore:

$$v_p^{max} = M_p F_p^{max} \tag{9.89}$$

If the velocity of the boundary with the attached pore were to exceed v_p^{max}, then separation will occur. The limiting condition for separation can therefore be written as:

$$v_b = v_p^{max} \tag{9.90}$$

Substituting for v_b from Equation 9.84, we obtain:

$$\frac{M_b M_p}{N_A M_b + M_p} F_b = M_p F_p^{max} \tag{9.91}$$

Putting $N_A \approx 1/X^2$ and substituting for the other parameters from Table 9.5, Equation 9.91 gives:

$$G_{sep} = \left(\frac{\pi r}{X^2} + \frac{D_s \delta_s \delta_{gb}}{D_a r^3} \right)^{-1} \tag{9.92}$$

where G_{sep} is the grain size when the boundary separates from the pore. Assuming that $X \approx G$, Equation 9.92 can be written as:

$$\left(\frac{D_s \delta_s \delta_{gb}}{D_a r^3} \right) G_{sep}^2 - G_{sep} + \pi r = 0 \tag{9.93}$$

The solution to this quadratic equation determines the separation curve shown in Figure 9.40. Pore–boundary separation is predicted to occur only for certain ratios of grain size to pore size. Brook also considered the effect of a segregated impurity (solute) on the boundary mobility. The mobility limited by solute drag is lower than the intrinsic mobility, allowing larger pores to migrate with the boundary before separation. The separation region is therefore shifted to larger grain and pore sizes.

Possible trajectories for sintering can be identified on the basis of the G vs. $2r$ diagram. As outlined earlier, grain growth and pore coalescence often contribute to the coarsening of porous ceramics, particularly in the later stages of sintering, resulting in an increase in both the average grain size and the average pore size. For this situation, a possible trajectory for the microstructural evolution runs diagonally upward from left to right.

A modification of the G vs. $2r$ map was used by Berry and Harmer [69] to show the sintering trajectories for Al_2O_3 on a grain size G vs. density ρ map (Figure 9.41). The influence of dopants and other variables can now be discussed in terms of their effects on the separation region and the sintering trajectory. For the achievement of high density during sintering, the separation region must be avoided. According to Figure 9.40 and Figure 9.41, the use of a dopant that reduces the boundary mobility has the effect of shifting the separation region to higher grain sizes, thereby making it easier for the trajectory to bypass the separation region. Flattening of the G vs. ρ trajectory

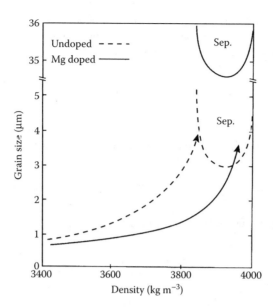

FIGURE 9.41 Grain size–density map for Al$_2$O$_3$, illustrating the effect of raising the surface diffusion coefficient by a factor of 4, reducing the lattice diffusion coefficient by a factor of 2, and reducing the grain boundary mobility by a factor of 34. This has the effect of flattening the grain size–density trajectory and raising the separation region to larger grain sizes, thereby making it possible to sinter to full density. (From Berry, K.A. and Harmer, M.P., Effect of MgO solute on microstructure development in Al$_2$O$_3$, *J. Am. Ceram. Soc.*, 69, 143, 1986. With permission.)

by enhancing the ratio of the densification rate to the grain growth rate is also effective. This can be achieved by the use of a dopant that lowers the boundary mobility or by selecting a heating schedule that enhances the densifying mechanisms relative to the coarsening mechanisms.

Although the microstructural maps provide a valuable method for representing the interaction between densification and coarsening, the idealized geometry of the assumed microstructure (a spherical pore on an isolated boundary of uniform curvature) means that the maps have limited applicability to actual final stage microstructures. Factors such as grain size distribution, pore size distribution, number of pores, and dihedral angle will affect the simple relationships between pore size and boundary curvature assumed for separation [70,71]. The statistical nature of the separation process must also be recognized. Separation of only a fraction of the pores from the boundary is sufficient to cause abnormal grain growth, as opposed to the assumption in the simple analysis that all pores separate from the grain boundary.

9.4 LIQUID-PHASE SINTERING

In many ceramics, the formation of a liquid phase is commonly used to assist the sintering process. The purpose is usually to enhance densification rates, achieve accelerated grain growth, or to produce specific grain boundary properties. Enhanced densification results from (1) enhanced rearrangement of the particles, and (2) enhanced matter transport through the liquid phase. Figure 9.42 shows a sketch of an idealized two-sphere model in which the microstructural aspects of liquid-phase sintering are compared with those of solid-state sintering. In liquid-phase sintering, if, as we assume, the liquid wets and spreads to cover the solid surfaces, the particles will be separated by a liquid bridge. Friction between the particles is reduced, so they rearrange more easily under the compressive capillary stress exerted by the liquid. In solid-state sintering by, for example, grain boundary diffusion, the diffusivity is equal to $(D_{gb}\delta_{gb})$, whereas in liquid-phase

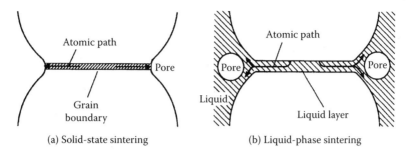

(a) Solid-state sintering   (b) Liquid-phase sintering

FIGURE 9.42 Sketch of an idealized two-sphere model comparing the microstructural aspects of (a) solid-state sintering with (b) liquid-phase sintering.

sintering, it is equal to $(D_L\delta_L)$, where D_L, is the diffusion coefficient of the solute atoms in the liquid, and δ_L is the width of the liquid bridge. Because $\delta_L \gg \delta_{gb}$ and $D_L \gg D_{gb}$, the liquid provides a path for enhanced matter transport.

Assuming as earlier that the liquid wets and spreads over the solid surfaces, the solid–vapor interface will be eliminated and pores will form in the liquid. The reduction of the liquid–vapor interfacial area provides a driving force for densification. For a spherical pore of radius r in a liquid, the pressure difference across the curved surface is given by the equation of Young and Laplace:

$$\Delta p = -\frac{2\gamma_{LV}}{r} \tag{9.94}$$

where γ_{LV} is the specific surface energy of the liquid–vapor interface. The pressure in the liquid is lower than that in the pore and this generates a compressive capillary stress on the particles. This compressive stress due to the liquid is equivalent to placing the system under an external hydrostatic pressure, the magnitude of which is given by Equation 9.94. Taking $\gamma_{LV} \approx 1$ J/m^2 and $r \approx 0.5$ μm gives $p \approx 4$ MPa. Pressures of this magnitude can provide an appreciable driving force for sintering.

In ceramics, the liquid is commonly formed by using an additive phase that forms a eutectic liquid with a small amount of the powder on heating. The liquid should have a low contact angle for good wetting of the solid, an appreciable solubility for the solid, and a reasonably low viscosity for rapid matter transport through the liquid. The amount of liquid formed during sintering is often small, typically less than a few volume percent, which can make precise control of the liquid composition difficult. The distribution of the liquid phase and of the resulting solidified phases produced on cooling after densification is critical to achieving the required properties of the sintered material.

Liquid-phase sintering is particularly effective for ceramics such as Si_3N_4 and SiC that have a high degree of covalent bonding and are therefore difficult to densify by solid-state sintering. The process is also important when the use of solid-state sintering is too expensive or requires too high a fabrication temperature. However, the enhanced densification rates achieved by liquid-forming additives are only of interest if the properties of the fabricated ceramic remain within the required limits. The liquid phase used to promote sintering commonly remains as a glassy intergranular phase that may degrade high-temperature mechanical properties such as creep and fatigue resistance.

Some examples of ceramic liquid-phase sintering systems and their applications are given in Table 9.6. When sufficient liquid is present and interfacial energies are fairly uniform, the microstructure may consist of individual, rounded grains dispersed in the liquid. Such a microstructure is commonly seen in metallic systems, and in some ceramic systems such as MgO/silicate (Figure 9.43a). In many ceramic systems, angular, prismatic, or elongated grain shapes are more commonly found as a result of liquid-phase sintering (Figure 9.43b–d).

TABLE 9.6
Some Examples of Ceramic Liquid-Phase Sintering Systems and Their Applications

Ceramic system	Additive content (wt%)	Application
Al_2O_3 + talc	~5	Electrical insulators
$MgO + CaO\ SiO_2$	<3	Refractories
$ZnO + Bi_2O_3$	2–3	Electrical varistors
$BaTiO_3 + TiO_2$	<1	Dielectrics
$Si_3N_4 + MgO$	5–10	Structural ceramics
$Si_3N_4 + Y_2O_3–Al_2O_3$	5–10	Structural ceramics
$SiC + Y_2O_3–Al_2O_3$	5–10	Structural ceramics
$WC + Ni$	~10	Cutting tools

9.4.1 Stages of Liquid-Phase Sintering

Liquid-phase sintering is generally regarded as proceeding in a sequence of three dominant stages [72]: (1) *rearrangement* of the solid phase and redistribution of the liquid, driven by capillary stress gradients, (2) densification and shape accommodation of the solid phase involving *solution precipitation*, (3) *final densification* driven by residual porosity in the liquid phase (Figure 9.44). The extent to which each stage influences densification depends on the volume fraction of liquid, so there are many variants in this conceptual picture. For high liquid content, complete densification can be achieved by rearrangement alone, whereas at the low liquid contents common for many systems, the solid skeleton inhibits densification, so solution-precipitation and final-stage sintering are required to achieve further densification.

Stage 1: Rearrangement and Liquid Redistribution
As the temperature of the green body is raised, a limited amount of solid-state sintering may occur in some systems prior to the formation of the liquid. With the formation of the liquid phase, capillary pressure gradients can be significant, reaching a few MPa when submicron liquid-phase surface curvatures are present (Equation 9.94), sufficient to cause rapid liquid flow and particle rearrangement when the effective viscosity of the compact is still low. The pressure gradients will cause liquid to flow from regions with large liquid pores to regions of smaller liquid pores and, as a consequence, from the surface to the interior of the compact. This principle leads to a sequential pore filling [73]. Even though liquid redistribution is relatively easy, it is desirable to start with a compact that is as homogeneous as practical. The same concerns about homogeneity of the starting compact apply to liquid-phase sintering as they do to solid-state sintering. The liquid-forming additive should be distributed as homogeneously as possible in the green compact. Precoating the particles with the liquid-forming phase using fluidized bed vapor deposition or chemical precipitation from solution (Chapter 2) can be used. A second stage rearrangement, slower than the first, in which the liquid phase is thought to penetrate the grain boundaries of dense, multigrain particles, has also been recognized. As described later, partial or complete grain boundary penetration is dictated by the dihedral angle.

Stage 2: Solution-Precipitation
As densification by rearrangement slows, effects dependent on the solid solubility in the liquid and the diffusivity in the liquid dominate, giving the second stage termed *solution-precipitation*. The solid dissolves at the solid-liquid interfaces with a higher chemical potential, diffuses through the liquid and precipitates on the particles at other sites with a lower chemical potential. In one model [74], referred to as *contact flattening*, dissolution occurs at the wetted contact area between the

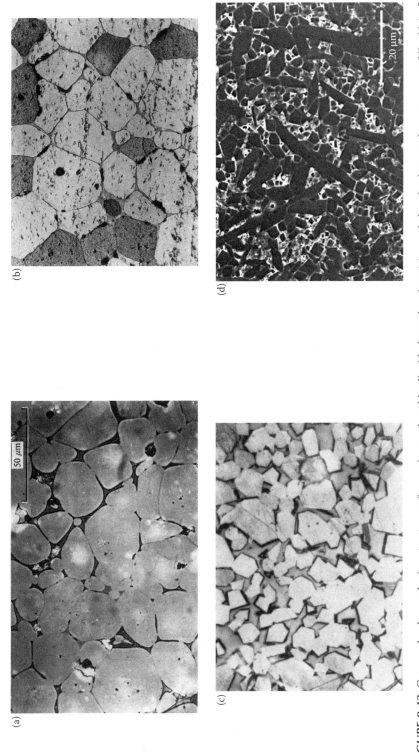

FIGURE 9.43 Commonly observed microstructures of ceramics produced by liquid-phase sintering: (a) rounded grains in a moderate amount of liquid (> 5 vol%), (b) grains with flat contact surfaces in a low volume fraction of liquid (<2–5 vol%), (c) prismatic grains dictated by anisotropic interfacial energy with a moderate to high liquid volume (>10 vol%), (d) elongated grains with flat or curved sides resulting from anisotropic interfacial energy with a low liquid content (<2–5 vol%).

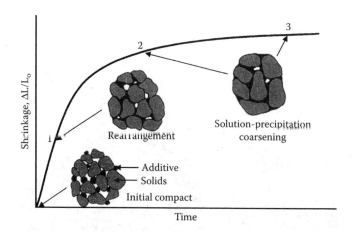

FIGURE 9.44 Schematic evolution of a powder compact during liquid-phase sintering. The three dominant stages overlap significantly.

particles where the capillary stress leads to a higher chemical potential, and precipitation occurs at sites away from the contact area (Figure 9.45a). In another model [75], involving Ostwald ripening, which is expected to be more applicable to systems with a wide distribution of particle sizes, dissolution occurs at the surfaces of the small particles and precipitation occurs at the surfaces of the large particles (Figure 9.45b). The models lead to densification equations for the shrinkage $\Delta L/L_o$ as a function of time t, where $\Delta L/L_o$ is proportional to $t^{1/3}$ when diffusion through the liquid is rate controlling, or $t^{1/2}$ when the solid–liquid phase boundary reaction is rate controlling.

The models further assume that the grain boundary liquid phase does not change composition or structure, but this is not always the case. There are now several examples where the composition of the grain boundary film is found to be different from that of the liquid in the adjacent triple junction or even from boundary to boundary [76,77], as shown in Figure 9.46. It is quite likely that such is more the rule than the exception. Densification by the solution–precipitation mechanism is also accompanied by changes in the shape of the grains. For a small amount of liquid, the grains develop flat faces and assume the shape of a polyhedron to achieve more efficient packing, a process described as *grain shape accommodation* (see Figure 9.43b).

Stage 3: Final Densification

The final stage of liquid-phase sintering is controlled by the densification of the solid particulate skeletal network. The process is slow because of the large diffusion distances in the coarsened structure and the rigid skeleton of contacting solid grains. Ostwald ripening dominates the final stage and the residual pores become larger if they contain trapped gas, leading to compact swelling. Coarsening is accompanied by grain shape accommodation, which allows more efficient packing of the grains. Liquid may be released from the more efficiently packed regions and it may flow into the isolated pores, leading to densification, or to partial expulsion of the liquid from the article.

9.4.2 KINETIC AND THERMODYNAMIC FACTORS

Successful liquid-phase sintering is dependent on several kinetic and thermodynamic factors, as well as processing factors.

9.4.2.1 Contact Angle

The degree of wetting is characterized by the contact angle θ that depends on the various interfacial energies for the solid–liquid–vapor system and is usually referred to a droplet of liquid on a flat

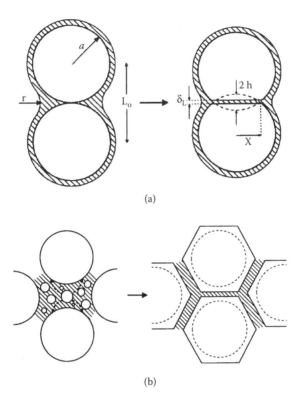

(a)

(b)

FIGURE 9.45 (a) Idealized two-sphere model for densification by contact flattening. (b) Schematic diagram illustrating densification accompanied by Ostwald ripening. Grain shape accommodation can also occur when the liquid volume fraction is low.

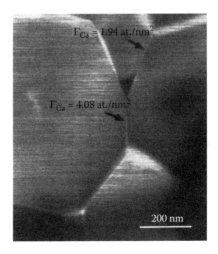

FIGURE 9.46 STEM image showing two neighboring grain boundary films with remarkably different calcium excess in calcia-doped silicon nitride. (Gu, H. et al., Dopant distribution in grain-boundary films in calcia-doped silicon nitride ceramics, *J. Am. Ceram. Soc.*, 81, 3125, 1998. With permission.)

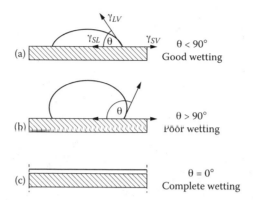

FIGURE 9.47 Wetting behavior between a liquid and a solid showing (a) good wetting, (b) poor wetting, (c) complete wetting for a liquid with a contact angle of θ.

solid surface (Figure 9.47). Though this does not represent full equilibrium, it is often cited as such, and is given by:

$$\gamma_{SV} = \gamma_{SL} + \gamma_{LV}\cos\theta \qquad (9.95)$$

where γ_{LV}, γ_{SV}, and γ_{SL} are the specific energies of the liquid–vapor, solid–vapor, and solid–liquid interfaces, respectively. Low contact angles are required for good wetting of the solid phase. For large contact angles, the liquid phase can actually produce repulsive forces rather than attractive forces between the particles [1], so further densification can only proceed by grain growth.

The γ_{LV} values for many inorganic melts, such as silicates, are often in the range of 0.1–0.5 J/m^2, with a value of ~0.3 J/m^2 commonly cited for molten silicates, and the surface energies of liquid metals and metal oxides as high as 2 J/m^2. The change in surface energy with temperature for melts such as silicates is not well documented. Compositional changes also modify the surface energy. For example, the surface energy of Fe_xO–SiO_2 melts at 1300°C decreases approximately linearly, from ~0.55 J/m^2 at zero wt% SiO_2 to 0.35 J/m^2 at 45 wt% SiO_2 [78].

The geometries shown in Figure 9.47 ignore the complication that for effective liquid-phase sintering, the solid must have some solubility in the liquid. This solubility effect leads to detailed wetting geometries or to liquid distribution that requires modification of Equation 9.95. Complete equilibrium must include a force balance in both dimensions as well as the attainment of constant curvature surfaces. These effects have been examined by Cannon et al. [79].

9.4.2.2 Dihedral Angle

The dihedral angle is related to the solid–liquid interfacial tension, γ_{SL}, and the grain boundary tension, γ_{gb}, by the force balance (Figure 9.48):

$$2\gamma_{SL}\cos(\psi/2) = \gamma_{gb} \qquad (9.96)$$

For $\gamma_{gb}/\gamma_{SL} > 2$, the liquid completely penetrates the grain boundary, whereas for $\gamma_{gb}/\gamma_{SL} < 2$, only partial penetration of the boundary is achieved, so solid-state processes can be significant.

9.4.2.3 Effects of Gravity

Because of the low effective viscosity of powder compacts containing a liquid phase, the weight of the system itself can lead to significant creep and distortion. Judicious support for large articles

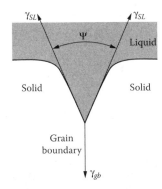

FIGURE 9.48 Dihedral angle ψ for a liquid at a grain boundary.

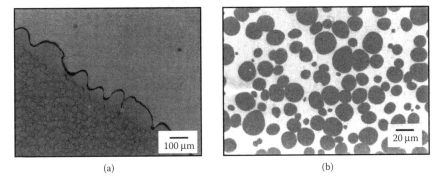

FIGURE 9.49 Micrographs of W(15.4 wt% Ni + 6.6 wt% Fe) after liquid-phase sintering for 1 min in microgravity at 1507°C: (a) low-magnification view of sample edge and wavy solidified liquid, (b) high-magnification view of sample interior showing grain agglomeration. (Courtesy of R.M. German.)

is thus necessary. An additional problem is the possibility of liquid redistribution, leading to particle settling or liquid drainage. This is particularly important for metallic systems, when a significant amount of liquid is present and a large density difference exits between the solid and the liquid. The effects have been examined under microgravity conditions. An interesting observation is that solids suspended in a liquid still tend to agglomerate (Figure 9.49). As this is a metallic system, surface charge effects are not expected to be present. It is possible that the metal particles are propelled as a result of uneven dissolution and precipitation rates along their surfaces, leading to collision and joining. It is possible that a liquid film may be present at the contact areas between the particles, but this has not yet been determined.

9.4.2.4 Grain Boundary Films

Initially, when there is a significant amount of liquid between the grains, the thickness of the grain boundary is determined by kinetic factors. The thinning rate of a liquid layer at a grain boundary of fixed cross section due to viscous flow under a compressive capillary force has been calculated to be proportional to h^5, where h is the instantaneous layer thickness [80]. This predicts an infinite time for the layer thickness to reach zero. However, when the film gets very thin, i.e., in the nanometer range, effects other than viscous flow become dominant, including charge interactions between surfaces [81], as well as structural and chemical forces [82]. The thickness of the grain boundary film appears to have an equilibrium value from one boundary to another, in any given material (Figure 9.50), regardless of the volume fraction of the solidified liquid phase (glass), with the excess glass located elsewhere, such as three- and four-grain junctions.

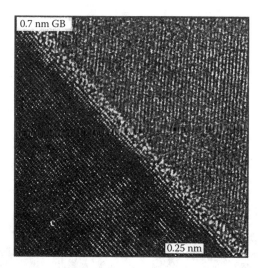

FIGURE 9.50 Amorphous grain boundary film with a constant equilibrium thickness in SiC(Al + B + C). (Courtesy of L.C. De Jonghe.)

Grain boundary films in ceramics are not just restricted to the thin equilibrium films. In a variety of other ceramics, thicker glass films (10 nm to several microns) are also observed. These thicker films represent a different regime of behavior and may vary in thickness with the volume fraction of liquid and from one boundary to another in a given sample of material.

9.4.2.5 Volume Fraction of Liquid Phase

The distribution of the liquid phase depends on the dihedral angle and the relative amount of liquid phase, as well as on the porosity. If sufficient liquid is present, initial rearrangement leads to a fully dense material. The relative amount of liquid and solid at this condition depends on the rearranged density of the particulate solid. For example, if the solid could rearrange to approach a dense random packing, then ~35 vol% of liquid would fill all the void space without further densification by other mechanisms. Such large volume fractions of liquid are often used in porcelains, and in cemented carbides. In the case of clayware and porcelains, the liquid phases are molten silicates which remain as glass after cooling [83]. This gives the ceramic ware a glassy appearance, and such ceramics are referred to as *vitrified*.

9.4.3 PHASE DIAGRAMS IN LIQUID-PHASE SINTERING

Phase diagrams serve as a useful guide for the selection of liquid-producing additives and the heating conditions in liquid-phase sintering. The diagrams give the phases present under equilibrium conditions, but the reaction kinetics during liquid-phase sintering are often too fast for equilibrium to be achieved. For a binary system consisting of a major component (or base B) and a liquid-producing additive A, Figure 9.51 shows an idealized binary phase diagram that indicates the desirable composition and temperature characteristics for liquid-phase sintering. A desirable feature is a large difference in melting temperature between the eutectic and the base B. The composition of the system should also be chosen away from the eutectic composition so that the liquid volume increases slowly with temperature, preventing the sudden formation of all of the liquid at or near the eutectic temperature. A typical sintering temperature would be chosen somewhat above the eutectic temperature with a composition in the (L + S₂) region.

Zinc oxide is liquid-phase sintered with ~0.5 mol % of Bi_2O_3 (and small additions of other oxides) for the production of varistors. The phase diagram for the $ZnO(Bi_2O_3)$ system (Figure 9.52)

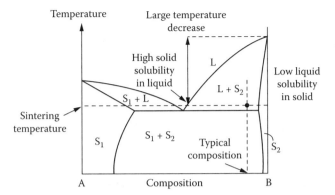

FIGURE 9.51 Model binary phase diagram showing the composition and sintering temperature associated with liquid-phase sintering in the L + S$_2$ phase field. The favorable characteristics for liquid-phase sintering include a suppression of the melting temperature, high solid solubility in the liquid, and low liquid solubility in the solid.

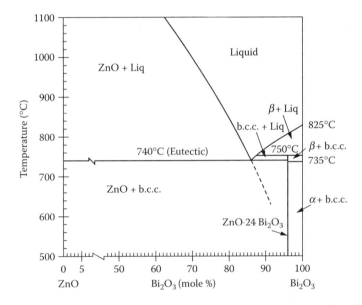

FIGURE 9.52 ZnO–Bi$_2$O$_3$ phase diagram showing limited solid solubility of Bi$_2$O$_3$ in ZnO and the formation of an eutectic containing 86 mol% Bi$_2$O$_3$ at 740°C. (From Safronov, G.M. et al., *Russ. J. Inorg. Chem.*, 16, 460, 1971.)

shows the formation of a Bi-rich liquid phase above the eutectic temperature of 740°C [84]. Samples that have been quenched from the sintering temperature show a liquid film that completely penetrates the grain boundaries, indicating a zero dihedral angle. During slow cooling, the liquid and solid composition change, with precipitation of the principal phase, leading to a substantial increase in the dihedral angle and to a nonwetting configuration [85].

9.5 SINTERING PRACTICE

A wide variety of techniques have been developed to obtain dense ceramics with a desired microstructure and phase composition. In general, these methods involve a manipulation of the heating schedule and, in some cases, the use of an applied pressure. Heating schedules can be simple, as

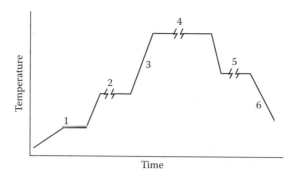

FIGURE 9.53 Generalized heating schedule.

in isothermal sintering, or have a complex temperature-time relationship, as in rate-controlled sintering. Pressure may be applied either uniaxially, with or without a die, or by the surrounding gas. Control of the sintering atmosphere is also important, and precise control of oxygen or nitrogen partial pressure as a function of temperature may in some cases be beneficial or essential. Insoluble gases trapped in closed pores may obstruct the final stage of densification or lead to post-densification swelling, and, in these cases, a change of sintering atmosphere or vacuum sintering is indicated. The practice of sintering further includes a control of the particle characteristics, green compact structure, and consideration of chemistry as a function of processing conditions.

9.5.1 HEATING SCHEDULES

A general heating schedule is shown in Figure 9.53. Binder burnout, removal of volatiles such as water, and conversion of additives such as organometallics or polymers take place in stage 1. Typically, the first hold temperature is, at most, a few hundred degrees Celsius. The heat-up rate should be carefully controlled, otherwise it is quite possible that rapid heating will cause boiling and evaporation of organic additives, leading to specimen bloating or even shattering. Stage 2 can be included to promote chemical homogenization or reaction of powder components. Stage 3 represents the heat up to the isothermal sintering stage 4, during which the majority of the densification and microstructure development takes place, which is then followed by a cool down. An additional hold in stage 5, prior to the final cool down of stage 6, may also be included to relieve internal stresses or allow for precipitation or other reactions.

9.5.1.1 Isothermal Sintering

In isothermal sintering, perhaps better called isothermal stage sintering, the temperature is increased monotonically to a sintering hold temperature (typically 0.5 to 0.8 of the melting temperature for solid-state sintering, or somewhat above the eutectic melting temperature for liquid-phase sintering), and lowered to room temperature afterwards. The holding time is long compared to the heat-up time. This is the most common heating schedule. The heat-up times are limited by the sample size and by the thermal characteristics of the furnace. Heat-up times for large bodies can stretch over many hours, to avoid temperature gradients that could lead to cracking or to the formation of an outer dense layer on an incompletely densified core, as would result from differential densification. During the heat-up phase of the isothermal sintering, significant densification and microstructural changes can take place, which are often ignored in sintering studies. The isothermal sintering is chosen to achieve the required final density within some reasonable time. Higher sintering temperature leads to faster densification, but the grain growth rate also increases. If the temperature is too high, abnormal grain growth may occur, limiting the final density. Figure 9.54 shows qualitatively some possible sintering profiles for different temperatures. The intended service temperature of the

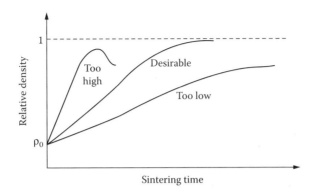

FIGURE 9.54 Schematic sintering curves illustrating the selection of an optimum isothermal sintering temperature.

material must also be taken into account. As a a general rule, sintering should be carried out significantly above the service temperature.

9.5.1.2 Constant Heating–Rate Sintering

In this case, the sample is heated to a specified temperature at a controlled heating rate, and immediately cooled. Constant heating rate experiments may actually be simpler to analyze theoretically than the "isothermal" ones, because strict isothermal sintering is not possible. In practice, the use of constant heating rates is also limited by the sample size. High heating rates are most useful in laboratory studies. Manipulation of the heating rate can be useful when crystallization or chemical reactions take place during densification [86–88]. For example, an aluminosilicate glass showed only limited sinterability when heated at 0.2°C/min because crystallization occurred prior to the glass reaching full density, but heating at 2°C/min delayed the onset of crystallization, allowing the glass to be sintered to full density [86]. The effect of heating rate on the sintering of polycrystalline ZnO was studied by Chu et al. [89]. In general, higher heating rates lead to a finer grain size. For a wide range of heating rates (0.5–15°C/min), plots of the relative density vs. temperature cluster in a fairly narrow band (Figure 9.55), and the derivative of the strain with respect to temperature for this range of constant heating rates falls on the same master curve. Similar observations were made for Al_2O_3 [90].

9.5.1.3 Multistage Sintering

Multistage sintering is frequently used in practice, sometimes introducing extended temperature plateaus or more complex temperature-time sequences in the heating schedule, with the purpose of achieving either specific chemical or microstructural features. An example of staged sintering, involving two separate peak temperatures in which the temperature is reduced between the stages is in the sintering of the ion-conducting ceramic sodium beta-alumina [91]. Optimum results with respect to microstructure and strength were achieved by reducing the sintering temperature by about 150°C after reaching the first peak temperature (about 1500°C).

A reduction in grain size and an increase in uniformity were also found for MgO, Al_2O_3, and ZnO, using a two-stage sintering technique [92,93]. The first stage consisted of an extended hold at a temperature where shrinkage was insignificant (0.5 to 1%) after 48 h. This temperature is readily found by monitoring the sintering in a dilatometer at a constant heating rate. At the holding temperature, limited coarsening occurs by surface diffusion or vapor transport which appears to improve the microstructural homogeneity of the porous powder compact. Figure 9.56 shows the change in pore size distribution after limited coarsening for an Al_2O_3 powder compact. Excessive

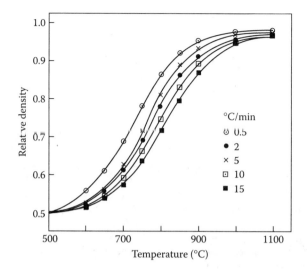

FIGURE 9.55 Relative density vs. temperature for ZnO powder compacts with the same initial density (0.50 ± 0.01), sintered at constant heating rates of 0.5 to 15°C/min. (From Chu, M.-Y., Rahaman, M.N., and Brook, R.J., Effect of heating rate on sintering and coarsening, *J. Am. Ceram. Soc.*, 74, 1217, 1991.)

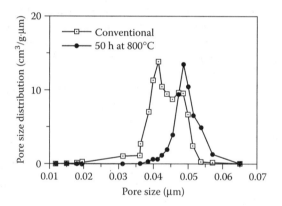

FIGURE 9.56 Pore size distribution determined by mercury porosimetry for Al_2O_3 powder compacts heated at 10°C/min to 800°C (conventional) and for 50 h at 800°C, showing that the distribution narrows for the compact heated for 50 h at 800°C. (From Lin, F.J.T., De Jonghe, L.C., and Rahaman, M.N., Microstructure refinement of sintered alumina by a two-step sintering technique, *J. Am. Ceram. Soc.*, 80, 2269, 1997.)

coarsening in this stage should be avoided. Microstructures of the compacts sintered by the two-stage technique and conventionally (without the coarsening step) are shown in Figure 9.57.

9.5.1.4 Rate-Controlled Sintering

In rate-controlled sintering, the densification rate is coupled with the temperature control of the sintering furnace in such a way as to keep the densification rate constant or at a limited value [94,95]. The result is a fairly complicated temperature history that at times approaches the staged sintering processes. Even though the underlying mechanisms are not fully understood, beneficial effects on microstructure have been reported in a few cases, including where chemical reactions occur during densification [96,97]. Manipulation of the sintering temperature is necessarily limited by the thermal impedance of the sintering furnace and by the size of the sample.

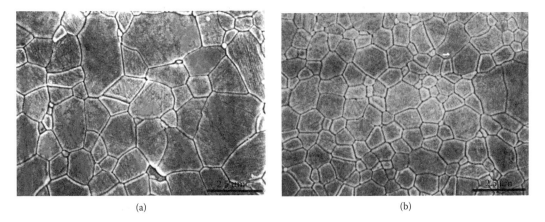

(a)   (b)

FIGURE 9.57 Scanning electron micrographs of Al_2O_3 compacts after sintering at 4°C/min to 1450°C: (a) conventional sintering (relative density = 0.98; grain size = 1.5 ± 0.2 μm), (b) two-step sintering with a pre-coarsening step of 50 h at 800°C (relative density = 0.99; grain size = 1.2 ± 0.1 μm). (From Lin, F.J.T., De Jonghe, L.C., and Rahaman, M.N., Microstructure refinement of sintered alumina by a two-step sintering technique, *J. Am. Ceram. Soc.*, 80, 2269, 1997.)

9.5.1.5 Fast Firing

In fast firing, the heating schedule is manipulated to enhance the densification rate relative to the coarsening rate [98], making the attainment of high density and fine grain size more favorable. The most common situation is one where the activation energy for densification is greater than that for coarsening (Figure 9.58), so at higher temperatures the densification rate is faster than the coarsening rate. The faster the sample is heated through the low temperature region where the ratio of densification rate to coarsening rate is unfavorable, the better the expected result. The argument therefore is for the use of rapid heating and short sintering times at high temperature. Fast firing has been shown to be effective for Al_2O_3 [99], a system for which the activation energy for the densification mechanism (lattice diffusion) is known to be greater than that for the coarsening mechanism (surface diffusion) and for $BaTiO_3$ [100] but not for MgO. Figure 9.59 shows density

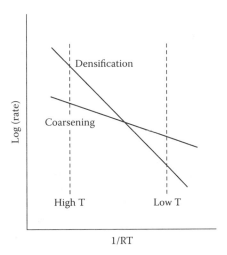

FIGURE 9.58 Under conditions where the densification mechanism has a higher activation energy than the coarsening mechanism, rapid heating to high firing temperatures (fast firing) can be beneficial for the achievement of high density with fine grain size.

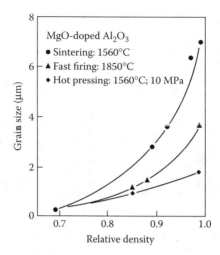

FIGURE 9.59 Grain size vs. density trajectories for Al_2O_3 doped with 200 ppm MgO fabricated by hot pressing, conventional sintering, and fast firing. (From Harmer, M.P. and Brook, R.J., Fast firing — microstructural benefits, *J. Br. Ceram. Soc.*, 80, 147, 1981.)

vs. grain size trajectories for fast firing, sintering, and hot pressing of MgO-doped Al_2O_3. The process can be particularly effective for thin walled tubes, in rapid zone sintering where the ceramic piece is moved through a hot zone.

9.5.1.6 Microwave Sintering

Since the 1970s, there has been a growing interest in the use of microwaves for heating and sintering ceramics. Microwave heating is fundamentally different from conventional heating in which electrical resistance furnaces are typically used: heat is generated internally by interaction of the microwaves with the atoms, ions, and molecules of the material (Figure 9.60). The method is effective for heating more complex ceramic forms rapidly [101]. Heating rates in excess of 1000°C/min can be achieved, and significantly enhanced densification rates have been reported. As an example, Figure 9.61 compares densification data for MgO-doped Al_2O_3 in conventional and microwave sintering [102]. The mechanisms responsible for the enhanced densification are as yet unclear.

Microwave sintering of ceramics is quite straightforward. The ceramic body, usually contained in nonabsorbing or weakly absorbing insulation such as loose, nonconducting powder, is placed

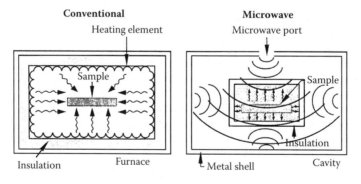

FIGURE 9.60 Heating patterns in conventional and microwave furnaces. (From Sutton, W.H., *Am. Ceram. Soc. Bull.*, 68, 376, 1989. With permission.)

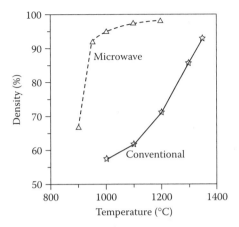

FIGURE 9.61 Relative density vs. temperature for MgO-doped Al₂O₃ powder compacts during sintering by microwave heating and by conventional heating. (From Janney, M.A. and Kimrey, H.D., Diffusion-controlled processes in microwave-fired oxide ceramics, *Mater. Res. Soc. Symp. Proc.*, 189, 215, 1991. With permission.)

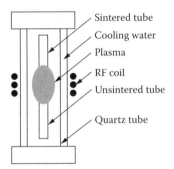

FIGURE 9.62 Schematic of plasma sintering apparatus.

within a microwave cavity. It is possible to use a simple consumer microwave oven to achieve densification if the ceramic body is properly insulated. The shape of the ceramic body affects the local heating rates significantly and achieving sufficiently uniform heating can be difficult [103]. The microwave frequency plays a significant role in the temperature gradients that can develop within the ceramic body. High frequencies tend to heat the exterior of the sample more than the interior, and a combination of frequencies ranging between about 2.5 and 85 GHz has been proposed as providing more uniform heating [104]. Continuous microwave sintering has also been reported, in which samples are passed through the apparatus [105].

A variant of microwave sintering is one where the microwaves ignite a plasma that surrounds the ceramic part. Very high sintering rates can be achieved this way. A schematic of the method is shown in Figure 9.62. Full densification was achieved for tubular specimens by translating the tube at a rate of 25 mm/min through a plasma of about 50 mm in height [106]. It is, as yet, not clear whether the plasma itself enhances the densification rates in addition to the high temperature created by the plasma.

9.5.1.7 Plasma-Assisted Sintering

Several attempts have been made at increasing the heating rates, giving rise to terms such as superfast or ultrafast sintering. One method, referred to as *spark plasma sintering*, involves passing a direct current pulse through a powder compact contained in a graphite die, under an applied

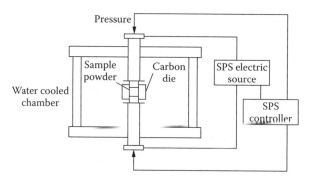

FIGURE 9.63 Schematic of spark plasma sintering (SPS) equipment.

pressure of 30 to 50 MPa (Figure 9.63). Specimen temperatures are difficult to assess in this method, and are usually measured using an optical pyrometer focused on the graphite die wall. Both the die and the specimen are heated by the current pulse. Heating rates in excess of 600°C/min have been reported. The high heating rates are thought to be caused, in part, by spark discharges generated in the voids between the particles. Remarkably high densification rates may be achieved under such conditions, with minimal grain growth. This approach is particularly useful in producing dense ceramic bodies from nanoscale powders [107,108]. The specimen shapes that can be prepared by spark plasma sintering are limited to simple shapes such as discs that can be contained in the compression die. A related method is a thermal explosive forming, in which a reactive mixture of components, e.g., Ti and C, is heated up in a die, and then ignited by passing an electrical current pulse. The process may perhaps be regarded as a self-propagating high-temperature synthesis under pressure. Formation of dense boride, nitride and carbide composites has been reported [109].

9.5.2 Pressure-Assisted Sintering

There are three principal methods of pressure-assisted sintering: hot pressing, hot isostatic pressing, and sinter forging, of which hot pressing is the most widely used. A key advantage is the ability to significantly enhance the densification rate relative to the coarsening rate (see Equation 9.45), thereby guaranteeing, in most cases, the attainment of high density with fine grain size.

9.5.2.1 Hot Pressing

Hot pressing is a convenient laboratory method for preparing dense samples [110]. Heat and pressure are applied to a sample contained typically in a high-strength graphite die, at applied pressures of 25–50 MPa. Other die materials such as aluminum oxide and silicon carbide, or dies confined by an outer metal or fiber-wound mantle, are also used in rare cases. Reaction of the graphite die with the sample can be decreased by spray coating the inside of the die with boron nitride, if the processing temperature remains below about 1350°C. Above that temperature the boron nitride can react with graphite. Lining the graphite die with graphite foil is also useful for prolonging the life of the die. Multiple plate-shaped samples may be hot pressed together, using graphite spacers. A method that allows for shape flexibility was developed by Lange and Terwilliger [111], in which the sample is packed in coarse powder in the hot-pressing die, developing a roughly isostatic pressure on the work piece.

A schematic of the standard process is shown in Figure 9.64. In a typical process, a moderate pressure (10–20 MPa) is applied from the start and, upon reaching the hot-pressing temperature, the pressure is increased to the required value. Some guidance for selecting the appropriate hot-pressing temperature may be obtained from the constant heating rate sintering curve for the powder compact, but some trial and error is also required. An applied pressure is usually maintained

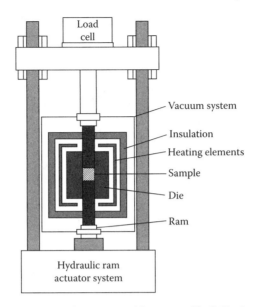

FIGURE 9.64 Schematic of the hot-pressing process. (Courtesy of L.C. De Jonghe.)

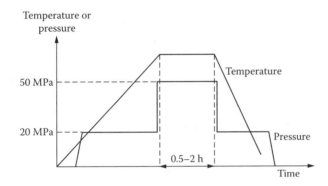

FIGURE 9.65 Pressure–temperature schedule for hot pressing with a high-strength graphite die.

during the cool-down period as well. A schematic of the thermomechanical treatment is shown in Figure 9.65.

Hot pressing is less effective for very fine powder that is well compacted prior to hot pressing. For submicron powders, the sintering stress can be significantly greater than the pressure that the hot-pressing die can tolerate, so the applied stress appears less effective. Benefits may still be derived from the applied pressure when it can assist in the rearrangement process or in the collapse of large pores. In case full densification is difficult to achieve, additives similar to those in sintering can be used. As a result of the significant uniaxial strain in hot pressing, texture may develop in the finished part. Trapping of insoluble gases in residual pores may cause swelling of the ceramic at elevated service temperatures. This can be avoided by hot pressing in vacuum.

Reactive hot pressing has also been used successfully [112, 113]. Densification is more readily achieved in this case compared to sintering. Related to reactive hot pressing is hydrothermal hot pressing. In this method, powders are compacted under hydrothermal conditions. The powder is essentially compacted inside an autoclave at temperatures between 100 and 350°C. The ram design should allow for fluid to leave the sample. Materials such as hydroxyapatite [114] and amorphous titania [115] have been compacted this way.

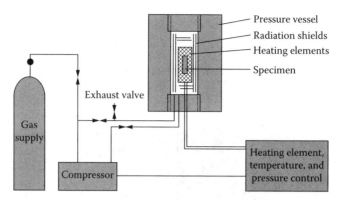

FIGURE 9.66 Schematic of hot-isostatic-pressing equipment. (Courtesy of L.C. De Jonghe.)

9.5.2.2 Sinter Forging

Sinter forging is similar to hot pressing, but without confining the sample in a die. Uniaxial strains are, as a result, significantly larger than in hot pressing [116]. The method has been used effectively to produce microstructures with elongated, aligned grains (Figure 9.27b) in layer-structured ferro-electrics such as bismuth titanate [44]. For fine powders, use can be made of possible superplastic deformation modes during sinter forging [117].

9.5.2.3 Hot Isostatic Pressing (HIPing)

In this method [118], developed around 1955, the preconsolidated powder is tightly enclosed in a glass or metal container, sealed under vacuum and positioned in the pressure vessel (Figure 9.66). Alternatively, the sample can be pre-densified to closed porosity by sintering, after which a can is not needed in subsequent HIPing. A compressor introduces inert gas pressure, and the sample is heated to the sintering temperature, which may be up to 2000°C. During this time the gas pressure rises further, to as much as 30,000 psi, and the container collapses around the sample, transmitting the isostatic pressure to the sample. Commercial hot isostatic presses may have internal chamber diameters approaching 1 m. Heating elements are typically graphite, molybdenum, tungsten, or tantalum. The quality of the ceramics produced by hot isostatic pressing is perhaps the highest obtainable by any other pressure-assisted method, since externally heated dies cannot withstand the pressure that can be applied by HIPing.

9.5.3 PARTICLE AND GREEN COMPACT CHARACTERISTICS

Although the sintering behavior of real powders is considerably more complex than that assumed in the models, the sintering theories clearly indicate the key parameters that must be controlled to optimize sintering. The particle size and particle packing of the green body are key factors, but characteristics such as size distribution, shape, and structure of the particles, can also exert a significant influence.

9.5.3.1 Particle Size

The densification rate of polycrystalline powder compacts depends strongly on particle size (see Equation 9.37), and the reduction of the particle size provides an important method for speeding up sintering. This is illustrated by the data shown in Figure 9.67 for CeO_2 powder compacts with three different particle sizes [119]. Because of packing difficulties, the observed sintering rate is lower than that predicted by the models but the enhancement of the rate with decreasing particle size is still considerable. A consequence of the enhanced sintering rate is the use of lower sintering

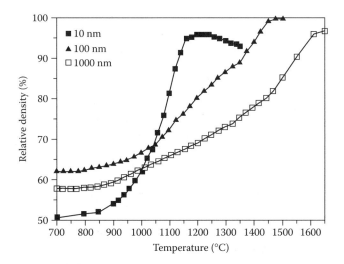

FIGURE 9.67 Effect of particle size on the sintering of CeO₂ powder compacts during constant heating rate sintering (10°C/min) in a oxygen atmosphere. (Zhou, Y.-C. and Rahaman, M.N., Hydrothermal synthesis and sintering of ultra-fine CeO₂ powders, *J. Mater. Res.*, 8, 1680, 1993.)

temperatures. Nanoscale particles (less than 50 to 100 nm in size) commonly exhibit large reductions in the sintering temperature. The data in Figure 9.67 indicate that isothermal sintering of 10-nm CeO_2 particles can be achieved at less than ~1150°C, compared to 1500 to 1600°C for 1-μm particles. Unfortunately, finer particles are more prone to agglomeration, so the green density is often lower that that for coarser particles, and the shrinkage during sintering is larger. Contamination can also be a problem due to the large surface area of the powder. The removal of surface impurities, such as hydroxyl groups, from the surface of nanoscale particles can also be difficult. Decomposition during sintering can lead to trapped gases in the pores which can limit the final density, as seen in Figure 9.67 for the 10-nm particles.

9.5.3.2 Particle Size Distribution

The solid-state sintering models generally assume that the particles are monodisperse but a particle size distribution can have significant effects on sintering. Coble [120] modeled the initial stage sintering of a linear array of particles with different sizes and found that an equivalent particle size, considerably smaller than the average size, can be used to account for the shrinkage. The sintering rates of binary mixtures of particles were intermediate between those for the end-member sizes. A simple rule of mixtures has in fact been proposed to describe the sintering data of binary mixtures of alumina powders where the coarser phase is relatively inactive [121,122]. Commonly, differential densification between the coarser phase and the finer phase coupled with interactions between the particles in the coarser phase can severely inhibit densification [1].

The use of mixtures of discrete particle sizes or a wide distribution of particle size can result in an increase in the packing density (because the finer particles fit into the interstices of the larger particles), so the shrinkage required for complete densification is reduced. This reduction in the shrinkage is important in industrial sintering of large objects. Figure 9.68 shows that increasing the width of the particle size distribution has a benefit in the early stages of sintering [123], presumably due to the presence of the fine powders with the higher driving force for sintering. The behavior in the later stages of sintering depends critically on the particle packing of the green body. For heterogeneous packing, an increase in the width of the particle size distribution is expected to enhance the detrimental effects of differential densification and grain growth (due to the enhanced driving force arising from the size difference), so the final density may decrease. However, if the

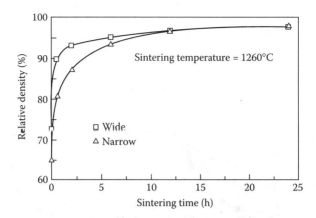

FIGURE 9.68 Effect of particle size distribution on the sintering of slip-cast Al_2O_3 powder compacts prepared from wide and narrow size distribution powders having the same median particle size. (From Yeh, T.S. and Sacks, M.D., Effect of green microstructure on sintering of alumina, *Ceram. Trans.*, 7, 309, 1990. With permission.)

particle packing is homogeneous, with small pores of narrow size distribution, a high final density can be achieved regardless of the width of the initial particle size distribution (Figure 9.68), emphasizing the importance of the pore characteristics in sintering.

9.5.3.3 Particle Shape and Particle Structure

Particle shape influences sintering primarily through its effect on the packing of the green body. Deviation from the spherical or equiaxial shape leads to a reduction in the packing density and packing homogeneity, resulting in a reduction of densification. Compacts of acicular (or elongated) particles, however, can be sintered to high density if the particles are aligned and the packing is homogeneous, as demonstrated for acicular Fe_2O_3 particles [124]. Some solids maintain or develop faceted shapes during sintering, but the effects of shape are often complicated by other factors, such as packing and composition.

Particles used in sintering are commonly dense, discrete units that can either be polycrystalline or a single crystal. Polycrystalline particles of Y_2O_3-stabilized ZrO_2 have been observed to sinter faster than single-crystal particles and this has been attributed to the larger number of grain boundaries contributing to mass transport [125]. Spherical, monodisperse particles synthesized by the Stober process consist of an aggregate of much finer particles (see Figure 2.20a). The influence of this particle substructure on sintering has been well characterized by Edelson and Glaeser [62]. For example, heating TiO_2 to 700°C leads to sintering of the fine particles in the spheres and conversion to rutile, so the spheres consist of 1 to 3 grains by the time the spheres start to sinter at ~1000°C.

9.5.3.4 Particle Packing

For enhanced sintering rates and the attainment of high density, the particles must be homogeneously packed with a high packing density. These packing characteristics produce fine, uniform pores with a low pore coordination number corresponding to the shrinking pore in Figure 9.8. Furthermore, the number of particle contacts is maximized, providing many grain boundaries and short diffusion paths for rapid matter transport into the fine pores. Rhodes [126] provided a clear demonstration of these principles in a study of the sintering of Y_2O_3 stabilized ZrO_2. Using a fine powder (crystallite size ~10 nm), he prepared a suspension and allowed the agglomerates to settle. The fine particles in the supernatant were then used to prepare compacts by gravitational settling in a centrifuge.

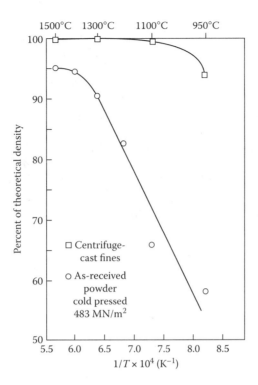

FIGURE 9.69 Effect of particle packing on the sintering of Y_2O_3-stabilized ZrO_2 powder compacts after 1 h at various temperatures. The compact with the more homogeneous green microstructure, formed by centrifugal casting of fine, agglomerate-free particles, reaches a significantly higher density than a compact formed by die pressing the agglomerated, as-received powder. (From Rhodes, W.W., Agglomerate and particle size effects on sintering of yttria-stabilized zirconia, *J. Am. Ceram. Soc.*, 64, 19, 1981. With permission.)

After drying, the compact had a density of 0.74 of the theoretical density. Sintering for 1 h at 1100°C produced almost complete densification, whereas compacts prepared by die pressing of the as-received, agglomerated powder reached a relative density of only 0.95 after sintering for 1 h at 1500°C (Figure 9.69).

Barringer and Bowen [127] developed a processing approach based on the uniform packing of monodisperse, spherical particles. Though rapid sintering was achieved at considerably lower temperatures, the compacts could not be sintered to full density (the residual porosity was ~5%). A problem is that the uniform consolidation of monodispere particles leads to the formation of small regions with three-dimensional, ordered packing (typical of the packing in crystals), referred to as *domains*, which are separated by packing flaws (voids) at the domain boundaries. The faster sintering of the ordered regions, when compared to the domain boundaries, leads to differential stresses that cause enlargement of the voids. These voids correspond to the growing or stable pores in Figure 9.8.

Liniger and Raj [128] have suggested that dense, random packing of particles with a distribution of sizes may provide an alternative ideal packing geometry for enhanced sintering. When compared to the ordered packing of monodisperse particles, the random structure would be less densely packed than the domains but the fluctuations in density should be less severe, leading to homogeneous sintering and the reduction of flaw generation due to differential sintering. Homogeneously packed Al_2O_3 green bodies formed by colloidal processing of fine powders have in fact been sintered to almost full density at temperatures as low as ~1200°C [123,129].

It is important to appreciate the implications of deviating from the ideal structure because most practical sintering, particularly for industrial applications, are not performed with an ideal system

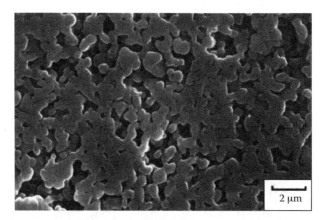

FIGURE 9.70 Differential densification during the sintering of an Al_2O_3 powder compact formed by die pressing, leading to the development of denser and less dense regions.

of homogeneous, densely packed, fine particles. A key problem is differential densification, which may lead to enlargement of the voids (Figure 9.70) or even the generation of crack-like voids in the less dense regions (Figure 1.23). The large voids correspond to the growing pore in Figure 9.8, so they limit densification. Another problem is that the denser regions can support grain growth. Enhanced grain growth starts at an earlier point during sintering, increasing the probability for the initiation of abnormal grain growth.

A relevant question is whether the effects of differential densification can be corrected or reversed during sintering. Lange [130] proposed that a limited amount of normal grain growth is beneficial for sintering in that it reduces the pore coordination number of the pores. This means that grain growth can convert the growing pore with a high coordination number in Figure 9.8 to a shrinking pore with low coordination number. Whereas support for Lange's suggestion is provided by the sintering studies of Lin et al. [131] on heterogeneously packed MgO powder, grain growth can, however, provide only a short-term benefit because the increasing diffusion distance eventually produces a significant lowering of the sintering rate [132]. In general, the effects of heterogeneous packing can be alleviated to some extent during sintering, but the benefits may not be as compelling as may be required in most systems. A more realistic approach would be based on optimizing the particle packing in the green body.

9.5.3.5 Effect of Green Density

Higher green density leads to a reduction in the shrinkage required to reach a given sintered density. Several studies have shown a correlation between the sintered density and the green density for the same sintering conditions. The sintered density is observed to decrease with decreasing green density for green density values below 55 to 60% of the theoretical [133,134]. Commonly, compacts with green densities lower than 40 to 45% of the theoretical are difficult to sinter to high density. Increasing green density is found to delay the onset of enhanced grain growth during the later stages of sintering [135]. These trends in the data can be explained in terms of the higher probability for the occurrence of pores with large coordination numbers with decreasing green density, coupled with the enhancement of differential densification. In general, the benefits of higher green density for sintering are clear, so optimization of the green density would form a useful approach. However, for some powders, particularly some nanoscale powders, a low green density does not necessarily limit the ability to reach a high sintered density (Figure 9.71). Compacts of nanoscale CeO_2 powders with a green density as low as 20% have been sintered to almost full density [136].

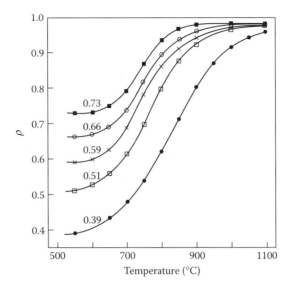

FIGURE 9.71 Effect of green density on the sintering of ZnO powder compacts during constant heating rate sintering (5°C/min to 1100°C), showing that for this powder, the final density is almost independent of the green density. (From Rahaman, M.N., De Jonghe, L.C., and Chu, M.-Y., Effect of green density on densification and creep during sintering, *J. Am. Ceram. Soc.*, 74, 514, 1991.)

9.5.4 CHEMISTRY CONSIDERATIONS

The sintering atmosphere can often have a decisive effect on the ability to reach a target micro-structure. Atmosphere is not directly considered in the sintering models, but the theories can provide a basis for understanding the phenomena arising from the atmosphere conditions. Both physical phenomena (e.g., the atmospheric gas trapped in the pores) and chemical phenomena (e.g., volatility, ionic oxidation state, and defect chemistry) are involved.

9.5.4.1 Gases in Pores

In the final stage of sintering, gas from the atmosphere (or from volatile species in the solid) is trapped in the pores when the pores pinch off and become isolated. Further sintering is influenced by the solubility of the gas atoms in the solid, as has been demonstrated by Coble [137] in the sintering of Al_2O_3. Coble found that MgO-doped Al_2O_3 can be sintered to theoretical density in vacuum O_2 or H_2, which can diffuse out to the surface of the solid, but not in air, N_2, He or Ar, which have a limited solubility in Al_2O_3.

When the gas has a high solubility in the solid, the densification rate is unaffected by the gas trapped in the pores because rapid diffusion through the lattice or along the grain boundaries can occur during shrinkage. For an insoluble gas, as shrinkage of the isolated pores takes place, the gas is compressed and its pressure increases. Shrinkage stops when the gas pressure becomes equal to the driving force for sintering. Assuming spherical pores, the limiting final density is reached when the gas pressure p_i in the pores balances the sintering driving force:

$$p_i = 2\gamma_{SV} / r \tag{9.97}$$

where γ_{SV} is the specific energy of the solid–vapor interface, and r is the pore radius when sintering stops. Applying the gas law $p_1 V_1 = p_2 V_2$ to the initial situation (the pores just become isolated) and to the limiting situation (shrinkage stops) gives:

$$p_oN\left(\frac{4}{3}\pi r_o^3\right) = p_iN\left(\frac{4}{3}\pi r^3\right) \tag{9.98}$$

where N is the number of pores per unit volume, p_0 is the pressure of the sintering atmosphere, and r_0 is the radius of the pores when they become isolated. Substituting for p_i from Equation 9.97 gives:

$$p_o r_o^3 = 2\gamma_{SV}r^2 \tag{9.99}$$

Assuming no coalescence of the pores, the porosity P_0 when the pores become isolated and the limiting final porosity P_f are related by:

$$\frac{P_o}{P_f} = \frac{r_o^3}{r^3} \tag{9.100}$$

Substituting for r from Equation 9.99, after some rearrangement, Equation 9.100 gives:

$$P_f = P_o\left(\frac{p_o r_o}{2\gamma_{SV}}\right)^{3/2} \tag{9.101}$$

According to Equation 9.101, the limiting density is controlled essentially by the pressure of the gaseous atmosphere and by the radius of the pores when they become isolated. Homogeneous packing of fine particles (giving fine pores) and, if practical, sintering in vacuum, will improve the final density.

When coarsening occurs with a slightly soluble or insoluble gas trapped in the pores, pore coalescence leads to a phenomenon referred to as *bloating*, whereby the density of the body starts to decrease, as illustrated in Figure 9.72. Bloating is also a common occurrence when relatively dense ceramics, previously hot pressed in a graphite die (or furnace), are annealed in an oxidizing atmosphere. Oxidation of carbonaceous impurities leads to the generation of a gas that provides the pressure for the expansion.

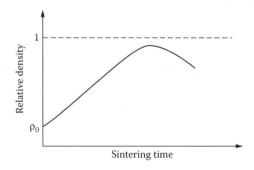

FIGURE 9.72 Diagram illustrating the phenomenon of bloating produced by coarsening when the pores are filled with an insoluble or slightly soluble gas. The density decreases with time after reaching a maximum value.

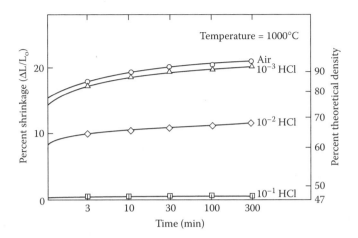

FIGURE 9.73 Shrinkage vs. time for Fe_2O_3 powder compacts sintered at 1000°C in different partial pressures of HCl. (From Readey, D. W. Vapor transport and sintering, *Ceramic Trans.*, 7, 86, 1990. With permission.)

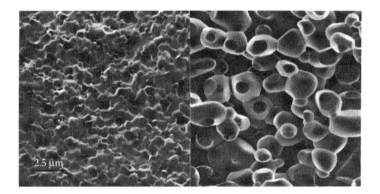

FIGURE 9.74 Scanning electron micrographs of fractured surfaces of Fe_2O_3 powder compacts sintered for 5 h at 1200°C in air (left) and 10% HCl (right). (From Readey, D.W., Vapor transport and sintering, *Ceram. Trans.*, 7, 86, 1990. With permission.)

9.5.4.2 Vapor Transport

Readey [138] has demonstrated how enhanced vapor transport produced, for example, by sintering in a reactive atmosphere, can provide another key process variable for microstructure control. For Fe_2O_3, vapor transport can be enhanced by the introduction of HCl gas into the atmosphere:

$$Fe_2O_3(s) + 6HCl(g) \rightarrow Fe_2Cl_6(g) + 3H_2O(g) \qquad (9.102)$$

Densification decreases as the HCl pressure and the amount of vapor transport increases, and almost stops when the HCl pressure reaches values greater than 10^{-2} MPa (Figure 9.73). As shown in Figure 9.74, the decrease in densification is caused by particle coarsening.

9.5.4.3 Volatilization and Decomposition

For some powder compositions (e.g., sodium β-alumina and lead-based ferroelectric ceramics), evaporation of volatile components can occur during sintering, making it difficult to control the composition and the microstructure of the sintered material. Evaporation of PbO in lead-based electroceramics, can be represented as:

$$PbO(s) \rightleftarrows PbO(g) \rightleftarrows Pb(g) + 1/2O_2 \qquad (9.103)$$

Lead is poisonous so, in addition to controlling lead loss, the evaporated lead must be contained. In practice, this is achieved by surrounding the sample with lead-based powder compositions [139], such as $PbZrO_3$ and PbO for lead–lanthanum–zirconium–titanate (PLZT), to provide a positive vapor pressure in a closed alumina crucible (Figure 9.75). With the controlled atmosphere apparatus, PLZT can be sintered to full density, yielding materials with a high degree of transparency.

Silicon nitride shows significant decomposition at the high temperatures (1700–1800°C) required for its densification. The decomposition reaction can be written as:

$$Si_3N_4 \rightleftarrows 3Si(g) + 2N_2(g) \qquad (9.104)$$

The vapor pressure of the N_2 gas generated by the decomposition is ~0.1 atm at the sintering temperature, so considerable weight loss can occur if the decomposition is not controlled. One solution is to surround the sample with a powder of the same composition in a closed graphite crucible under N_2 gas at atmospheric pressure. A better method involves raising the N_2 gas pressure in the sintering atmosphere (to values of 10 to 20 atm or higher) so that the reaction in Equation 9.105 is driven to the left [140].

9.5.4.4 Oxidation State

The sintering atmosphere influences the oxidation state of certain cations, particularly those of the transition elements (e.g., chromium), and control of the oxidation state has been shown to have a significant effect on the densification of chromium-containing oxides [141,142]. Figure 9.76 shows the porosity of three chromites as a function of oxygen partial pressure, p_{O_2}, in the sintering atmosphere after sintering for 1 h at temperatures ranging from 1600 to 1740°C. The chromites show little densification if the p_{O_2} is greater than ~10^{-8} atm, but relative densities of ~99% can be obtained by sintering in a p_{O_2} of ~10^{-12} atm when the Cr ion is stabilized in its trivalent state (as Cr_2O_3). At higher p_{O_2}, Cr_2O_3 becomes unstable and vaporizes as CrO_2 or CrO_3. The high vapor pressure enhances vapor transport, leading to neck growth and coarsening, but little densification. The dramatic effects of oxidation state control observed for the chromites have not been repeated in other ceramic systems, so the approach sees limited applicability.

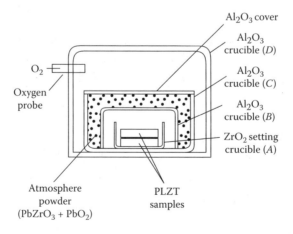

FIGURE 9.75 Apparatus for the sintering of lead–lanthanum–zirconium–titanate (PLZT) in controlled atmosphere. (From Snow, G.S., Improvements in atmosphere sintering of transparent PLZT ceramics, *J. Am. Ceram. Soc.*, 56, 479, 1973. With permission.)

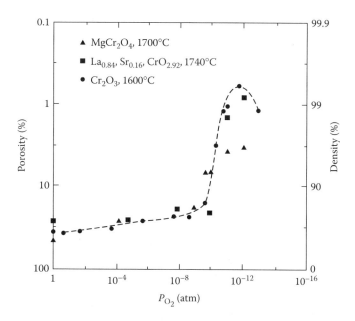

FIGURE 9.76 Final porosity vs. oxygen partial pressure in the atmosphere for sintered chromites. (From Anderson, H.U., Influence of oxygen activity on the sintering of MgCr$_2$O$_4$, *J. Amer. Ceram. Soc.*, 57, 34, 1974. Courtesy of H.U. Anderson.)

Atmosphere control is particularly important in the sintering of ferrites for magnetic devices, such as manganese zinc ferrite and nickel–zinc ferrite [143]. In addition to limiting the evaporation of zinc, the atmospheric conditions must also give the right concentration of ferrous iron in the sintered ferrite for achieving the desired magnetic properties. This is commonly done by first sintering in an atmosphere of high p_{O_2} (0.3–1.0 atm) to minimize the evaporation of zinc, followed by annealing at a lower temperature in a low p_{O_2} atmosphere (10–100 ppm) to give the desired concentration of ferrous iron (Figure 9.77).

9.5.4.5 Defect Chemistry and Stoichiometry

The sintering behavior of the highly stoichiometric oxides (e.g., Al$_2$O$_3$ and ZrO$_2$) shows almost no dependence on the p_{O_2} in the atmosphere. For other ceramic oxides, particularly the transition metal oxides, the p_{O_2} can change the concentration and type of lattice defects and, hence, the stoichiometry of the compound. The sintering of these oxides would therefore be expected to depend significantly on p_{O_2} of the atmosphere. Oxygen ions have a significantly larger ionic radius than most cations, so they are often less mobile and can control the rate of sintering. Thus an atmosphere that produces an increase in the oxygen vacancy concentration may enhance sintering. In practice, it is often difficult to separate the influence of defect chemistry and stoichiometry from the influence of changes in the nondensifying mechanisms (e.g., vapor transport and surface diffusion) also caused by the change in the atmosphere.

Densification of NiFe$_2$O$_4$ and NiAl$_2$O$_4$ powders is severely retarded for compositions with an excess of Fe or Al, and this behavior was explained in terms of a reduction in the oxygen vacancy concentration [144]. Figure 9.78 shows densification and grain growth data for Fe$_3$O$_4$ as a function of p_{O_2} at three different temperatures [145]. Although the type and concentration of the lattice defects may be expected to vary over the wide p_{O_2} range, the sintered density is insensitive to the atmosphere. However, p_{O_2} has a significant effect on the grain growth. High p_{O_2} leads to small grain size whereas the grain size is much larger at low p_{O_2}. Although the data cannot be fully explained, it was suggested that oxygen lattice diffusion controlled the pore mobility. Oxygen lattice

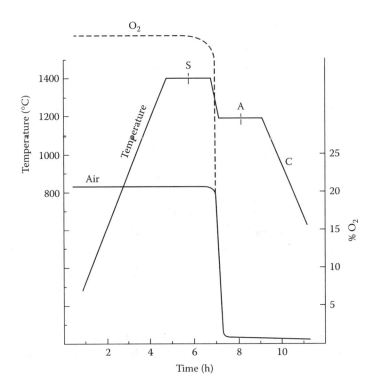

FIGURE 9.77 Schematic firing cycle for manganese–zinc ferrites. (From Reynolds, T., III, Firing, in *Treatise on Materials Science and Technology*, Vol. 9, Wang, F.F.Y., Ed., Academic Press, New York, 1976, p. 199. With permission.)

diffusion is faster at lower p_{O_2}, so an increase in the pore mobility will give a faster grain growth rate if the boundary mobility is controlled by the pore motion.

9.6 CONCLUDING REMARKS

In this chapter, we discussed some of the fundamental aspects of sintering. The elimination of porosity due to matter transport is driven by the reduction in surface free energy. The mechanisms of sintering and idealized models for analyzing sintering were considered. Because of the drastic simplifications, the models can provide only a qualitative understanding of the sintering of real powder compacts. Some aspects of microstructure development were also discussed, particularly from the point of view of grain growth and pore evolution. Simple models provided qualitative predictive results, as well as some indications of how to control the microstructure during sintering. The importance of the particle packing in the green body was very evident. Control of normal grain growth and the avoidance of abnormal grain growth were found to be essential for the production of ceramics with high density and fine grain size. Some basic aspects of liquid-phase sintering were also discussed. We reviewed some of the practical methods used for achieving the required microstructure during sintering, such as the use of controlled heating schedules, pressure-assisted sintering, and control of the sintering atmosphere. The effects of particle size and particle packing of the green body on sintering were also discussed.

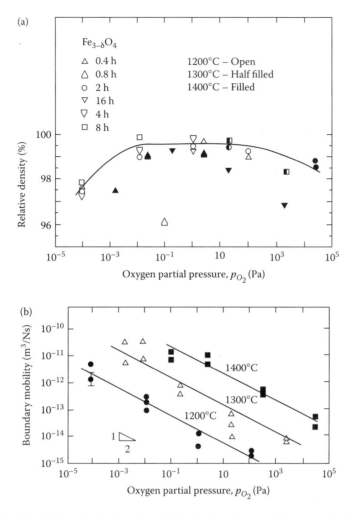

FIGURE 9.78 Data for (a) the relative density and (b) the boundary mobility as a function of the oxygen partial pressure in the atmosphere during the sintering of Fe_3O_4 powder compacts under the conditions of time and temperature indicated. (From Yan, M.F., Grain growth in Fe_3O_4, *J. Am. Ceram. Soc.*, 63, 443, 1980. With permission.)

PROBLEMS

9.1

   a. Assuming isotropic shrinkage, derive the following equation relating the relative density ρ to the shrinkage $\Delta L/L_o$ of a powder compact:

$$\rho = \rho_o / (1 - \Delta L / L_o)^3$$

   where ρ_o is the initial relative density and L_o is the initial length of the compact.

   b. Derive the following relationship between the densification strain and the volumetric strain:

$$\Delta\rho / \rho = - \Delta V / V_o$$

where V_o is the initial volume of the compact.

9.2

Assuming that a powder has a surface energy of 1 J/m², estimate the maximum amount of energy that is available for densification for spherical particles with a diameter of 1 μm compacted to a green density of 60% of the theoretical density. Assume that there is no grain growth and that the grain boundary energy is 0.5 J/m².

9.3

Estimate the difference in equilibrium vapor pressure for two spherical particles, one with a diameter of 0.1 μm and the other with a diameter of 10 μm, assuming a temperature of 500°C, an atomic radius of 10^{-10} m and a surface energy of 1 J/m². Discuss what would happen if the two particles were placed in the same closed box at 500°C.

9.4

Consider a hypothetical oxide with a dihedral angle of 150°. If the dihedral angle is changed to 90° but the surface energy remains the same, would the oxide densify more readily or less readily? Explain why.

9.5

Assuming a constant grain boundary width of 0.5 nm, plot the fractional volume of a polycrystalline solid occupied by the grain boundaries as a function of the grain size in the range of 10 nm to 10 μm. To simplify the calculation, the grains may be assumed to have a cubic shape.

9.6

Sintering is a continuous process but in the models, the process is divided into stages. Explain why. For idealized solid-state sintering, make a table for the three stages of sintering, and for each stage, give the following: the geometrical model, the approximate density range of applicability, and the mechanisms of sintering. Indicate on the table whether the mechanism is densifying or nondensifying, and the source and sink for matter transport.

9.7

For the two-sphere sintering model, derive the equations given in Figure 9.14 for the radius of curvature of the neck surface, r, the surface area of the neck A, and the volume of the material transported into the neck V.

9.8

Derive the equations given in Table 9.2 for sintering of two spheres by (a) lattice diffusion, (b) vapor transport, and (c) surface diffusion.

9.9

The lattice diffusion coefficient for Al^{3+} ions in Al_2O_3 is 4.0×10^{-14} cm²/sec at 1400°C and the activation energy is 580 kJ/mol. Assuming that sintering is controlled by lattice diffusion of Al^{3+} ions, estimate the initial rate of sintering for an Al_2O_3 powder compact of 1 μm particles at 1300°C.

9.10

A ZnO powder compact is formed from particles with an average size of 3 µm. Assuming that densification occurs by a lattice diffusion mechanism with an activation energy of 250 kJ/mol, estimate the factor by which the densification rate will change if: (a) the particle size is reduced to 0.3 µm, (b) the compact is hot pressed under an applied pressure of 40 MPa, and (c) the sintering temperature is raised from 1000 to 1200°C. The specific surface energy of ZnO can be assumed to be 1 J/m^2.

9.11

Zinc oxide has a fairly high vapor pressure for commonly used sintering temperatures above half the melting point, so coarsening due to vapor transport can reduce the densification rate. Discuss how the changes in the particle size, applied pressure and temperature described in Problem 9.10 will influence the rate of vapor transport.

9.12

Distinguish between normal and abnormal grain growth. Why it is important to control normal grain growth and avoid abnormal grain growth for the achievement of high density during sintering?

9.13

For a dense, pure polycrystalline ZnO in which the grain growth follows normal, parabolic kinetics, the average grain size after annealing for 120 min at 1200°C is found to be 5 µm. Annealing for 60 min at 1400°C gives an average grain size of 11 µm. If the average grain size at time = 0 is 2 µm, estimate what the average grain size will be after annealing for 30 min at 1600°C.

9.14

A dense, pure polycrystalline MgO has an average grain size of 1 µm and the grain growth follows normal, parabolic kinetics. Estimate what the average grain size will be after 5 h annealing at 1700°C, given that the grain boundary diffusion coefficient is $D_{gb} = 10^{-8}$ cm^2/sec, the grain boundary width $\delta_{gb} = 0.5$ nm, and the grain boundary energy $\gamma_{gb} = 0.5$ J/m^2.

If some SiO$_2$ is present, leading to the formation of a continuous, liquid film with a thickness of 50 nm at the grain boundaries, estimate the change in the boundary mobility if the diffusion coefficient through the liquid D_L is (1) equal to D_{gb} and (2) equal to $100 D_{gb}$.

9.15

Consider the grain boundary between two spherical particles with approximately the same radius. In one case, both particles are single crystalline. In the other case, one particle is polycrystalline and the second particle single crystalline. Compare the grain growth phenomena that may be expected to occur in the two cases.

9.16

Compare the densification of a homogeneously packed powder compact of 5-µm single-crystalline particles with the densification of a compact of 5-µm agglomerates consisting of 0.5-µm single-crystalline particles. Assume that the particles have the same chemical composition.

9.17

Compare the conditions that control the stability of a pore in a polycrystalline ceramic with those for a pore in a glass.

9.18

A dense Al_2O_3 ceramic (grain size $= 1\ \mu m$) contains 10 vol% of fine ZrO_2 particles that are uniformly distributed at the grain boundaries. If the particle size of the ZrO_2 is $0.2\ \mu m$, and the grain boundary energy is $0.5\ J/m^2$, estimate the maximum pinning force that the ZrO_2 particles can exert on unit area of the grain boundary. If the ceramic is annealed at a high enough temperature (e.g., 1600°C), discuss the main features of the grain growth phenomena that may be expected to occur.

9.19

According to the sintering equations, grain growth during sintering leads to a reduction of the densification rate. Can grain growth actually lead to densification of ceramic powder compacts? Explain your answer.

9.20

Consider a polycrystalline ceramic with pores of various sizes situated at the grain boundaries and within the grains, as sketched schematically as follows:

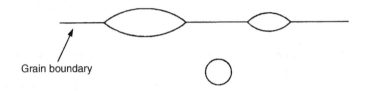

Grain boundary

a. Use arrows to sketch the possible atomic transport paths from the grain boundaries to the pores.
b. Which pore will disappear first? Which pore will disappear last? Explain.
c. If the three pores are close enough so that the diffusion distances between the pores are relatively small, will there be any matter transported between the pores? Use arrows to show the direction of matter transport and explain.

9.21

For each of the stages associated with classic liquid-phase sintering, give the name of the stage, the dominant processes, and the main microstructural changes.

9.22

Determine the range of contact angles for the following conditions: (a) $\gamma_{LV} = \gamma_{SV} < \gamma_{SL}$; (b) $\gamma_{LV} > \gamma_{SV} > \gamma_{SL}$; (c) $\gamma_{LV} > \gamma_{SV} = \gamma_{SL}$.

9.23

Given the following interfacial energies (in units of J/m^2) for liquid-phase sintering: $\gamma_{SV} = 1.0$; $\gamma_{LV} = 0.8$; γ_{ss} (or γ_{gb}) $= 0.75$; and $\gamma_{SL} = 0.35$, determine: (a) the solid–liquid contact angle and (b) whether the liquid will penetrate the grain boundaries.

9.24

State the relationship for the dihedral angle ψ is terms of the grain boundary energy γ_{gb} and the solid–liquid interfacial energy γ_{SL} for a polycrystalline solid in contact with a liquid.

Consider two spheres with the same radius a in contact during liquid-phase sintering. Derive an expression for the equilibrium radius of the neck X formed between the spheres when the dihedral angle is ψ. Using the relationship between the dihedral angle and the interfacial energies, plot a graph of the ratio X/a vs. γ_{gb}/γ_{SL}.

9.25

Wetting of the solid by the liquid phase is a necessary criterion for efficient liquid-phase sintering. Discuss the wetting phenomena that are generally relevant to the liquid-phase sintering of ceramic systems.

9.26

Briefly discuss whether each of the following factors is likely to promote shrinkage or swelling during the initial stage of liquid-phase sintering: (a) high solid solubility in the liquid; (b) low liquid solubility in the solid; (c) large contact angle; (d) small dihedral angle; (e) high green density of the compact; (f) large particle size of the liquid-producing additive.

9.27

Discuss how each of the following variables is expected to influence the densification and micro-structural evolution during liquid-phase sintering of ZnO: (a) the composition of the liquid-producing additive, (b) the volume fraction of the liquid phase, (c) the sintering temperature, (d) the particle size of the ZnO powder, and (e) the application of an external pressure.

9.28

Plot the limiting porosity in a powder compact as a function of the grain size (in the range of 0.1 to 100 μm) when sintering is carried out in an insoluble gas at atmospheric pressure. Assume that the pore size is one third the grain size and that the specific surface energy of the solid–vapor interface is 1 J/m^2.

9.29

When sintered in an oxygen atmosphere at 1700°C for a few hours, an MgO-doped Al$_2$O$_3$ powder compact (starting particle size ≈0.2 μm) reaches theoretical density with an average grain size of ~10 μm. However, when sintered under similar conditions in air (1 atm pressure), the compact reaches a limiting density of 95% of the theoretical. At the end of the intermediate stage when the pores become pinched off, the average grain size and average pore size are found to be 2 μm and 0.5 μm, respectively. Estimate the average grain size and average pore size when the sample sintered in air reaches its limiting density, stating any assumptions that you make.

Further sintering of the sample in air leads to grain growth with pore coalescence, with the pores reaching an average size of 4 μm. Will the porosity be different? Explain. If the porosity is different, estimate how different.

9.30

A student is given the task of sintering a fine-grained BaTiO$_3$ powder (average particle size ≈ 0.1 μm; purity > 99.9%) to a density greater than 98% of the theoretical value and with an average grain size not greater than 1 μm. Discuss how the student should attempt to accomplish the task.

REFERENCES

1. Rahaman, M.N., *Ceramic Processing and Sintering*, 2nd ed., Marcel Dekker, New York, 1995.
2. German, R.M., *Sintering Theory and Practice*, John Wiley & Sons, New York, 1996.
3. De Jonghe, L.C. and Rahaman, M.N., Sintering of ceramics, in *Handbook of Advanced Ceramics*, Vol. 1: Materials Science, Somiya, S., Aldinger, F., Claussen, N., Spriggs, R.M., Uehino, K., Koumoto, K., and Kaneno, M., Eds., Elsevier, New York, 2003, chap. 4.
4. De Jonghe, L.C. and Rahaman, M.N., Sintering stress of homogeneous and heterogeneous powder compacts, *Acta Metall.*, 36, 223, 1988.
5. Raj, R., Analysis of the sintering pressure, *J. Am. Ceram. Soc.*, 70, C210, 1987.
6. Handwerker, C.A., Dynys, J.M., Cannon, R.M., and Coble, R.L., Dihedral angles in magnesia and alumina: distributions from surface thermal grooves, *J. Am. Ceram. Soc.*, 73, 1371, 1990.
7. Kingery, W.D. and Francois, B., Sintering of crystalline oxides, I. Interactions between grain boundaries and pores, in *Sintering and Related Phenomena*, Kuczynski, G.C., Hooton, N.A., and Gibbon, G.F., Eds., Gordon and Breach, New York, 1967, p. 471.
8. Coble, R.L. and Cannon, R.M., Current paradigms in powder processing, *Mater. Sci. Res.*, 11, 151, 1978.
9. (a) Coble, R.L., Sintering crystalline solids. I. Intermediate and final state diffusion models, *J. Appl. Phys.*, 32, 787, 1961.
   (b) Coble, R.L., Sintering crystalline solids. II. Experimental test of diffusion models in powder compacts, *J. Appl. Phys.*, 32, 793, 1961.
10. Frenkel, J., Viscous flow of crystalline bodies under the action of surface tension, *J. Phys. (Moscow)*, 9, 385, 1945.
11. Exner, H.E., Principles of single phase sintering, *Rev. Powder Metall. Phys. Ceram.*, 1, 1, 1979.
12. Coblenz, W.S., Dynys, J.M., Cannon, R.M., and Coble, R.L., Initial stage sintering equations: a critical analysis and assessment, in *Sintering Processes*, Kuczynski, G.C., Ed., Plenum, New York, 1980, p. 141.
13. Ashby, M.F., A first report on sintering diagrams, *Acta Metall.*, 22, 275, 1974.
14. Swinkels F.B. and Ashby, M.F., A second report on sintering diagrams, *Acta Metall.*, 29, 259, 1981.
15. (a) Scherer, G.W., Sintering of low-density glasses: I, theory, *J. Am. Ceram. Soc.*, 60, 236, 1977.
    (b) Scherer, G.W. and Bachman, D.L., Sintering of low-density glasses: II, experimental study, *J. Am. Ceram. Soc.*, 60, 239, 1977.
16. Mackenzie, J.K. and Shuttleworth, R., Phenomenological theory of sintering, *Proc. Phys. Soc. (London)*, 62, 833, 1949.
17. Herring, C., Effect of change of scale on sintering phenomena, *J. Appl. Phys.*, 21, 301, 1950.
18. Coble, R.L., Diffusion models for hot pressing with surface energy and pressure effects as driving forces, *J. Appl. Phys.*, 41, 4798, 1970.
19. Bross, P. and Exner H.E., Computer simulation of sintering processes, *Acta Metall.*, 27, 1013, 1979.
20. Svoboda, J. and Reidel H., New solution describing the formation of interparticle necks in solid-state sintering, *Acta Metall. Mater.*, 43, 1, 1995.
21. Jagota, A. and Dawson, P.R., Simulation of viscous sintering of two particles, *J. Am. Ceram. Soc.*, 73, 173, 1990.
22. Jagota, A., Simulation of the viscous sintering of coated particles, *J. Am. Ceram. Soc.*, 77, 2237, 1994.
23. Brook, R.J., Controlled grain growth, in *Treatise on Materials Science and Technology*, Wang, F.F.Y., Ed., Academic Press, New York, 1976.
24. Atkinson, H.V., Theories of normal grain growth in pure single phase systems, *Acta Metall.*, 36, 469, 1988.
25. Burke, J.E. and Turnbull, D., Recrystallization and grain growth, *Prog. Metal Phys.*, 3, 220, 1952.
26. Anderson, M.P., Srolovitz, D.J., Grest, G.S., and Sahni, P.S., Computer simulation of grain growth — I. Kinetics, *Acta Metall.*, 32, 783, 1984.
27. Srolovitz, D.J., Anderson, M.P., Sahni, P.S., and Grest, G.S., Computer simulation of grain growth — II. Grain size distribution, topology, and local dynamics, *Acta Metall.*, 32, 793, 1984.
28. Hillert, M., On the theory of normal and abnormal grain growth, *Acta Metall.*, 13, 227, 1965.
29. Srolovitz, D.J., Grest, G.S., and Anderson, M.P., Computer simulation of grain growth — V. Abnormal grain growth, *Acta Metall.*, 33, 2233, 1985.

30. Thompson, C.V., Frost, H.J., and Spaepen, F., The relative rates of secondary and normal grain growth, *Acta Metall.*, 35, 887, 1987.

31. Rollet, A.D., Srolovitz, D.J., and Anderson, M.P., Simulation and theory of abnormal grain growth — anisotropic grain boundary energies and mobilities, *Acta Metall.*, 37, 1227, 1989.

32. Yang, W., Chen, L., and Messing, G.L., Computer simulation of anisotropic grain growth, *Mater. Sci. Eng. A* 195, 179, 1995.

33. Kunaver, U. and Kolar D., Three-dimensional computer simulation of anisotropic grain growth in ceramics, *Acta Mater.*, 46, 4629, 1998.

34. Harmer, M.P., A history of the role of MgO in the sintering of α-Al_2O_3, *Ceram. Trans.*, 7, 13, 1990.

35. Kaysser, W.A., Sprissler, M., Handwerker, C.A., and Blendell, J.E., Effect of liquid phase on the morphology of grain growth in alumina, *J. Am. Ceram. Soc.*, 70, 339, 1987.

36. Song, H. and Coble, R.L., Origin and growth kinetics of plate-like abnormal grains in liquid-phase-sintered alumina, *J. Am. Ceram. Soc.*, 73, 2077, 1990.

37. Horn, D.S. and Messing, G.L., Anisotropic grain growth in TiO_2-doped alumina, *Mater. Sci. Eng. A* 195, 169, 1995.

38. Powers, J.D. and Glaeser, A.M., Titanium effects on sintering and grain growth of alumina, in *Sintering Technology*, German, R.M., Messing, G.L., and Cornwall, R.G., Eds., Marcel Dekker, New York, 1996, p. 333.

39. Becher, P.F., Microstructural design of toughened ceramics, *J. Am. Ceram. Soc.*, 74, 255, 1991.

40. Cao, J.J., Moberly-Chan, W.J., De Jonghe, L.C., Gilbert, C.J., and Ritchie, R.O., In situ toughened silicon carbide with Al-B-C additions, *J. Am. Ceram. Soc.*, 79, 461, 1996.

41. Huang, T., Rahaman, M.N., Mah, T.-I., and Parthasarathay, T.A., Effect of SiO_2 and Y_2O_3 additives on the anisotropic grain growth of dense mullite, *J. Mater. Res.*, 15, 718, 2000.

42. Seabaugh, M.M., Kersht, I.H., and Messing G.L., Texture development by templated grain growth in liquid-phase-sintered α-alumina. *J. Am. Ceram. Soc.*, 80, 1181, 1997.

43. Horn, J.A., Zhang, S.C., Selvaraj, U., Messing, G.L., and Trolier-McKinstry, S., Templated grain growth of textured bismuth titanate, *J. Am. Ceram. Soc.*, 82, 921, 1999.

44. Patwardhan, J.S. and Rahaman, M.N., Compositional effects on densification and microstructural evolution of bismuth titanate, *J. Mater. Sci.*, 39, 133, 2004.

45. Yamamoto, T. and Sakuma, T., Fabrication of barium titanate single crystals by solid-state grain growth, *J. Am. Ceram. Soc.*, 77, 1107, 1994.

46. Li, T., Scotch, A.M., Chan, H.M., Harmer, M.P., Park, S.-E., Shrout, T.R., and Michael, J.R., Single crystals of $Pb(Mg_{1/3}Nb_{2/3})O_3$–35 mol% $PbTiO_3$ from polycrystalline precursors, *J. Am. Ceram. Soc.*, 81, 244, 1998.

47. Greenwood, G.W., The growth of dispersed precipitates in solutions, *Acta Metall.*, 4, 243, 1956.

48. Wagner, C., Theory of Ostwald ripening, *Z. Electrochem.*, 65, 581, 1961.

49. Lifshitsz, I.M. and Slyozov, V.V., The kinetics of precipitation from supersaturated solutions, *Phys. Chem. Solids*, 19, 35, 1961.

50. Ardell, A.J., The effect of volume fraction on particle coarsening: theoretical considerations, *Acta Metall.*, 20, 61, 1972.

51. (a) Enomoto, Y., Kawasaki, K., and Tokuyama, M., Computer modeling of Ostwald ripening, *Acta Metall.*, 35, 904, 1987.
    (b) Enomoto, Y., Kawasaki, K., and Tokuyama, M., The time dependent behavior of the Ostwald ripening for the finite volume fraction, *Acta Metall.*, 35, 915, 1987.

52. Coble, R.L., Transparent Alumina and Method of Preparation, U.S. Patent # 3, 026, 210, 1962.

53. (a) Kingery, W.D., Plausible concepts necessary and sufficient for interpretation of ceramic grain boundary phenomena: I, grain-boundary characteristics, structure and electrostatic potential, *J. Am. Ceram. Soc.*, 57, 1, 1974.
    (b) Kingery, W.D., Plausible concepts necessary and sufficient for interpretation of ceramic grain boundary phenomena: II, solute segregation, grain boundary diffusion, and general discussion, *J. Am. Ceram. Soc.*, 57, 74, 1974.

54. Cahn, J.W., The impurity drag effect in grain boundary motion, *Acta Metall.*, 10, 789, 1962.

55. Yan, M.F., Cannon, R.M., and Bowen, H.K., Grain boundary migration in ceramics, in *Ceramic Microstructures '76*, Fulrath, R.M. and Pask, J.A., Eds., Westview Press, Boulder, CO, 1977, p. 276.

56. Glaeser, A.M., Bowen, H.K., and Cannon, R.M., Grain-boundary migration in LiF: I, mobility measurements, *J. Am. Ceram. Soc.*, 69, 119, 1986.

57. Chen, P.-L. and Chen, I.-W., Role of defect interaction in boundary mobility and cation diffusivity of CeO$_2$, *J. Am. Ceram. Soc.*, 77, 2289, 1994.

58. Rahaman, M.N. and Zhou, Y.-C., Effect of dopants on the sintering of ultra-fine CeO$_2$ powder, *J. Eur. Ceram. Soc.*, 15, 939, 1995.

59. Smith, C.S., Grains, phases, and interfaces: an interpretation of microstructure, *Trans. AIME*, 175, 15, 1948.

60. (a) Stearns, L.C., and Harmer, M.P., Particle-inhibited grain growth in Al$_2$O$_3$-SiC: I, experimental results, *J. Am. Ceram. Soc.*, 79, 3013, 1996.
    (b) Stearns, L.C. and Harmer, M.P., Particle-inhibited grain growth in Al$_2$O$_3$-SiC: II, equilibrium and kinetic analyses, *J. Am. Ceram. Soc.*, 79, 3020, 1996.

61. Greskovich, C. and Lay, K.W., Grain growth in very porous Al$_2$O$_3$ compacts, *J. Am. Ceram. Soc.*, 55, 142, 1972.

62. Edelson, L.H. and Glaeser, A.M., Role of particle substructure in the sintering of monosized titania, *J. Am. Ceram. Soc.*, 71, 225, 1988.

63. Kingery, W.D. and Francois, B., Grain growth in porous compacts, *J. Am. Ceram. Soc.*, 48, 546, 1965.

64. Rhines, F. and DeHoff, R., Channel network decay in sintering, *Mater. Sci. Res.*, 16, 49, 1983.

65. Coleman, S.C. and Beeré, W.B., The sintering of open and closed porosity in UO$_2$, *Philos. Mag.*, 31, 1403, 1975.

66. Shewmon, P.G., Movement of small inclusions in solids by a temperature gradient, *Trans. AIME*, 230, 1134, 1964.

67. Hsueh, C.H., Evans, A.G., and Coble, R.L., Microstructure development during final/initial stage sintering: I, pore/grain boundary separation, *Acta Metall.*, 30, 1269, 1982.

68. Brook, R.J., Pore-grain boundary interactions and grain growth, *J. Am. Ceram. Soc.*, 52, 56, 1969.

69. Berry, K.A. and Harmer, M.P., Effect of MgO solute on microstructure development in Al$_2$O$_3$, *J. Am. Ceram. Soc.*, 69, 143, 1986.

70. Carpay, F.M.A., Discontinuous grain growth and pore drag, *J. Am. Ceram. Soc.*, 60, 82, 1977.

71. Handwerker, C.A., Cannon, R.M., and Coble, R.L., Final stage sintering of MgO, *Adv. Ceram.*, 10, 619, 1984.

72. German, R.M., *Liquid Phase Sintering*, Plenum Press, New York, 1985.

73. Kwon, O.J. and Yoon, D.N., Closure of isolated pores in liquid-phase sintering of W–Ni, *Int. J. Powder Technol.*, 17, 127, 1981.

74. Kingery, W.D., Densification during sintering in the presence of a liquid phase. I. Theory, *J. Appl. Phys.*, 30, 301, 1959.

75. Yoon, D.N. and Huppmann, W.J., Grain growth and densification during liquid-phase sintering of W–Ni, *Acta Metall.*, 27, 693, 1979.

76. Gu, H., Pan, X., Cannon, R.M., and Rühle, M., Dopant distribution in grain-boundary films in calcia-doped silicon nitride ceramics, *J. Am. Ceram. Soc.*, 81, 3125, 1998.

77. Brydson, R., Chen, S.-C., Riley, F.L., Milne, S.J., Pan, X., and Rühle, M., Microstructure and chemistry of intergranular glassy films in liquid-phase-sintered alumina, *J. Am. Ceram. Soc.*, 81, 369, 1998.

78. Mills, K.C. and Keene, B.J., Physical properties of BOS slags, *Int. Mater. Rev.*, 32, 1, 1987.

79. Cannon, R.M., Saiz, E., Tomsia, A.P., and Carter, W.C., Reactive wetting taxonomy, *Mater. Res. Soc. Symp. Proc.*, 357, 279, 1995.

80. Lange, F.F., Liquid-phase sintering: are liquids squeezed out from between compressed particles? *J. Am. Ceram. Soc.*, 65, C23, 1982.

81. Clarke, D.R., Shaw, T.M., Philipse, A.P., and Horn, R.G., Possible electrical double layer contribution to the equilibrium thickness of intergranular glass films in polycrystalline ceramics, *J. Am. Ceram. Soc.*, 76, 1201, 1993.

82. Wang, H. and Chiang, Y.-M., Thermodynamic stability of intergranular amorphous films in bismuth-doped zinc oxide, *J. Am. Ceram. Soc.*, 81, 89, 1998.

83. Cambier, F. and Leriche, A., Vitrification, in *Materials Science and Technology*, Vol. 17B, Cahn, R.W., Haasen, P., and Kramer, E.J., Eds., VCH, New York, 1996, chap. 15.

84. Safronov, G.M., Batog, V.N., Stepanyuk, T.V., and Fedorov, P.M., Equilibrium diagram of the bismuth oxide–zinc oxide system, *Russ. J. Inorg. Chem.*, 16, 460, 1971.

85. Chiang, Y.-M., Birnie, D.P., III, and Kingery, W.D., *Physical Ceramics*, John Wiley & Sons, New York, 1997, p. 428.

86. Panda, P.C., Mobley, W.M., and Raj, R., Effect of heating rate on the relative rates of sintering and crystallization in glass, *J. Am. Ceram. Soc.*, 72, 2361, 1989.

87. Keddie, J.L., Braun, P.V., and Giannelis, E.P., Interrelationship between densification, crystallization, and chemical evolution in sol-gel-titania thin films, *J. Am. Ceram. Soc.*, 77, 1592, 1994.

88. Boccaccini, A.R., Stumpfe, W., Taplin, D.M.R., and Ponton, C.B., Densification and crystallization of glass powder compacts during constant heating rate sintering, *Mater. Sci. Eng. A* 219, 26, 1996.

89. Chu, M.-Y., Rahaman, M.N., and Brook, R.J., Effect of heating rate on sintering and coarsening, *J. Am. Ceram. Soc.*, 74, 1217, 1991.

90. Lange, F.F., Approach to reliable powder processing, *Ceram. Trans.*, 1, 1069, 1988.

91. Duncan, J.H. and Bugden, W.G., Two-peak firing of beta-alumina, *Proc. Br. Ceram. Soc.*, 31, 221, 1981.

92. Lin, F.J.T., De Jonghe, L.C., and Rahaman, M.N., Microstructure refinement of sintered alumina by a two-step sintering technique, *J. Am. Ceram. Soc.*, 80, 2269, 1997.

93. Chu, M.-Y., De Jonghe, L.C., Lin, M.F.K., and Lin, F.J.T., Pre-coarsening to improve microstructure and sintering of powder compacts, *J. Am. Ceram. Soc.*, 74, 2902, 1991.

94. Huckabee, M.L., Hare, T.M., and Palmour, H., III, Rate-controlled sintering as a processing method, in *Processing of Crystalline Ceramics*, Palmour, H., III, Davis, R.F., and Hare, R.T., Eds., Plenum Press, New York, 1978, p. 205.

95. Huckabee, M.L., Paisley, M.J., Russell, R.L., and Palmour, H., III, RCS — taking the mystery out of densification profiles, *J. Am. Ceram. Soc.*, 73, 82, 1994.

96. Agarwal, G., Speyer, R.F., and Hackenberger, W.S., Microstructural development of ZnO using a rate-controlled sintering dilatometer, *J. Mater. Res.*, 11, 671, 1996.

97. Ragulya, A.V., Rate-controlled synthesis and sintering of nanocrystalline barium titanate powder. *Nanostruct. Mater.*, 10, 349. 1998.

98. Brook, R.J., Fabrication principles for the production of ceramics with superior mechanical properties, *Proc. Br. Ceram. Soc.*, 32, 7, 1982.

99. Harmer, M.P. and Brook, R.J., Fast firing — microstructural benefits, *J. Br. Ceram. Soc.*, 80, 147, 1981.

100. Mostaghaci, H. and Brook, R.J., Production of dense and fine grain size $BaTiO_3$ by fast firing, *Trans. Br. Ceram. Soc.*, 82, 167, 1983.

101. Johnson, D.L., Ultra-rapid sintering, *Mater. Sci. Res.*, 10, 243, 1984.

102. Janney, M.A. and Kimrey, H.D., Diffusion-controlled processes in microwave-fired oxide ceramics, *Mater. Res. Soc. Symp. Proc.*, 189, 215, 1991.

103. Birnboim, A. and Carmel, Y., Simulation of microwave sintering of ceramic bodies with complex geometry, *J. Am. Ceram. Soc.*, 82, 3024, 1999.

104. Birnboim, A., Gershon, D., Calame, J., Birman, A., Carmel, Y., Rodgers, J., Levush, B., Bykov, Y.V., Eremeev, A.G., Holoptsev, V.V., Semonov, V.E., Dadon, D., Martin, P.L., Rosen, M., and Hutcheon, R., Comparative study of microwave sintering of zinc oxide at 2.45, 30, and 83 GHz, *J. Am. Ceram. Soc.*, 81, 1493, 1998.

105. Cheng, J., Agrawal, D., Roy, R., and Jayan, P.S., Continuous microwave sintering of alumina abrasive grits, *J. Mater. Process. Technol.*, 108, 26, 2000.

106. Henrichsen, M., Hwang, J., Dravid, V.P., and Johnson, D.L., Ultra-rapid phase conversion in beta-alumina tubes, *J. Am. Ceram. Soc.*, 83, 2861, 2000.

107. Gao, L., Shen, Z., Miyamoto, H., and Nygren, M., Superfast densification of oxide/oxide ceramic composites, *J. Am. Ceram. Soc.*, 82, 1061, 1999.

108. Takeuchi, T., Tabuchi, M., and Kageyama, H., Preparation of dense $BaTiO_3$ ceramics with submicrometer grains by spark plasma sintering, *J. Am. Ceram. Soc.*, 82, 939, 1999.

109. Gutmanas, E. and Gotman, I., Dense high-temperature ceramics by thermal explosion under pressure, *J. Eur. Ceram. Soc.*, 19, 2381, 1999.

110. Vasilos, T. and Spriggs, R.M., Pressure sintering of ceramics, in *Progress in Ceramic Science*, Vol. 4, Burke, J.E., Ed., Pergamon Press, New York, 1966, p. 97.

111. Lange, F.F. and Terwilliger, G.R., The powder vehicle hot-pressing technique, *Bull. Am. Ceram. Soc.*, 52, 563, 1973.

112. Zhang, G.-J., Deng, Z.-Y., Kondo, N., Yang, J.-F., and Ohji, T., Reactive hot pressing of ZrB_2-SiC composites, *J. Am. Ceram. Soc.*, 83, 2330, 2000.

113. Wen, G., Li, S.B., Zhang, B.S., and Guo, Z.X., Processing of in situ toughened B-W-C composites by reaction hot pressing of B₄C and WC, *Scripta Mater.*, 43, 853, 2000.

114. Hosoi, K., Hashida, T., Takahashi, H., Yamasaki, N., and Korenaga, T., New processing technique for hydroxyapatite ceramics by the hydrothermal hot-pressing method, *J. Am. Ceram. Soc.*, 79, 2271, 1996.

115. Yanagisawa, K., Ioku, K., and Yamasaki, N., Formation of anatase porous ceramics by hydrothermal hot-pressing of amorphous titania spheres, *J. Am. Ceram. Soc.*, 80, 1303, 1997.

116. He, Y.J., Winnubst, A.J.A., Verweij, A., and Burggraaf, A.J., Sinter-forging of zirconia toughened alumina, *J. Mater. Sci.*, 29, 6505, 1994.

117. Kondo, N., Suzuki, Y., and Ohji, T., Superplastic sinter-forging of silicon nitride with anisotropic microstructure formation, *J. Am. Ceram. Soc.*, 82, 1067, 1999.

118. Saller, H., Paprocki, S., Dayton, R., and Hodge, E., A Method of Bonding, Can. Patent # 680160, 1964.

119. Zhou, Y.-C. and Rahaman, M.N., Hydrothermal synthesis and sintering of ultra-fine CeO₂ powders, *J. Mater. Res.*, 8, 1680, 1993.

120. Coble, R.L., Effects of particle size distribution in initial stage sintering, *J. Am. Ceram. Soc.*, 56, 461, 1973.

121. Onoda, G.Y. Jr. and Messing, G.L., Packing and sintering relations for binary powders, *Mater. Sci. Res.*, 11, 99, 1978.

122. Smith, J.P. and Messing, G.L., Sintering of bimodally distributed alumina powders, *J. Am. Ceram. Soc.*, 67, 238, 1984.

123. Yeh, T.S. and Sacks, M.D., Effect of green microstructure on sintering of alumina, *Ceram. Trans.*, 7, 309, 1990.

124. Yamaguchi T. and Kosha, H., Sintering of acicular Fe₂O₃ particles, *J. Am. Ceram. Soc.*, 64, C84, 1981.

125. Slamovich E.B. and Lange, F.F., Densification behavior of single-crystal and polycrystalline spherical particles of zirconia, *J. Am. Ceram. Soc.*, 73, 3368, 1990.

126. Rhodes, W.W., Agglomerate and particle size effects on sintering of yttria-stabilized zirconia, *J. Am. Ceram. Soc.*, 64, 19, 1981.

127. (a) Barringer, E.A. and Bowen, H.K., Formation, packing, and sintering of monodisperse TiO₂ powders, *J. Am. Ceram. Soc.*, 65, C199, 1982.
(b) Barringer, E.A. and Bowen, H.K., Effect of particle packing on the sintered microstructure, *Appl. Phys. A*, 45, 271, 1988.

128. Liniger, E. and Raj, R., Packing and sintering of two-dimensional structures made from bimodal particle size distributions, *J. Am. Ceram. Soc.*, 70, 843, 1987.

129. Cesarano, J. III and Aksay, I.A., Processing of highly concentrated aqueous α-alumina suspensions stabilized with polyelectrolytes, *J. Am. Ceram. Soc.* 71, 1062, 1988.

130. Lange, F.F., Sinterability of agglomerated powders, *J. Am. Ceram. Soc.*, 67, 83, 1984.

131. Lin, M., De Jonghe L.C., and Rahaman, M.N., Creep-sintering of MgO powder compacts, *J. Am. Ceram. Soc.*, 70, 360, 1987.

132. Harmer, M.P. and Zhao, J., Effect of pores on microstructure development, *Mater. Sci. Res.*, 21, 455, 1987.

133. Rahaman, M.N., De Jonghe, L.C., and Chu, M.-Y., Effect of green density on densification and creep during sintering, *J. Am. Ceram. Soc.*, 74, 514, 1991.

134. Bruch, C.A., Sintering kinetics for the high density alumina process, *Am. Ceram. Soc. Bull.*, 41, 799, 1962.

135. Occhionero, M.A. and Halloran, J.W., The influence of green density on sintering, *Mater. Sci. Res.*, 16, 89, 1984.

136. Chen, P.-L. and Chen, I-W., Sintering of fine oxide powders: II, sintering mechanisms, *J. Am. Ceram. Soc.*, 80, 637, 1997.

137. Coble, R.L., Sintering of alumina: effect of atmosphere, *J. Am. Ceram. Soc.*, 45, 123, 1962.

138. Readey, D.W., Vapor transport and sintering, *Ceram. Trans.*, 7, 86, 1990.

139. Snow, G.S., Improvements in atmosphere sintering of transparent PLZT ceramics, *J. Am. Ceram. Soc.*, 56, 479, 1973.

140. Mitomo, M., Pressure sintering of Si₃N₄, *J. Mater. Sci.*, 11, 1103, 1976.

141. Ownby, P.D., Oxidation state control of volatile species in sintering, *Mater. Sci. Res.*, 6, 431, 1973.

142. Anderson, H.U., Influence of oxygen activity on the sintering of MgCr₂O₄, *J. Am. Ceram. Soc.*, 57, 34, 1974.

143. Reynolds, T., III, Firing, in *Treatise on Materials Science and Technology*, Vol. 9, Wang, F.F.Y., Ed., Academic Press, New York, 1976, p. 199.

144. Reijnen, P., Nonstoichiometry and sintering of ionic solids, in *Reactivity of Solids*, Mitchell, J.W. et al., Eds., John Wiley & Sons, New York, 1969, p. 99.

145. Yan, M.F., Grain growth in Fe_3O_4, *J. Am. Ceram. Soc.*, 63, 443, 1980.

Appendix A: Physical Constants

Velocity of light, c	2.998×10^8 m s^{-1}
Permittivity of vacuum, ε_o	8.854×10^{-12} F m^{-1}
Permeability of vacuum, $\mu_o = 1/\varepsilon_o c^2$	1.257×10^{-6} H m^{-1}
Elementary charge, e	1.602×10^{-19} C
Planck constant, h	6.626×10^{-34} J s
Avogadro number, N_A	6.022×10^{23} mol^{-1}
Atomic mass unit, $m_u = 10^{-3}/N_A$	1.661×10^{-27} kg
Mass of electron, m_e	9.110×10^{-31} kg
Mass of proton, m_p	1.673×10^{-27} kg
Mass of neutron, m_n	1.675×10^{-27} kg
Faraday constant, $F = N_A e$	9.649×10^4 C mol^{-1}
Rydberg constant, R_∞	1.097×10^7 m^{-1}
Bohr magneton, μ_B	9.274×10^{-24} J T^{-1}
Gas constant, R	8.314 J K^{-1} mol^{-1}
Boltzmann constant, $k = R/N_A$	1.381×10^{-23} J K^{-1}
Gravitational constant, G	6.67×10^{-11} N m^2 kg^{-2}
Stefan–Boltzmann constant, σ	5.670×10^{-8} W m^{-2} K^{-4}
Standard volume of ideal gas, V_o	22.414×10^{-3} m^3 mol^{-1}
Acceleration due to gravity, g (at sea level and zero degree latitude)	9.780 m s^{-2}

Appendix B:
SI Units — Names and Symbols

Quantity	Unit	Symbol	Relation to Other Units
Base Units			
Length	Meter	m	
Mass	Kilogram	kg	
Time	Second	s	
Electric current	Ampere	A	
Temperature	Kelvin	K	
Amount of substance	Mole	mol	
Luminous intensity	Candela	cd	
Supplementary Units			
Plane angle	Radian	rad	
Solid angle	Steradian	sr	
Derived Units with Special Names			
Frequency	Hertz	Hz	s^{-1}
Temperature	Degree Celsius	°C	$T(°C) = T(K) - 273.2$
Force	Newton	N	$kg\ m\ s^{-2}$
Pressure and stress	Pascal	Pa	$N\ m^{-2}$
Energy	Joule	J	$N\ m$
Power	Watt	W	$J\ s^{-1}$
Electric charge	Coulomb	C	$A\ s$
Potential	Volt	V	$J\ C^{-1}$
Resistance	Ohm	Ω	$V\ A^{-1}$
Capacitance	Farad	F	$C\ V^{-1}$
Magnetic flux	Weber	Wb	$V\ s$
Flux density	Tesla	T	$Wb\ m^{-2}$
Inductance	Henry	H	$V\ s\ A^{-1}$
Luminous flux	Lumen	lm	$cd\ sr$

Appendix C:
Conversion of Units

Length	1 micron (μm) = 10^{-6} m
	1 Ångström (Å) = 10^{-10} m
	1 inch (in) = 25.4 mm
Mass	1 pound (lb) = 0.454 kg
Volume	1 liter (l) = 10^{-3} m^{-3}
Density	1 gm cm^{-3} = 10^3 kg m^{-3}
Force	1 dyne = 10^{-5} N
Angle	$1°$ = 0.01745 rad
Pressure and stress	1 lb in.$^{-2}$ (psi) = 6.89×10^3 Pa
	1 bar = 10^5 Pa
	1 atmosphere (atm) = 1.013×10^5 Pa
	1 torr = 1 mm Hg = 133.32 Pa
Energy	1 erg = 10^{-7} J
	1 calorie = 4.1868 J
	1 electron volt (eV) = 1.6022×10^{-19} J
Viscosity	
Dynamic	1 poise = 0.1 Pa s
Kinematic	1 stokes = 10^{-4} m^2 s^{-1}

Decimal Fractions and Multiples

Fraction	Prefix	Symbol	Multiple	Prefix	Symbol
10^{-3}	Milli	m	10^3	Kilo	k
10^{-6}	Micro	μ	10^6	Mega	M
10^{-9}	Nano	n	10^9	Giga	G
10^{-12}	Pico	p	10^{12}	Tera	T
10^{-15}	Femto	f	10^{15}	Peta	P
10^{-18}	Atto	a	10^{18}	Exa	E

Appendix D:
Aperture Size of U.S. Standard Wire Mesh Sieves (ASTM E 11:87)

Aperture Size (mm or μm)	Sieve Number	Aperture Size (μm)	Sieve Number
5.6 mm	3.5	300	50
4.75	4	250	60
4.00	5	212	70
3.35	6	180	80
2.80	7	150	100
2.36	8	125	120
2.00	10	106	140
1.70	12	90	170
1.40	14	75	200
1.18	16	63	230
1.00	18	53	270
850 μm	20	45	325
710	25	38	400
600	30	32	450
500	35	25	500
425	40	20	635
355	45		

Appendix E:
Densities and Melting Points of Some Elements, Ceramics, and Minerals

Chemical Formula	Density (g/cm³)	Melting Point (°C)	Common Names
Al	2.70	660	Aluminum
AlN	3.255	2000(d)	Aluminum nitride
Al_2O_3	3.986	2053	Alumina; corundum
$Al_2O_3 \cdot H_2O$	3.44	—	Aluminum monohydrate; boehmite
	3.40	—	Diapsore
$Al_2O_3 \cdot 3H_2O$	2.53	—	Aluminum trihydroxide; bayerite
	2.42	—	Gibbsite
$Al_2O_3 \cdot SiO_2$	3.25	—	Sillimanite
	3.59	—	Kyanite
	3.15	—	Aandalusite
$3Al_2O_3 \cdot 2SiO_2$	3.16	1850	Mullite
$Al_2O_3 \cdot 2SiO_2 \cdot 2H_2O$	2.65	—	Kaolinte
$Al_2O_3 \cdot 2SiO_2 \cdot 4H_2O$	2.62	—	Halloysite
$Al_2O_3 \cdot 4SiO_2 \cdot 2H_2O$	2.78	—	Pyrophyllite
$Al_2O_3 \cdot TiO_2$	3.70	1200(d)	Aluminum titanate
B_4C	2.50	2350	Boron carbide
BN	2.270	2500(s)	Boron nitride (hexagonal)
	3.470		Boron nitride (cubic)
B_2O_3	2.55	450(d)	Boric oxide
BaO	5.72	1972	Barium oxide (cubic)
$BaO \cdot Al_2O_3 \cdot 2SiO_2$	3.25	—	Barium feldspar; celsian
$BaTiO_3$	6.02	1625	Barium titanate
BeO	3.01	2577	Beryllium oxide; bromelite
C	3.513	4400	Diamond
	2.2	4480	Graphite
$CaCO_3$	2.71	900(d)	Calcium carbonate; calcite
	2.83	825(d)	Aragonite
$CaCO_3 \cdot MgCO_3$	2.86	750(d)	Dolomite
CaF_2	3.18	1418	Calcium fluoride; fluorite
CaO	3.34	2898	Calcium oxide; lime
$Ca(OH)_2$	2.2	580(d)	Calcium hydroxide; slaked lime
$CaO \cdot Al_2O_3$	2.98	1605	Calcium aluminate
$CaO \cdot 2Al_2O_3$	2.91	1720	Calcium dialuminate
$CaO \cdot 6Al_2O_3$	3.69	1840	Calcium hexaluminate

Continued.

461

Chemical Formula	Density (g/cm³)	Melting Point (°C)	Common Names
$CaO \cdot Al_2O_3 \cdot 2SiO_2$	2.76	1551	Anorthite; lime feldspar
$CaO \cdot MgO \cdot 2SiO_2$	3.30	1391	Diopside
$2CaSO_4 \cdot H_2O$	—	160(d)	Plaster of Paris
$CaSO_4 \cdot 2H_2O$	2.32	160(d)	Gypsum
$CaO \cdot SiO_2$	2.92	1540	Calcium metasilicate; wollastonaite
$CaO \cdot TiO_2$	3.98	1975	Calcium metatitanate; perovskite
CeO_2	7.65	2400	Cerium (IV) oxide
Ce_2O_3	6.2	2210	Cerium (III) oxide
Cr_2O_3	5.22	2329	Chromium oxide
Cu	8.96	1085	Copper
CuO	6.30–6.45	1326	Cupric oxide; tenorite
Cu_2O	6.00	1235	Cuprous oxide; cuprite
FeO	6.0	1377	Ferrous oxide; wuestite
Fe_2O_3	5.25	1565	Ferric oxide; hematite
Fe_3O_4	5.17	1597	Ferroso-ferric oxide; magnetite
FeS_2	5.02	>600(d)	Iron disulfide; pyrite
$FeTiO_3$	4.72	~1470	Ilmenite
K_2O	2.35	350(d)	Potash
$K_2O \cdot Al_2O_3 \cdot 4SiO_2$	2.49	1686	Leucite
$K_2O \cdot Al_2O_3 \cdot 6SiO_2$	2.56	—	Orthoclase; microcline; potash feldspar
$K_2O \cdot 3Al_2O_3 \cdot 6SiO_2 \cdot 2H_2O$	2.80–2.90	—	Mica; muscovite
$LaCrO_3$	6.70		Lanthanum chromite
$LaMnO_3$	5.70		Lanthanum manganite
La_2O_3	6.51	2304	Lanthanum oxide; lanthana
LiF	2.64	848	Lithium fluoride
Li_2O	2.013	1570	Lithium oxide; lithia
$Li_2O \cdot Al_2O_3 \cdot 4SiO_2$	3.13	—	Spodumene (alpha)
	2.40	—	Spodumene (beta)
$Li_2O \cdot Al_2O_3 \cdot 8SiO_2$	2.42	—	Petalite
$MgCO_3$	3.05	540(d)	Magnesium carbonate; magnesite
MgO	3.581	2850	Magnesia; periclase
$MgO \cdot Al_2O_3$	3.55	2135	Magnesium aluminate; spinel
$2MgO \cdot 2Al_2O_3 \cdot 5SiO_2$	2.60	—	Cordierite (beta)
$Mg(OH)_2$	2.37	350(d)	Brucite
$MgO \cdot SiO_2$	3.19	1550(d)	Magenium metasilicate; clinostatite
$2MgO \cdot SiO_2$	3.21	1900	Magnesium orthosilicate; fosterite
$3MgO \cdot 4SiO_2 \cdot H_2O$	2.71	—	Talc; soapstone
$MoSi_2$	6.24	2030	Molybdenum disilicide
$NaCl$	2.17	800	Sodium chloride
Na_2O	2.27	1132(d)	Sodium oxide; soda
$Na_2O \cdot Al_2O_3 \cdot 2SiO_2$	2.61	1526	Nephelite
$Na_2O \cdot Al_2O_3 \cdot 4SiO_2$	3.34	1000	Jadeite
$Na_2O \cdot Al_2O_3 \cdot 6SiO_2$	2.63	1100	Albite; soda feldspar
$Na_2O \cdot SiO_2$	2.40	1088	Sodium metasilicate
$Na_2O \cdot 2SiO_2$		875	Sodium disilicate
Nb_2O_5	4.55	1500	Niobia
NiO	6.72	1955	Nickel oxide; bunsenite

Continued.

Chemical Formula	Density (g/cm³)	Melting Point (°C)	Common Names
PbO	9.35	—	Litharge (transforms to massicot)
	9.64	897	Massicot
PbO_2	9.64	300(d)	Lead dioxide; plattnerite
Pb_3O_4	8.92	830	Red lead
PbS	7.60	1114	Lead sulfide; galena
$PbTiO_3$	7.9	1290	Lead titanate
Si	2.329	1414	Silicon
SiC	3.217	2700	Silicon carbide
Si_3N_4	3.184	1900(d)	Silicon nitride
SiO_2	2.648	—	α-Quartz (transforms to β-quartz)
	2.533	—	β-Quartz (transforms to tridymite)
	2.265	—	Tridymite (transforms to cristobalite)
	2.334	1722	Cristobalite
	2.196	—	Silica (vitreous)
$SrTiO_3$	5.10	2080	Strontium titanate
TiB_2	4.52	3200	Titanium diboride
TiC	4.910	3000	Titanium carbide
TiO_2	4.23	1840	Titania; rutile
	3.89	—	Anatase (transforms to rutile)
	4.14	—	Brookite (transforms to rutile)
UO_2	10.97	2825	Uranium dioxide; urania
WC	15.6	2785	Tungsten carbide
Y_2O_3	5.03	2438	Yttrium oxide; yttria
$3Y_2O_3 \cdot 5Al_2O_3$	4.55	1970	Yttrium aluminum garnet
ZnO	5.606	1975	Zinc oxide; zincite
$ZnO \cdot Al_2O_3$	4.50		Zinc aluminate
ZnS	4.09	1700	Zinc sulfide; wurtzite
	4.04	1700	Sphalerite
ZrB_2	6.09	3245	Zirconium diboride
ZrO_2	5.56	—	Zirconia (monoclinic)
	6.10	—	Zirconia (tertragonal)
	5.68–5.91	2680	Zirconia (cubic)
$ZrO_2 \cdot SiO2$	4.60	2250	Zirconium orthosilicate; zircon

Note: d = decomposition; s = sublimation

Index

A

Acheson process, 57
Activation energy, 52, 386, 425
Activity, 48
Additives
    forming, 280–292
    grain growth control, 398–401
    liquid-phase sintering, 414
Adsorption from solution
    ions, 150–153, 283
    polyelectrolytes, 174–175
    polymers, 165–167
    surfactants, 284–286
Agglomerate, 22, 38
    definition, 98
    effect on packing, 274
    effect on sintering, 432
    hard, 98
    soft, 98
Aggregate, 99
Aging of gels, 14, 216
Alkoxide (*see* Metal alkoxide)
Alkyl ammonium acetate, 285
Aluminum oxide
    bayerite, 205, 235
    boehmite, 205, 235
    composite, 8, 10
    dihedral angle, 373
    extrusion, 318
    fast firing, 425
    gel, 234–236
    gelcasting, 313
    grain growth, 394–395
    microwave sintering, 426
    powders
        Bayer process, 73
        from solution, 67
    pressure casting, 307–308
    reaction bonding, 10
    sintering
        atmosphere, 435
        effect of particle packing, 433
        grain size–density map, 412
        role of MgO, 398, 411
        two-step (two-stage), 423–424
    slip casting, 303
    suspension, 174–176
    tape casting, 310
    translucent, 398, 435

Arrhenius relation, 51, 386
Atomic absorption spectroscopy, 122
Atomic emission spectroscopy, 122
Atmosphere
    effect on decomposition, 52
    effect on sintering, 435–440
        β-Al_2O_3, 437
        chromites, 438
        Fe_3O_4, 437
        ferrites, 439
        gases in pores, 435–436
        oxidation number, 438
        PLZT, 438
        Si_3N_4, 438
        swelling (bloating), 436
        translucent Al_2O_3, 435
        vapor transport, 437
Attractive surface forces, 142–148
Auger electron spectroscopy, 130–132

B

Ball milling, 41–45 (*see also* Comminution)
    agitated ball mill (attrition mill), 44
    dry ball milling, 42
    effect of dispersant, 42
    grinding media, 43
    milling rate, 43
    rate of grinding, 42–43
    tumbling ball mill, 43
    vibratory ball mill (vibro-mill), 44
    wet ball milling, 42
Barium titanate
    fast firing, 425
    powder synthesis, 69, 71
        coprecipitation, 69
        hydrothermal, 71
    single crystal conversion, 395
    tape casting, 310
Bayer process, 73
BET surface area, 114–116
Binder, 287–290
    cellulose, 287, 289
    natural, 287
    sodium silicate, 290
    synthetic, 288
    viscosity grade, 289, 290
    water-soluble, 288, 289
Binder burnout (*see* Thermal debinding)

RELATED TITLES

Sintering
Mohamed Rahaman
ISBN: 0849372860

Modern Ceramic Engineering Properties, Processing and Use in Design
David W. Richerson
ISBN: 0842786343

Properties of Glass-Forming Melts
David Pye, Innocent Joseph, and Angelo Montenero
ISBN: 1574446622

Introduction to Character Residual Stress by Neutron Diffraction
M.T. Hutchings, P.J. Withers, T.M. Holden, and Torben Lorentzen
ISBN: 0415310008

Ceramic Fabrication Technology
Roy Rice
ISBN: 0824708539

Ceramic Materials for Electronics, 3e
Relva Buchanan
ISBN: 0824740289